AAPG Memoir 86

Global Resource Estimates from Total Petroleum Systems

By

Thomas S. Ahlbrandt[1], Ronald R. Charpentier[1],
T. R. Klett[1], James W. Schmoker[1],
Christopher J. Schenk[1], and Gregory F. Ulmishek[1]

[1] U.S. Geological Survey, Denver, CO

Published by

The American Association of Petroleum Geologists
Tulsa, Oklahoma, U.S.A.

ISBN: 0-89181-367-5

AAPG Editor: Ernest A. Mancini
Geoscience Director: James B. Blankenship
Production: ProType Inc., Tulsa, Oklahoma

The cover illustration shows 3-D model simulation illustrating oil-migration pathways and accumulations (in green) in the petroleum drainage area containing Majnoon Field in Iraq and Jufeyr Field in Iran, which are located in the Zagros-Mesopotamian Cretaceous-Tertiary Total Petroleum System. The vitrinite reflectance of the Jurassic unit that sourced the oil is shown in shaded relief.

This and other AAPG publications are available from:

The AAPG Bookstore
P.O. Box 979
Tulsa, OK 74101-0979
U.S.A.
Telephone: 1-918-584-2555 or 1-800-364-AAPG (U.S.A. and Canada)
Fax: 1-918-560-2652 or 1-800-898-2274 (U.S.A. and Canada)
E-mail: bookstore@aapg.org
www.aapg.org

Geological Society Publishing House
Unit 7, Brassmill Enterprise Centre
Brassmill Lane, Bath BA13JN
U.K.
Telephone: +44-1225-445046
Fax: +44-1225-442836
E-mail: sales@geolsoc.org.uk
www.geolsoc.org.uk

Canadian Society of Petroleum Geologists
No. 160, 540 Fifth Avenue S.W.
Calgary, Alberta T2P 0M2
Canada
Telephone: 1-403-264-5610
Fax: 1-403-264-5898
E-mail: jaime.croft@cspg.org
www.cspg.org

Affiliated East-West Press Private Ltd.
G-1/16 Ansari Road, Darya Ganj
New Delhi 110 002
India
Telephone: +91-11-23279113
Fax: +91-11-23260538
E-mail: affiliat@vsnl.com

AAPG

wishes to thank the following

for their generous contributions

to

Global Resource Estimates from Total Petroleum Systems

by

Thomas S. Ahlbrandt, Ronald R. Charpentier,
T. R. Klett, James W. Schmoker, Christopher J. Schenk,
and Gregory F. Ulmishek

USGS Energy Resources Program

Contributions are applied toward the production costs of publication, thus directly reducing the book's purchase price and making the volume available to a larger readership.

About the Authors

Thomas S. Ahlbrandt received his B.A. (1969) and Ph.D. (1973) in geology at the University of Wyoming. He has 19 years of petroleum experience with ExxonMobil, BPAmoco, Amerada Hess, and independents, and an additional 20 years of service with the U.S. Geological Survey. He is currently the World Energy Project Chief for the U.S. Geological Survey in Denver and led the recent USGS World Petroleum Assessment 2000. Ahlbrandt's recognitions include Distinguished Lecturer of the American Association of Petroleum Geologists (AAPG, 2002–2003), the Distinguished Service Award from AAPG (2002), Outstanding Scientist from the Rocky Mountain Association of Geologists (1999), and Distinguished Alumnus of the University of Wyoming (2000). He has served on the Executive Committee of AAPG as Chairman of the House of Delegates (1995–1996).

Ronald R. Charpentier has worked with the U.S. Geological Survey oil and gas assessment teams as a mathematical geologist since 1979. He specializes in the development of quantitative methodologies to assess oil and gas resources, both conventional and unconventional. As a participant in many assessments of U.S. and world resources, he has also studied the petroleum systems of such diverse areas as the U.S. midcontinent and offshore West Africa. Charpentier received his B.S. in geology from the University of Kansas and an M.S. and Ph.D. in geology from the University of Wisconsin/Madison.

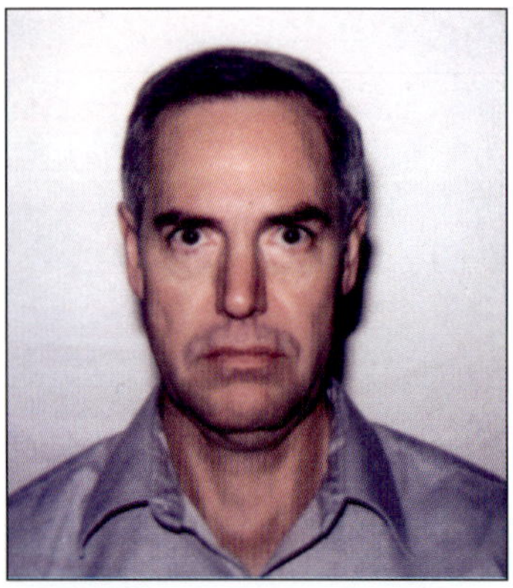

T. R. Klett is a research petroleum geologist with the U.S. Geological Survey (USGS). He is responsible for the assessment of petroleum resources of the world, as well as the development and implementation of resource-assessment methodologies. He serves on a panel of senior USGS research scientists, which oversees and reviews the assessment of both domestic and international petroleum resources. Klett holds a B.S. in geology from Ohio University, an M.S. from the University of Houston (Clear Lake), and a Ph.D. in geology/geochemistry from the Colorado School of Mines. He started his career as a mudlogger, working in and along the Gulf of Mexico. In 1984, he joined Exxon Production Research Company (now ExxonMobil Research Company). He has been with the USGS since 1991.

James W. Schmoker is an emeritus scientist at the U.S. Geological Survey, Denver, Colorado, where he worked for more than a decade on issues of petroleum resource assessment. He has contributed to the methodologies used by the USGS in a number of their more recent oil and gas assessments, including the World Petroleum Assessment 2000.

Christopher J. Schenk has been a Research Geologist with the Energy Program of the U.S. Geological Survey for 25 years. Schenk is presently Project Chief of the U.S. National Oil and Gas Assessment, with the goal of assessing the undiscovered oil and gas resource potential of the U.S. Schenk is also Assessment Coordinator for South and Central America for the U.S. Geological Survey World Energy Project. Research interests include sequence stratigraphy, reservoir characterization, and assessment of oil and gas resources.

Gregory F. Ulmishek is a research geologist with the USGS in Denver. After a 20-year career as an exploration and research geologist in the former USSR, he immigrated to the U.S. in 1980 and was employed by the Argonne National Laboratory. In 1987, he joined the USGS as a member of the World Energy Resource Program. Currently, Ulmishek is the Regional Coordinator for the Former Soviet Union and is involved in studies of basins in the Arctic. He has published more than 90 papers in scientific journals and books. Ulmishek received an M.S. degree in geology from the Moscow Petroleum Institute and a Ph.D. from the Institute of Geology and Exploration for Fossil Fuels in Moscow.

Acknowledgments

We thank the 38 organizations and members of the World Energy Consortium who greatly assisted us with data and identification of the total petroleum systems of the world and supporting data to aid in the definition and assessment of those petroleum systems. Geologic maps were compiled by Feliks M. Persits and geologic data by Sandra J. Lindquist, both contractors to the USGS on the World Energy Project. We thank Katharine L. Varnes who extensively reviewed and edited this manuscript, and Margarita V. Zyrianova who provided technical support for illustrations and text. We also thank our colleagues in the World Energy Project, and Vito F. Nuccio, Central Energy Team Chief Scientist.

The cover of this volume is a 3-D model simulation illustrating oil-migration pathways and accumulations (in green) in the petroleum drainage area containing Majnoon Field in Iraq and Jufeyr Field in Iran, which are located in the Zagros-Mesopotamian Cretaceous-Tertiary Total Petroleum System. The vitrinite reflectance of the Jurassic unit that sourced the oil is shown in shaded relief. Courtesy of Douglas Steinshouer, Janet Pitman, and Steve Cazenave, USGS, Denver, Colorado (for details refer to Pitman et al., 2004).

We thank Beverly Molyneux, AAPG Special Publications Coordinator, for her efforts on this publication.

Table of Contents

About the Authors iv

Acknowledgments vi

I. Summary 1

II. Introduction 5

- A. Purpose and approach
- B. Commodities assessed
- C. Previous work and approaches

III. Results of Resource Assessment of Undiscovered Conventional Petroleum 31

- A. World level results, excluding the U.S.
  1. World level total grown petroleum endowment
  2. Comparison to previous USGS world assessments
- B. World level results, including U.S.
- C. Region level results
- D. Results and rankings by assessment unit, total petroleum system, and province

IV. Reserve Growth 53

V. Preliminary Examination of Geologic Relationships 57

- A. Geologic characterization of assessment units
  1. Source rocks:
     - a. Source rock age
     - b. Source rock character
     - c. Peak maturation age
  2. Reservoirs:
     - a. Reservoir rock age
     - b. Reservoir rock depositional environment
     - c. Reservoir rock lithology
  3. Seals: Seal rock lithology
  4. Traps: Trap type
- B. Other observations on geologic relationships

VI. Field Analyses 139

- A. Remaining large fields
- B. Field size-depth-pressure-temperature relationships

VII. Onshore, Offshore, and Country Allocations 145
A. Onshore versus offshore resources
B. Petroleum volumes by country

VIII. Discussion 153
A. Natural gas
B. North American natural gas supply concerns

IX. Conclusions 157

X. Appendices 159
1A. Assessment units ranked by mean (expected) undiscovered oil
1B. Assessment units ranked by mean (expected) undiscovered natural gas
1C. Assessment units ranked by conventional oil endowment
1D. Assessment units ranked by conventional natural gas endowment
1E. Assessment units ranked by future oil volume
1F. Assessment units ranked by future natural gas volume
2A. Total petroleum systems ranked by mean (expected) undiscovered oil
2B. Total petroleum systems ranked by mean (expected) undiscovered natural gas
2C. Total petroleum systems ranked by conventional oil endowment
2D. Total petroleum systems ranked by conventional natural gas endowment
2E. Total petroleum systems ranked by future oil volume
2F. Total petroleum systems ranked by future natural gas volume
3A. Provinces ranked by mean (expected) undiscovered oil
3B. Provinces ranked by mean (expected) undiscovered natural gas
3C. Provinces ranked by mean (expected) undiscovered total petroleum
3D. Provinces ranked by conventional oil endowment
3E. Provinces ranked by conventional natural gas endowment
3F. Provinces ranked by conventional total petroleum endowment
3G. Provinces ranked by future oil volume
3H. Provinces ranked by future natural gas volume
3I. Provinces ranked by total future petroleum volume
4. Volumes of crude oil, natural gas, and natural gas liquids and geologic attributes of assessment units identified in the World Petroleum Assessment 2000 (USGS, 2000).
5. Volumes of crude oil, natural gas, and natural gas liquids and geologic attributes of total petroleum systems containing assessment units identified in the World Petroleum Assessment 2000 (USGS, 2000).

6. Assessment units ranked by estimated size of their largest undiscovered oil field
7. Assessment units ranked by estimated size of their largest undiscovered natural gas field
8. Assessment units ranked by estimated number of undiscovered oil fields larger than 1024 MMBO
9. Assessment units ranked by estimated number of undiscovered natural gas fields larger than 1024 million BOE
10. Discovered and undiscovered petroleum volumes in assessed portions of countries

XI. Glossary 317

XII. References 319

I: Summary

The U.S. Geological Survey (USGS) recently assessed potential volumes of undiscovered technically recoverable conventional oil, natural gas, and natural gas liquids (NGL) of the world, exclusive of the United States (USGS, 2000). Using a geology based methodology, the USGS defined 149 total petroleum systems (TPS) and 246 assessment units (AU) in 128 oil and natural gas provinces worldwide, and quantitatively assessed undiscovered resources in each of the 246 AU. This assessment provides estimates of the quantities of conventional, technically recoverable oil, natural gas, and natural gas liquids, exclusive of the United States, that have the potential to be added to reserves in the next 30 yr (1995 to 2025). These amounts include estimates of undiscovered volumes and the estimated volume of reserve growth for discovered fields. The assessment results are shown in Table 1, and are summarized as follows: a mean of 649 billion bbl of undiscovered conventional oil with a range of 344–1107 billion bbl of oil; a mean of 4669 tcf of conventional natural gas with a range of 2299–8174 tcf, and a mean of 207 billion bbl of natural gas liquids with a range of 95–378 billion bbl of NGL, exclusive of resources in the United States. Adding these assessment results to estimates of undiscovered oil and natural gas resources of the United States, the cumulative production, remaining reserves, and reserve growth of provinces throughout the world, a total world endowment of conventional oil resource, was estimated by the USGS to be about 3 trillion barrels of oil equivalent (BOE). Using the same procedure, the total world endowment of conventional natural gas was estimated by the USGS to be about 15,000 tcf or 2.6 trillion BOE, and about one-third of a trillion bbl of NGL for a total world petroleum endowment of about 5.9 trillion BOE. This total endowment of 5.9 trillion BOE is for conventional resources only; not included are significant tar sand resources, oil shale resources, and unconventional natural gas resources worldwide including gas hydrates. The USGS estimated that about 1 trillion BOE has been consumed to date, or about 18% of the world's endowment of conventional oil and natural gas resources, leaving about 82% of the conventional oil and natural gas resources to be utilized or discovered.

Provinces of the world that the USGS estimated to have the greatest potential for undiscovered conventional oil resources range from provinces with well-established production histories such as the Middle East, the Siberian and Caspian provinces of the Former Soviet Union, and the Niger and Congo delta provinces of Africa, to provinces with significant geologic uncertainty and no production such as the Northeast Greenland Shelf. Provinces of the world that were estimated to have the greatest potential for volumes of undiscovered natural gas included the Western Siberian Basin Province, the shelves of the Barents and Kara Sea Provinces of the Former Soviet Union, the Rub al Khali Province of the Middle East, and the Norwegian Sea Province. In some provinces, such as the East Siberia Province and the Northwest Shelf Province of Australia, large volumes of discovered but stranded gas are known, and significant volumes of natural gas remain to be discovered. The locations of major natural gas total petroleum systems differ from the major oil total petroleum systems. Additional research, particularly on natural gas systems, is needed.

In defining the geology of 149 total petroleum systems and the 246 assessment units that were evaluated throughout the world, the USGS arrived at several generalizations regarding the geology of petroleum system elements (traps, source rocks, seals, reservoirs) for the conventional oil and natural gas resources of the world. Structural traps are estimated to contain the greatest volume of undiscovered oil and natural gas resource, less vol-

 DOI:10.1306/1061770M863175

ume is estimated to be in combination traps, and the least volume is in stratigraphic traps. For source rocks, marine mudstones are the dominant source rock type, and Mesozoic marine source rocks are the most significant in terms of both discovered and undiscovered volumes of oil and natural gas resources. For reservoirs, non-marine to marginal marine reservoirs contain the greatest volumes of conventional oil and natural gas, whereas deepwater clastic reservoirs are the least significant, despite recent large discoveries in these reservoirs. Clastic reservoirs will dominate future discoveries, whereas carbonate reservoirs contain significant volumes of known resources. For seals, shale seals are the most common in the total petroleum systems examined in this study, and evaporite seals are critical for the preservation of oil and natural gas in total petroleum systems with Paleozoic reservoirs. Migration of oil and natural gas from mature source rocks is estimated to be less than 12.4 mi (20 km), either vertical or lateral, in most of the world's petroleum systems. The major oil and natural gas accumulations of the world are depth sensitive, with oil and natural gas occurrence decreasing significantly with reservoir depth greater than 2.8 mi (4.5 km). The USGS analysis also indicates that half of the world's undiscovered oil and natural gas is offshore. Approximately 25% of the undiscovered volumes of oil and natural gas are estimated to be in Arctic provinces, making this area the next great exploration frontier.

The 8 years since 1995 are less than one-third of the 30-yr forecast span (1995–2025) for the assessment (USGS, 2000). During this period, 18% of estimated oil resources and 27% of the estimated natural gas resources have been added to reserves from new field discoveries and reserve growth for the provinces assessed in USGS (2000). If the entire world is considered, exclusive of the U.S. and Canada, these percentages increase to 23% of estimated oil and 31% of estimated natural gas in this timeframe. Estimated reserve growth is very significant with 26% of the potential oil volume and 52% of the estimated natural gas volume via reserve growth being realized (Klett et al., in press). Overall, about 10% of the estimated undiscovered oil and natural gas was added to reserves by new discoveries assessed in provinces (USGS, 2000). Results of these analyses suggest that estimates in USGS (2000) are somewhat reasonable assuming linear rates of reserves additions, and are moderate relative to recent higher estimates for both oil and natural gas.

TABLE 1. World level summary of petroleum estimates for undiscovered conventional petroleum and reserve growth for oil, natural gas, and natural gas liquids (NGL).

[BBOE = billions of barrels of oil equivalent. Six thousand ft^3 of gas equals one BOE. F95 represents a 95% chance of at least the amount tabulated. Other fractiles are defined similarly. Production and reserves normalized to January 1, 1996. Shading indicates not applicable.]

	Oil				Natural Gas					NGL			
	Billion Barrels				Trillion Cubic Feet				BBOE	Billion Barrels			
	F95	F50	F5	Mean	F95	F50	F5	Mean	Mean	F95	F50	F5	Mean
World (excluding United States)													
Undiscovered conventional	334	607	1107	649	2299	4333	8174	4669	778	95	189	378	207
Reserve growth (conventional)	192	612	1031	612	1049	3305	5543	3305	551	13	42	71	42
Remaining reserves*				859				4621	770				68
Cumulative production*				539				898	150				7
Total				2659				13,493	2249				324
United States**													
Undiscovered conventional	66		104	83	393		698	527	88	Combined with oil			
Reserve growth (conventional)				76				355	59	Combined with oil			
Remaining reserves				32				172	29	Combined with oil			
Cumulative production				171				854	142	Combined with oil			
Total				362				1908	318				
World Total (including United States)				3021				15,401	2567				

*World reserve and cumulative production data reflect only those parts of the world actually assessed and are from Petroconsultants (1996) and NRG Associates (1995).

**U.S. data from Gautier et al. (1996) and MMS (1996).

II: Introduction

A. PURPOSE AND APPROACH

Fossil fuels will remain an important source of energy for much of this century under most energy projections, although their abundance and potential shortages are debated. Their distribution, quantities, and availability will thus continue to be of paramount importance for world economies. Policy decisions made in the U.S. and elsewhere depend on detailed knowledge of petroleum resources and their distribution in the world. Therefore, petroleum resources are periodically reassessed, not just because new data become available and better geologic models are developed, but also because many non-geologic factors such as technologic advances, accessibility to markets, and geographic or societal constraints determine which part of the crustal abundance of petroleum will be economic and acceptable throughout some foreseeable future.

Detailed petroleum information was published in a recent world assessment of petroleum resources conducted by the U.S. Geological Survey (USGS) in their World Petroleum Assessment (WPA) 2000 (USGS, 2000). Two major components of world petroleum resources exclusive of the United States were assessed by the U.S. Geological Survey (USGS, 2000): undiscovered conventional resources and reserve (field) growth. This assessment provided estimates of the quantities of conventional technically recoverable oil, natural gas, and natural gas liquids (NGL), outside the United States, that have the potential to be added to reserves (Table 1). Many individuals and groups, such as the International Energy Agency (IEA 2000, 2001, 2002) and the Energy Information Administration of the Department of Energy (EIA, 2000, 2001a,b), Cavallo (2002), and Greene et al. (2003) use the USGS (2000) study as their prime reference for world petroleum resources.

In this report, reference to world resources or world endowment is understood to refer to those 128 geologic provinces of the world assessed in this study. Where petroleum volumes from the United States are included for a comprehensive discussion, they will be identified and represent the 1995 USGS national (U.S.) onshore and state waters assessment (USGS, 1995; Gautier et al., 1996) and the 1996 U.S. offshore assessment (MMS, 1996).

Undiscovered conventional petroleum resources, the first component of the USGS (2000), were assessed on the basis of geology and exploration and discovery history. Potential additions to reserves from reserve growth, the second component, were estimated for the USGS (2000) study only for the world as a whole using a factor developed from United States reserve growth and applied to world data. The significance of reserve growth for the world is increasingly recognized (Odell, 1973; Klett and Schmoker, 2003; Stark, 2003). However, regional reserve growth estimates allocated from the world totals have since been derived and are reported here. The latter portion of even a 30-yr time frame is difficult to predict because of continuous technologic progress and the rapidly evolving world economy.

This (USGS, 2000) is the fifth in a series of USGS world petroleum assessments (exclusive of the U.S.) that began in 1981. This assessment has a 30-yr forecast span, in contrast to previous USGS world studies that presented estimates of ultimate petroleum resources. The effective date of this study is January 1, 1996. Data for known volumes used in this study are from Petroconsultants (1996) and NRG Associates (1995) and are used with the permission of these companies. In later sections, a discussion of an updated version of this earlier Petroconsultants (1996) database is compared to the IHS Energy Group (formerly Petroconsultants) (2001) data to analyze the estimates.

 DOI:10.1306/1061771M863175

Petroleum resource assessments are only estimates, but accuracy may increase through time as knowledge, technology, and economics change. Every assessment by each assessing organization is different, and is a refinement of previous work. The current USGS world assessment (USGS, 2000) differs from the previous four USGS world assessments in terms of methodology, personnel, and documentation. The USGS (2000) assessment methodology was extensively evaluated prior to selection as the appropriate assessment method and model (Ahlbrandt et al., 1998).

The critical geologic controls on petroleum distribution are encompassed by the total petroleum system (TPS); this approach, rather than a play approach, was used in USGS (2000) for the first time by the USGS (Figure 1). TPS differs from the "petroleum system" of Magoon and Dow (1994) by the inclusion of undiscovered resources (Magoon and Schmoker, 2000). Each TPS was divided into assessment units (AU), the basic units assessed (Figure 1). The advantages of using a total petroleum systems approach is that it combines the elements of source, reservoir, trap, and seal with the tectonostratigraphic history of basin evolution and hydrocarbon migration, generation, and accumulation greatly enhancing the assessment of undiscovered hydrocarbon resources potential (Doust, 2003). For example, the area (also called pod) of active petroleum source rock is a critical element of the TPS (Figure 1). Summaries of individual TPS and AU used in USGS (2000) are available at the website noted at the end of this section. Overall, 159 TPS and 270 AU were identified in 96 countries and two jointly held areas in 128 geologic provinces (Table 2; Figure 2). Of these, 149 TPS (Table 2; Figure 3) and 246 AU (Table 2; Figure 4) were assessed. The undiscovered resources and endowments in the assessed TPS for oil (Figures 5, 6) and natural gas (Figures 7, 8) show the high concentration of resources in a relatively few TPS although the ranking of TPS by volume of resources is different for oil and natural gas. All estimates were made at the assessment unit level. These results were then aggregated to TPS and province levels and in turn allocated to the region and ultimately aggregated to the world as previously discussed.

For the current assessment, the world was divided into eight regions approximating the U.S. State Department regions, and each region was subdivided into geologic provinces (Figure 2). Ninety-five percent of the world's known petroleum was found to exist in 76 of these provinces exclusive of the U.S. based on data from Petroconsultants (1996) (now IHS Energy Group); these 76 were designated as priority provinces for assessment. Fifty-two other provinces were also examined (Klett et al., 1997; Klett et al., 2000a) (Figure 2). A geologic province was not comprehensively assessed unless assessment units covered the entire area. The assessed areas were those judged to be significant on a world scale in terms of reported (known) petroleum volumes, geologic potential for new petroleum discoveries, and political importance. Antarctica, as well as some other remote geologic provinces, were excluded based on the judgment that these areas were unlikely to contribute significant proved petroleum reserves in the 30-yr time frame.

Offshore areas of provinces originally were limited to water depths of 6562 ft (2000 m), but several were extended to water depths of 13,124 ft (4000 m) in recognition of rapidly developing deep-water drilling and production technology (Table 2; Figure 2).

Estimated volumes of undiscovered conventional resources and corresponding uncertainty of the world from this assessment are presented in Table 1, which also includes cumulative production, volumes of remaining reserves, and additions to reserves from reserve growth. A series of appendices provides estimates of oil and natural gas endowments; undiscovered and future oil, natural gas, and NGL volumes for each AU (Appendices 1a–1f). The estimated volumes of undiscovered resources of each AU were aggregated into estimates for each TPS (Appendices 2a–2f) which were then aggregated into estimates for each geologic province (Appendices 3a–3i). Volumes for geologic provinces were in turn aggregated to provide region and world volumes as shown in Table 1.

B. COMMODITIES ASSESSED

Commodities considered in this study are crude oil, natural gas, and NGL that can be produced from the subsurface through a well and that are trapped as discrete (conventional) accumulations. Because of reporting uncertainties, forecasts of undiscovered NGL volumes present in natural gas fields refer to all liquids, including crude oil as defined here.

The USGS definition of a conventional accumulation is based on geology. A conventional accumulation is a discrete entity and commonly bounded by a downdip water contact that is controlled by the buoyancy of petroleum in water. This definition does not incorporate economic or political factors such as deep water, remoteness, harsh climate, regulatory status, or engineering techniques.

Not included in this assessment are heavy oil (API gravity <15°); tar accumulations; oil shales; gas hydrates; natural gas dissolved in geopressured brines; coalbed methane; and low-permeability, fractured-shale or basin-center continuous accumulations. However, an effort was made to describe some of the continuous-type resources. Twenty-four AU that were not quantitatively assessed contain continuous-type resources, 17 of these were described in USGS (2000).

Assessment procedures focused on oil in oil fields and natural gas in natural gas fields. However, additional secondary products, such as dissolved natural gas (or NGL) in oil fields and liquids in natural gas fields, also contribute significantly to the world's total petroleum resource. Such components were included in this assessment by specifying ratios of these additional products at estimated minimum, median, and maximum values as part of the input data for each AU. These ratios were then used to estimate volumes of these undiscovered additional products.

Natural gas, as considered in this assessment, is a mixture of hydrocarbon gases (mainly methane) and nonhydrocarbon gases (such as carbon dioxide, hydrogen sulfide, and nitrogen). Estimated undiscovered natural gas volumes presented here are for natural gas and may include nonhydrocarbon gases.

For oil, data of environmental and economic interest include H_2S content, API gravity, and drilling and water depths; for natural gas such data include the nonhydrocarbon components, API gravity of liquids, and drilling and water depths. These data are included in archive files on the 4th CD-ROM of DDS-60 in USGS (2000).

C. PREVIOUS WORK AND APPROACHES

The USGS in 2000 assessed the conventional oil, natural gas, and natural gas liquids (NGL) resources of the world outside the United States (U.S.) that have the potential to be added to reserves within a 30-yr time frame (1995–2025). The assessment encompassed a 5-yr effort from 1995–2000 by 41 people comprising a 100 person-yr effort. The USGS undertook this world petroleum study in order to provide impartial, scientifically based petroleum-resource information essential to the economic and strategic security of the United States. Material found in Chapter AR of DDS-60 (Ahlbrandt et al., 2000) has been extended here, and documents the quantity, quality, and location of petroleum energy resources. This publication discusses the distribution of estimated undiscovered resources within TPS; geologic analyses of the significant TPS of the world; input data; and allocations of estimated undiscovered resources to countries and onshore and offshore areas. Additional data such as more than 14,000 supporting discovery history plots, geologic summaries for each AU, and summaries of methodology and regional geologic summaries are found on DDS-60. Individual TPS reports as well as DDS-60 may be found at http://energy.cr.usgs.gov/oilgas/wep/. To enlarge the maps printed in this Memoir, refer to the digital maps contained in DDS-60 at the above website.

Table 2. Master list of names and codes for regions (1st digit, e.g. **1**), provinces (2nd to 4th digits, e.g. 1**008**), total petroleum systems (5th and 6th digits, e.g. 1008**01**), and assessment units (7th and 8th digits, e.g. 100801**01**) that were assessed in USGS (2000) and used in this report.

TPS = Total Petroleum System

AU = Assessment Unit

CODE	NAME
1	**Former Soviet Union Region**
1008	**Timan-Pechora Basin Province**
100801	Domanik-Paleozoic TPS
10080101	Northwest Izhma-Pechora Depression AU
10080102	Main Basin Platform AU
10080103	Foredeep Basins AU
1009	**Dnieper-Donets Basin Province**
100901	Dnieper-Donets Paleozoic TPS
10090101	Carboniferous-Lower Permian Clastics AU
10090102	Devonian Synrift AU
10090103	Continuous Basin-Centered Gas Accumulation AU
1015	**Volga-Ural Region Province**
101501	Volga-Ural Domanik-Paleozoic TPS
10150101	Supra-Domanik Carbonates/Clastics AU
10150102	Lower Volga AU
10150103	Sub-Domanik Devonian Clastics AU
10150104	Self-Sourced Domanik AU
101502	Belsk Basin TPS
10150201	Permian Reefs/Thrust Folds AU
1016	**North Caspian Basin Province**
101601	Paleozoic North Caspian TPS
10160101	North and West Margins Subsalt Pinnacle Reefs AU
10160102	North and West Margins Subsalt Barrier Reefs AU
10160103	East and Southeast Margins Subsalt AU
10160104	South Margin Subsalt AU
10160105	Basin Center Subsalt AU
10160106	Suprasalt AU
1050	**South Barents Basin Province**
105001	South and North Barents Triassic-Jurassic TPS
10500101	Kolguyev Terrace AU
10500102	South Barents and Ludlov Saddle AU
10500103	North Barents AU
1060	**North Barents Basin (Province, assessed with Province 1050)**
1108	**Azov-Kuban Basin Province**
110801	Azov-Kuban Mesozoic-Cenozoic TPS
11080101	Tertiary Foredeep AU
11080102	Foreland Slope AU
11080103	Subsalt Jurassic AU
1109	**Middle Caspian Basin Province**
110901	Terek-Caspian TPS
11090101	Foldbelt-Foothills AU
11090102	Terek-Sunzha Subsalt Jurassic AU
11090103	Foreland Slope and Foredeep AU
110902	South Mangyshlak TPS
11090201	South Mangyshlak (Entire) AU
110903	Stavropol-Prikumsk TPS
11090301	Offshore Prikumsk Zone AU
11090302	Onshore Stavropol-Prikumsk AU
11090303	Central Caspian Offshore AU
1112	**South Caspian Basin Province**
111201	Oligocene-Miocene Maykop/Diatom TPS
11120101	Apsheron-Pribalkhan Zone AU
11120102	Lower Kura Depression and Adjacent Shelf AU
11120103	Gograndag-Okarem Zone and Adjacent Shelf AU
11120104	Central Offshore AU
11120105	Iran Onshore/Nearshore AU
1150	**North Ustyurt Basin Province**
115001	Buzachi Arch and Surrounding Areas Composite TPS
11500101	Mesozoic Sandstone Reservoirs AU
115002	North Ustyurt Jurassic TPS
11500201	Jurassic-Tertiary Reservoirs AU
115003	North Ustyurt Paleozoic TPS
11500301	Upper Paleozoic Carbonates AU
1154	**Amu-Darya Basin Province**
115401	Amu-Darya Jurassic-Cretaceous TPS
11540101	Northern and Western Areas AU
11540102	Karabil-Badkhyz (Southern Area) AU
11540103	Murgab Depression Suprasalt AU
11540104	Murgab Depression Subsalt AU
1174	**West Siberian Basin Province**
117401	Bazhenov-Neocomian TPS
11740101	Upper Jurassic-Cretaceous Sandstones AU
11740102	Self-Sourced Bazhenov Fractured Reservoirs AU
117402	Togur-Tyumen TPS
11740201	Pre-Upper Jurassic AU
117403	Northern West Siberian Mesozoic Composite TPS
11740301	Northern West Siberian Onshore Gas AU
11740302	South Kara Sea Offshore AU
1207	**Baykit Arch Province**
120701	Yenisey Foldbelt Riphean-Craton Margin Riphean TPS
12070101	Yenisey Foldbelt Riphean-Craton Margin Riphean AU

Table 2. *(Cont.)*

CODE	NAME
1209	**Angara-Lena Terrace (Province, assessed with Province 1210)**
1210	**Nepa-Botuoba Arch Province**
121001	Baikal-Patom Foldbelt Riphean-Craton Margin Vendian TPS
12100101	Baikal-Patom Foldbelt Riphean-Craton Margin Vendian AU
1211	**Cis-Patom Foredeep (Province, assessed with Province 1210)**
1322	**North Sakhalin Basin Province**
132201	North Sakhalin Neogene TPS
13220101	Onshore and Offshore Northeastern Shelf AU
2	**Middle East and North Africa Region**
2004	**Ma'Rib-Al Jawf/Masila Basin Province**
200401	Madbi Amran/Qishn TPS
20040101	Ma'Rib-Al Jawf/Shabwah/Masila AU
2014	**Ghaba Salt Basin Province**
201401	North Oman Huqf/'Q'-Haushi TPS
20140101	Ghaba-Makarem Combined Structural AU
2016	**Fahud Salt Basin Province**
201601	North Oman Huqf-Shu'aiba TPS
20160101	Fahud-Huqf Combined Structural AU
201602	Middle Cretaceous Natih TPS
20160201	Natih-Fiqa Structural/Stratigraphic AU
2019	**Rub Al Khali Basin Province**
201901	Cretaceous Thamama/Wasia TPS
20190101	Cretaceous Reservoirs in Northwest Desert Anticlines AU
20190102	Cretaceous Reservoirs in South Gulf Suprasalt Structural AU
20190103	Mesozoic/Tertiary Foredeep Fold and Thrust AU
201902	Jurassic Hanifa/Diyab-Arab TPS
20190201	Jurassic Reservoirs in Northwest Desert Anticlines AU
20190202	Jurassic Reservoirs in South Gulf Suprasalt/Qatar Arch Structural AU
201903	Silurian Qusaiba TPS
20190301	Khuff Carbonates in Salt Structures AU
20190302	Paleozoic Reservoirs AU
2020	**Interior Homocline-Central Arch Province**
2021	**Greater Ghawar Uplift Province**
202101	Central Arabia Qusaiba Paleozoic TPS
20210101	Central Arch Horst-Block Anticlinal Oil and Gas AU
20210102	North Gulf Salt Basin Structural Gas AU
202102	Arabian Sub-Basin Tuwaiq/Hanifa-Arab TPS
20210201	Horst-Block Anticlinal Oil AU
20210202	Salt-Involved Structural Oil AU
2022	**Qatar Arch Province**
2023	**Widyan Basin-Interior Platform Province**
202301	Paleozoic Qusaiba/Akkas/Abba/Mudawwara TPS
20230101	Horst/Graben-Related Oil and Gas AU
202302	Jurassic Gotnia/Barsarin/Sargelu/Najmah TPS
20230201	Platform Horst/Graben-Related Oil AU
20230202	Basinal Oil and Gas AU
2024	**Mesopotamian Foredeep Basin Province**
2029	**Wadi-Surhan Basin (Province, assessed with Province 2023)**
2030	**Zagros Fold Belt Province**
203001	Zagros-Mesopotamian Cretaceous-Tertiary TPS
20300101	Cretaceous Reservoirs AU
20300102	Tertiary Reservoirs AU
203002	Paleozoic-Permian/Triassic TPS
20300201	Northern Qatar Arch Extension AU
2043	**Sirte Basin Province**
204301	Sirte-Zelten TPS
20430101	Southeast Sirte Clastics AU
20430102	Central Sirte Carbonates AU
20430103	Offshore Sirte Hypothetical AU
20430104	Southeast Sirte Hypothetical AU
2048	**Pelagian Basin Province**
204801	Bou Dabbous-Tertiary TPS
20480101	Bou Dabbous-Tertiary Structural/Stratigraphic AU
204802	Jurassic-Cretaceous Composite TPS
20480201	Jurassic-Cretaceous Structural/Stratigraphic AU
2054	**Trias/Ghadames Basin Province**
205401	Tanezzuft-Oued Mya TPS
20540101	Tanezzuft-Oued Mya Structural/Stratigraphic AU
205402	Tanezzuft-Melrhir TPS
20540201	Tanezzuft-Melrhir Structural/Stratigraphic AU
205403	Tanezzuft-Ghadames TPS
20540301	Tanezzuft-Ghadames Structural/Stratigraphic AU
2056	**Illizi Basin Province**
205601	Tanezzuft-Illizi TPS
20560101	Tanezzuft-Illizi Structural/Stratigraphic AU
2058	**Grand Erg/Ahnet Basin Province**
205801	Tanezzuft-Timimoun TPS
20580101	Tanezzuft-Timimoun Structural/Stratigraphic AU
205802	Tanezzuft-Ahnet TPS
20580201	Tanezzuft-Ahnet Structural/Stratigraphic AU
205803	Tanezzuft-Sbaa TPS
20580301	Tanezzuft-Sbaa Structural/Stratigraphic AU
205804	Tanezzuft-Mouydir TPS
20580401	Tanezzuft-Mouydir Structural/Stratigraphic AU
205805	Tanezzuft-Benoud TPS
20580501	Tanezzuft-Benoud Structural/Stratigraphic AU
205806	Tanezzuft-Bechar/Abadla TPS
20580601	Tanezzuft-Bechar/Abadla Structural/Stratigraphic AU
2071	**Red Sea Basin Province**
207101	Sudr-Nubia TPS
20710101	Gulf of Suez Block-Fault Fairway AU
20710102	Gulf of Suez Qaa Plain AU
20710103	Southern Gulf of Suez AU
207102	Maqna TPS
20710201	Red Sea Coastal Block Faults AU
20710202	Red Sea Salt Basin AU

Table 2. *(Cont.)*

CODE	NAME
3	**Asia Pacific Region**
3115	**Junggar Basin Province**
311501	Lucaogou-Karamay/Ulho/Pindequan TPS
31150101	Upper Paleozoic/Lower Mesozoic Nonmarine Coarse Clastics AU
311502	Jurassic Coal-Jurassic/Tertiary TPS
31150201	Jurassic/Tertiary Fluvial and Lacustrine Sandstone AU
311503	Lucaogou/Jurassic Coal-Paleozoic/Mesozoic TPS
31150301	Continuous-Type (Basin-Centered) Gas AU
3127	**Bohaiwan Basin Province**
312701	Shahejie-Shahejie/Guantao/Wumishan TPS
31270101	Tertiary Lacustrine AU
31270102	Pre-Tertiary Buried Hills AU
312702	Carboniferous/Permian Coal-Paleozoic TPS
31270201	Coal-Sourced Gas
3128	**Ordos Basin Province**
312801	Yanchang-Yanan TPS
31280101	Jurassic/Triassic Fluvial and Lacustrine Sandstone AU
312802	Taiyuan/Shanxi-Majiagou/Shihezi TPS
31280201	Basin Centered Gas (Continuous) in Ordovician Carbonate and Permian Sandstone AU
31280202	Carboniferous and Permian Coalbed Methane (Continuous) AU
3142	**Sichuan Basin Province**
314201	Maokou/Longtang-Jialingjiang/Maokou/Huanglong TPS
31420101	Southeastern Fold Belt AU
31420102	Northwestern Depression/Foldbelt AU
31420103	Continuous Gas in Central Uplift/Northwestern Depression AU
314202	Daanzhai-Daanzhai/Lianggaoshan TPS
31420201	Jurassic Lacustrine AU
314203	Xujiahe-Xujiahe/Shaximiao TPS
31420301	Continuous Gas in Northwestern Depression/Central Uplift AU
314204	Cambrian/Silurian Marine Shale-Dengying/Lower Paleozoic TPS
31420401	Leshan-Longnusi Paleohigh AU
31420402	Lower Paleozoic of Southeastern Fold Belt AU
3144	**Songliao Basin Province**
314401	Qingshankou-Putaohua/Shaertu TPS
31440101	Subtle Traps AU
31440102	Anticlinal AU
314402	Jurassic Coal-Denglouku/Nongan TPS
31440201	Structural Traps AU
3154	**Tarim Basin Province**
315401	Ordovician/Jurassic-Phanerozoic TPS
31540101	Tarim Basin Excluding Marginal Foldbelts AU
31540102	Kuche (Northern) Foldbelt AU
31540103	Southwest Foldbelt AU
3701	**Baram Delta/Brunei-Sabah Basin Province**
370101	Brunei-Sabah TPS
37010101	Brunei-Sabah Deltaics AU
37010102	Brunei-Sabah Turbidites AU
3702	**Greater Sarawak Basin Province**
370201	Sarawak Basin TPS
37020101	Central Luconia AU
37020102	Balingian AU
370202	East Natuna TPS
37020201	East Natuna AU
3703	**Malay Basin Province**
370301	Oligocene-Miocene Lacustrine TPS
37030101	South Malay Lacustrine AU
37030102	North Malay Lacustrine AU
370302	Miocene Coaly Strata TPS
37030201	South Malay Coaly AU
3808	**Central Sumatra Basin Province**
380801	Brown Shale-Sihapas TPS
38080101	Pematang/Sihapas Siliciclastics AU
3817	**Kutei Basin Province**
381701	Kutei Basin TPS
38170101	Kutei Basin Deltaics AU
38170102	Kutei Basin Turbidites AU
38170103	Kutei Basin Fold and Thrust Belt AU
3822	**North Sumatra Basin Province**
382201	Bampo-Cenozoic TPS
38220101	North Sumatra AU
38220102	Mergui AU
3824	**Northwest Java Basin Province**
382401	Banuwati-Oligocene/Miocene TPS
38240101	Sunda/Asri AU
382402	Jatibarang/Talang Akar-Oligocene/Miocene TPS
38240201	Ardjuna AU
382403	Tertiary-Parigi TPS
38240301	Biogenic Gas AU
382404	Tertiary-Cenozoic TPS
38240401	Billiton Basin Hypothetical AU
3828	**South Sumatra Basin Province**
382801	Lahat/Talang Akar-Cenozoic TPS
38280101	South Sumatra AU
3910	**Bonaparte Gulf Basin Province**
391001	Milligans-Carboniferous/Permian TPS
39100101	Barnett AU
391002	Keyling/Hyland Bay-Permian TPS
39100201	Petrel AU
39100202	Vulcan Graben AU
391003	Jurassic/Early Cretaceous-Mesozoic TPS
39100301	Malita AU
3913	**Browse Basin Province**
391301	Late Jurassic/Early Cretaceous-Mesozoic TPS
39130101	Late Jurassic/Early Cretaceous-Mesozoic AU
3930	**Gippsland Basin Province**
393001	Latrobe TPS
39300101	Gippsland AU

Table 2. *(Cont.)*

CODE	NAME
3948	**Northwest Shelf Province**
394801	Dingo-Mungaroo/Barrow TPS
39480101	Dingo-Mungaroo/Barrow AU
394802	Locker-Mungaroo/Barrow TPS
39480201	Locker-Mungaroo/Barrow AU
4	**Europe Region**
4017	**Vestford-Helgeland Province**
401701	Upper Jurassic Spekk TPS
40170101	Halten Terrace-Trondelag Platform AU
40170102	Mid-Norway Continental Margin AU
4025	**North Sea Graben Province**
402501	Kimmeridgian Shales TPS
40250101	Viking Graben AU
40250102	Moray Firth AU
40250103	Central Graben AU
4035	**Northwest German Basin (Province, assessed with Province 4036)**
4036	**Anglo-Dutch Basin Province**
403601	Carboniferous-Rotliegend TPS
40360101	Southern Permian Basin-U.K. Onshore AU
40360102	Southern Permian Basin-Europe Onshore AU
40360103	Southern Permian Basin-Offshore AU
4047	**North Carpathian Basin Province**
404701	Isotopically Light Gas TPS
40470101	Foreland Basin AU
404702	Mesozoic/Paleogene Composite TPS
40470201	Deformed Belt AU
404703	Paleozoic Composite TPS
40470301	Paleozoic Reservoirs AU
4048	**Pannonian Basin Province**
404801	Greater Hungarian Plain Neogene TPS
40480101	Greater Hungarian Plain Basins AU
404802	Zala-Drava-Sava Mesozoic/Neogene TPS
40480201	Zala-Drava-Sava Basins AU
404803	Danube Neogene TPS
40480301	Danube Basin AU
404804	Transcarpathian Neogene TPS
40480401	Transcarpathian Basin AU
404805	Central Carpathian Paleogene TPS
40480501	Central Carpathian Paleogene Basin AU
404806	Hungarian Paleogene TPS
40480601	Hungarian Paleogene Basin AU
4057	**Transylvanian Basin Province**
405701	Transylvanian Composite TPS
40570101	Transylvanian Neogene Suprasalt Gas AU
40570102	Transylvanian Subsalt AU
4060	**Po Basin Province**
406001	Porto Garibaldi TPS
40600101	Neogene Flysch Gas AU
406002	Meride/Riva di Solto TPS
40600201	Thermal Triassic AU
4061	**Carpathian-Balkanian Basin Province**
406101	Moesian Platform Composite TPS
40610101	Moesian Platform AU
406102	Dysodile Schist-Tertiary TPS
40610201	Romania Flysch Zone AU
40610202	Romania Ploiesti Zone AU
4068	**Provence Basin Province**
406801	Pre-Messinian TPS
40680101	Subsalt AU
5	**North America Region**
5200	**East Greenland Rift Basins Province**
520001	Permian/Upper Jurassic Composite TPS
52000101	Northeast Greenland Shelf Rift Systems AU
5215	**Labrador-Newfoundland Shelf Province**
521501	Egret-Hibernia TPS
52150101	Jeanne d'Arc AU
521502	Mesozoic Composite TPS
52150201	Graben Systems Exclusive of Jeanne d'Arc AU
5243	**Alberta Basin Province**
524301	Keg River-Keg River TPS
52430101	Keg River Gas AU
52430102	Keg River Oil and Gas AU
524302	Duvernay-Leduc TPS
52430201	Leduc Gas AU
52430202	Leduc Oil and Gas AU
524303	Exshaw-Rundle TPS
52430301	Exshaw-Rundle Gas AU
52430302	Exshaw-Rundle Oil and Gas AU
524304	Combined Triassic/Jurassic TPS
52430401	Combined Triassic/Jurassic Gas AU
52430402	Combined Triassic/Jurassic Oil and Gas AU
524305	Mannville-Upper Mannville TPS
52430501	Mannville Gas AU
52430502	Mannville Continuous (Basin-Center) Gas AU
524306	Second White Specks-Cardium TPS
52430601	Second White Specks-Cardium Gas AU
52430602	Second White Specks-Cardium Oil and Gas AU
52430603	Upper Cretaceous Eastern Shallow Gas AU
5244	**Williston Basin, Canada Province**
524401	Yeoman TPS
52440101	Yeoman Oil AU
524402	Brighthölme TPS
52440201	Brightholme Oil AU
524403	Bakken TPS
52440301	Bakken Sandstone AU
524404	Lodgepole TPS
52440401	Lodgepole Oil AU
5245	**Rocky Mountain Deformed Belt (Province, assessed with Province 5243)**
5301	**Tampico-Misantla Basin (Province, assessed with Province 5305)**
5302	**Veracruz Basin (Province, assessed with Province 5305)**

Table 2. *(Cont.)*

CODE	NAME
5304	**Saline-Comalcalco Basin (Province, assessed with Province 5305)**
5305	**Villahermosa Uplift Province**
530501	Pimienta-Tamabra TPS
53050101	El Abra-Like Reef and Backreef Limestone AU
53050102	Tamabra-Like Debris-Flow-Breccia Limestone Overlying Evaporites AU
53050103	Tamabra-Like Debris-Flow-Breccia Limestone and Overlying Strata AU
53050104	Tamabra-Like Debris-Flow-Breccia Limestone of the Golden Lane AU
53050105	Tamaulipas-Like Basinal Limestone and Tertiary Strata Without Underlying Evaporites AU
53050106	Tamaulipas-Like Basinal Limestone and Tertiary Strata Overlying Evaporites AU
53050107	Tamabra-Like Debris-Flow-Breccia Limestone North of Campeche AU
6	**Central and South America Region**
6021	**Guyana-Suriname Basin Province**
602101	Cenomanian-Turonian TPS
60210101	Late Cretaceous-Tertiary Turbidites AU
60210102	Cretaceous Carbonates AU
60210103	Late Cretaceous-Tertiary Nearshore Sandstones AU
6022	**Foz do Amazonas Basin Province**
602201	Neogene TPS
60220101	Amazon Delta and Submarine Fan AU
60220102	Offshore Gas Hydrates AU
6029	**Sergipe-Alagoas Basin Province**
602901	Neocomian to Turonian Composite TPS
60290101	Western Pre-Aptian Reservoirs AU
60290102	Late Cretaceous-Tertiary Deep-Water Sandstones AU
6034	**Espirito Santo Basin Province**
603401	Cretaceous Composite TPS
60340101	Espirito Santo Shelf AU
60340102	Late Cretaceous-Tertiary Slide Blocks and Turbidites AU
60340103	Abrolhos Sub-Volcanic Structures AU
6035	**Campos Basin Province**
603501	Lagoa Feia-Carapebus TPS
60350101	Late Cretaceous-Tertiary Turbidites AU
60350102	Cretaceous Carbonates AU
60350103	Salt Dome Province Tertiary Sandstones AU
6036	**Santos Basin Province**
603601	Guaratiba-Guaruja (Cretaceous) Composite TPS
60360101	Santos Shelf AU
60360102	Salt-Structured Deep-Water Sandstones AU
6037	**Pelotas Basin Province**
603701	Cenomanian-Turonian-Tertiary Composite TPS
60370101	Pelotas Platform and Basin AU
60370102	Offshore Gas Hydrates AU
6041	**Putumayo-Oriente-Maranon Basin Province**
604101	Mesozoic-Cenozoic TPS
60410101	Hollin-Napo AU
60410102	Basin-Center Gas AU
604102	Paleozoic TPS
60410201	Ene AU
6045	**Santa Cruz-Tarija Basin Province**
604501	Los Monos-Machareti TPS
60450101	Sub-Andean Fold and Thrust Belt AU
60450102	Foreland Basins AU
60450103	Foreland Central Chaco High AU
6055	**Neuquen Basin Province**
605501	Neuquen Hybrid TPS
60550101	Neuquen Extensional Structures AU
60550102	Neuquen Foothills Structure AU
60550103	Dorsal de Neuquen Structure AU
6058	**San Jorge Basin Province**
605801	D-129 TPS
60580101	San Jorge Extensional Structures AU
60580102	San Bernardo Fold Belt Structures AU
605802	Aguada Bandera TPS
60580201	Hypothetical Basin-Centered Gas AU
6059	**Magallanes Basin Province**
605901	Lower Inoceramus TPS
60590101	Magallanes Extensional Structures AU
60590102	Andean Fold Belt Structures AU
6060	**Falklands Plateau Province**
606001	Neocomian Lacustrine TPS
60600101	North Falklands Basin AU
606002	Lower Cretaceous TPS
60600201	South Falklands Basin AU
6063	**Malvinas Basin Province**
606301	Lower Cretaceous Marine TPS
60630101	Malvinas Extensional Structures AU
6081	**Talara Basin Province**
608101	Cretaceous-Tertiary TPS
60810101	Cretaceous-Paleogene Basin AU
6083	**Progreso Basin Province**
608301	Neogene TPS
60830101	Neogene Pull-Apart Basin AU
608302	Cretaceous-Paleogene TPS
60830201	Cretaceous-Paleogene Santa Elena Block AU
6090	**Middle Magdalena Province**
609001	La Luna-La Paz TPS
60900101	Northern AU
60900102	Southern AU
60900103	Eastern AU
60900104	La Luna and Older AU
6096	**Llanos Basin Province**
609601	Gacheta-Mirador TPS
60960101	Central AU
60960102	Peripheral AU

Table 2. *(Cont.)*

CODE	NAME
6098	**East Venezuela Basin Province**
609801	Querecual TPS
60980101	East Venezuela Fold and Thrust Belt AU
60980102	Guarico Sub-Basin AU
60980103	Maturin Sub-Basin AU
60980104	Orinoco Heavy Oil and Tar Belt AU
609802	Upper Cretaceous/Tertiary TPS
60980201	Trinidad Basins AU
60980202	Orinoco Delta and Offshore AU
6099	**Maracaibo Basin Province**
609901	La Luna/Maracaibo TPS
60990101	Main Maracaibo Basin AU
60990102	Southwest Maracaibo Basin Fold Belt AU
60990103	Basin-Centered Gas AU
6103	**Tobago Trough Province**
610301	Lower Cruse TPS
61030101	Carupano Basin Gas AU
6107	**Lesser Antilles Deformed Belt Province**
610701	Tobago Trough Paleogene TPS
61070101	Inner Forearc Deformation Belt AU
6117	**Greater Antilles Deformed Belt Province**
611701	Upper Jurassic-Neocomian TPS
61170101	North Cuba Fold and Thrust Belt AU
7	**Sub-Saharan Africa and Antarctica Region**
7013	**Senegal Province**
701301	Cretaceous-Tertiary Composite TPS
70130101	Coastal Plain and Offshore AU
7183	**Gulf of Guinea Province**
718301	Cretaceous Composite TPS
71830101	Coastal Plain and Offshore AU
7192	**Niger Delta Province**
719201	Tertiary Niger Delta (Agbada/Akata) TPS
71920101	Agbada Reservoirs AU
71920102	Akata Reservoirs AU
7203	**West-Central Coastal Province**
720301	Melania-Gamba TPS
72030101	Gabon Subsalt AU
720302	Azile-Senonian TPS
72030201	Gabon Suprasalt AU
720303	Congo Delta Composite TPS
72030301	Central Congo Delta and Carbonate Platform AU
72030302	Central Congo Turbidites AU
720304	Cuanza Composite TPS
72030401	Cuanza-Namibe AU
7303	**Orange River Coastal Province**
730301	Cretaceous Composite TPS
73030101	Offshore AU
8	**South Asia Region**
8026	**Kohat-Potwar Province**
802601	Patala-Namal TPS
80260101	Kohat-Potwar Intrathrust Basin AU
8034	**Assam Province**
803401	Sylhet-Kopili/Barail-Tipam Composite TPS
80340101	Sylhet-Kopili/Barail-Tipam Composite AU
8042	**Indus Province**
804201	Sembar-Goru/Ghazij TPS
80420101	Greater Indus Foreland and Foldbelt AU
80420102	Indus Fan AU
8043	**Bombay Province**
804301	Eocene-Miocene Composite TPS
80430101	Eocene-Miocene Bombay Shelf AU
80430102	Eocene-Miocene Cambay Deltaic AU
8047	**Ganges-Brahmaputra Delta Province**
804701	Permian Coal TPS
80470101	Coalbed Gas AU
804702	Jalangi-Sylhet/Burdwan Composite TPS
80470201	Western Shelf and Slope AU
804703	Jenam/Bhuban-Bokabil TPS
80470301	Central Basin AU
80470302	Eastern Fold Belt AU
8048	**Irrawaddy Province**
804801	Eocene to Miocene Composite TPS
80480101	Central Burma Basin AU
80480102	Irrawaddy-Andaman AU

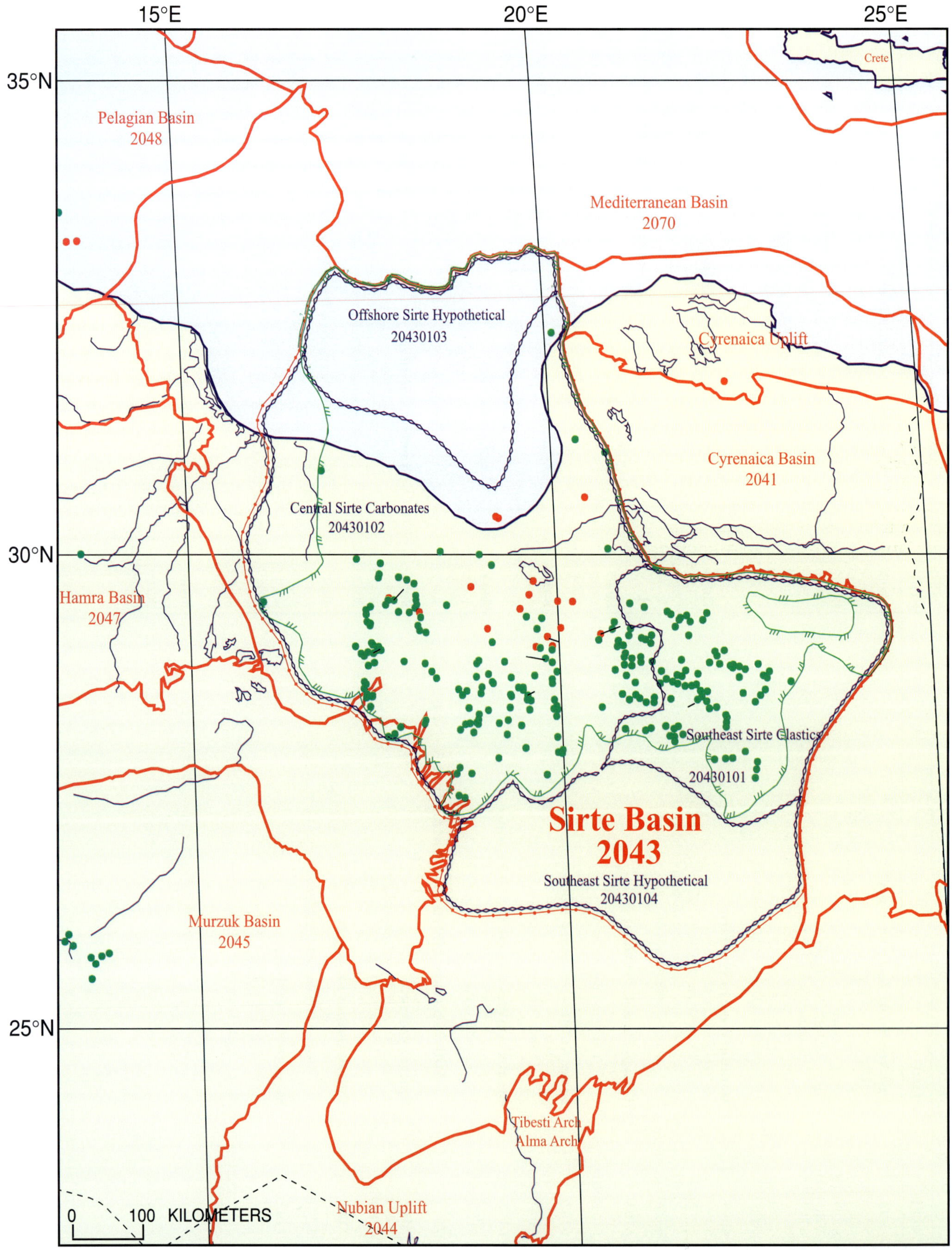

Projection: Robinson. Central meridian: 0

EXPLANATION

- Hydrography
- Shoreline
- Geologic province name and boundary
- Country boundary
- Natural Gas field centerpoint
- Oil field centerpoint

Annotated Fields with magnitude > 100,000 mbo or mmscfg

ASSESSMENT DATA

- Assessment units boundary and name
- Total petroleum system boundary
- K Area of active source rocks boundaryó Ticks indicate side of their presence; queried where extent uncertain

FIGURE 1. Sirte-Zelten Total Petroleum System showing the boundaries of the total petroleum system, area (pods) of active (thermally mature) source rocks, four assessment units, and oil and natural gas field centerpoints (for fields exceeding 100 million BOE). Note the 200-m bathymetric isobath forms a boundary between Offshore Sirte Hypothetical Assessment Unit (20430103) and Central Sirte Carbonates Assessment Unit (20430102). The 2000-m bathymetric isobath forms the northern boundary of the province, total petroleum system, and Offshore Sirte Hypothetical Assessment Unit 20430103 (from Ahlbrandt, 2001).

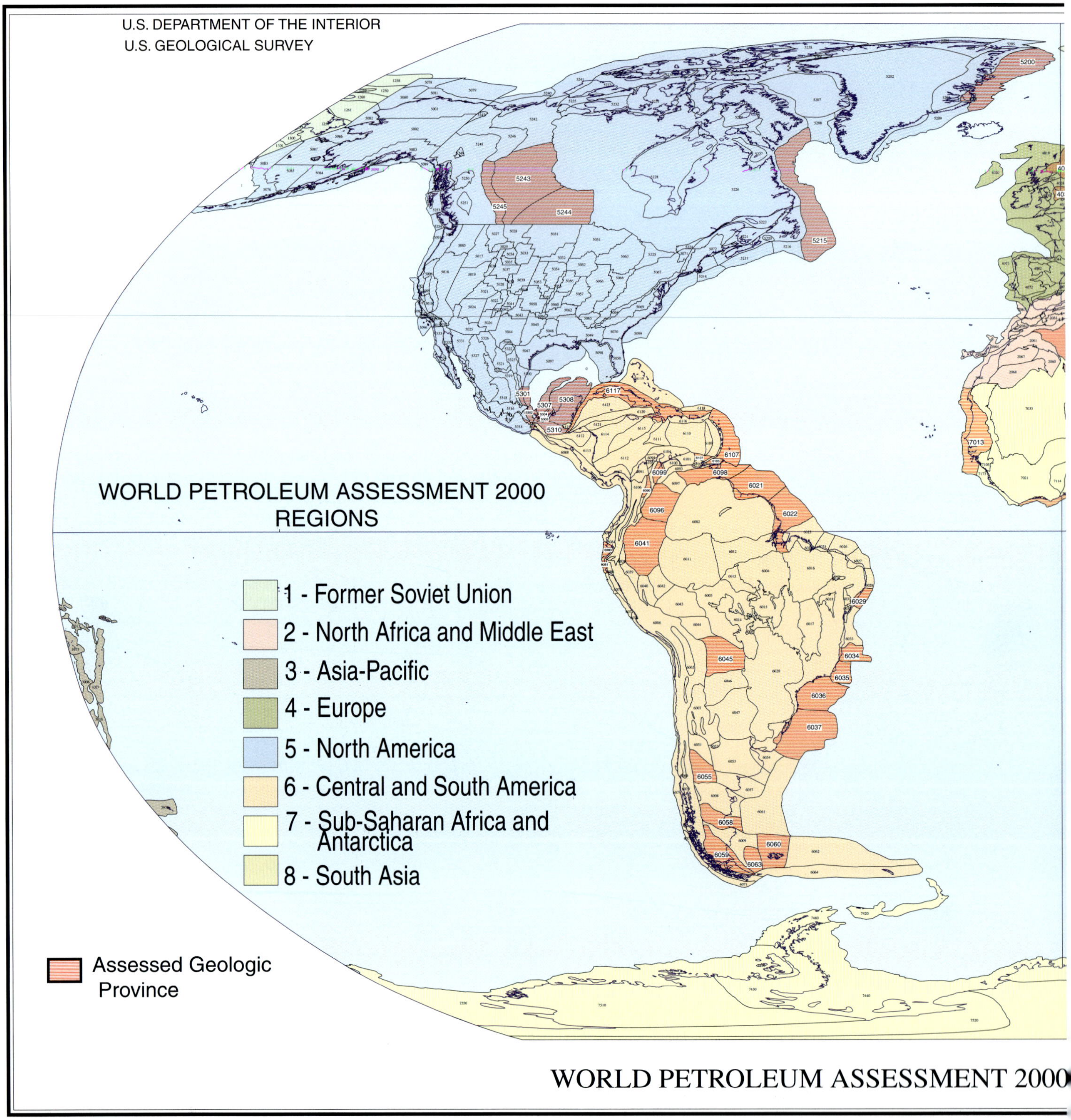

FIGURE 2. Geologic provinces identified by number for USGS World Petroleum Assessment 2000 (USGS, 2000). Assessed provinces in red. Note region numbers and names. In subsequent maps, assessed geologic provinces are highlighted by a heavy outline and are colored but not numbered. Within these assessed provinces, total petroleum systems (TPS) or assessment units (AU) relevant to that specific map are patterned and numbered with the respective 6- or 8-digit USGS code (Table 2). Unassessed provinces retain their 4-digit province code.

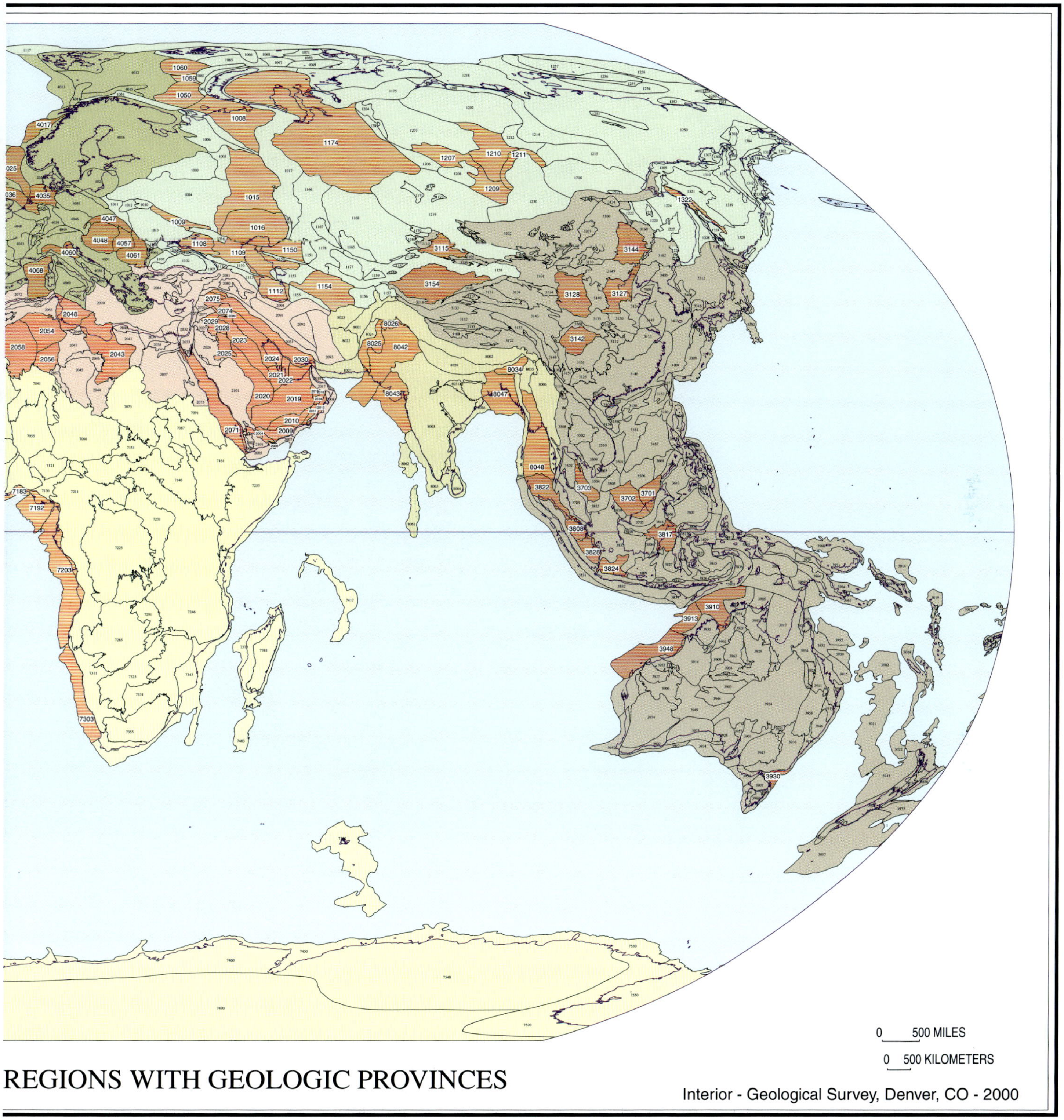

Projection Robinson
Central meridian 0

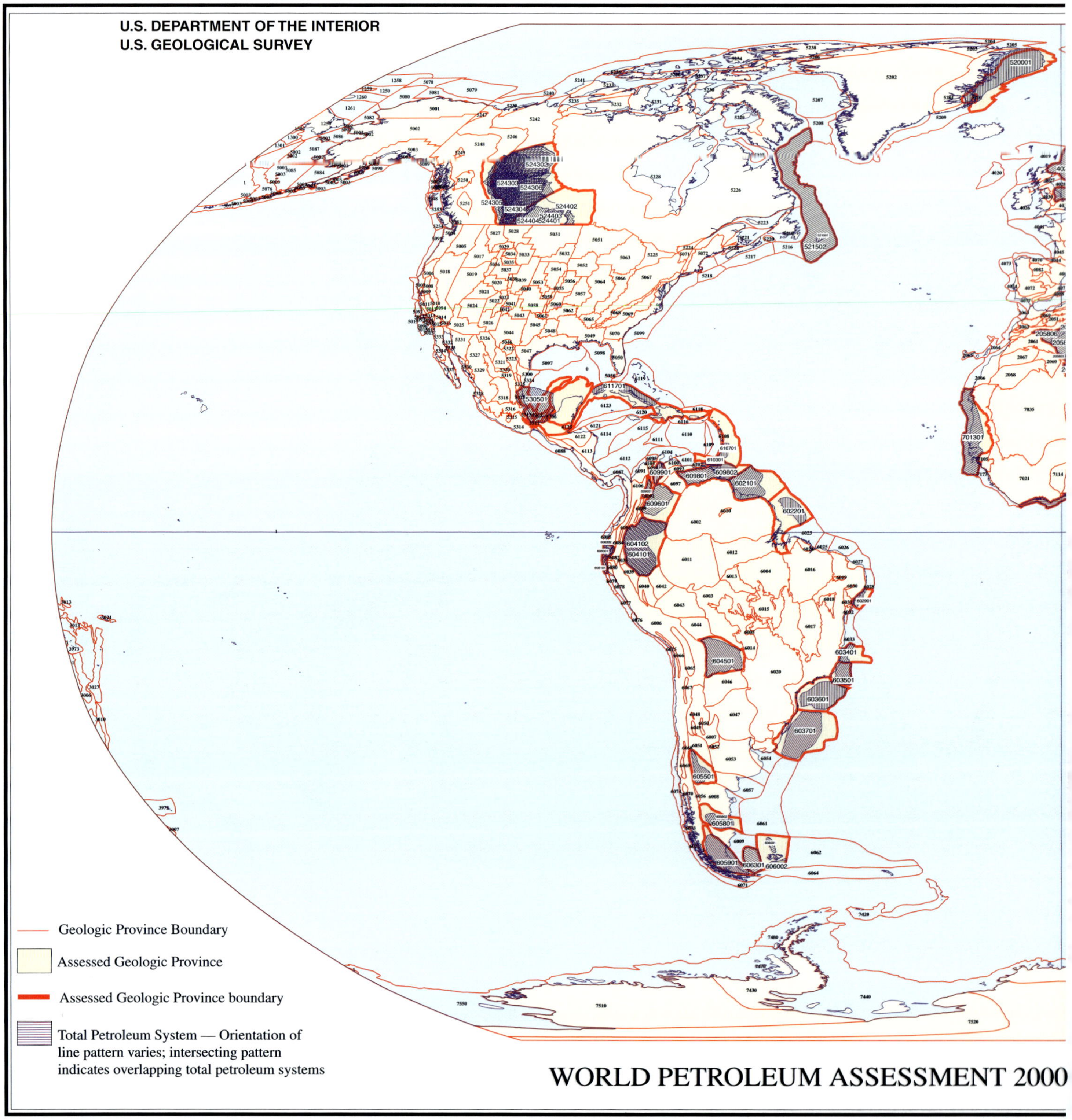

FIGURE 3. Total petroleum systems, identified by number. Those TPS assessed for USGS World Petroleum Assessment 2000 (USGS, 2000), are shaded in blue. The U.S. was not reassessed in this study; U.S. assessments by Gautier et al. (1996) and the MMS (1996) are included later in this report for comparative purposes.

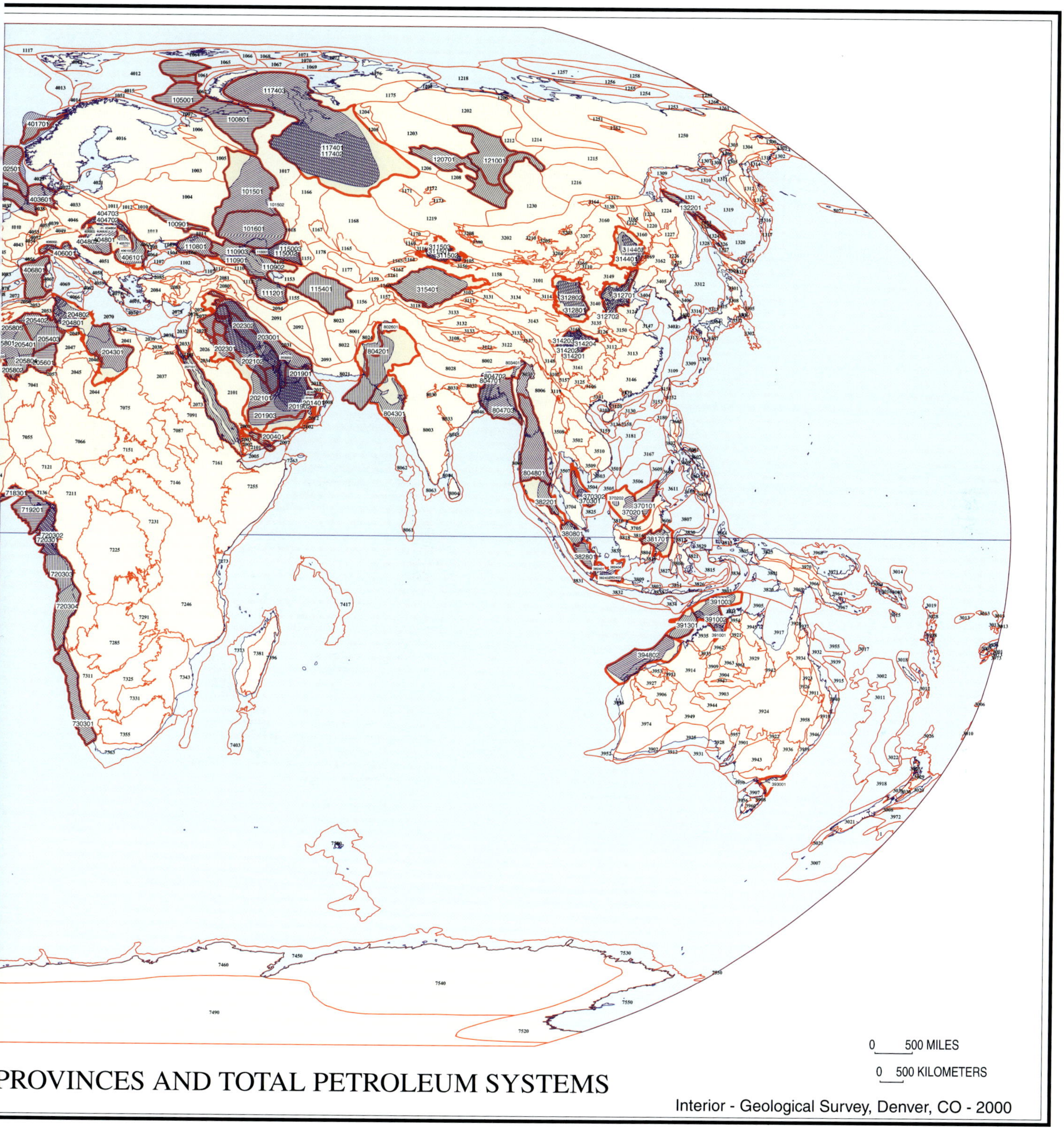

Projection Robinson
Central meridian 0

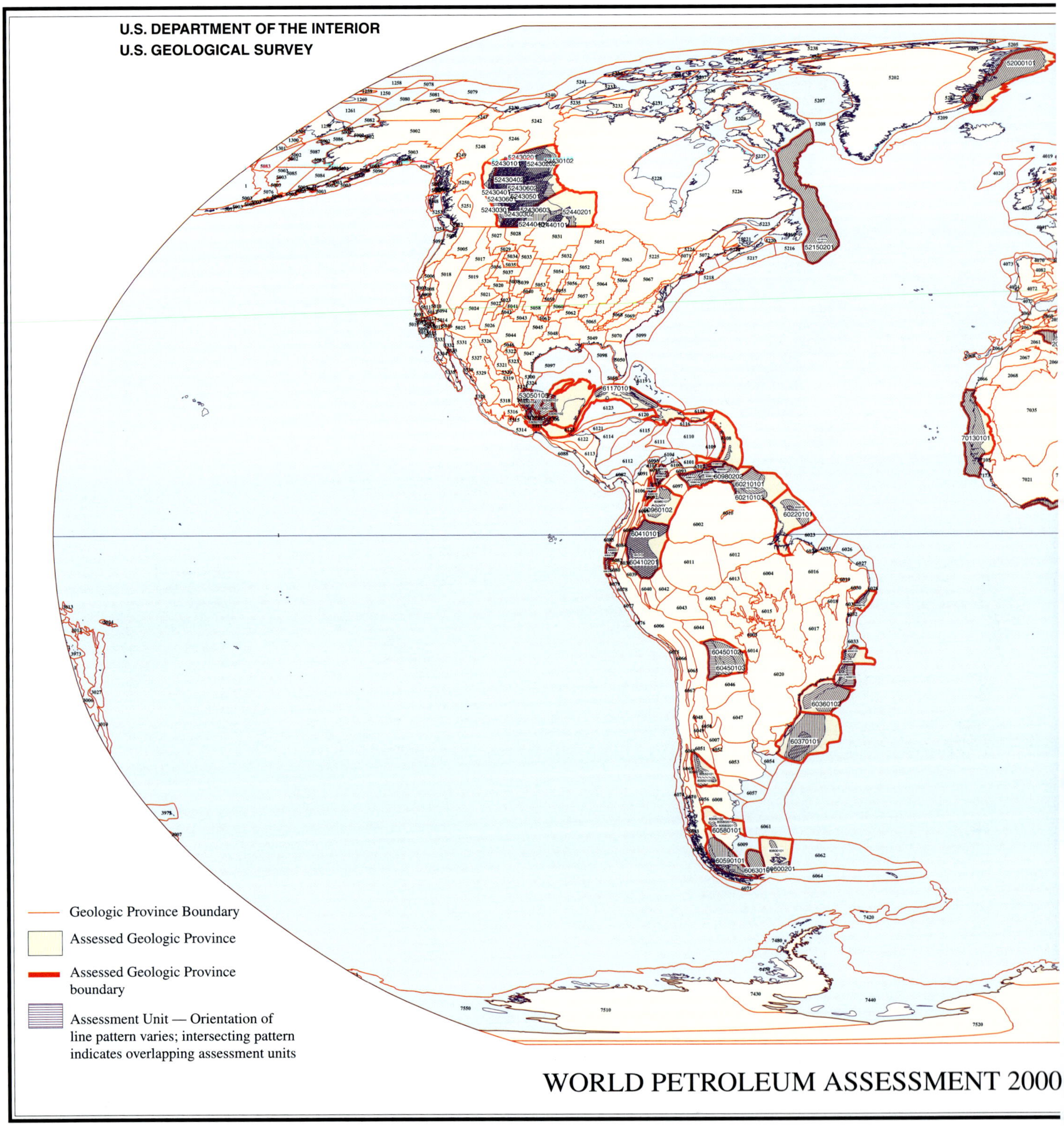

Figure 4. Assessment units of total petroleum systems evaluated for USGS World Petroleum Assessment 2000 (USGS, 2000). Assessment units are shaded in blue.

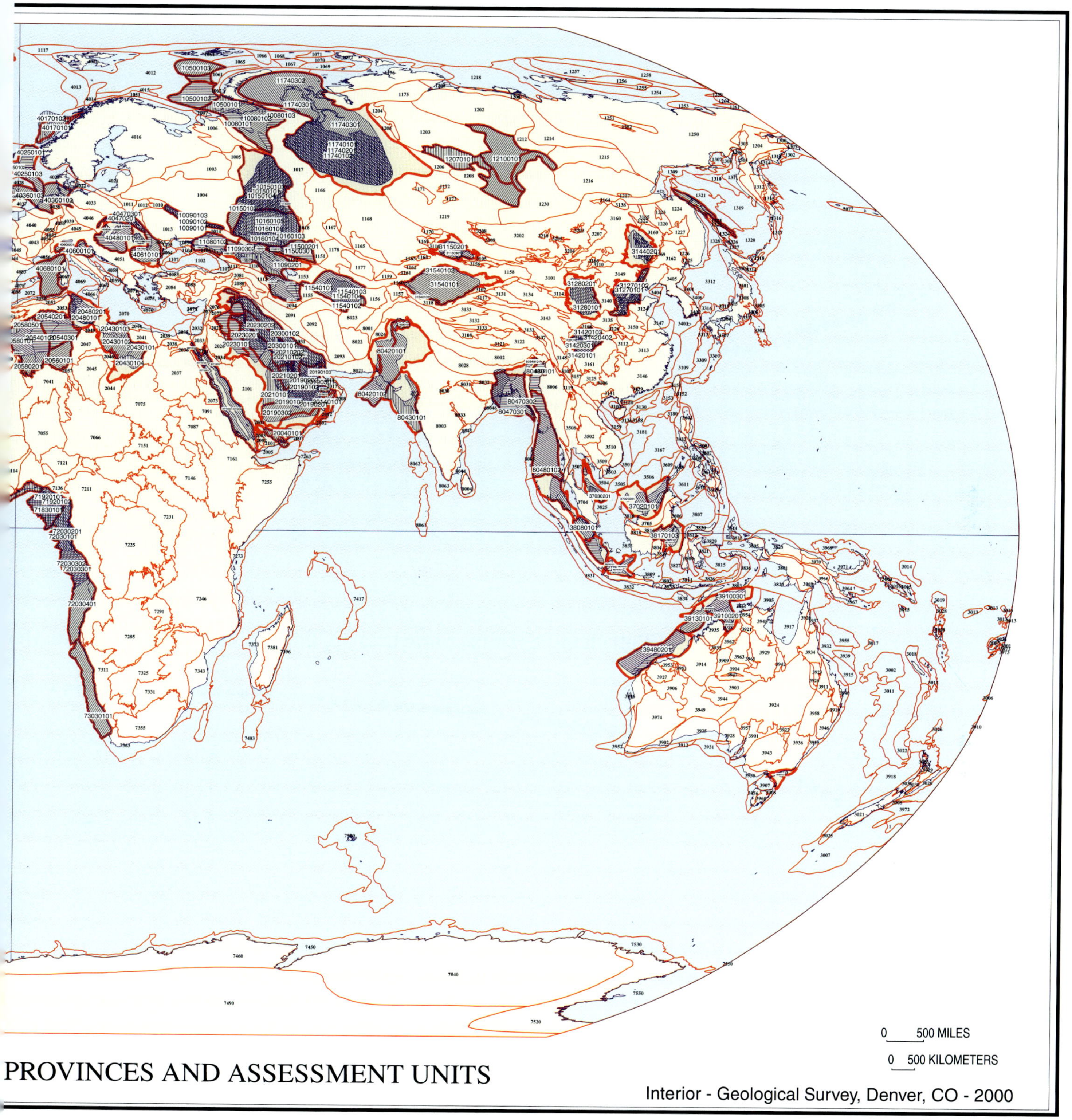

Projection Robinson
Central meridian 0

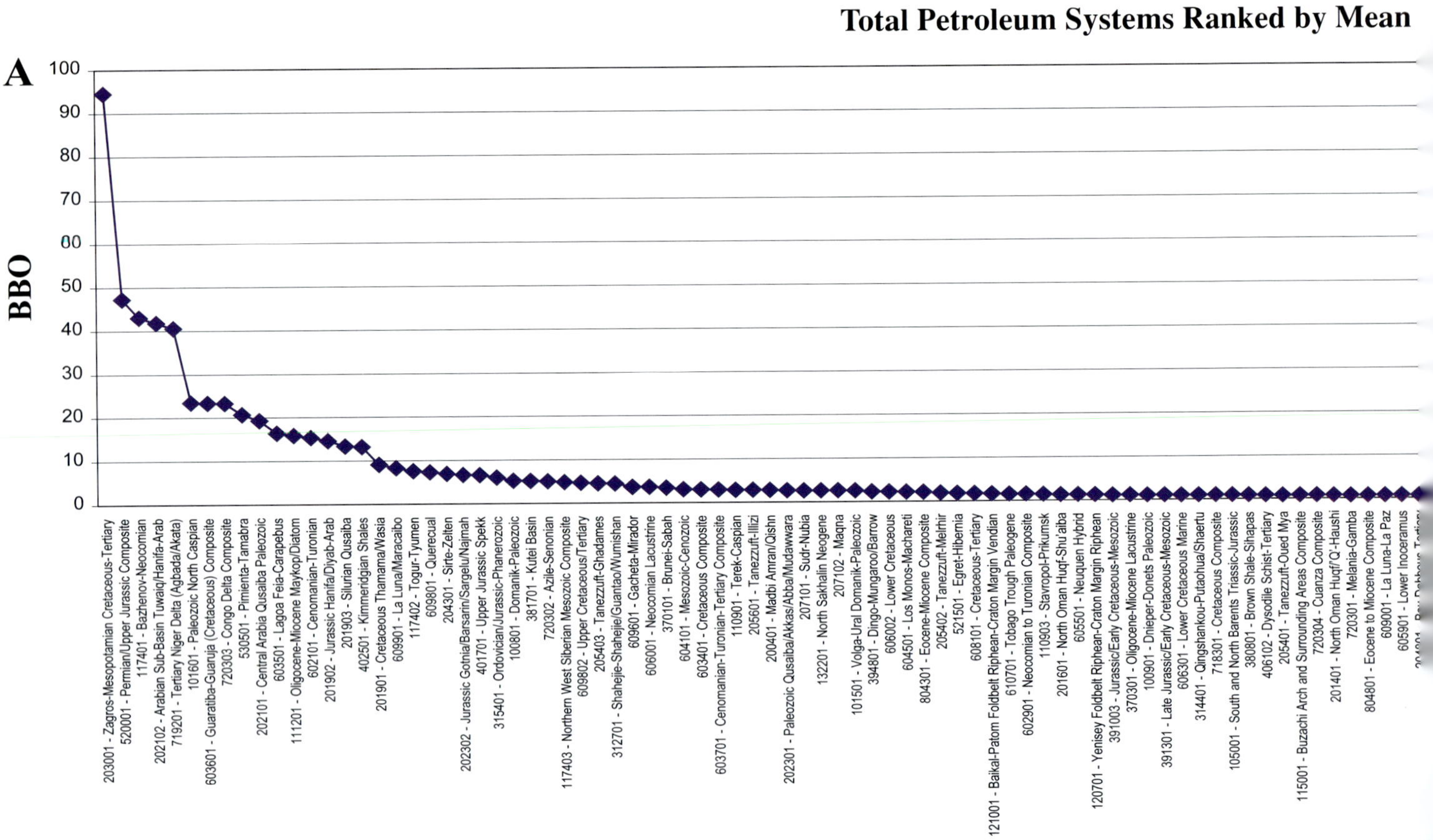

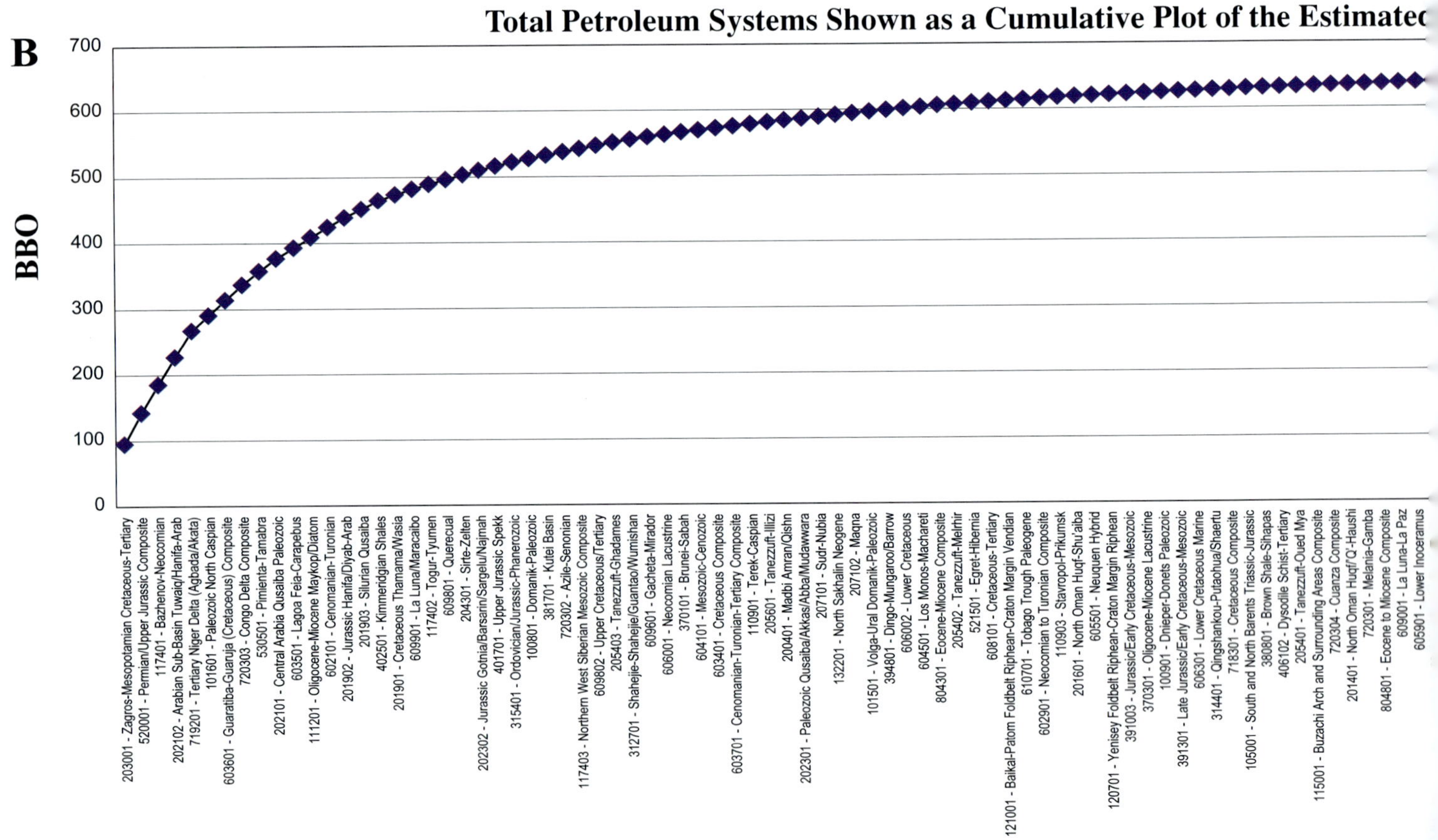

FIGURE 5. Total petroleum systems of the world A) ranked individually in order of size of estimated mean undiscovered volume of oil and B) shown as a cumulative plot of the estimated mean undiscovered oil volume of the same ranked TPS. The names and code numbers of the systems are shown; the first digit of the code refers to the Region, the following five digits the Province and TPS. BBO = billion barrels of oil.

Undiscovered Oil Endowment (BBO) as assessed in USGS (2000)

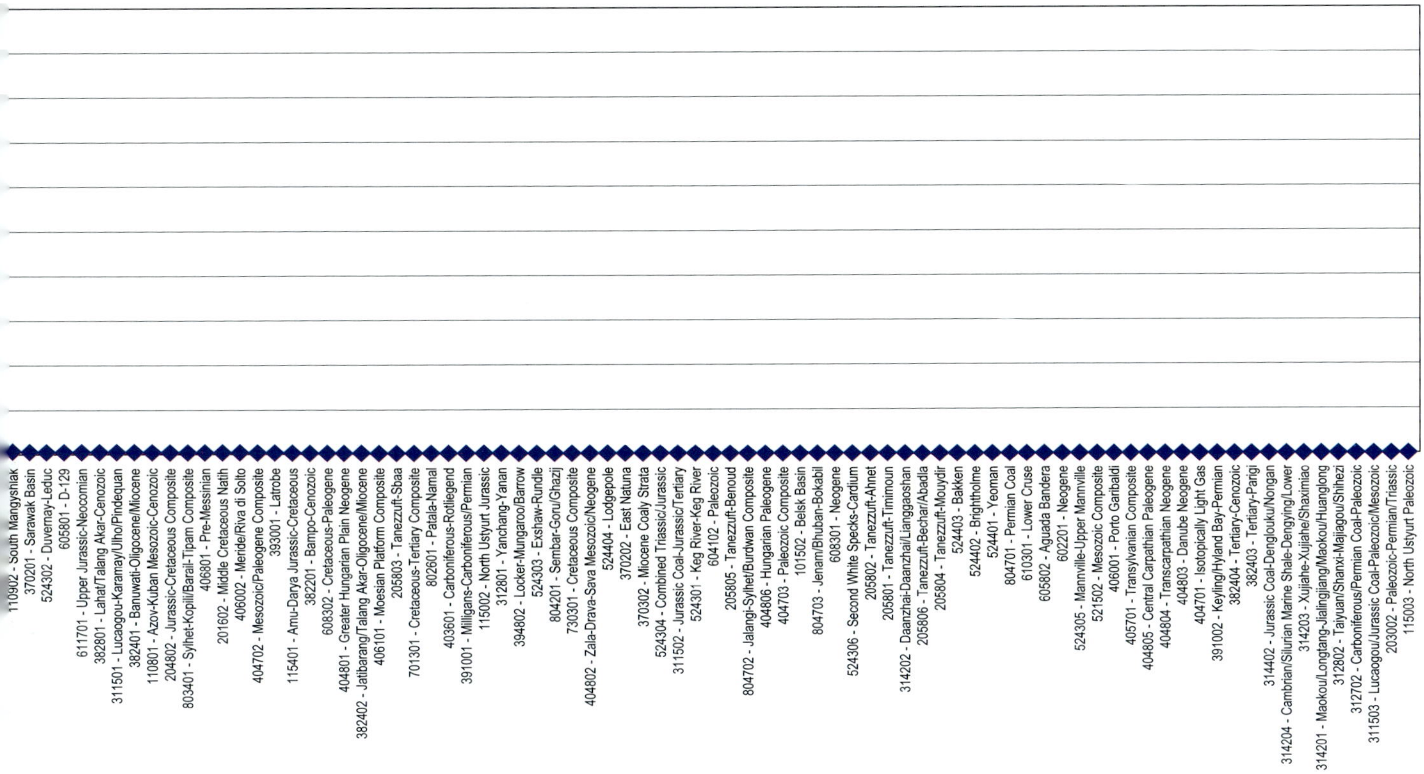

Mean Undiscovered Oil Volume (BBO) as assessed in USGS (2000)

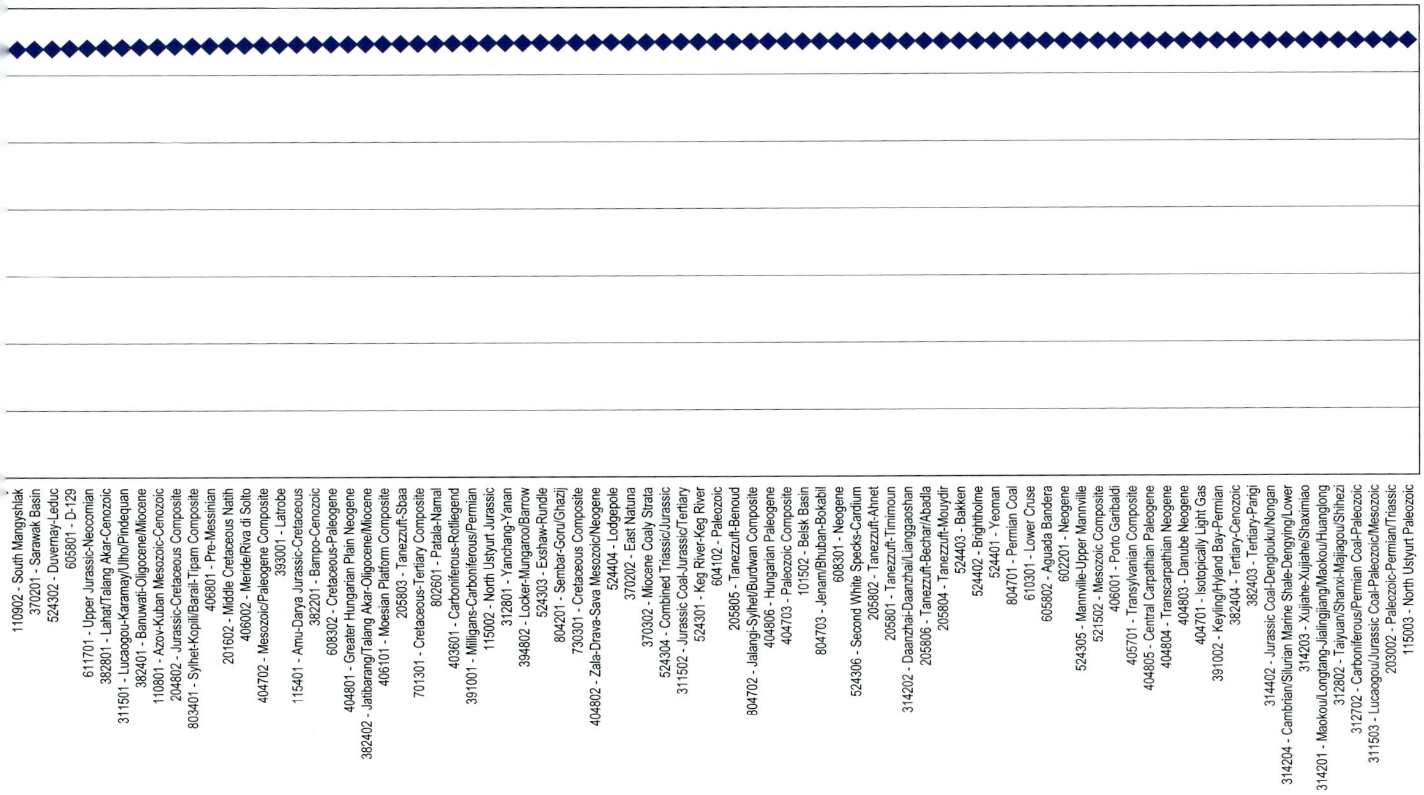

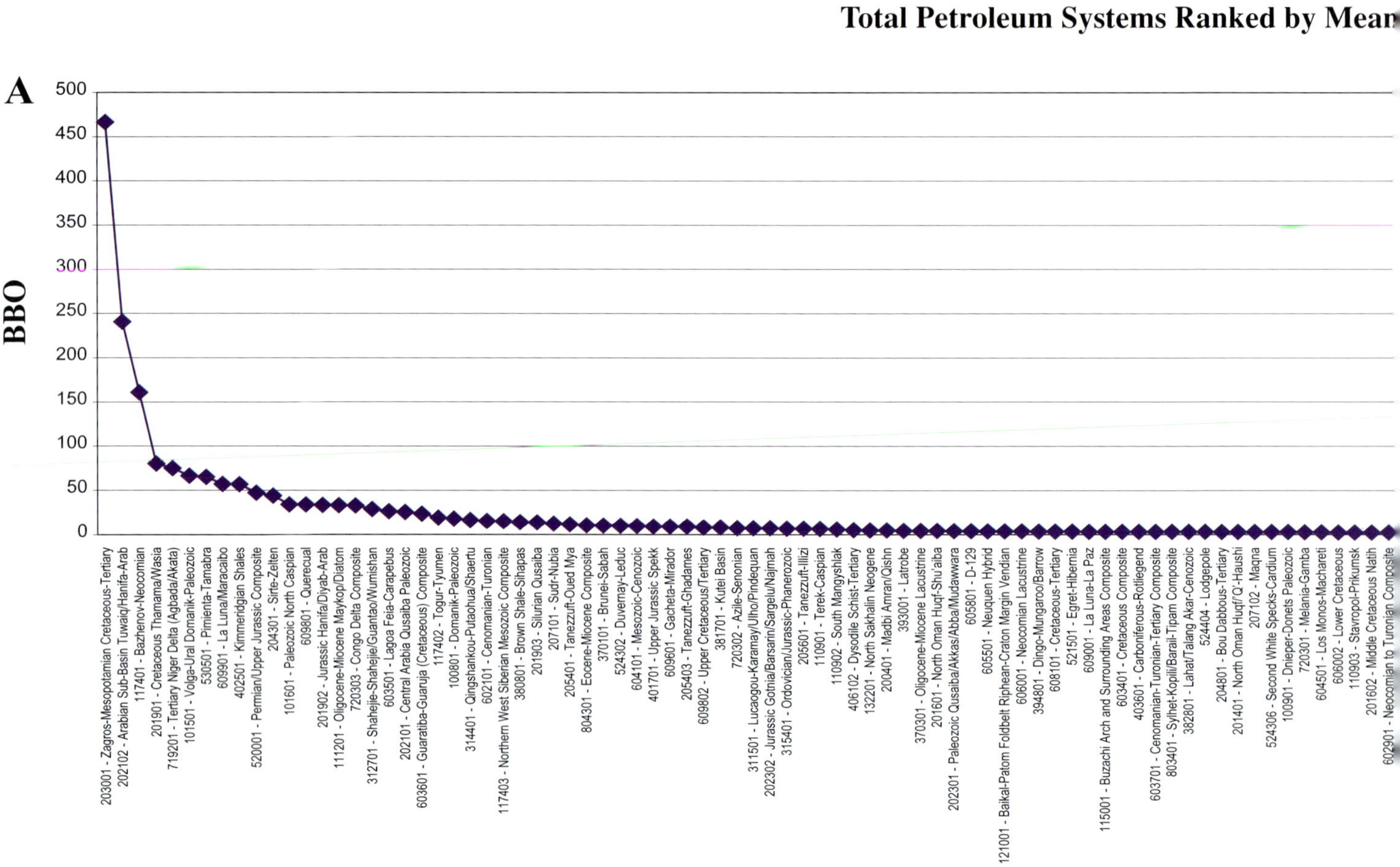

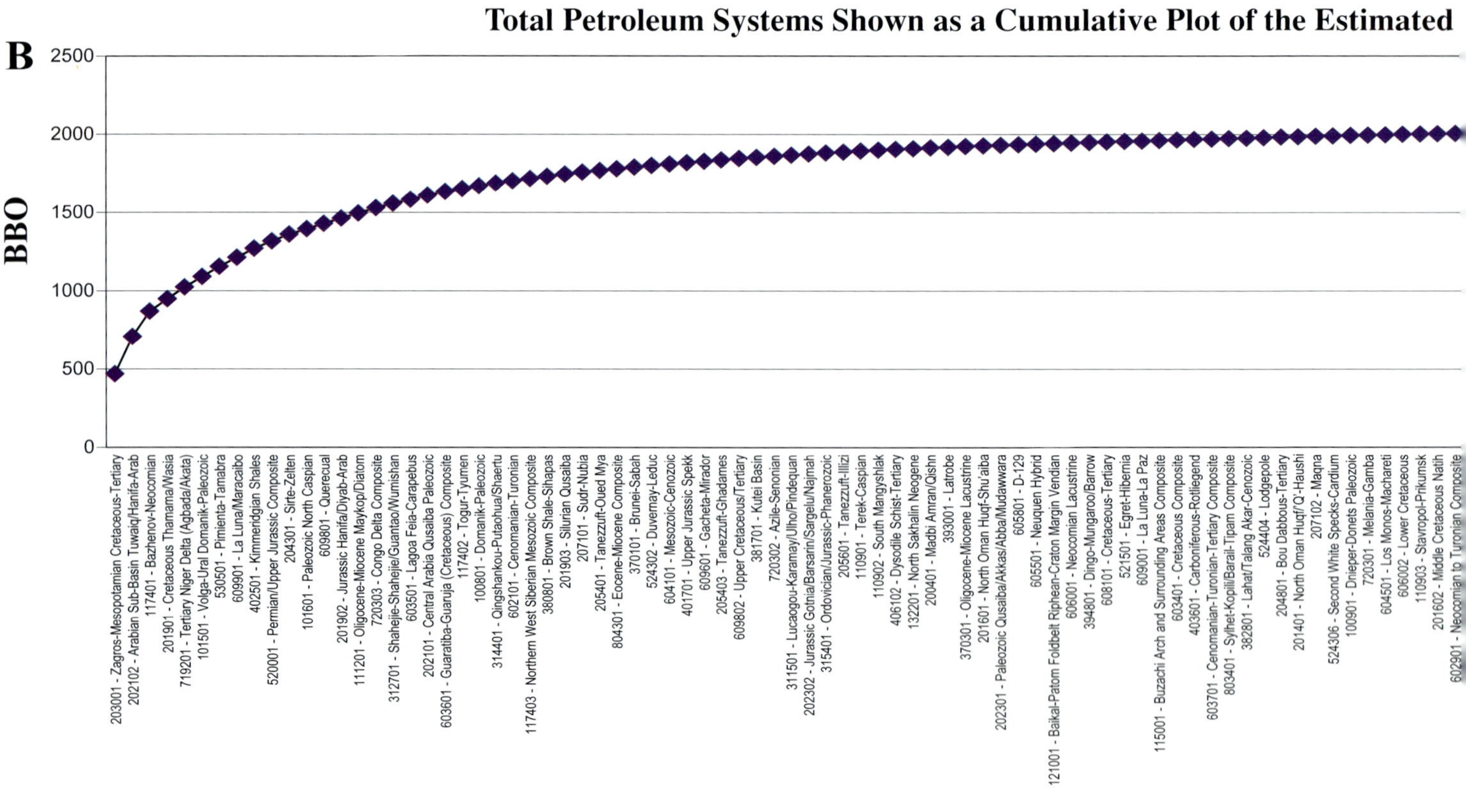

FIGURE 6. Total petroleum systems of the world A) ranked individually in order of size of estimated mean endowment of oil and B) shown as a cumulative plot of the estimated mean endowment of oil of the same ranked TPS. The names and code numbers of the systems are shown; the first digit of the code refers to the Region, the following five digits the Province and TPS. BBO = billion barrels of oil; Endowment = cumulative production plus remaining reserves plus mean estimate of undiscovered resources.

Oil Endowment (BBO) as assessed in USGS (2000)

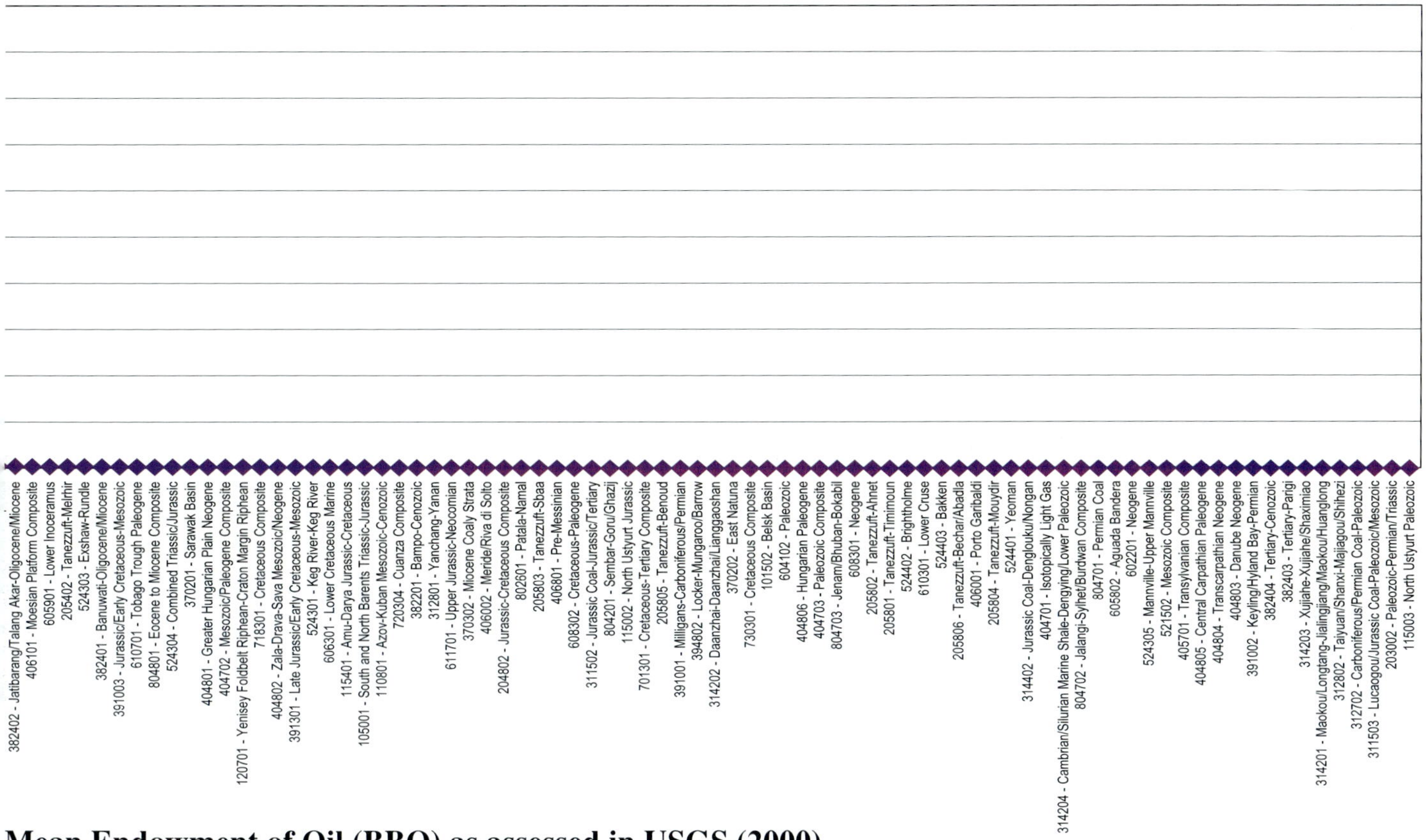

Mean Endowment of Oil (BBO) as assessed in USGS (2000)

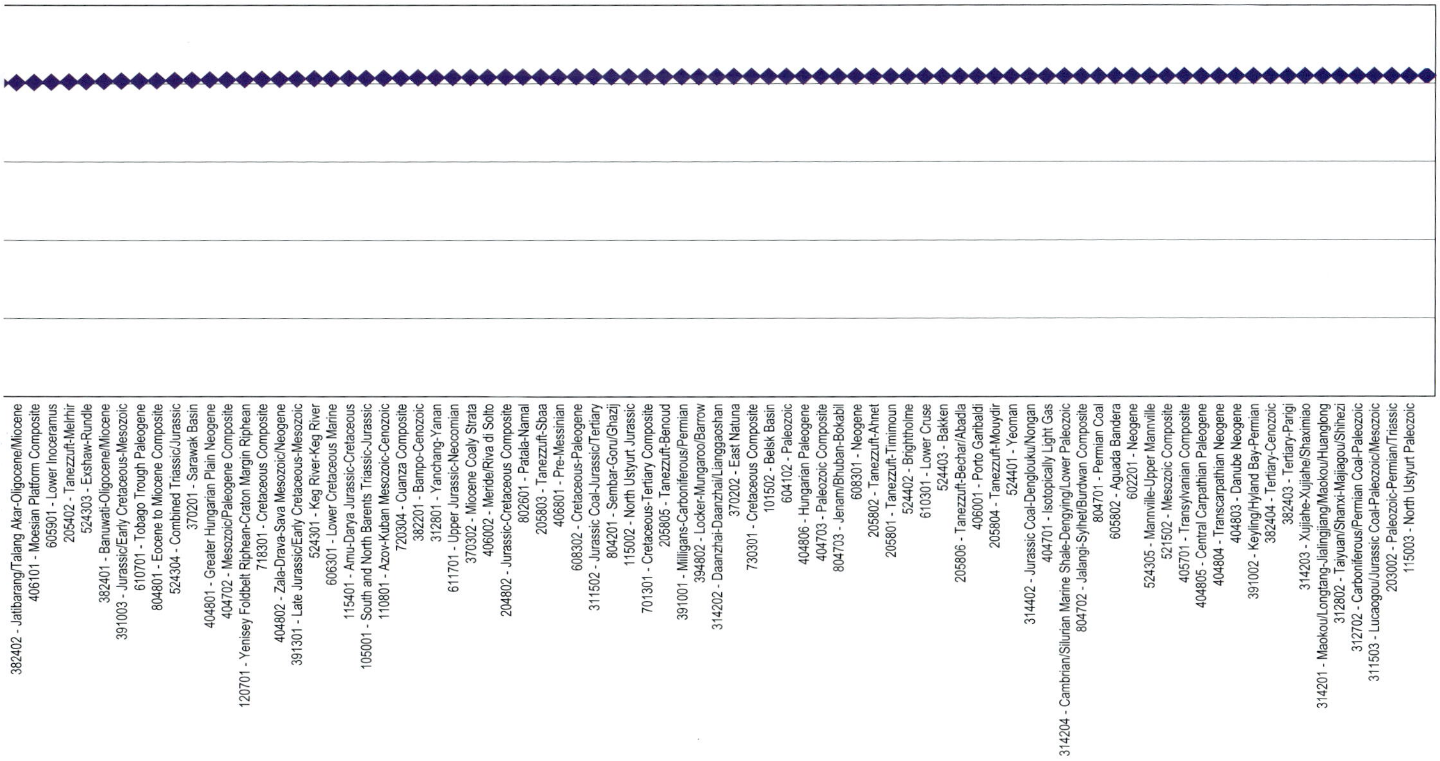

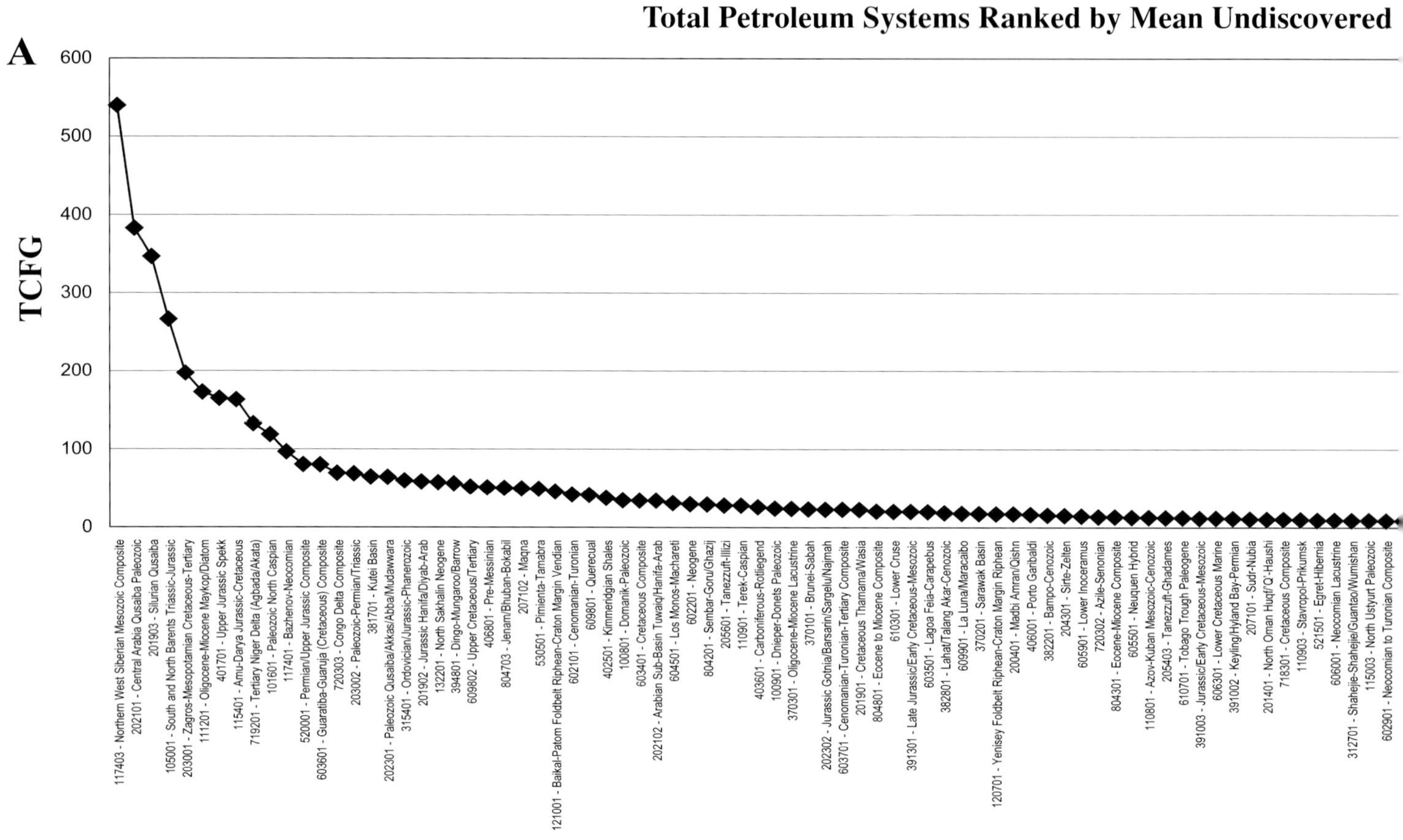

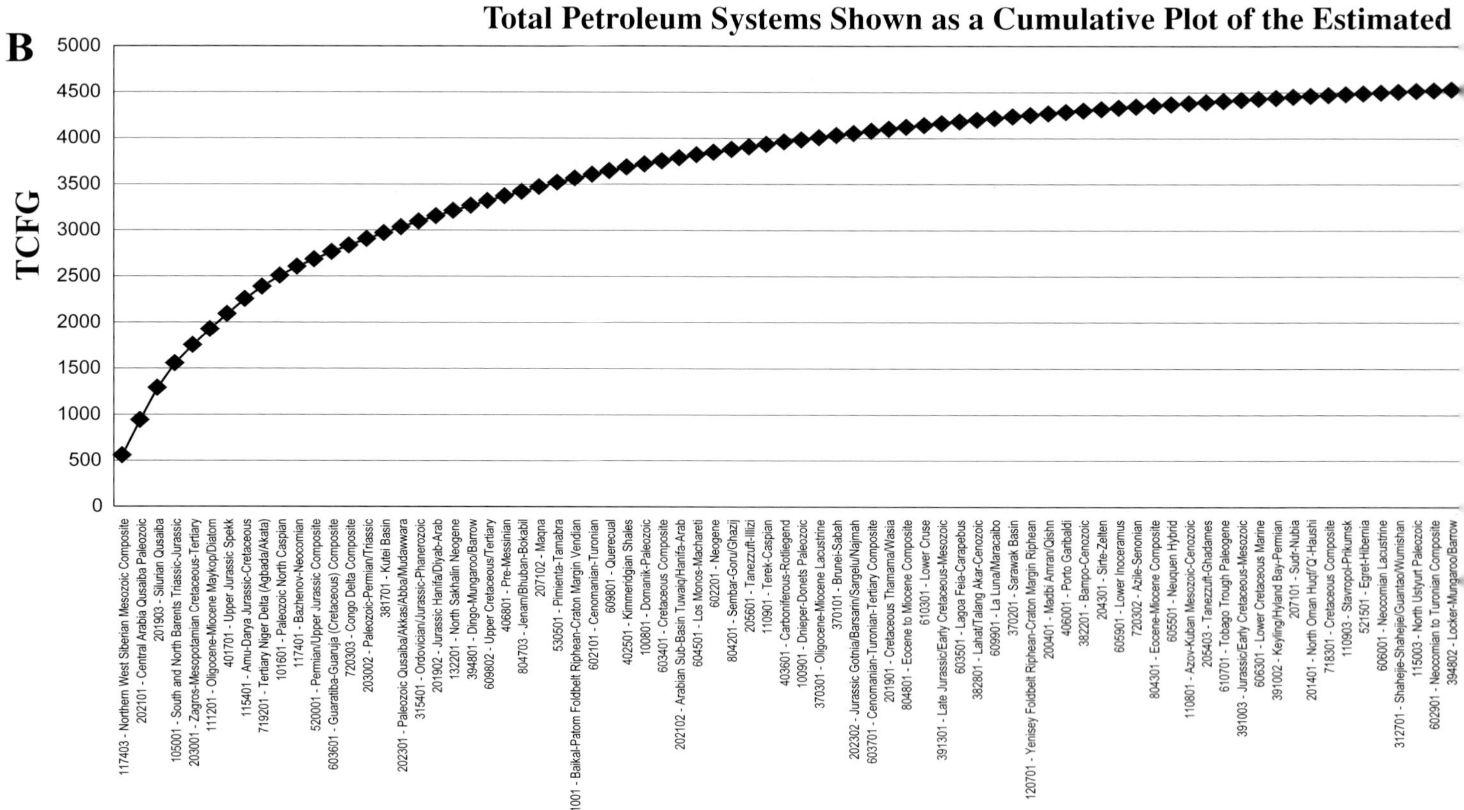

FIGURE 7. Total petroleum systems of the world A) ranked individually in order of size of estimated mean undiscovered natural gas and B) shown as a cumulative plot of the estimated mean undiscovered natural gas volume of the same ranked TPS. The names and code numbers of the systems are shown; the first digit of the code refers to the Region, the following five digits the Province and TPS. TCFG = cubic feet of natural gas.

Natural Gas (TCFG) as assessed in USGS (2000)

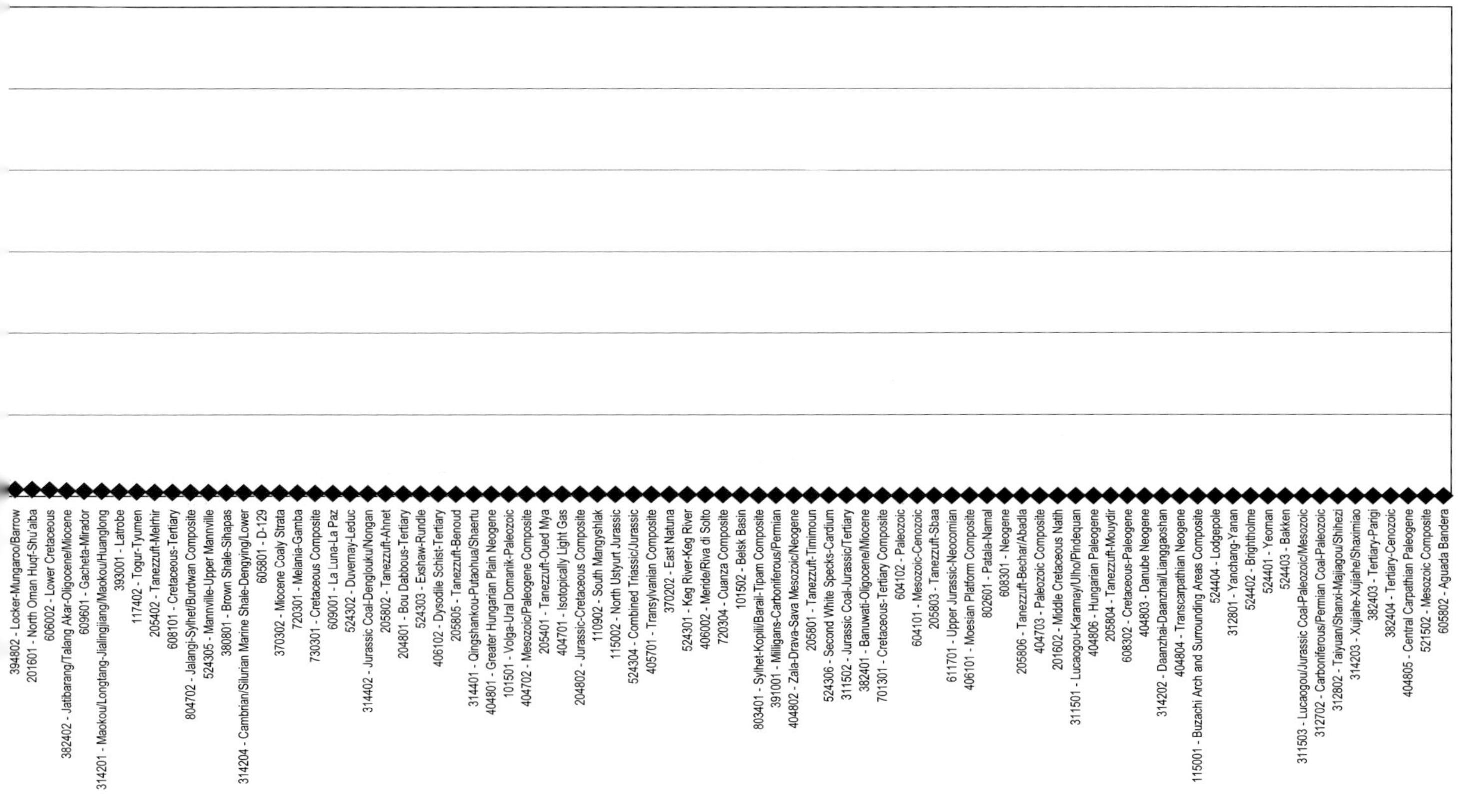

Mean Undiscovered Natural Gas Volume (TCFG) as assessed in USGS (2000)

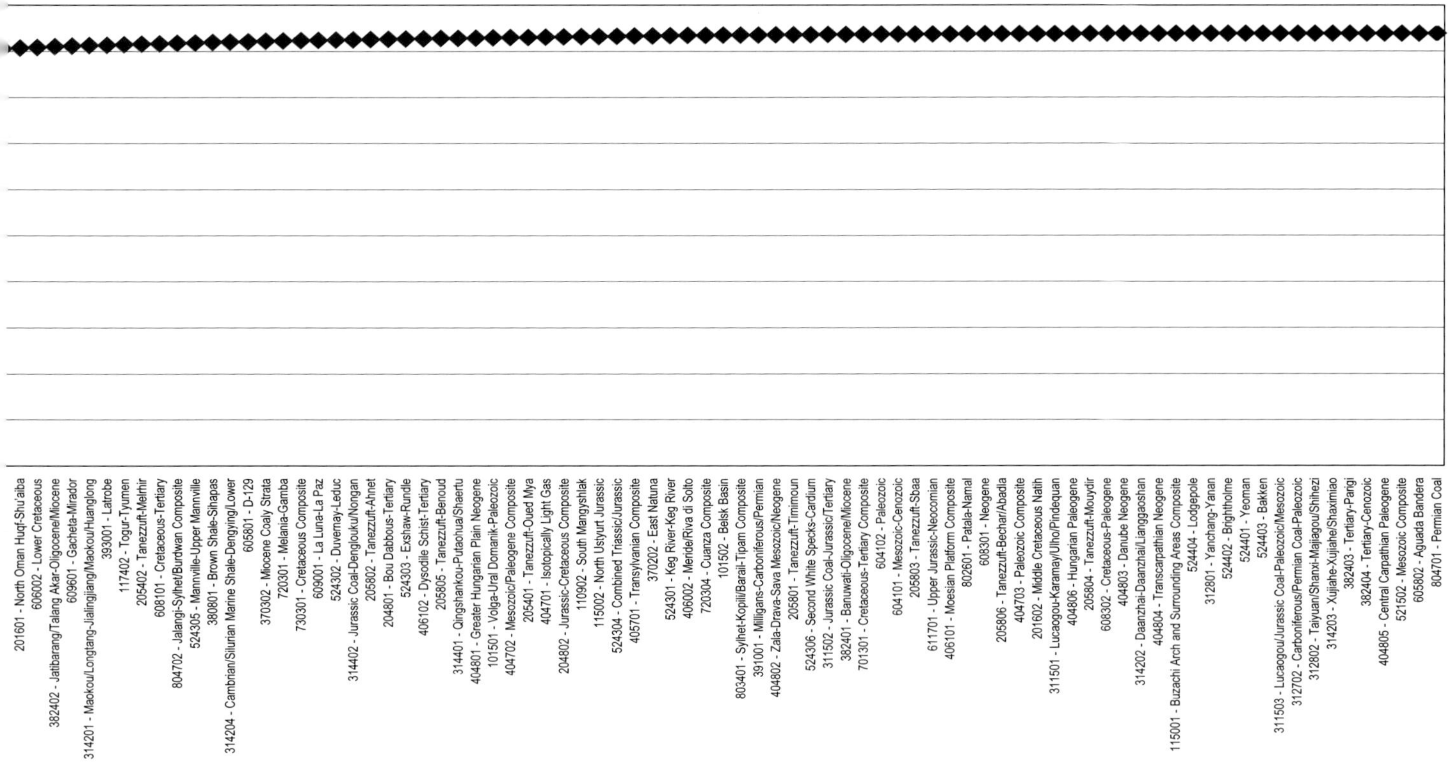

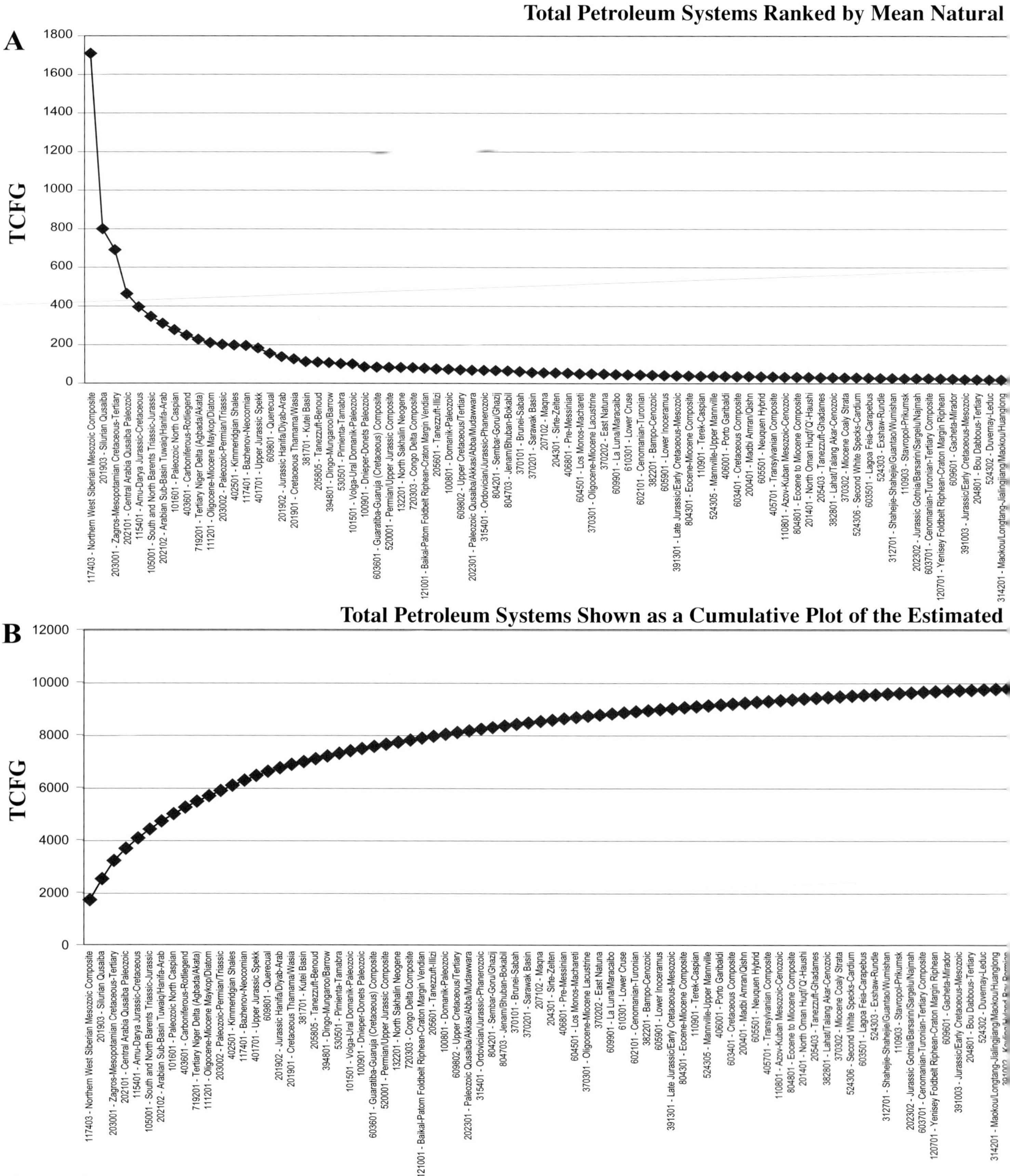

FIGURE 8. Total petroleum systems of the world. A) ranked individually in order of size of estimated mean endowment of natural gas and B) shown as a cumulative plot of the estimated mean endowment of natural gas of the same ranked TPS. The names and code numbers of the systems are shown; the first digit of the code refers to the Region, the following five digits the Province and TPS. TCFG = cubic feet of natural gas; Endowment = cumulative production plus remaining reserves plus mean estimate of undiscovered resources.

Gas Endowment (TCFG) as assessed in USGS (2000)

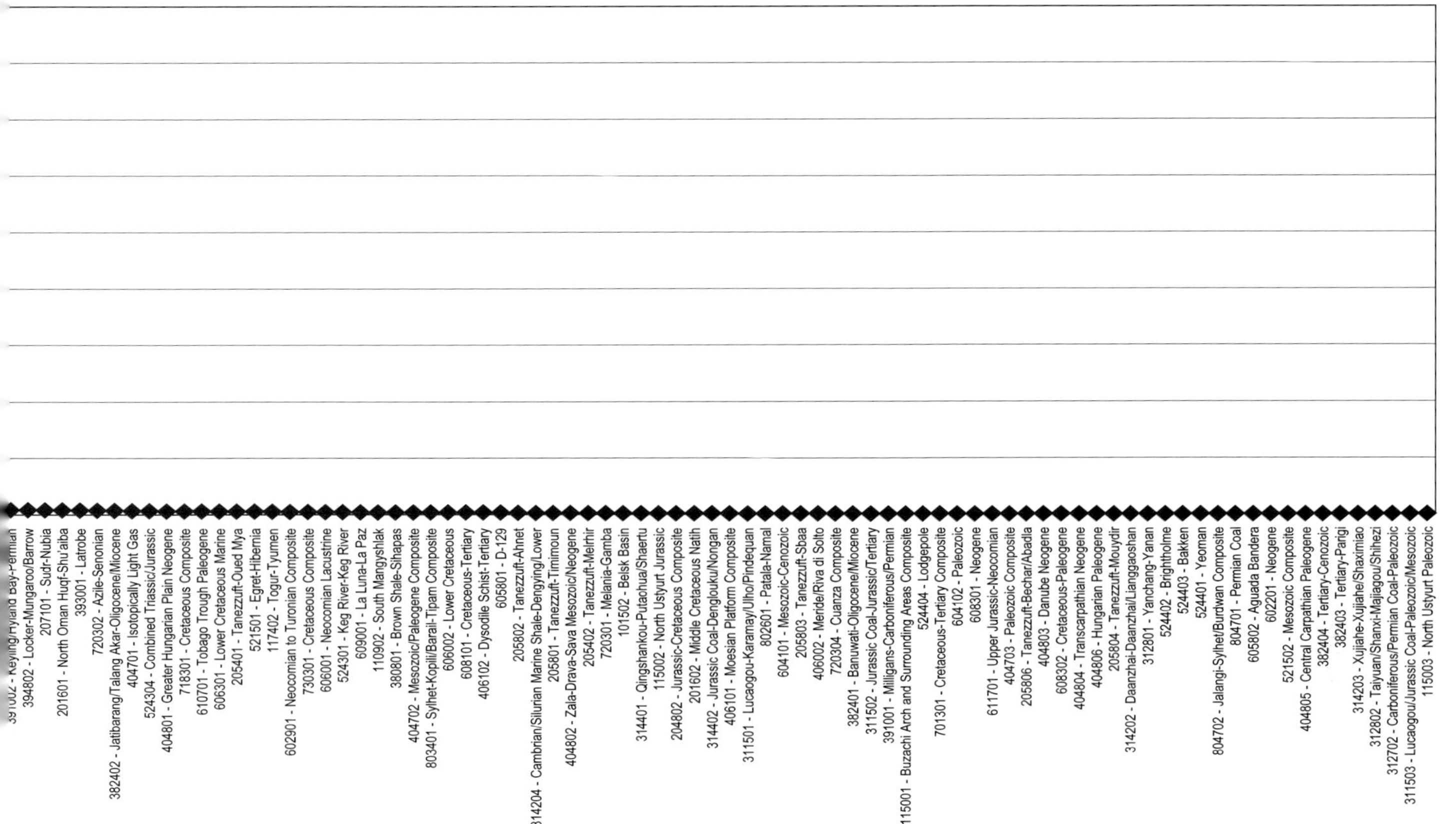

Mean Endowment of Natural Gas (TCFG) as assessed in USGS (2000)

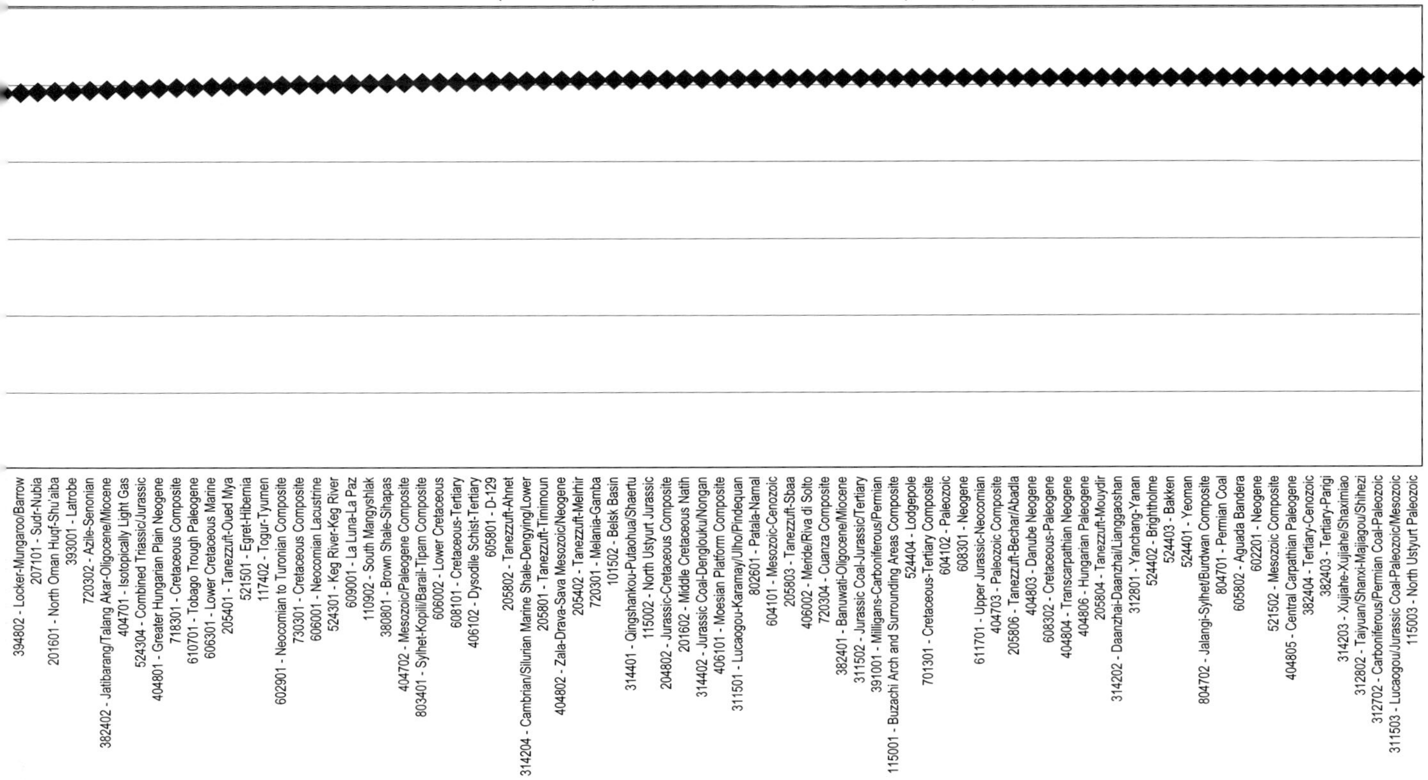

III. Results of Resource Assessment of Undiscovered Conventional Petroleum

A. WORLD LEVEL RESULTS, EXCLUDING THE U.S.

Volumes (statistical mean estimates) of undiscovered conventional oil, natural gas, and NGL (exclusive of the U.S.) as assessed in USGS (2000) are:

- 649 billion bbl of oil (Figure 9)
- 4669 tcf of natural gas (778 billion BOE) (Figure 10)
- 207 billion bbl of natural gas liquids (NGL) (Figure 11)

Table 1 summarizes the estimates of the ranges of volumes of undiscovered conventional oil, natural gas, and NGL for the world both exclusive and inclusive of U.S. petroleum resources. Lognormal probability distributions for the three commodities, exclusive of the U.S., are shown in Figures 9, 10, and 11. These estimates demonstrate that a significant volume of undiscovered petroleum is yet to be found in the world.

The estimates of undiscovered resources of individual assessment units were aggregated into TPS, provinces, regions, and finally the world. Perfect positive dependency among estimates was assumed at the AU, TPS, province, country, and region levels. In aggregating the eight region-level sets of estimates to the world level, a positive dependency of 0.5 was assumed. Dependencies assumed for aggregations are described in USGS (2000) by Charpentier et al. (2000).

1. World level total grown petroleum endowment

In developing the World Petroleum Assessment 2000 (USGS, 2000), the potential for discovering conventional oil and natural gas accumulations in hypothetical and frontier assessment units, as well as in established areas already known to be rich in petroleum, was examined. In addition to undiscovered oil and natural gas, other components, including cumulative production, remaining reserves, reserve growth, and undiscovered resources for oil (Figure 12) and natural gas (Figure 13) were considered. In this more comprehensive perspective, undiscovered conventional oil represents only 25% of the world's grown conventional oil endowment (excluding the U.S.) (Figure 14). The 3 other components of the world's grown conventional oil (cumulative production, remaining reserves, and reserve growth) contribute 20%, 32%, and 23%, respectively (Figure 14). In other words, 75% of the world's oil identified in this assessment (excluding the U.S.) has already been discovered indicating significant maturity of world exploration. If cumulative production is not considered, since it has already been consumed, then the remaining resources components on a percentage basis can be compared (Figure 15).

Figure 13 shows that natural gas is a less exploited commodity, on a world scale, than oil. Only 7% of the world's grown conventional natural gas endowment, excluding the U.S., has been produced; if the U.S. is included then 11.4% of the natural gas endowment has been used (Ahlbrandt, 2002). Undiscovered conventional natural gas represents about 34% of the world's (including the U.S.) grown natural gas endowment (Figure 13). Even for natural gas, 66.3% of the world's total grown conventional natural gas endowment (including the U.S.) identified in the World Petroleum Assessment 2000 (USGS, 2000) has been discovered.

The U.S. accounts for about 12% of the grown conventional endowments of both oil and natural gas for the entire world (Table 1). The United States has produced about 45% of its oil and natural gas endowments and carries only 9% as remaining reserves (Figure 16).

Each of the mean undiscovered resource estimates exceeds the world's cumulative production (excluding the U.S.) of these commodities. However, remaining reserves

 DOI:10.1306/1061773M863175

plus estimated reserve growth exceed the undiscovered volumes for oil and natural gas, but not for NGL (Table 1; Figures 14, 15).

2. Comparison to previous USGS world assessments

Comparison of assessments is not always meaningful because many factors are inevitably different. The previous USGS world assessment by Masters et al. (1994), also reported in Masters et al. (1997), utilized production data to January 1, 1993, and did not formally assess additional secondary products or reserve growth nor did it assess deep-water areas (generally deeper than 1640.5 ft [500 m]). Rather, the 1994 USGS study used a Delphi approach based upon consensus of geologists. The current assessment (USGS, 2000) used production data to January 1, 1996, assessed secondary products, and incorporated a much different methodology related to TPS and AU. Additionally, the two studies addressed sets of geologic provinces with different boundaries. Both assessments, however, were based on geology.

The comparison with Masters et al. (1994, 1997) is made here (excluding the U.S.) of estimates of undiscovered resources. Overall, the total petroleum resource estimated in 2000 is about 6% greater than that of the 1994 assessment; 1634 billion BOE versus 1544 billion BOE. However, if individual commodities are compared, further differences between the two assessments become apparent, as shown in Table 3.

B. WORLD LEVEL RESULTS, INCLUDING U.S.

The United States was not reassessed in this study. However, estimates for the U.S. made by Gautier et al. (1996) and the MMS (1996) are included here for purposes of comparison. The mean estimate of undiscovered conventional petroleum volume for this assessment, for oil, natural gas, and NGL taken together, is 9.1% higher than the previous assessment by Masters et al. (1994, 1997) (1650 billion BOE versus 1805 billion BOE). With respect to the three commodities assessed, undiscovered oil is about 26% higher than the previous assessment, undiscovered natural gas is about 10% lower, and NGL is about 103% higher (Figure 17).

C. REGION LEVEL RESULTS

Figures 18, 19, and 20 summarize the estimated mean undiscovered conventional volumes of oil, natural gas, and NGL for the eight USGS regions of the world (excluding the U.S. from Region 5).

For these regional comparisons (Figures 18, 19, and 20), it must be stressed that not every geologic province in the world was assessed. There are 309 provinces in the world that contain some petroleum. Only 128 of them were assessed as discussed by Klett et al. (1997, 2000a). Thus, relative proportions of undiscovered petroleum resources among regions could differ somewhat from those shown if such additional geographic areas had been assessed.

Table 4 ranks the Regions by decreasing volume of estimated undiscovered oil and undiscovered natural gas.

The overall dominance of Regions 1 (Former Soviet Union) and 2 (Middle East and North Africa) for undiscovered conventional petroleum resources is apparent. However, the assessment demonstrated that 47% of the undiscovered oil is estimated to exist in the other six regions of the world (Figure 18), reflecting particularly the deep-water oil exploration potential in Region 6 (offshore eastern South America) and Region 7 (offshore western Africa). Thirty-six percent of the undiscovered, conventional natural gas is in regions other than Region 1 and Region 2 (Figure 19). Undiscovered NGL are dominated by Region 1 (Former Soviet Union) and Region 2 (Middle East and North Africa) (Figure 20); and much of the remaining resources will come from offshore. Estimated NGL resources were increased dramatically in USGS (2000) by determining phase-associated products carefully at the AU level rather than by assuming a world average as done by Masters et al. (1994).

Adding the U.S. estimated undiscovered conventional resources (Gautier et al., 1996; MMS, 1996) increases Region 5's mean undiscovered liquids (oil and NGL) by 83 billion bbl of oil and its mean undiscovered natural gas by 528 tcf (88 billion BOE). Inclusion of the U.S. estimates of undiscovered resources in Region 5 causes that Region's ranking to move from fifth to second in the world in terms of undiscovered conventional oil, and from seventh to third in terms of undiscovered conventional natural gas.

A comparison of the USGS 1994 world assessment (Masters et al., 1994, 1997) and this assessment for the eight regions of the world (exclusive of the U.S.) is shown for oil (Figure 21) and for natural gas (Figure 22). The sets of geologic provinces assessed in the 1994 and 2000 studies differ somewhat as do definitions of the regions themselves. The comparisons shown in Figures 21 and 22 have been adjusted as much as possible for differences in regional boundaries.

D. RESULTS AND RANKINGS BY ASSESSMENT UNIT, TOTAL PETROLEUM SYSTEM, AND PROVINCE

The volumetric ranking for each AU is shown in Appendices 1a–1f; for each TPS in appendices 2a–2f; and for provinces in Appendices 3a–3i. The appendices for AU, TPS, and provinces follow a progression from undiscovered resources for oil followed by natural gas, then the endowment (cumulative production, remaining reserves, and mean undiscovered resource estimate—note "known" volumes are the sum of cumulative production and remaining reserves). Future volumes (remaining reserves combined with the mean undiscovered resource estimate) by commodity are then given at the AU, TPS, and province level. At the province level an additional ranking of undiscovered resources for estimated total petroleum (oil, natural gas, and NGL) is given in Appendix 3c and the estimated endowment of total petroleum is given in Appendix 3f.

The progression from estimates of undiscovered resources, the focus of the USGS (2000) assessment, to estimated endowment is useful for a more comprehensive view of the petroleum endowment at the AU, TPS, and provinces levels. Some AU, TPS, and provinces have a significant discovery history and their endowment may be larger, thus their endowment ranking may be different from the ranking by undiscovered resources alone. The aforementioned appendices allow comparison of past, present, or future petroleum potential at all levels and permit world comparisons. TPS rankings for undiscovered resources and endowment were highlighted for oil in Figures 5 and 6 and for natural gas in Figures 7 and 8. These plots demonstrate the large concentration of petroleum resources in relatively few total petroleum systems. However, the TPS have different rankings for oil and natural gas at province, AU, and TPS levels and illustrate geographic and volumetric differences between oil and natural gas total petroleum systems.

Data for known volumes used in this study are from Petroconsultants (1996) (now IHS Energy Group) and NRG Associates (1995) and are used with the permission of these companies. Reported volumes of individual fields were summed to provide the cumulative production and known volumes for each AU, TPS, province, and region. Only fields assigned to an AU were used in this study.

Conventional petroleum endowment and future petroleum volumes for TPS, provinces, and regions are necessarily approximate. The mean undiscovered volumes were estimated for each AU that was assessed (Figure 4), whereas known volumes were reported for all AU. Therefore, the endowment or future volume of a TPS, province, or region will contain the known volume, but not the undiscovered volume, of an unassessed AU. Twenty-four of the 270 AU were not quantitatively assessed.

Determination of the major commodity of a field is based on the ratio of natural gas endowment to oil endowment as ft^3 of natural gas per bbl of oil (ft^3/bbl of oil). The major commodity is oil if this ratio is less than 20,000 ft^3/bbl of oil and natural gas if this ratio is equal to or greater than 20,000 ft^3/bbl of oil.

The undiscovered petroleum resources are concentrated in relatively few TPS and AU. Likewise, the distributions for TPS (Figures 5, 6, 7, 8) and provinces are similarly concentrated (Figures 23, 24, 25). Eighty percent of the undiscovered oil (approximately 500 billion bbl of oil) occurs in approximately 16% of the AU, TPS (Figures 5, 6, 27), and provinces (Figures 23, 26); and 80% of the undiscovered natural gas (approximately 3700 tcf) is in about 22% of the AU, TPS (Figures 7, 8, 29), and provinces (Figures 24, 28). Eighty percent of the total undiscovered petroleum (approximately 1300 billion BOE) is present in approximately 21% of the provinces that contain assessed AU (Figure 2). The map distribution for mean undiscovered volumes of oil provinces is shown in Figure 26 and in TPS's shown in Figure 27. Similar plots for undiscovered natural gas are shown in Figures 28 and 29. These distributions are similar to that for discovered petroleum volumes (Klett et al., 1997; Klett et al., 2000a).

The rankings show that the greatest estimated volumes of undiscovered conventional oil are in the Middle East, northeast Greenland Shelf, the West Siberian and Caspian areas of the Former Soviet Union, and the Niger and Congo delta areas of Africa. Significant new undiscovered oil resource potential was identified in a number of areas with no significant production history, such as northeast Greenland and offshore Suriname (Figures 2, 5; Appendices 1a, 2a, 3a). Areas that contain the greatest estimated volumes of undiscovered conventional natural gas are in the West Siberia Basin, Barents and Kara Seas shelves of the Former Soviet Union, the Middle East, and the Norwegian Sea. A number of areas were identified that may contain significant additional undiscovered natural gas resources where large discoveries have been made but remain undeveloped; e.g. East Siberia and the Northwest Shelf of Australia (Figures 2, 7; Appendices 1b, 2b, 3b).

TABLE 3. Volumes of undiscovered world petroleum, by commodity, from this assessment and the previous USGS assessment. BBO = billion barrels of oil; BBOE = billion barrels of oil equivalent; BBNGL = billion barrels of natural gas liquids.

Commodity (excluding U.S.)	Masters et al., 1994	USGS, 2000	% Change
Oil	549 BBO	649 BBO	+18
Natural Gas	915 BBOE	778 BBOE	-15
NGL	90 BBNGL	207 BBNGL	+130
Total	1544 BBOE	1634 BBOE	+6

TABLE 4. Volumes of undiscovered oil and natural gas for assessed areas, by region. Volumes in billion barrels of oil, trillion cubic feet of natural gas, or billion barrels of oil equivalent (BBOE).

Regions (ranked in descending order of estimated undiscovered oil)	Billions of Barrels of Oil	Regions (ranked in descending order of estimated undiscovered natural gas)	Trillions of Cubic Feet of Natural Gas (BBOE)
Region 2 (Middle East and North Africa)	230	Region 1 (Former Soviet Union)	1611 (269)
Region 1 (Former Soviet Union)	116	Region 2 (Middle East and North Africa)	1370 (228)
Region 6 (Central and South America)	105	Region 6 (Central and South America)	487 (81)
Region 7 (Sub-Saharan Africa and Antarctica)	72	Region 3 (Asia Pacific)	379 (63)
Region 5 (North America exclusive of the U.S.)	70	Region 4 (Europe)	312 (52)
Region 3 (Asia-Pacific)	30	Region 7 (Sub-Saharan Africa and Antarctica)	235 (39)
Region 4 (Europe)	22	Region 5 (North America exclusive of the U.S.)	155 (26)
Region 8 (South Asia)	4	Region 8 (South Asia)	120 (20)

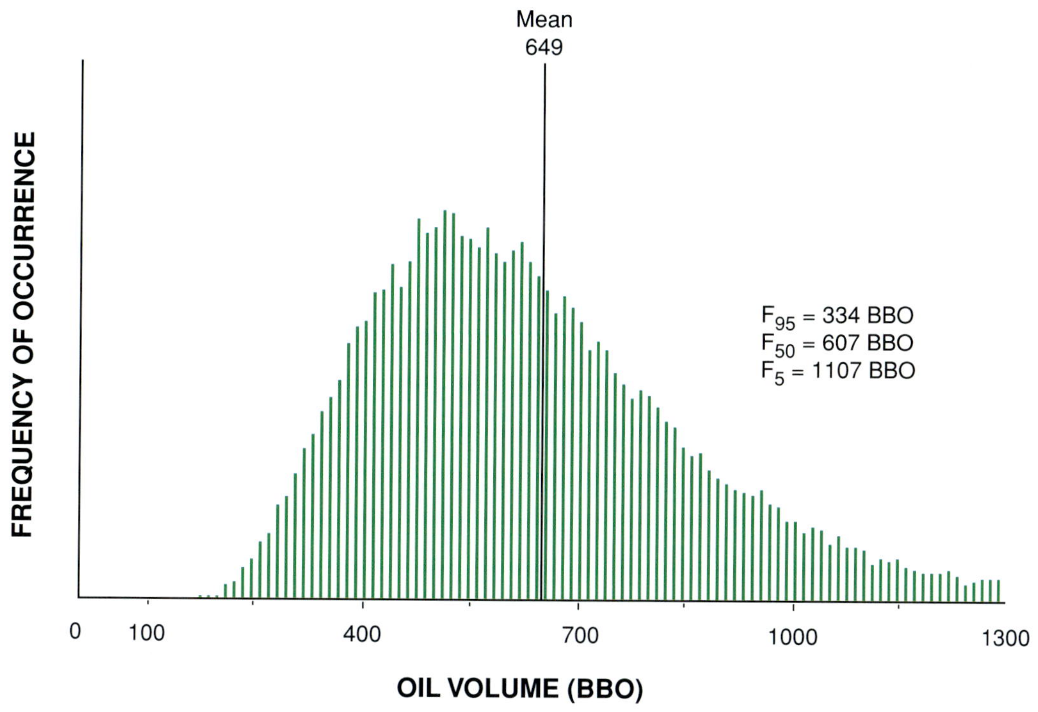

FIGURE 9. Graphical depiction of the forecast for world undiscovered oil resources (exclusive of the U.S.), with uncertainty expressed in the form of a lognormal probability distribution. BBO = billion barrels of oil. F_{95} (the 95th fractile) represents a 95% chance of the occurrence of at least the amount tabulated. Other fractiles are defined similarly.

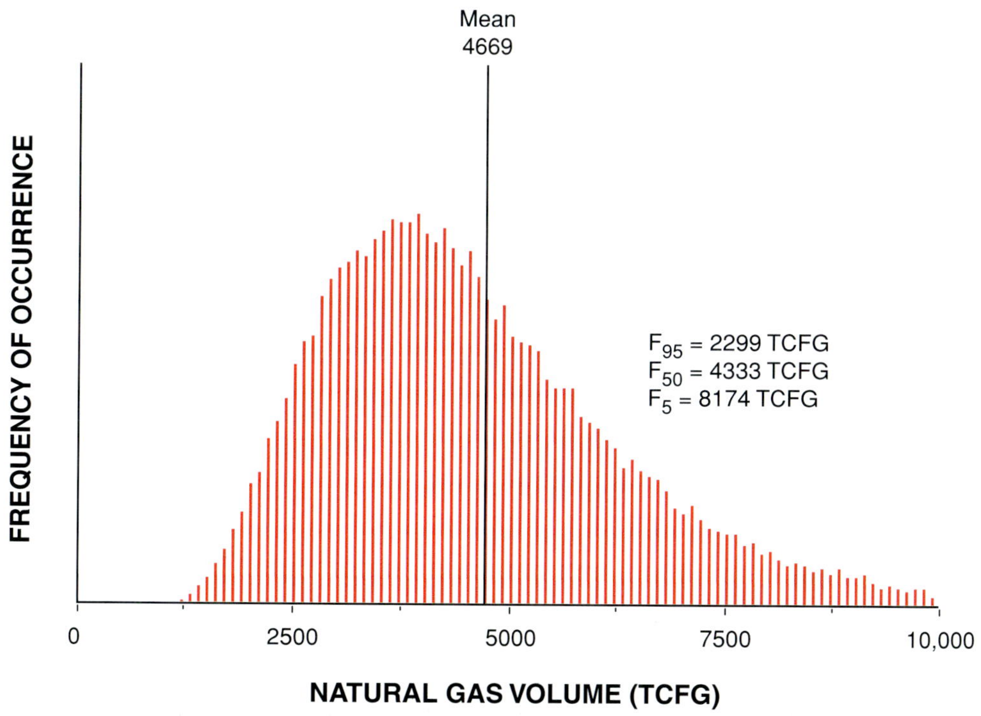

FIGURE 10. Graphical depiction of the forecast for world undiscovered natural gas resources (exclusive of the U.S.), with uncertainty expressed in the form of a lognormal probability distribution. TCFG = trillion cubic feet of natural gas. F_{95} (the 95th fractile) represents a 95% chance of the occurrence of at least the amount tabulated. Other fractiles are defined similarly.

Figure 11. Graphical depiction of the forecast for world undiscovered natural gas liquids (NGL) resources (exclusive of the U.S.), with uncertainty expressed in the form of a lognormal probability distribution. BBNGL = billion barrels of NGL. F_{95} (the 95th fractile) represents a 95% chance of the occurrence of at least the amount tabulated. Other fractiles are defined similarly.

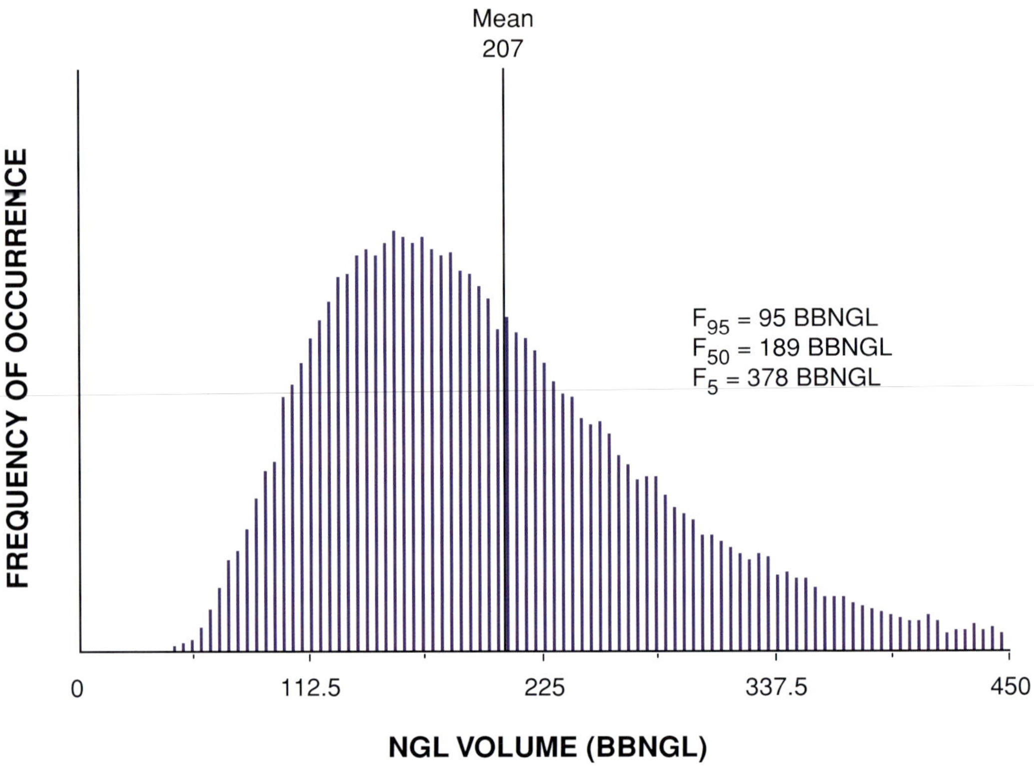

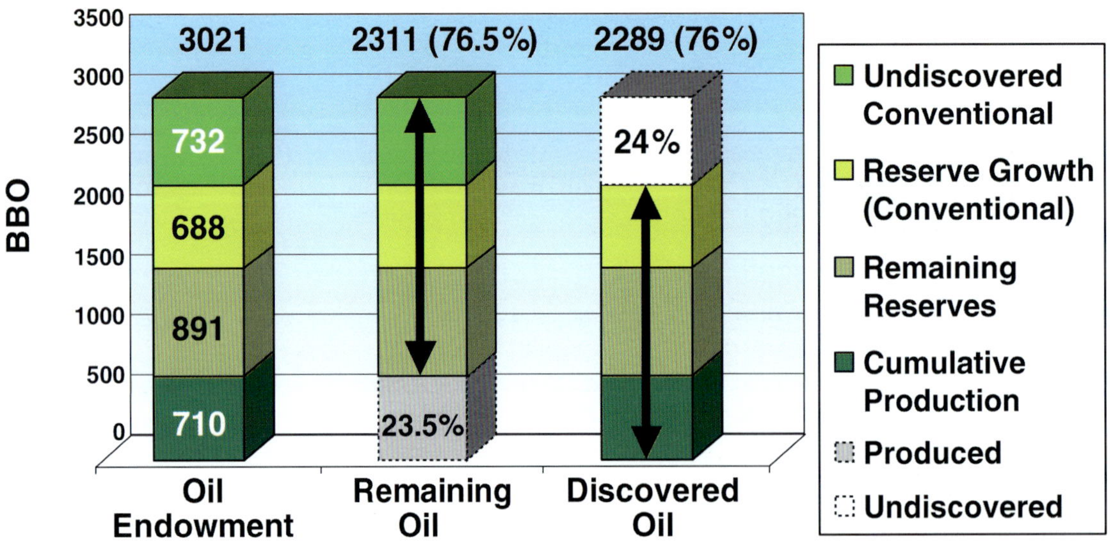

Figure 12. World's grown conventional oil endowment. Estimates include the 128 world provinces assessed by the USGS (2000) and the U.S. provinces assessed by the USGS (1995) and MMS (1996) as of January 1, 1996. Grown oil endowment includes the four components shown on the right side of diagram. BBO = billion barrels of oil.

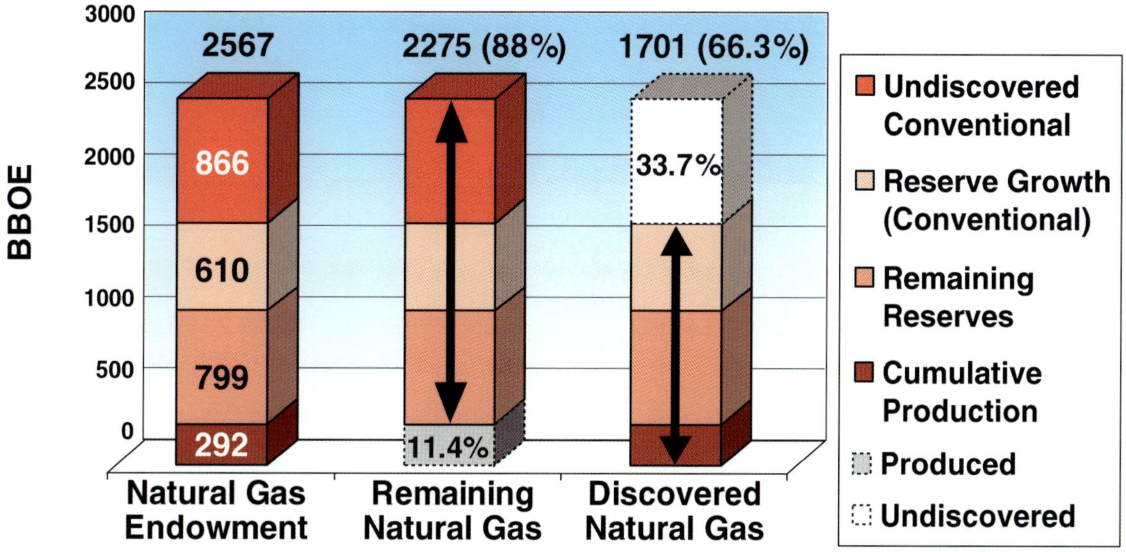

FIGURE 13. World's grown conventional natural gas endowment. Estimates include the 128 world provinces (USGS, 2000) and the U.S. provinces assessed by the USGS (1995) and MMS (1996) as of January 1, 1996. Grown natural gas endowment includes the four components shown on the right side of diagram. BBOE = billion barrels of oil equivalent.

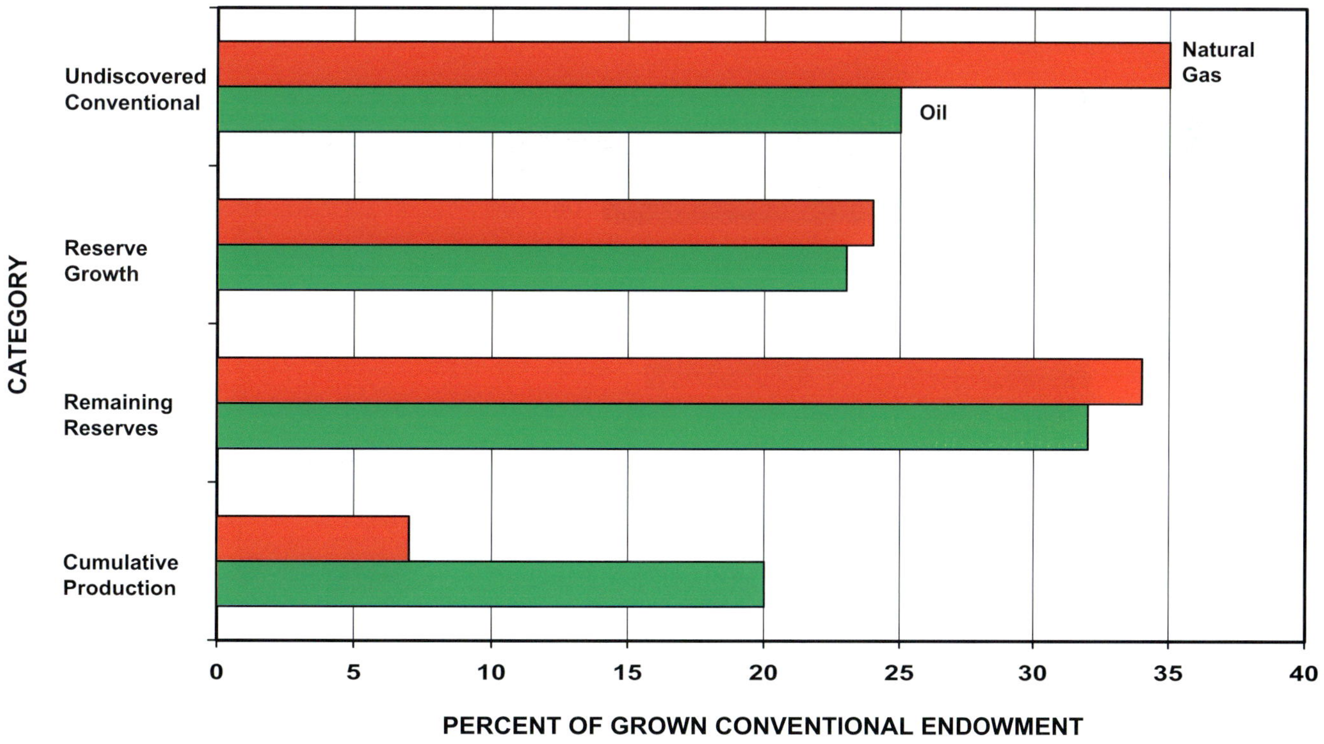

FIGURE 14. Percent of world's grown conventional endowments of oil and natural gas (excluding the U.S.) composed of cumulative production, remaining reserves, reserve growth, and undiscovered conventional volumes.

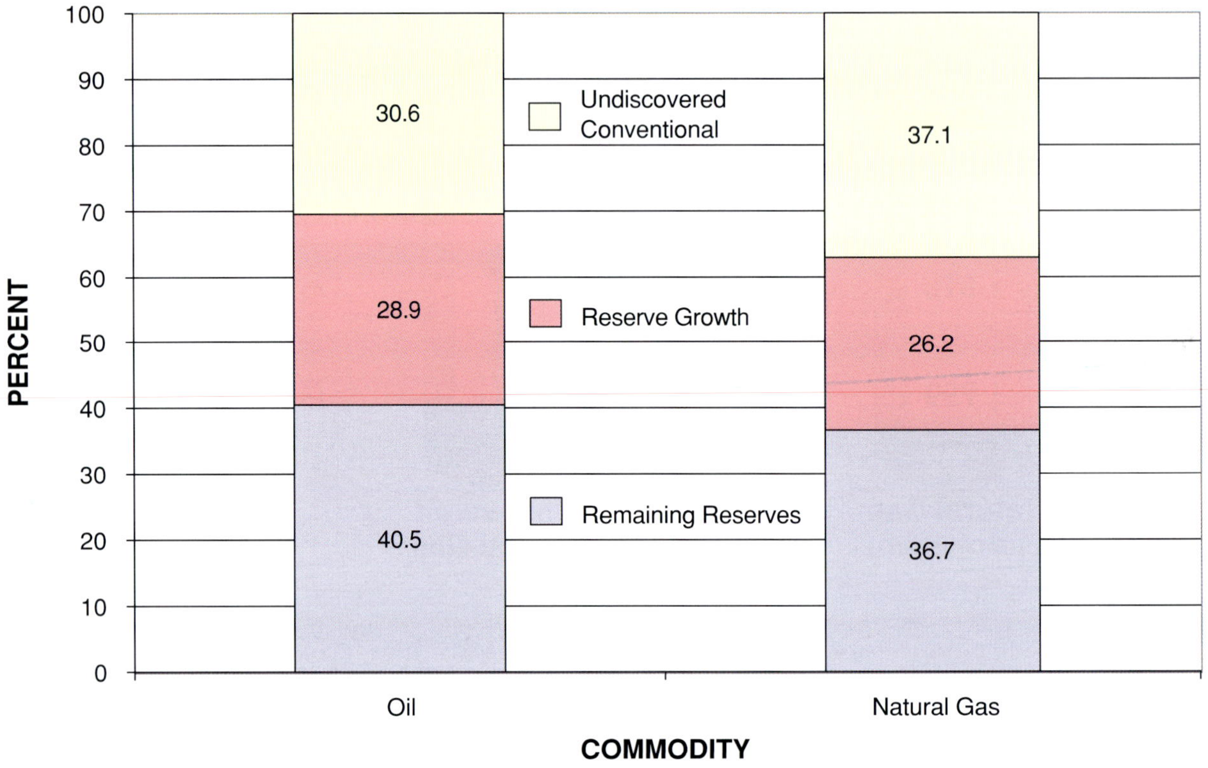

FIGURE 15. Comparison of world's remaining oil and natural gas reserves, estimated mean reserve growth, and estimated mean undiscovered conventional resources (excluding the U.S.) for oil and natural gas.

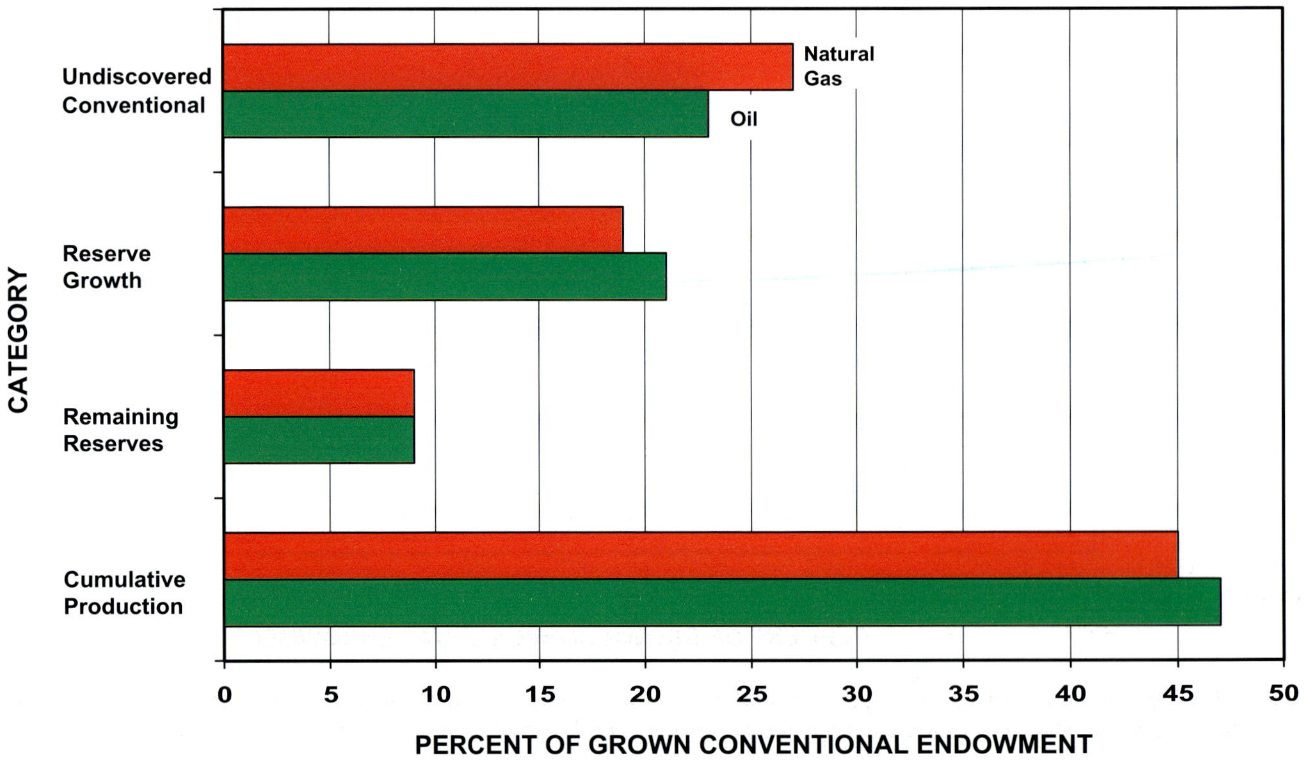

FIGURE 16. Percent of U.S. grown conventional endowments of oil and natural gas composed of cumulative production, remaining reserves, reserve growth, and undiscovered conventional volumes.

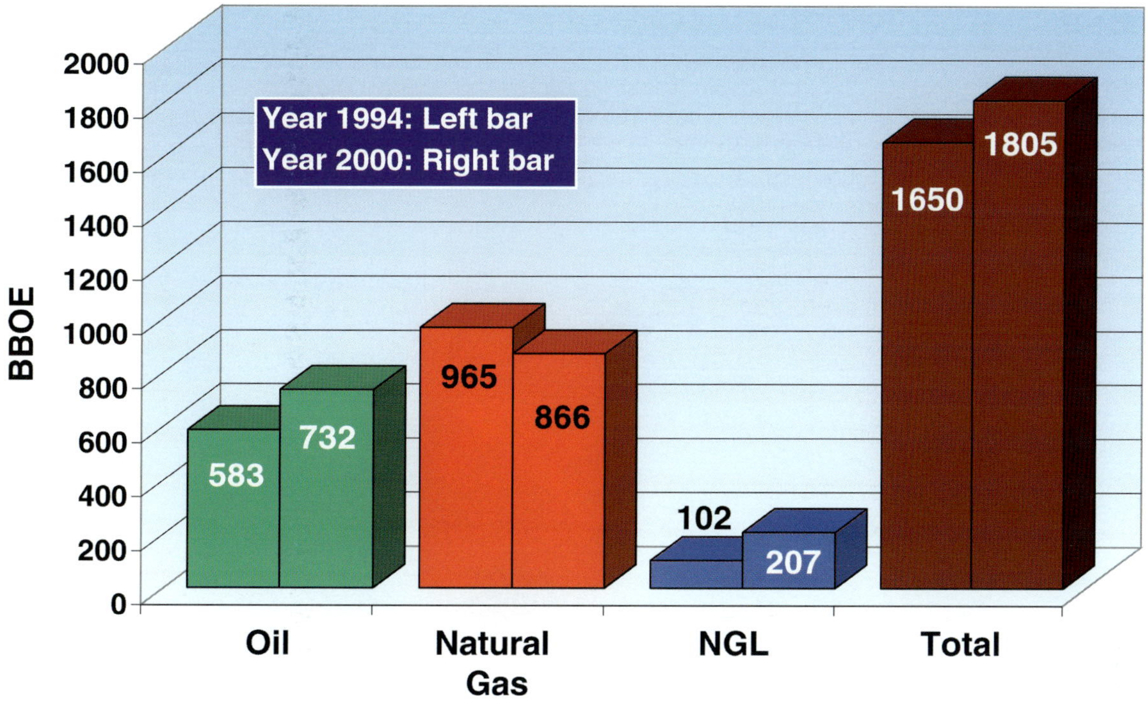

FIGURE 17. Mean estimates of the world's undiscovered conventional resources including the U.S. from 1994 USGS assessment (Masters et al., 1994, 1997) compared to the current assessment of the world (USGS, 2000). BBOE = billion barrels of oil equivalent.

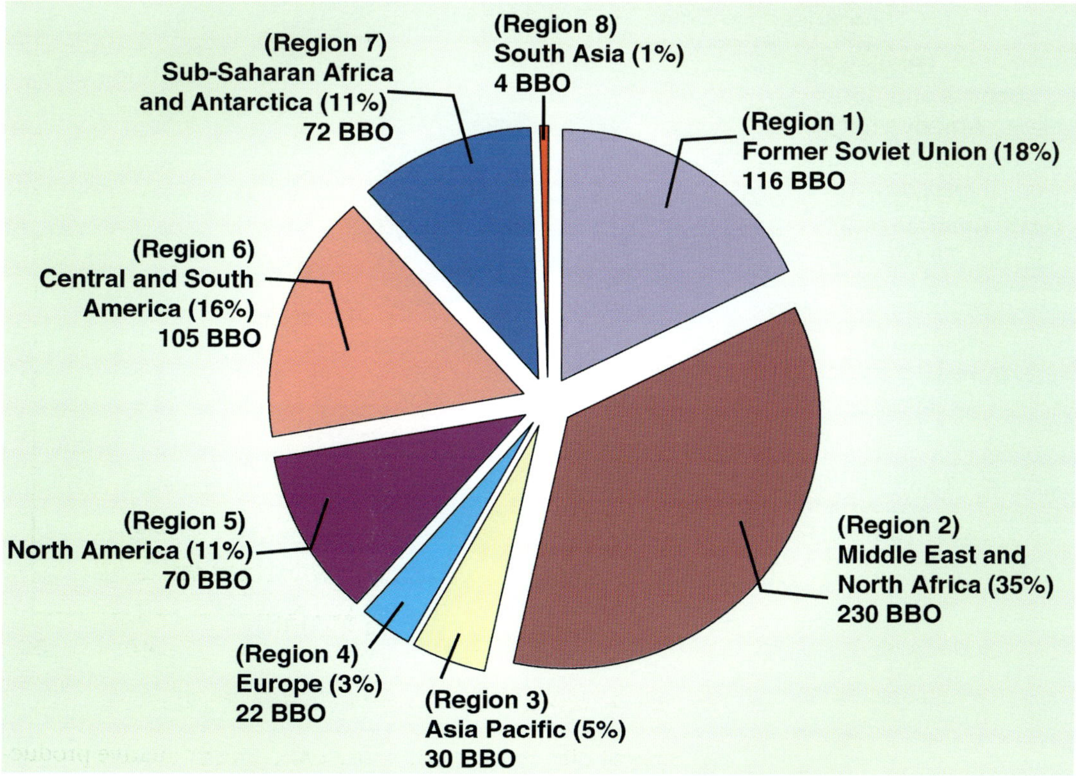

FIGURE 18. Estimates of mean undiscovered conventional oil (excluding the U.S.) for the eight USGS regions of the world. BBO = billion barrels of oil.

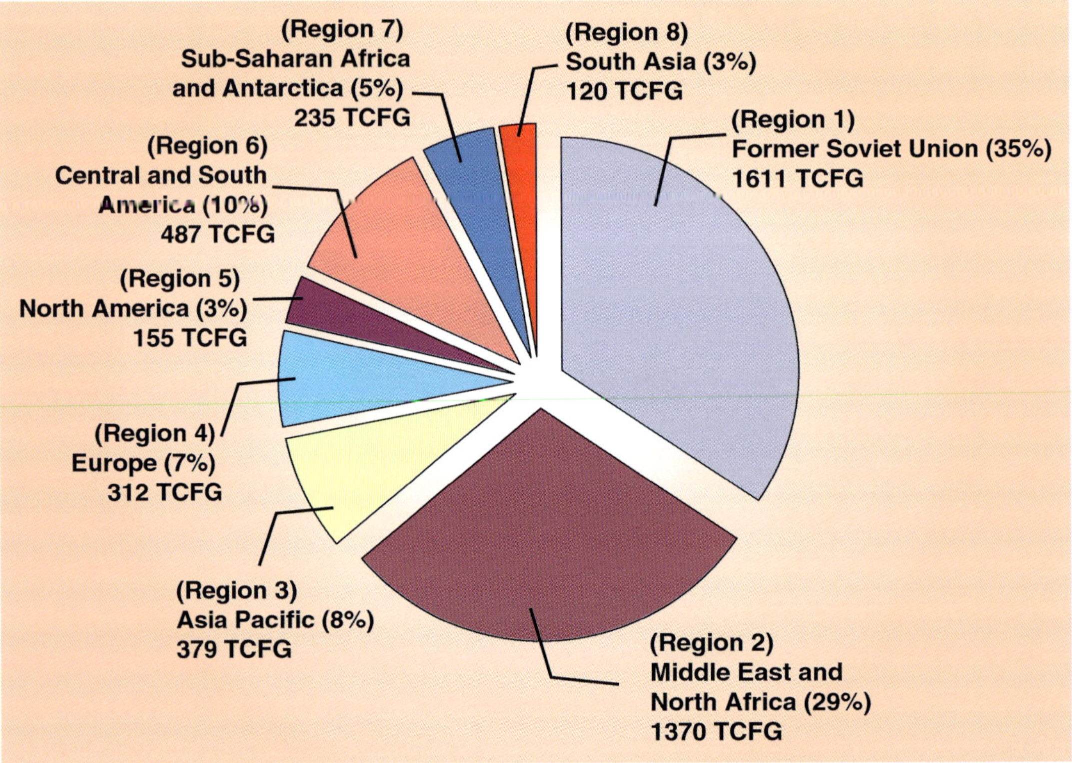

FIGURE 19. Estimates of mean undiscovered conventional natural gas (excluding the U.S.) for the eight USGS regions of the world. TCFG = trillion cubic feet of natural gas.

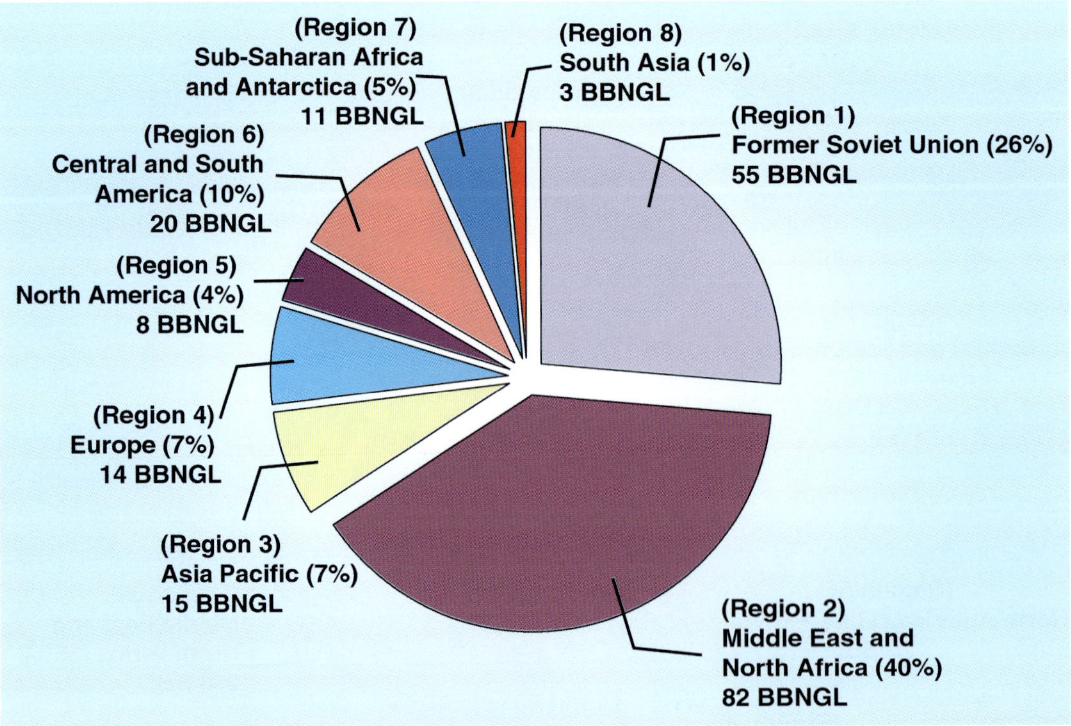

FIGURE 20. Estimates of mean undiscovered conventional natural gas liquids (NGL) (excluding the U.S.) for the eight USGS regions of the world. BBNGL = billion barrels of NGL.

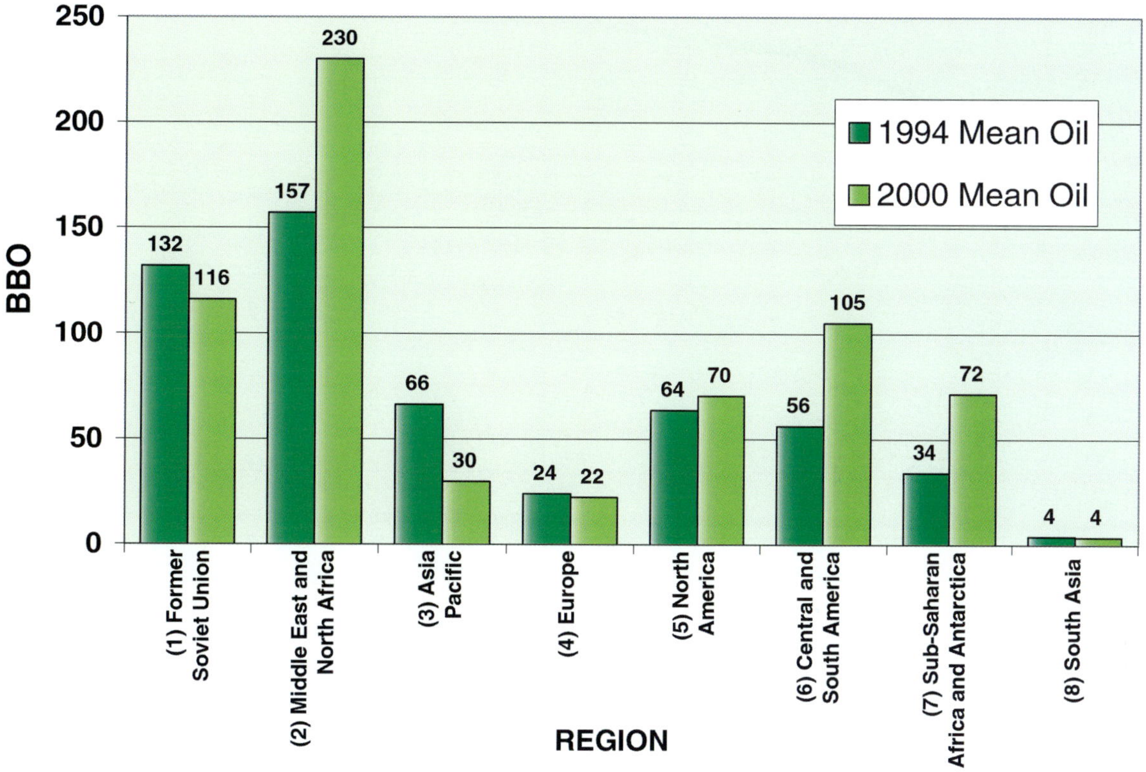

Figure 21. Comparison of the 1994 and 2000 USGS world petroleum assessments (excluding the U.S.) of estimated undiscovered, conventional mean oil resources for the eight USGS regions of the world. Data from Masters et al. (1994, 1997) and USGS (2000). BBO = billion barrels of oil.

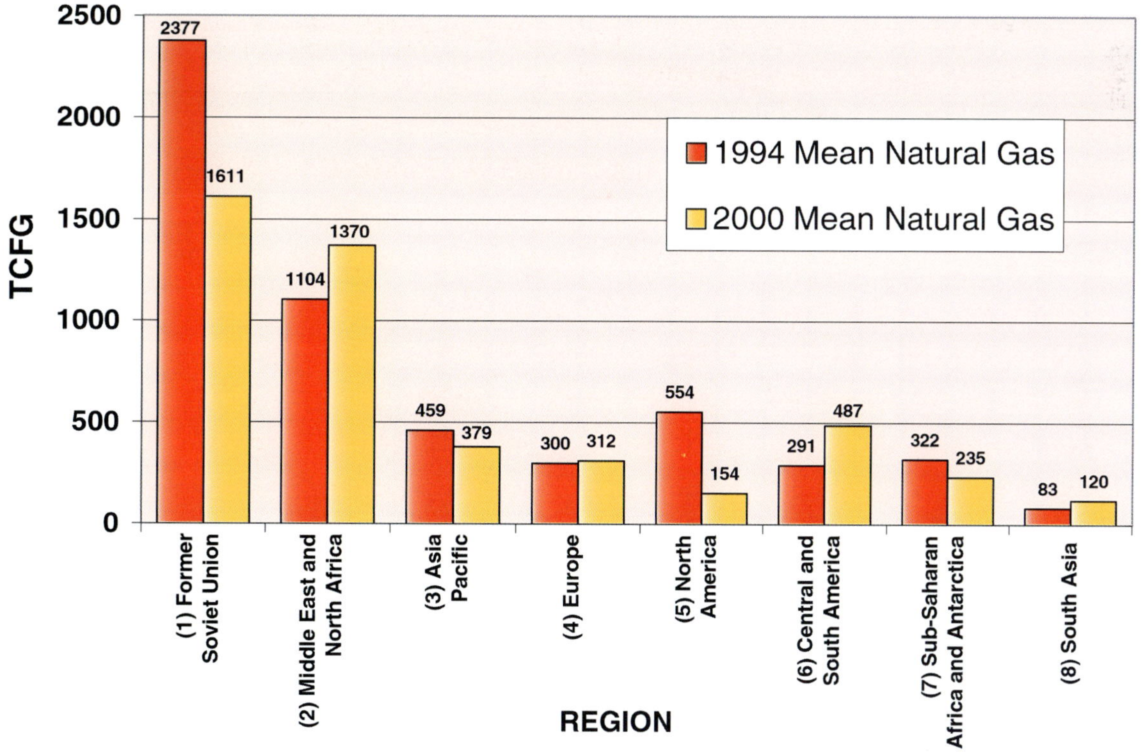

Figure 22. Comparison of the 1994 and 2000 USGS world petroleum assessments (excluding the U.S.) of estimated undiscovered, conventional mean natural gas resources for the eight USGS regions of the world. Data from Masters et al. (1994, 1997) and USGS (2000). TCFG = trillion cubic feet of natural gas.

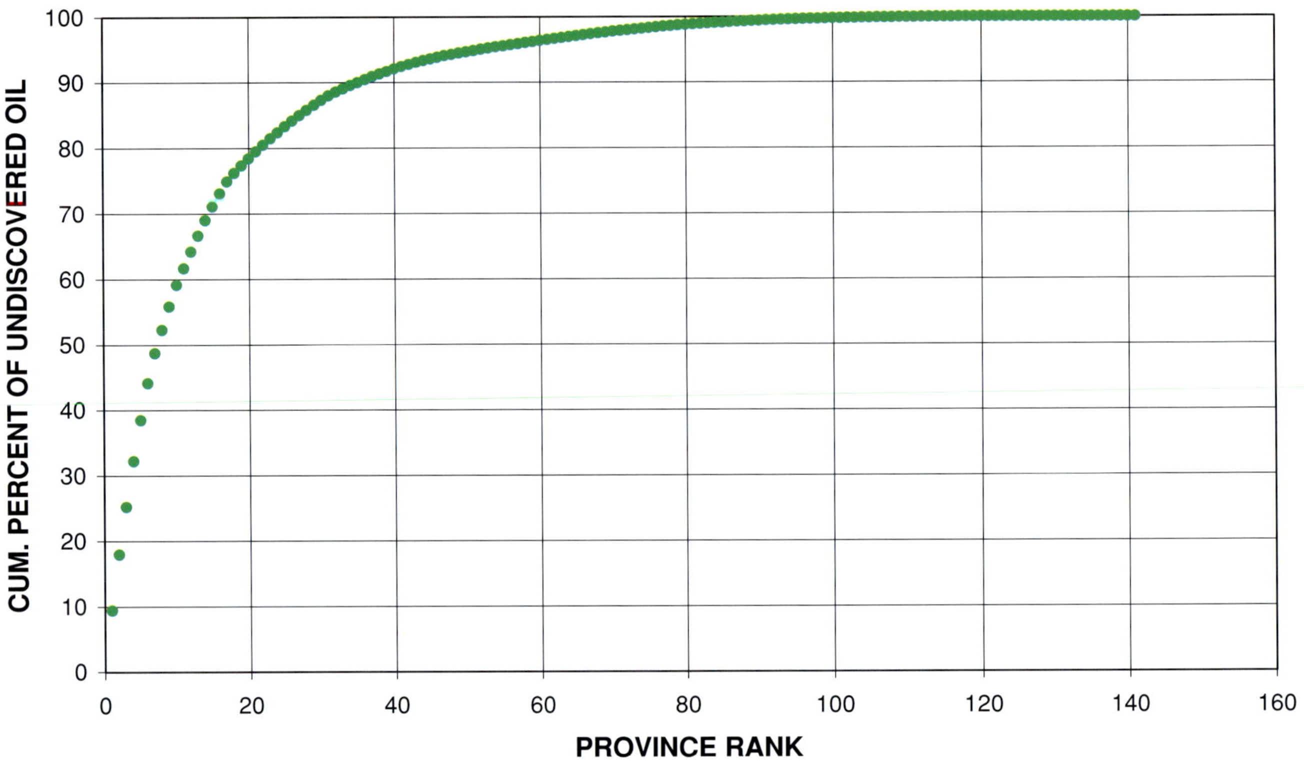

Figure 23. Cumulative percent of estimated undiscovered oil volume, with provinces ranked by decreasing volume.

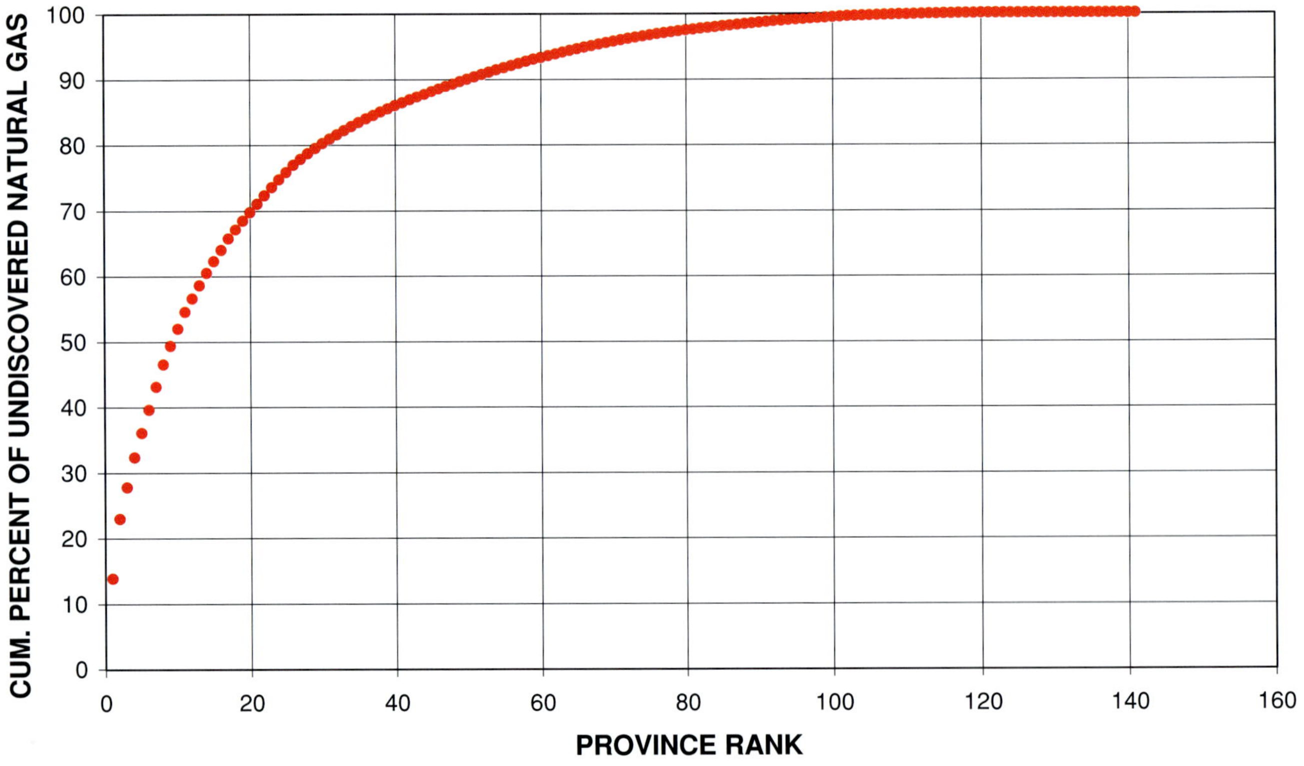

Figure 24. Cumulative percent of estimated undiscovered natural gas volume, with provinces ranked by decreasing volume.

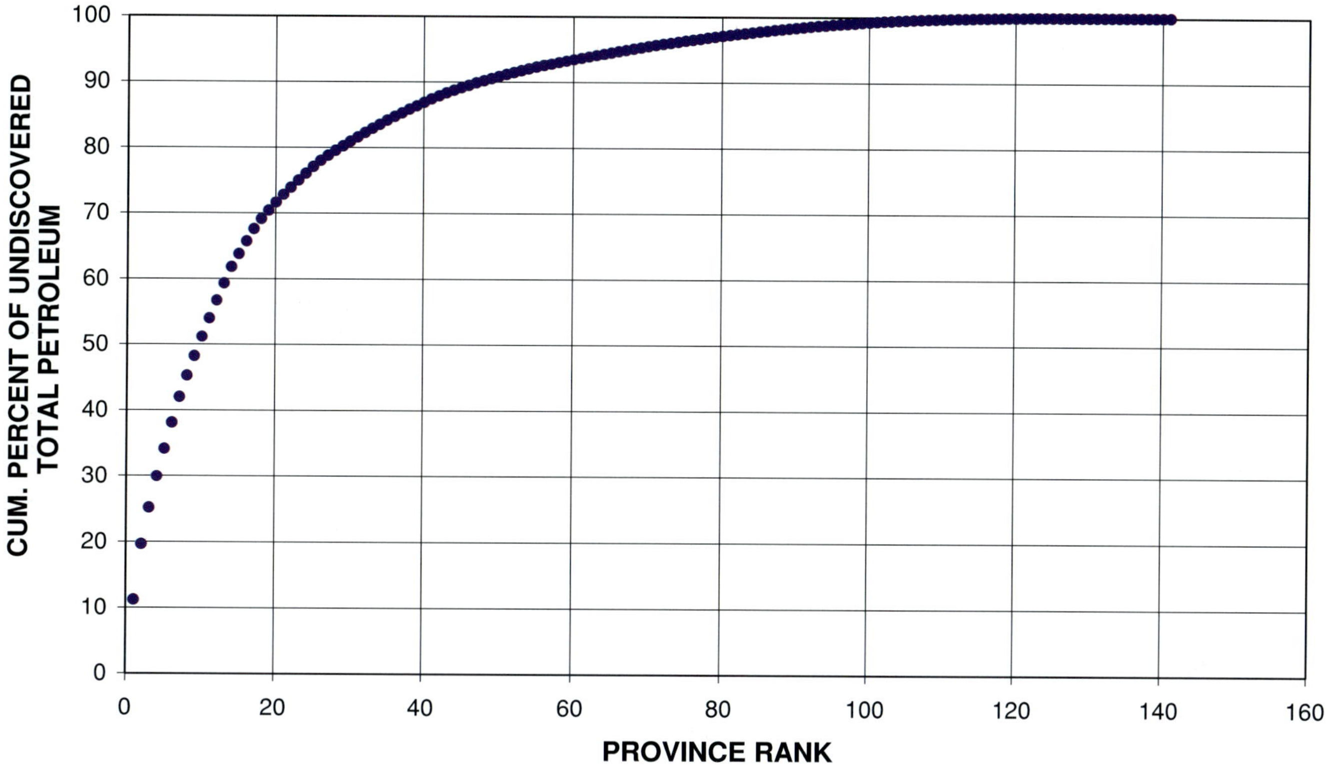

FIGURE 25. Cumulative percent of estimated total undiscovered petroleum volume, with provinces ranked by decreasing volume.

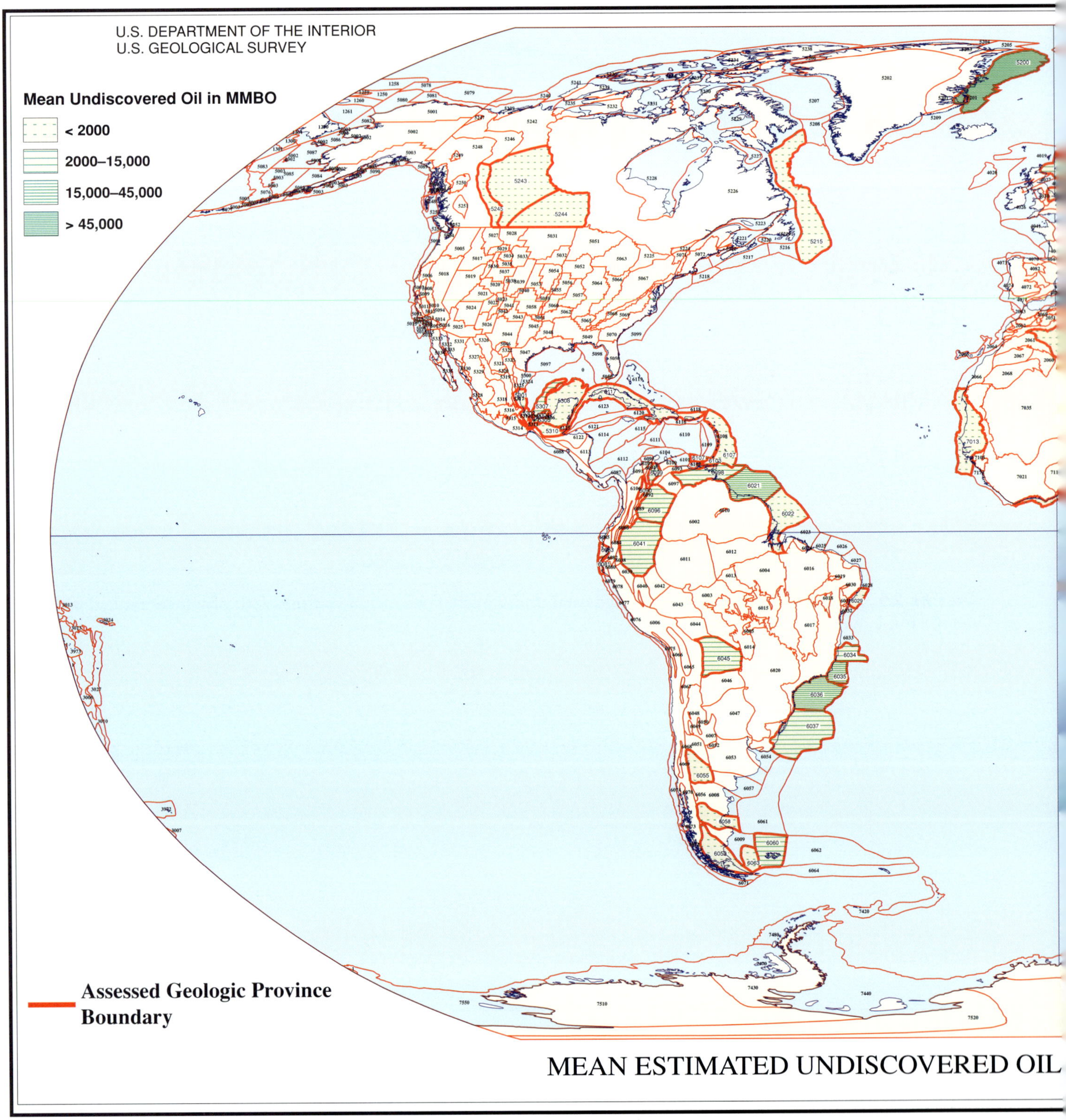

FIGURE 26. Map showing ranked categories of assessed geologic provinces by mean estimated undiscovered oil volume. Categories are <2000 MMBO; 2000–15,000 MMBO; 15,000–45,000 MMBO; >45,000 MMBO. MMBO = million barrels of oil.

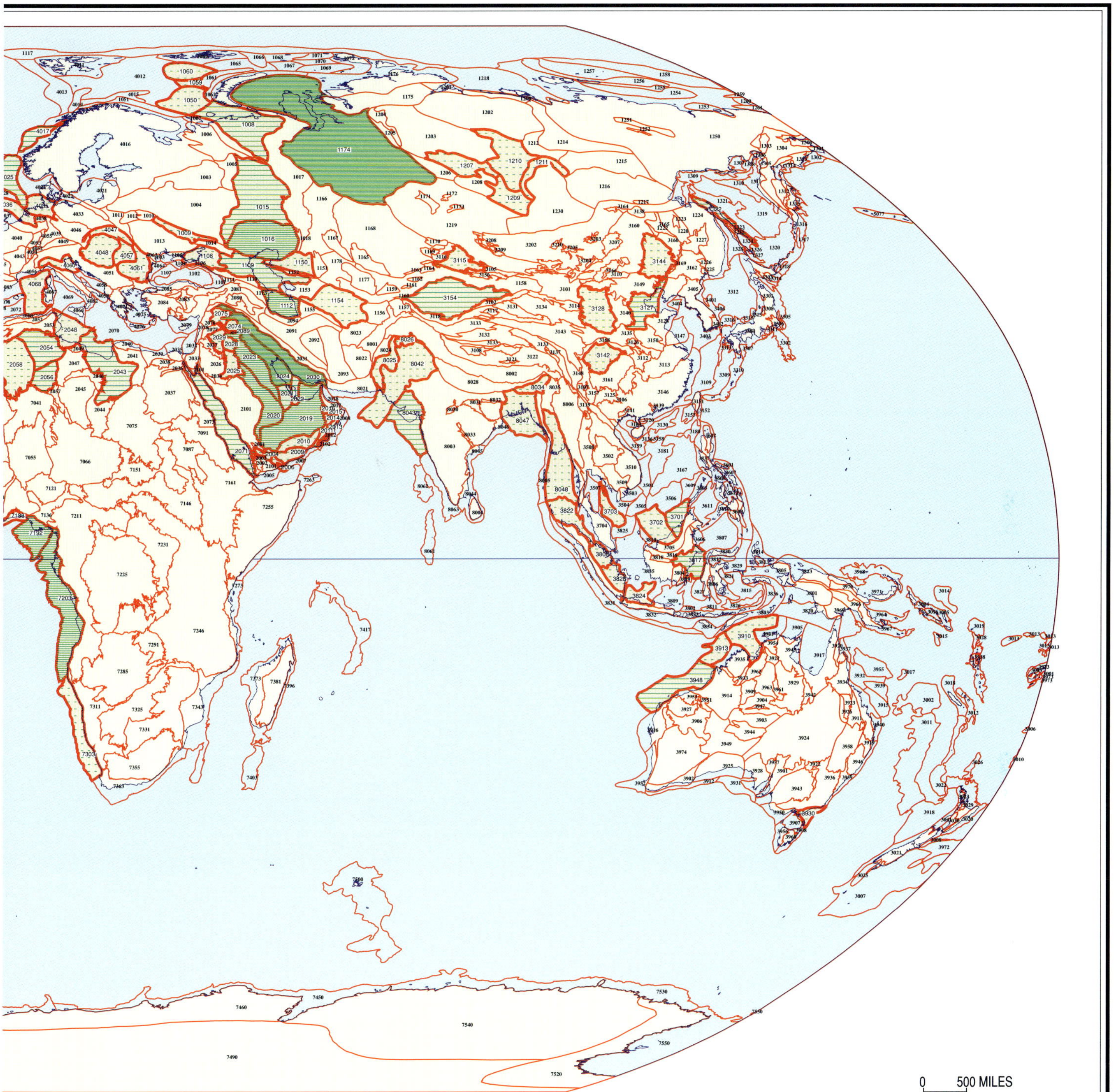

VOLUME (GEOLOGIC PROVINCE LEVEL)

Projection Robinson
Central meridian 0

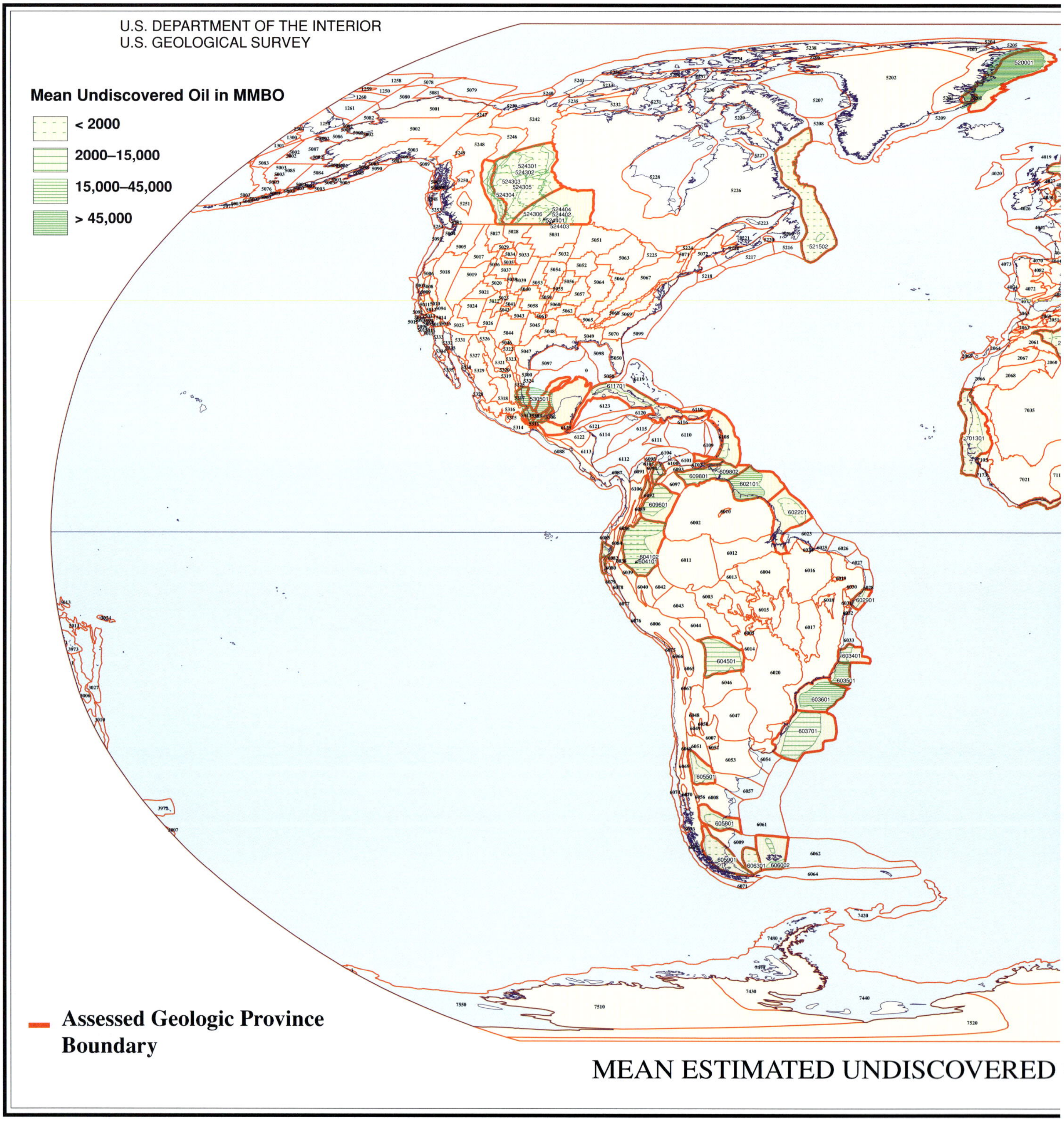

FIGURE 27. Map showing ranked categories of assessed total petroleum systems (TPS) by mean estimated undiscovered oil volume. Categories are <2000 MMBO; 2000–15,000 MMBO; 15,000–45,000 MMBO; >45,000 MMBO. MMBO = million barrels of oil.

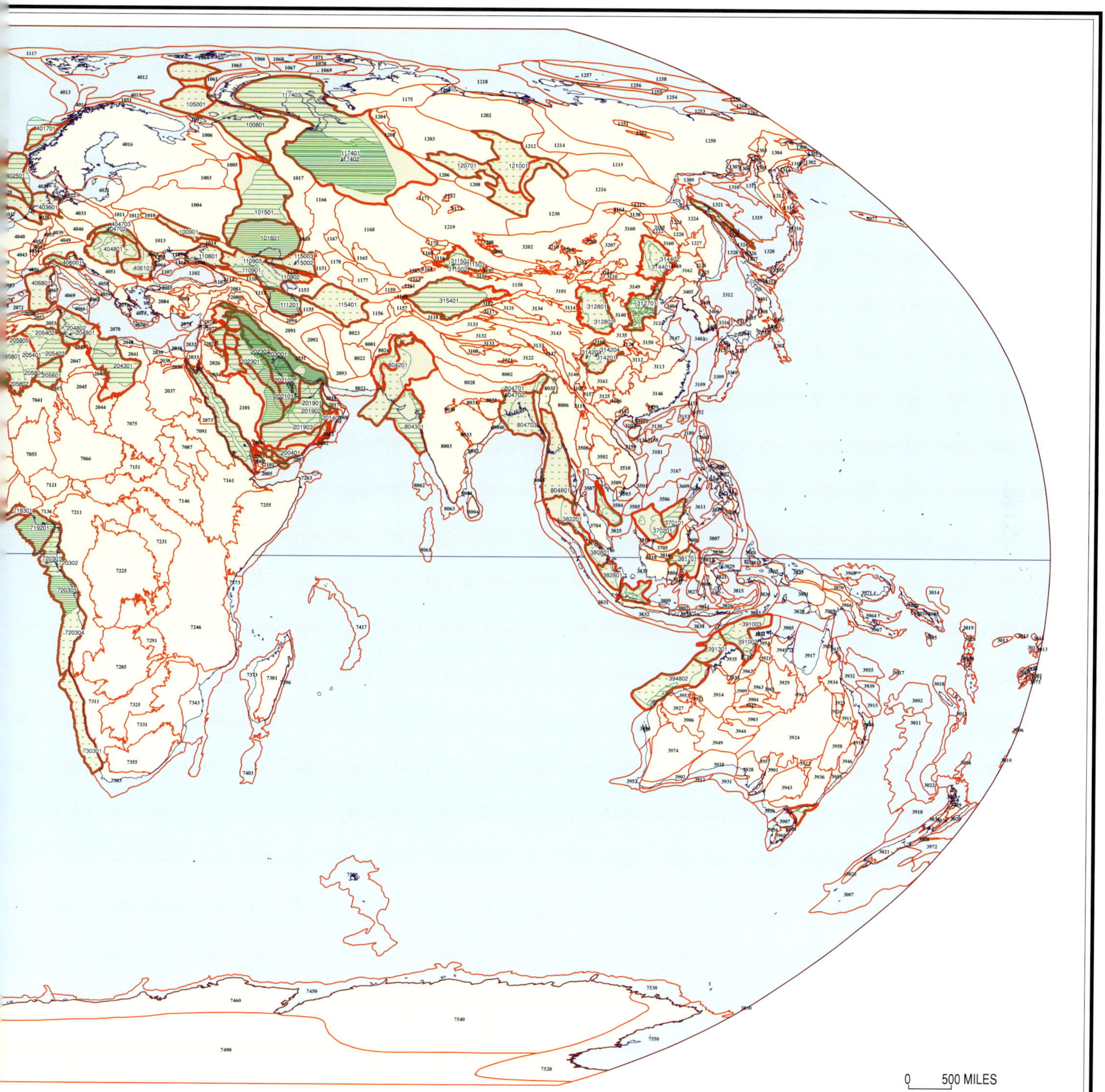

OIL VOLUME (TPS LEVEL)

Projection Robinson
Central meridian 0

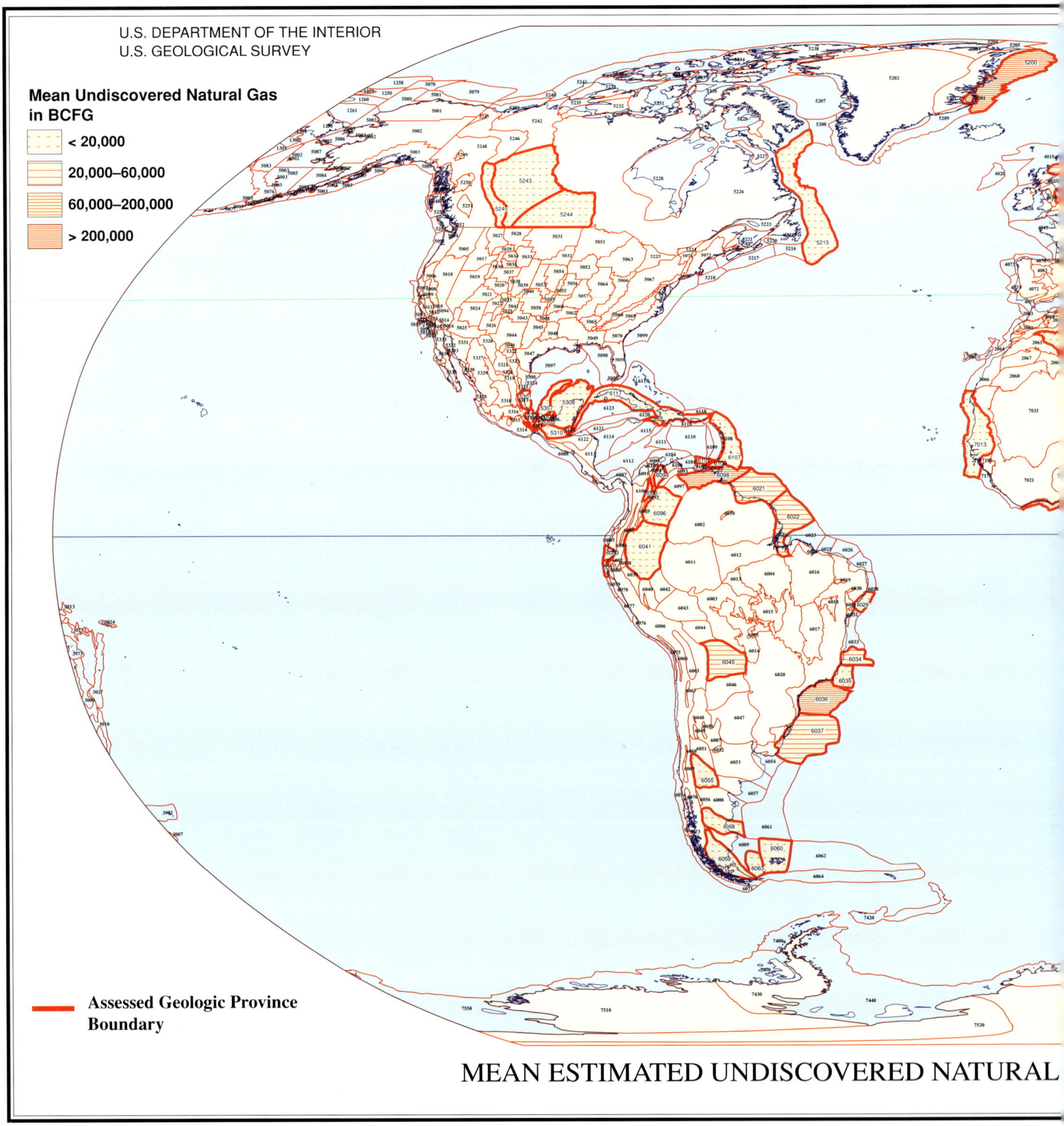

FIGURE 28. Map showing ranked categories of assessed geologic provinces by mean estimated undiscovered natural gas volume. Categories are <20,000 BCFG; 20,000–60,000 BCFG; 60,000–200,000 BCFG; >200,000 BCFG. BCFG = billion cubic feet of gas.

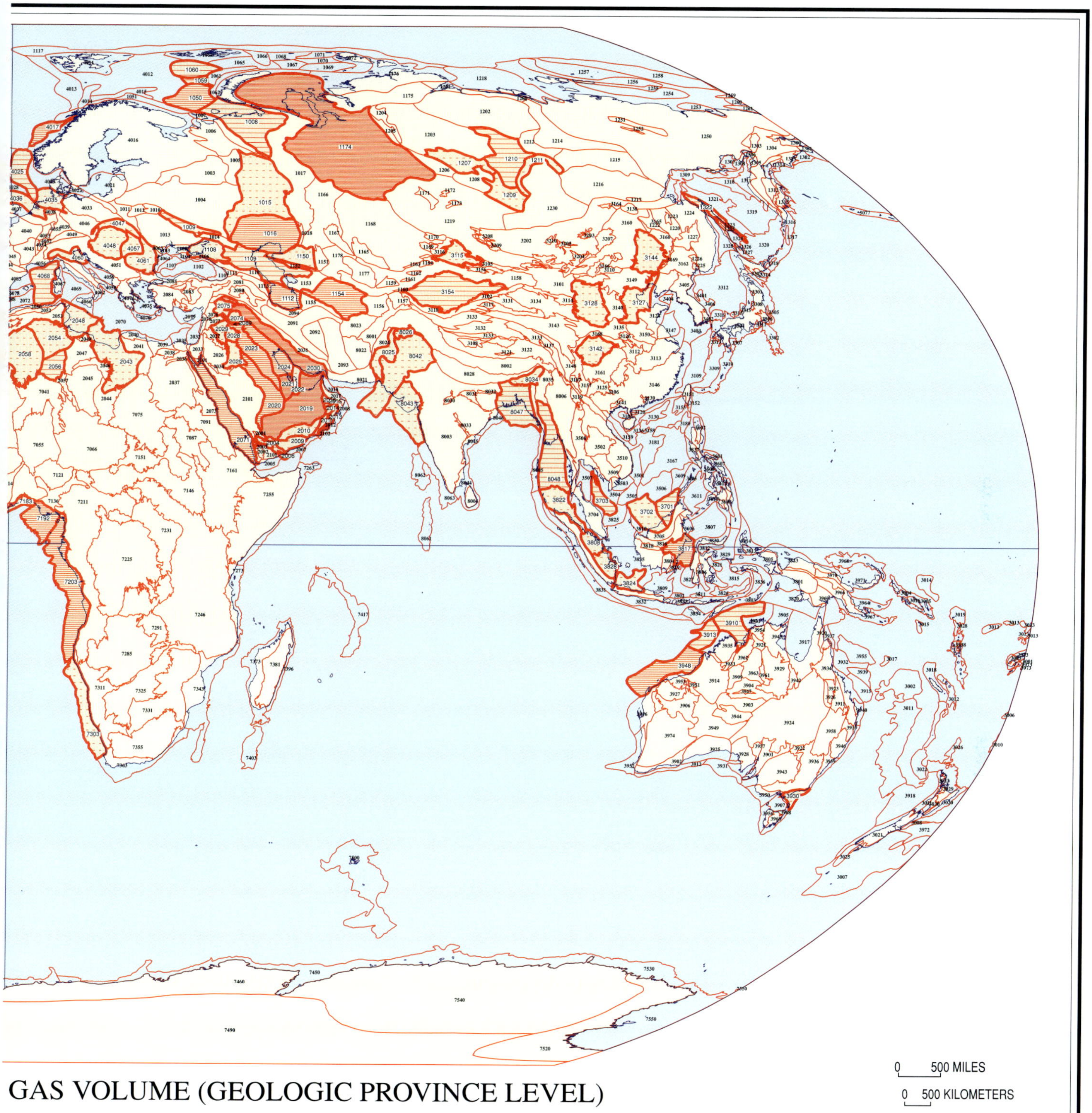

GAS VOLUME (GEOLOGIC PROVINCE LEVEL)

Projection Robinson
Central meridian 0

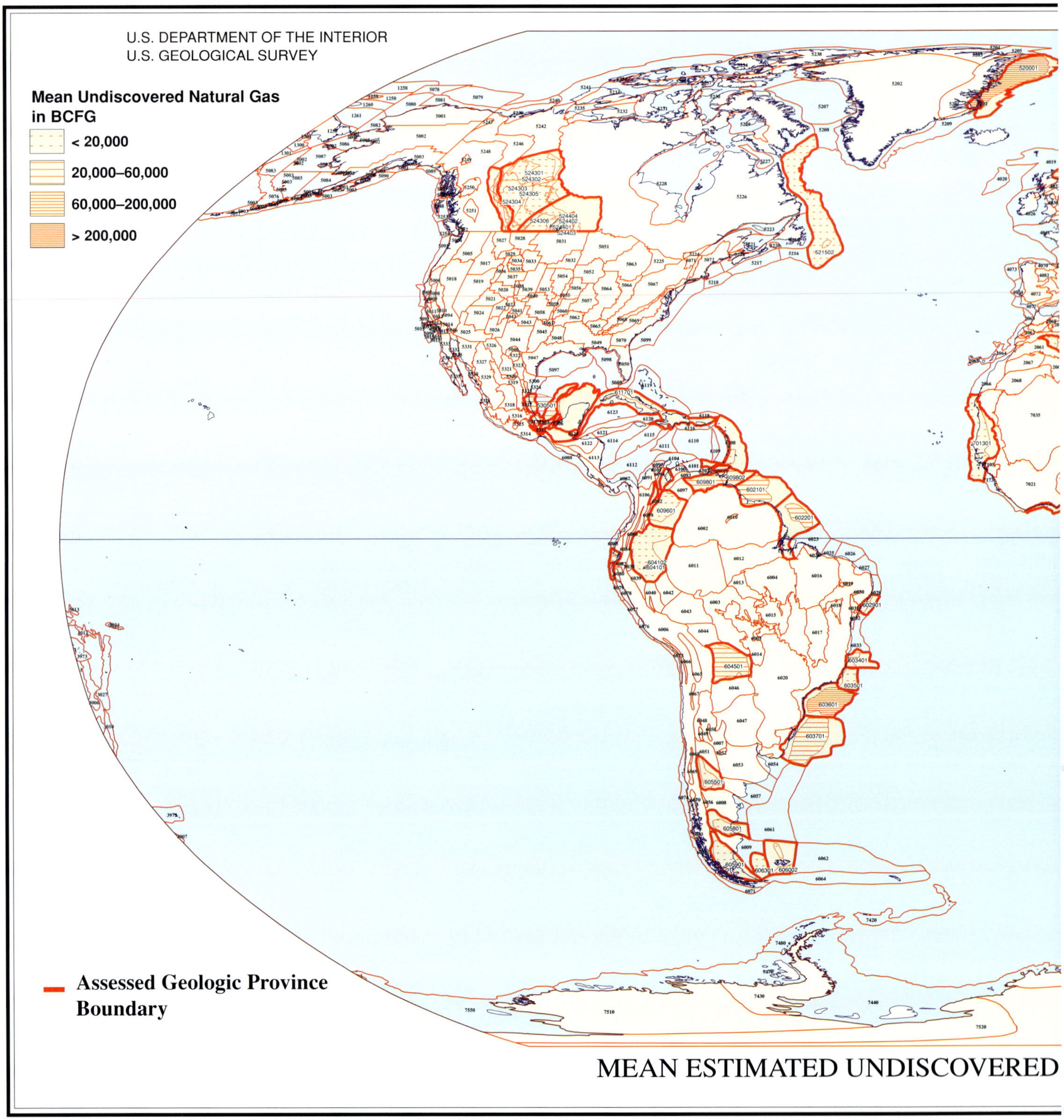

FIGURE 29. Map showing ranked categories of assessed total petroleum systems (TPS) by mean estimated undiscovered natural gas volume. Categories are <20,000 BCFG; 20,000–60,000 BCFG; 60,000–200,000 BCFG; >200,000 BCFG. BCFG = billion cubic feet of gas.

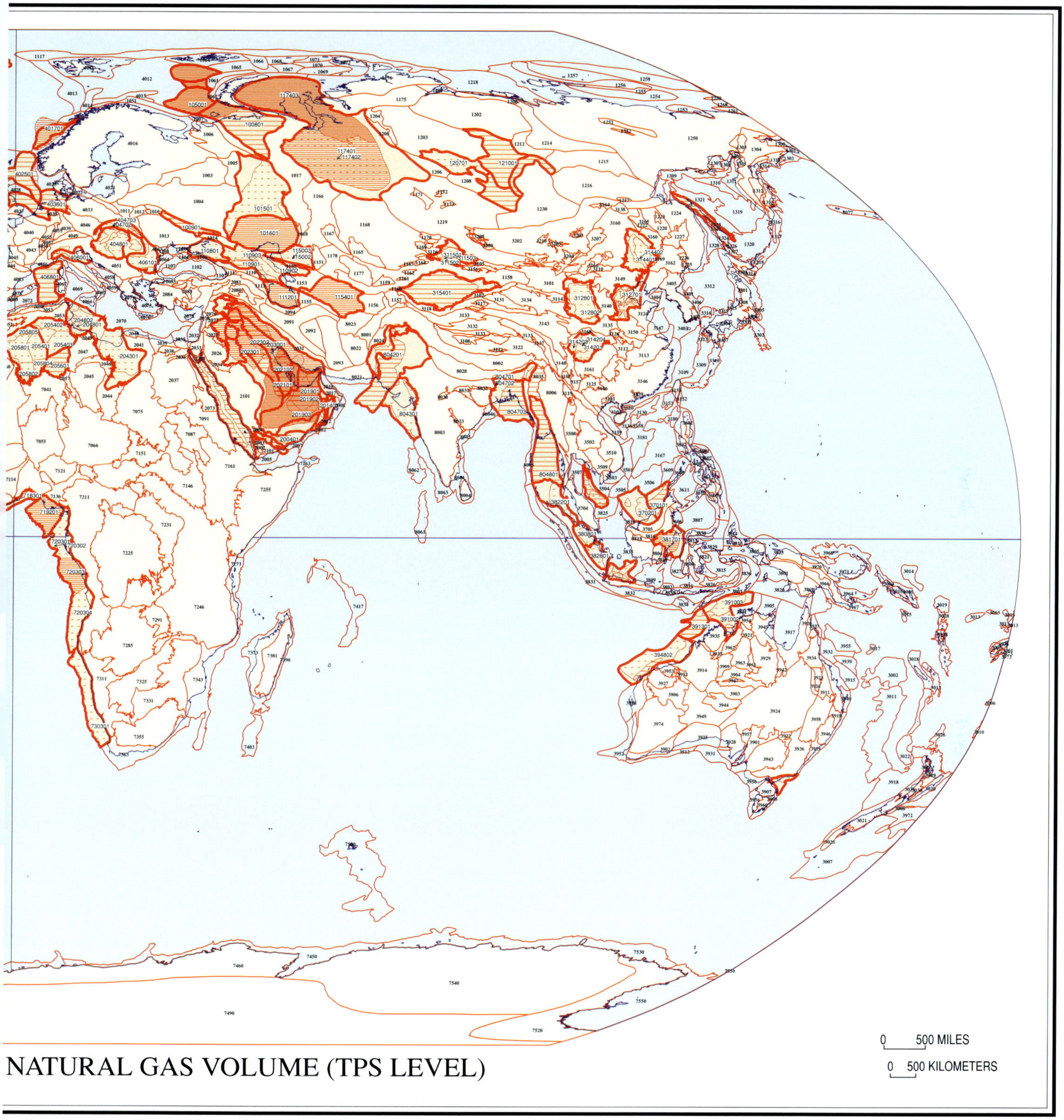

Projection Robinson
Central meridian 0

IV. Reserve Growth

Reserve growth refers to the increases, in estimated sizes of discovered fields, that can occur through time as oil and natural gas fields are developed and produced. In the U.S., which is one of the most intensely explored countries in the world, reserve growth is widely considered to be a major component of remaining oil and natural gas resources (Arrington, 1960; Root and Mast, 1993; Attanasi and Root, 1994; Root et al., 1995; Schmoker and Attanasi, 1996, 1997; Attanas and Schmoker, 1997; Attanasi et al., 1999; Verma, 2000; Klett, 2003; Klett and Schmoker, 2003). A forecast of world potential reserve growth was therefore a necessary element of the World Petroleum Assessment 2000 (USGS, 2000). The forecast (Table 1) shows that the estimated potential additions to reserves from reserve growth are nearly as large as the estimated undiscovered resource volumes (Schmoker and Klett, 2000). Unlike World Petroleum Assessment 2000 (USGS, 2000), allocations of the world reserve growth were made in this study for the eight regions of the world for liquid petroleum, oil and natural gas liquids, (Figure 30), and natural gas (Figure 31).

The technical approach to this forecast of reserve growth and detailed results are documented by Schmoker and Klett (2000), and Klett and Schmoker (2003). The potential reserve growth for world oil (excluding the U.S.) is estimated to range from 192 billion bbl of oil at a 95% probability to as much as 1031 billion bbl of oil at a 5% probability, with a mean of 612 billion bbl of oil (Table 1; Figure 32). These estimates, as well as those for natural gas and NGL, reflect a 30-yr forecast span (1995 to 2025).

The world potential reserve growth (excluding the U.S.) for natural gas is estimated to range from 1049 tcf (95% probability) to 5543 tcf (5% probability), with a mean of 3305 tcf (Table 1; Figure 33). World potential reserve growth (excluding the U.S.) for NGL is estimated to range from 13 billion bbl of NGL (95% probability) to 71 billion bbl of NGL (5% probability), with a mean of 42 billion bbl of NGL (Table 1; Figure 34).

These forecasts indicate that reserve growth could prove to be a major component of the world's future petroleum resources. As illustrated by Figures 14 and 15, the forecasts made in the World Petroleum Assessment 2000 (USGS, 2000) for the reserve growth of oil and for undiscovered oil resources (excluding the U.S.) are roughly equal. The world has been explored for oil to the point that the reserve growth of fields already discovered may provide nearly as much new oil in the next 30 yr as will come from discoveries of conventional fields.

Worldwide, natural gas is a commodity that is less mature in terms of exploration than oil. Reserve growth is predicted to provide somewhat less additional natural gas in the next 30 yr than will come from conventional new-field discoveries (26% versus 37% of future resources, respectively; Figure 14). Nevertheless, the current assessment indicates that slightly more than one-fourth of the future natural gas resources of the world (excluding the U.S.) reside in the reserve growth category (Figure 13).

International production and reserves, and resource databases supplied by proprietary sources such as IHS Energy Group (formerly Petroconsultants) or by organizations or publishing groups such as the *Oil & Gas Journal*, *World Oil*, or individual company analysis such as those annually reported by BP (British Petroleum) (2001) and Shell (2001) were used in this study. Reported, also called known, reserves include proved reserves, but additionally are said to include the median or F_{50} estimate of probable reserves. Because the definition of probable reserves is less restrictive than proved reserves, reserve growth should be somewhat lower for a world analysis relative to a U.S. analysis.

Potential additions to reserves from reserve growth were estimated for the first time by the USGS and were

 DOI:10.1306/1061774M863175

estimated at the world level only, but regional allocations of this total were made for oil and natural gas (Figures 30, 31). Reserve growth for the world has also been estimated recently by others (Schollnberger, 1998; Edwards, 1997, 2001, 2002 a, b; IEA, 1998, 2000, 2001; Stark, 2003).

The World Petroleum Assessment 2000 (USGS, 2000) has a 30-yr forecast span (1995–2025), in contrast to previous USGS assessments that presented estimates of ultimate petroleum resources without a time constraint. The latter portion of even a 30-yr time frame is difficult to predict because of continuous technological progress and a rapidly evolving world economy.

Because of reserve growth, the announced size of a newly discovered field does not equal its ultimate size. A newly discovered field is more likely than not to increase in size during its productive life. Thus, the size estimates of recent new field discoveries are commonly conservative.

The ranges of the forecasts for the world's reserve growth are large with respect to undiscovered resource volumes. These large ranges reflect the fact that worldwide patterns of reserve growth are poorly understood and difficult to establish. For the world as a whole, a set of uniform reporting requirements does not exist; criteria for the estimation of remaining reserves differ widely; and the technical, economic, and political conditions that drive the reserve-growth process vary from country to country. For these reasons, forecasts of reserve growth for entities smaller than the region were not developed as part of this assessment, but individual studies have demonstrated in greater detail the phenomenon of reserve growth in the Volga Ural and West Siberian basin provinces (Verma et al., 2000; Verma and Ulmishek, 2003), the North Sea (Klett and Gautier, 2003), and Iraq (Verma et al., 2004).

Future analyses of the patterns of reserve growth at regional and local levels offer a way to improve such assessments. Regional level reserve growth analyses are reported here based upon an allocation of the estimated world reserve growth reported rather than determination of individual regions.

The 8 years since 1995 are less than one-third of the 30-yr forecast span (1995–2025) for the assessment. During this period, 18% of estimated oil resources and 27% of the estimated natural gas resources have been added to reserves from new field discoveries and reserve growth for the provinces assessed in USGS (2000). Estimated reserve growth is very significant with 26% of the potential oil volume and 52% of the estimated natural gas volume via reserve growth being realized (Klett et al., in press). Overall, about 10% of the estimated undiscovered oil and natural gas was added to reserves by new discoveries in USGS (2000) assessed provinces. Reserve growth added roughly three times the oil and natural gas volumes than did new field discoveries (i.e. undiscovered resources as of 1995), demonstrating the volumetric significance of reserve growth for the world.

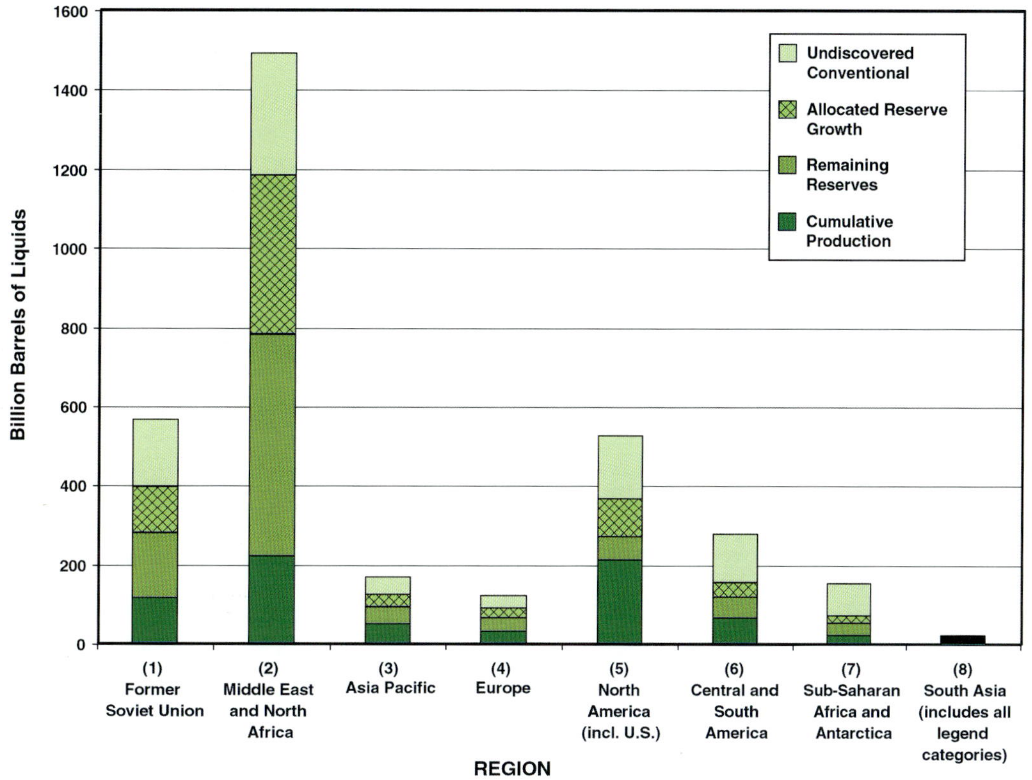

FIGURE 30. Grown endowment (cumulative production, remaining reserves, mean estimate of allocated reserve growth, and mean estimate of undiscovered resources) of conventional liquid (oil and natural gas liquids) for the eight regions of the world as defined by the World Petroleum Assessment 2000 (USGS, 2000).

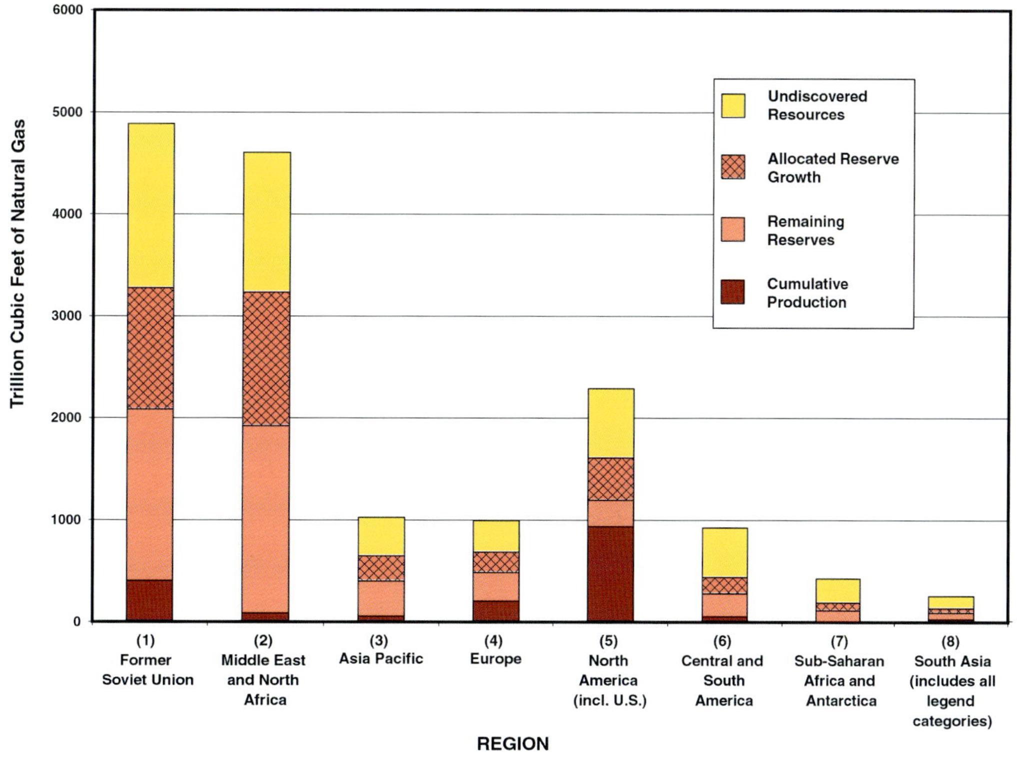

FIGURE 31. Grown endowment (cumulative production, remaining reserves, mean estimate of allocated reserve growth, and mean estimate of undiscovered resources) of conventional natural gas for the eight regions of the world as defined by the World Petroleum Assessment 2000 (USGS, 2000).

Figure 32. Forecast for world potential reserve growth of oil (exclusive of the U.S.), with uncertainty expressed in the form of a triangular probability distribution. F_{95} (the 95th fractile) represents a 95% chance of the occurrence of at least the amount tabulated. Other fractiles are defined similarly. BBO = billion barrels of oil. Forecast span is 30 yr (1995–2025).

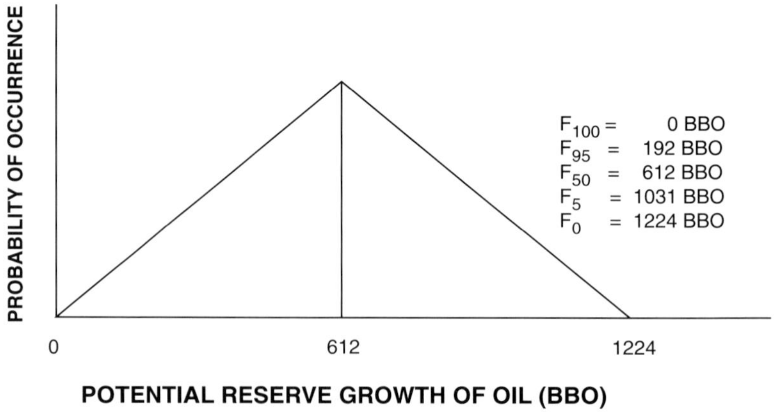

Figure 33. Forecast for world potential reserve growth of natural gas (exclusive of the U.S.), with uncertainty expressed in the form of a triangular probability distribution. F_{95} (the 95th fractile) represents a 95% chance of the occurrence of at least the amount tabulated. Other fractiles are defined similarly. TCFG = trillion cubic feet of natural gas. Forecast span is 30 yr (1995–2025).

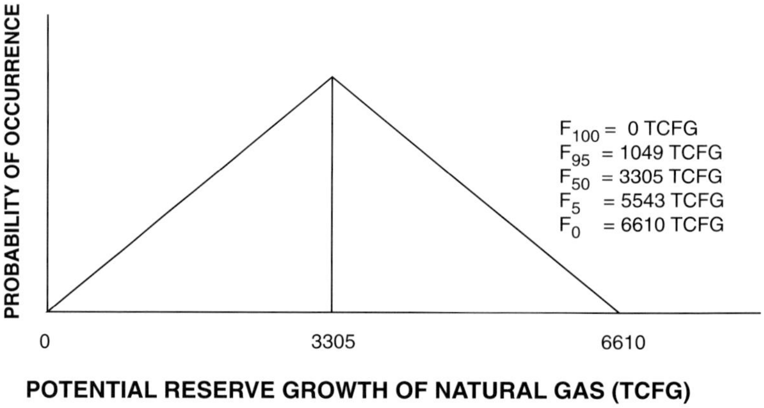

Figure 34. Forecast for world potential reserve growth of natural gas liquids (NGL) (exclusive of the U.S.), with uncertainty expressed in the form of a triangular probability distribution. F_{95} (the 95th fractile) represents a 95% chance of the occurrence of at least the amount tabulated. Other fractiles are defined similarly. BBNGL = billion barrels of NGL. Forecast span is 30 yr (1995–2025).

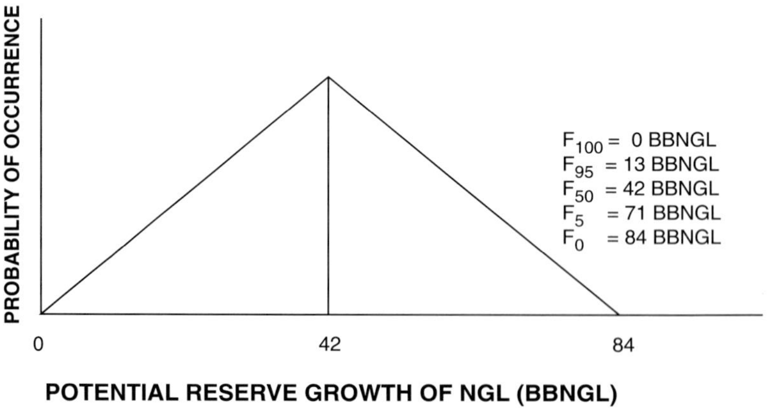

V. Preliminary Examination of Geologic Relationships

Although there have been individual analyses of source rock intervals and some elements of the total petroleum system (Klemme, 1986; Tissot et al., 1987; Klemme and Ulmishek, 1991; Perrodon, 1992; Magoon and Dow, 1994), the USGS (2000) study of world TPS provides many new geologic insights, petroleum volumes associated with the TPS, and a significant database for further studies. Some of these insights are discussed briefly for the elements of the TPS, and supporting data are shown in a series of Figures, maps, and Appendices.

In this study a large body of geologic information was assembled and synthesized for the TPS of the world included in the assessment. Eight geologic parameters from the geologic summaries for each of the assessment units (AU) were recorded in a spreadsheet (Appendix 4) and for each TPS (an aggregation of AU into their TPS) in Appendix 5. The geologic parameters are (1) source rock age, (2) source rock character, (3) peak maturation age, (4) reservoir rock age, (5) reservoir rock depositional environment, (6) reservoir rock lithology, (7) seal lithology, and (8) trap type.

The data from the eight geologic parameters were examined in terms of frequency of occurrence in the world TPS and their correlation with petroleum resource volumes. Total petroleum resource volumes, in billion BOE, are further distinguished as (1) known (cumulative production plus remaining reserves, [data from Petroconsultants, 1996]); (2) undiscovered volumes (data from USGS, 2000); and (3) conventional endowment, which consists of the sum of known reserves (cumulative production and remaining reserves) and undiscovered resources but does not include reserve growth estimates. World reserve growth estimates were not allocated below the regional level.

A. GEOLOGIC CHARACTERIZATION OF ASSESSMENT UNITS

1. Source Rocks:

a. Source rock age

Source rocks occur throughout the sedimentary rock record (Figures 35, 36, 37, 38, 39, 40); much petroleum has apparently been lost from older petroleum systems because, although such older source rocks are numerically abundant, the resources associated with them are volumetrically less significant than in younger systems (Figures 41, 42). Mesozoic source rocks (particularly Jurassic–Cretaceous) are the most important volumetrically in terms of conventional endowment, followed by Cenozoic source rock.

The numerical distribution of AU by specific age of source rock is shown in Figure 41, and the mean petroleum volumes of AU grouped by specific age of source rock are shown in Figure 42. Four maps (Figures 37, 38, 39, 40) illustrate the distribution of AU having source rock ages of Precambrian, Paleozoic, Mesozoic, and Cenozoic. Most TPS with Paleozoic source rocks are primarily on the ancient cratons, whereas TPS with Mesozoic and Cenozoic source rocks are largely on the passive and active continental margins and in foreland basins of the Alpine foldbelts.

b. Source rock character

Although there are many opinions as to the relative roles of source rock types, this compilation demonstrates that marine source rocks with Type II kerogen are by far the dominant source rock type (Figures 43, 44).

 DOI:10.1306/1061775M863175

The number of AU grouped by character of source rock, defined as coaly, lacustrine, shallow marine, and deep marine, is shown in Figure 43, and the estimated mean petroleum volumes of TPS grouped by depositional environment of source rock are shown in Figure 44. Four maps (Figures 45, 46, 47, 48) show the distribution of AU having coaly, lacustrine, shallow marine, and deep marine source rocks, respectively.

Shallow and deep marine source rocks are present in TPS with source rocks of all ages that are designated in this study. Coaly and lacustrine source rocks are more characteristic of TPS with Mesozoic and Cenozoic source rocks. It should be noted that there is no clear definition of deep marine versus shallow marine source rocks. Generally, deep marine source rocks were commonly deposited under anoxic conditions; however, anoxic to disoxic conditions of deposition are characteristic of some shallow marine source rocks.

Appendix 4 documents volumes and geologic attributes assigned to assessment units shown on the maps, and Appendix 5 shows the same for the TPS. The dominance of marine source rocks, particularly shallow marine source rocks, for both known (cumulative production plus remaining reserves) and undiscovered resources is clearly demonstrated.

c. Peak maturation age

The number of AU grouped by the age at which peak maturation of source rocks occurred is shown in Figure 49, and the mean petroleum volumes of TPS grouped by age at which peak maturation of source rocks occurred are shown in Figure 50. These data and plots demonstrate that young total petroleum systems, i.e., Cenozoic TPS, are volumetrically dominant for oil, and less so for natural gas. Three maps (Figures 51, 52, 53) illustrate the distribution of AU having peak maturation coinciding with Paleozoic, Mesozoic, and Cenozoic ages.

Apparently many total petroleum systems of the world, in which maturation of source rocks occurred early in geologic history, lost their petroleum due to subsequent tectonic processes and associated breaching of traps. Much of early generated oil and natural gas that has been preserved is associated with salt seals.

2. Reservoirs:

a. Reservoir rock age

The number of AU grouped by age of reservoir rock is shown in Figure 54, and the mean petroleum volumes of AU grouped by age of oldest reservoir rock in the AU are shown in Figure 55 and Appendix 4. Four maps (Figures 56, 57, 58, 59) show the distribution of AU having reservoir rock ages of Precambrian, Paleozoic, Mesozoic, and Cenozoic. A detailed breakdown of the number of AU by oldest reservoir rock age is shown in Figure 60, and a detailed breakdown of petroleum volumes by reservoir rock age is shown in Figure 61. The common occurrence of reservoir rocks at era boundaries is seen in Figures 60 and 61. This concentration of reservoirs adjacent to era boundaries is apparent and is supportive of the concept of petroleum concentration along inter-regional unconformities (Schlee, 1984).

b. Reservoir rock depositional environment

The number of AU grouped by depositional environment of reservoir rock is shown in Figure 62, and the mean petroleum volumes of AU grouped by depositional environment of reservoir rock are shown in Figure 63. Four maps (Figures 64, 65, 66, 67) illustrate the distribution of AU having reservoir rock depositional environments characterized as continental, paralic, shallow marine, and deep marine. Despite enormous recent success with deepwater reservoirs, volumetrically they are the least significant of those considered whereas non-marine to shallow-marine-reservoirs dominate (Figures 62, 63). This may reflect, in part, less mature exploration of deepwater reservoirs; however, it is doubtful that the relationships shown in Figure 63 will substantially change in the future.

The large conventional petroleum endowment related to non-marine and non-marine to marine reservoir rocks deposited in continental, paralic, and nearshore marine environments is apparent from Figure 63 in spite of paucity of source rocks (except lacustrine) in these environments. This emphasizes the commonality of vertical migration of petroleum in geologic sequences. The geographic distribution of dominantly continental, paralic, and shallow and deep marine reservoir rocks by AU is shown in Figures 64, 65, 66 and 67.

c. Reservoir rock lithology

The number of AU grouped by lithology of reservoir rock is shown in Figure 68, and the mean petroleum volumes of AU grouped by lithology of reservoir rock are shown in Figure 69. Both numerically and volumetrically, siliciclastic reservoirs are dominant (Figures 68, 69). Three maps (Figures 70, 71, 72) show the distribution of AU having primary reservoir lithology of siliciclastic, carbonate, and mixed siliciclastic and carbonate, respectively.

Siliciclastic reservoirs are abundant on all continents, whereas carbonate reservoirs are largely limited to the northern group of continents, which were located in low latitudes in Paleozoic time. They are also present in the Mesozoic sequences of northernmost Gondwana (North Africa and Arabian Peninsula) that also were located in low latitudes. Future discoveries will be largely from siliciclastic reservoirs as shown for both undiscovered oil and natural gas resources in Figure 69.

3. Seal: *Seal rock lithology*

The importance of the lithology of both regional and local seal rocks has long been recognized in petroleum systems (Downey, 1994). This study documents the number and volume of resources related to different seal lithologies.

The numerical distribution of AU by lithology of seal rock is shown in Figure 73, and the mean petroleum volumes of AU, grouped by lithology of seal rock, for oil and natural gas are shown in Figures 74, 75, and 76. Three maps (Figures 77, 78, and 79) illustrate the distribution of AU having seal rock lithologies of dominantly shale, dominantly evaporites, and mixed shale and evaporites respectively. Shale seals are numerically and volumetrically dominant as shown in Figures 73 and 74.

Evaporites (particularly salt) are very effective long-term seals, and evaporite seals are a critical preservation component of older Paleozoic assessment units (Figures 75, 76).

A large part of both discovered reserves and estimated undiscovered resources of AU with source rocks that reached maturity during Paleozoic and Mesozoic times is controlled by evaporite seals (Figure 75). Evaporite seals are especially important in terms of petroleum volumes in the extensional and passive tectonic settings; in the compressional setting the integrity of evaporite seals is commonly breached. Younger AU in which maturation of source rocks was reached during Mesozoic to Cenozoic and Cenozoic times have dominantly siliciclastic seals.

4. Traps: *Trap type*

The number of AU grouped by type of trap is shown in Figure 80, and the estimated mean petroleum volumes of AU grouped by trap type and tectonic settings are shown in Figures 81, 82, and 83. Six maps (Figures 84, 85, 86, 87, 88, 89) show the distribution of AU having a primary trap type characterized as stratigraphic, paleogeomorphic, compressional structures, extensional structures, non-tectonic structures, and combination structural-stratigraphic, respectively.

These data demonstrate that petroleum is trapped in many ways but no single trap type dominates in terms of petroleum volume. Compressional structural traps contain the largest volume of hydrocarbons, although not a majority, followed by extensional, stratigraphic, non-tectonic, structural-stratigraphic, and paleogeomorphic trap types (Figure 80). In general, structural traps contain the dominant volumes of both discovered and undiscovered hydrocarbons (Figures 82, 83). However, the exploration maturity of stratigraphic traps in the great majority of the world TPS is far less than that of structural traps.

The locations of AU containing stratigraphic traps, paleogeomorphic traps, compressional traps, and extensional traps are shown respectively in Figures 84, 85, 86, and 87. AU's containing non-tectonic structural traps are shown in Figure 88 and combination structural-stratigraphic traps in Figure 89. In most AU, traps containing a substantial part of discovered and undiscovered resources belong to more than one type and thus, these AU are shown on two or more maps in Figures 84–89.

B. OTHER OBSERVATIONS ON GEOLOGIC RELATIONSHIPS

Data were presented in the previous section with relatively little discussion or interpretation. Perusal of the figures and the tables in the Appendix, however, reveals additional interesting patterns and trends in and among the factors that exert basic geologic controls on world petroleum resources. Such data might serve to revise or confirm some long-held qualitative assumptions, and almost certainly offer insights into paths for future research.

Most of the total petroleum systems in the world are dominated by vertical migration or limited lateral migration (less than 12.4 mi [20 km]) from the mature source rock area. There are many differences between oil and natural gas total petroleum systems. Of the top ten oil total petroleum systems, in terms of total endowment (Appendix 2c), only two are also found in the top 10 natural gas petroleum system endowments (Appendix 2d). Geographically and geologically, the dominant oil total petroleum systems are different from the dominant natural gas total petroleum systems. These differences are also seen at the assessment unit (Appendices 1a–1f), total petroleum system (Appendices 2a–2f), and province (Appendices 3a–3i) levels.

Although some geoscientists argue that thermogenic natural gas is generated nearly simultaneously with oil in an active petroleum system, the data shown in Appendices 1a–1f, 2a–2f, 3a–3i, 4, and 5 suggest a more complicated story because the major oil and natural gas TPS at any level of consideration are not geographically the same. It would seem that natural gas total petroleum systems and their assessment units reflect sufficient differences from oil total petroleum systems to warrant entirely separate study particularly in light of increasing emphasis on natural gas. Many major conventional natural gas systems are closely linked to large unconventional (continuous) natural gas resources, and this further emphasizes the need for greater study of the natural gas systems. Unconventional (continuous) oil and natural gas systems are little studied throughout the world compared to conventional petroleum resources. Moreover, the natural gas endowment of 2.57 trillion BOE seems under-represented when compared to the 3.0 trillion bbl oil endowment. Natural gas generation is far more ubiquitous in nature than oil generation, and large quantities of natural gas are generated in many types of geologic formations that are devoid or almost devoid of effective oil-source rocks. For example, in the U.S. the two commodities, on a BOE basis, are nearly equivalent. The gap between oil and natural gas resources on the world scale may reflect several things: 1) There is a less complete natural gas database relative to the oil database. 2) Many discovered natural gas resources are undeveloped due to infrastructure limitations such as gas pipelines, resulting in considerable known but underutilized or stranded gas. Further, the reported field sizes for these natural gas resources are likely substantially underestimated because the fields are not being developed. 3) A fundamental question is whether natural gas total petroleum systems are under-represented due to economically or geographically biased sampling. For instance, the seemingly gas-prone Arctic provinces might rebalance the oil-to-natural gas relationship once they are explored and developed.

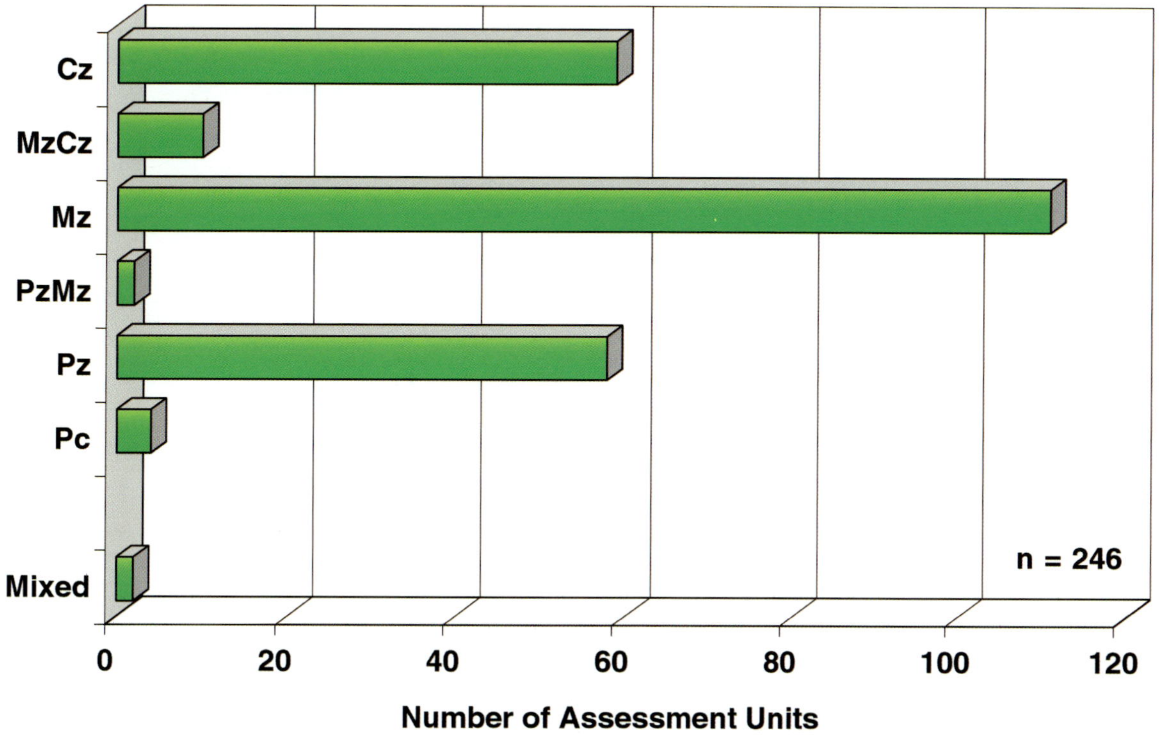

Figure 35. Numerical distribution of 246 assessment units by age of source rock. Pc = Precambrian; Pz = Paleozoic; PzMz = Paleozoic–Mesozoic; Mz = Mesozoic; MzCz = Mesozoic–Cenozoic; Cz = Cenozoic; Mixed = source rocks of multiple ages.

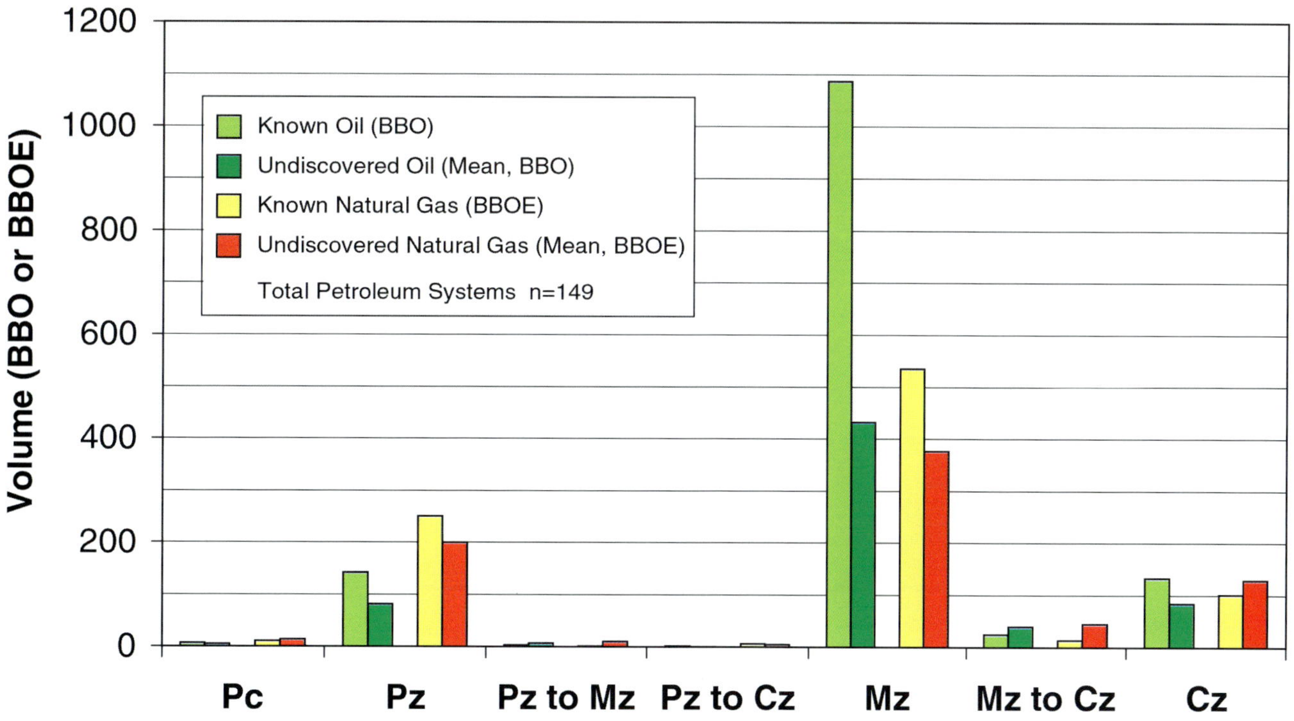

Figure 36. Distribution of mean estimated petroleum volumes of total petroleum systems by age of source rock. Known volume = cumulative production plus remaining reserves; undiscovered volume = estimated in this study. Pc = Precambrian; Pz = Paleozoic; Pz to Mz = Paleozoic–Mesozoic; Pz to Cz = Paleozoic–Cenozoic; Mz = Mesozoic; Mz to Cz = Mesozoic–Cenozoic; Cz = Cenozoic. BBO = billion barrels of oil; BBOE = billion barrels of oil equivalent.

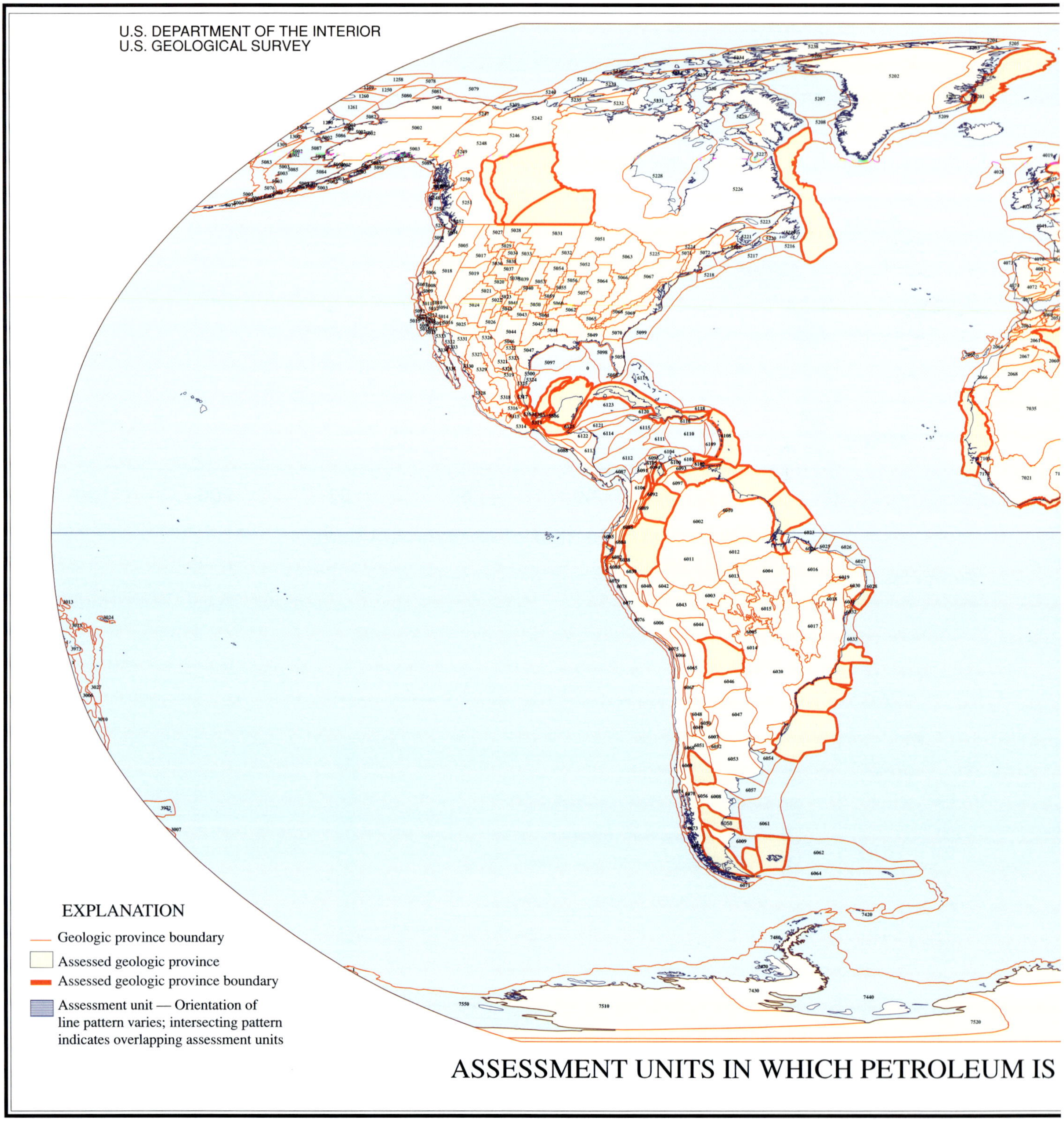

FIGURE 37. Map showing assessment units in which petroleum is derived from Precambrian source rocks.

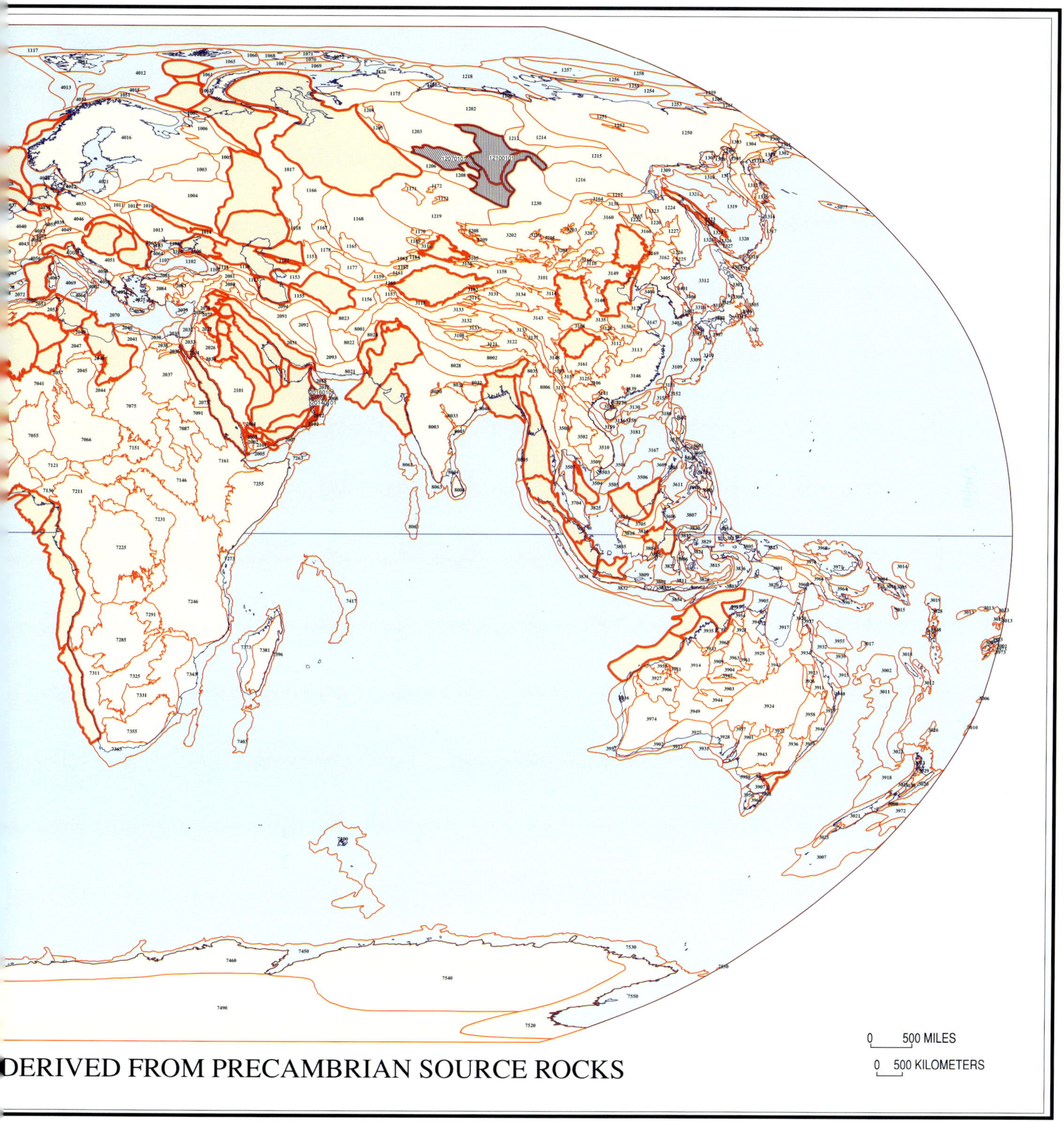

Projection Robinson
Central meridian 0

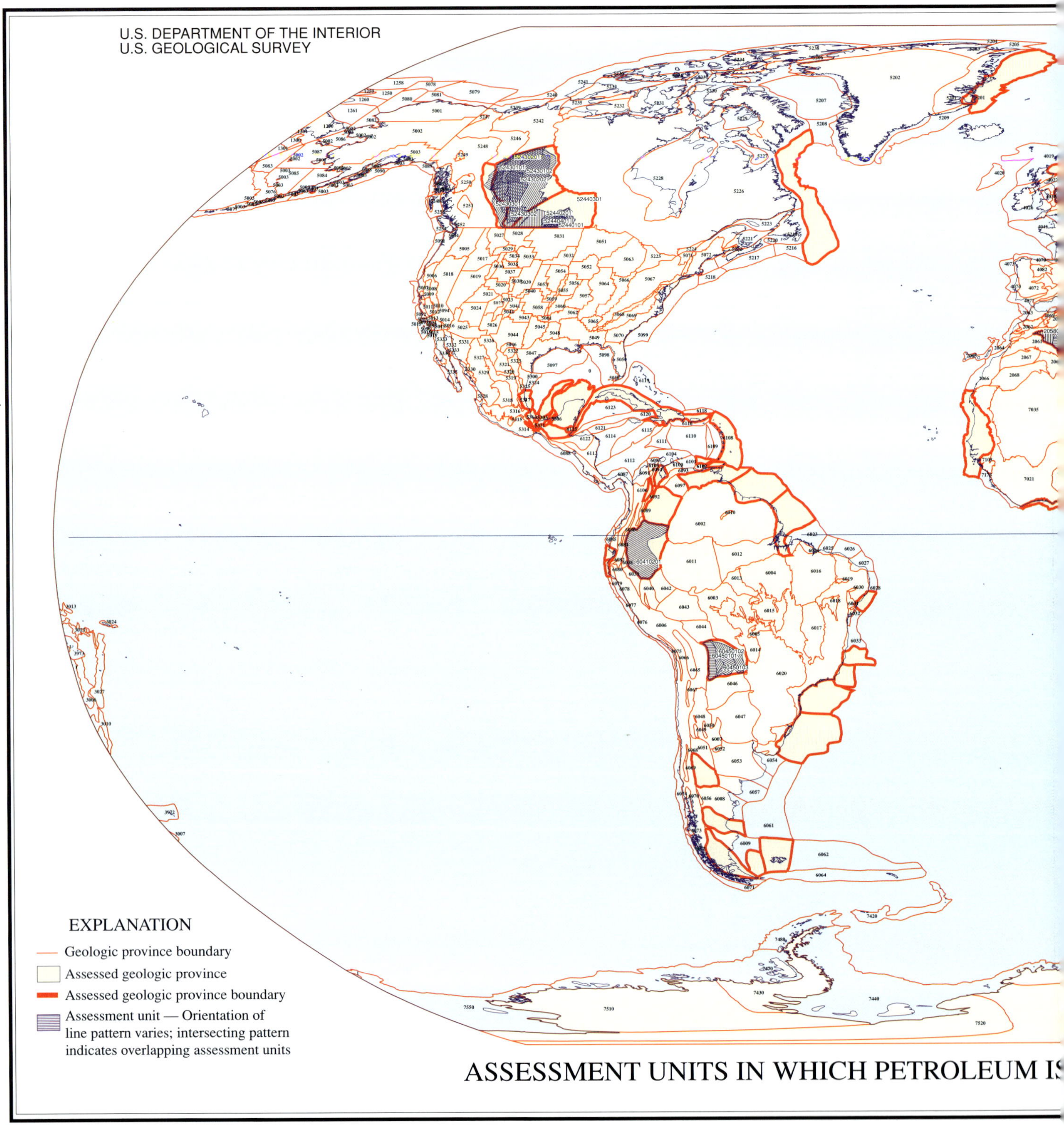

FIGURE 38. Map showing assessment units in which petroleum is derived from Paleozoic source rocks.

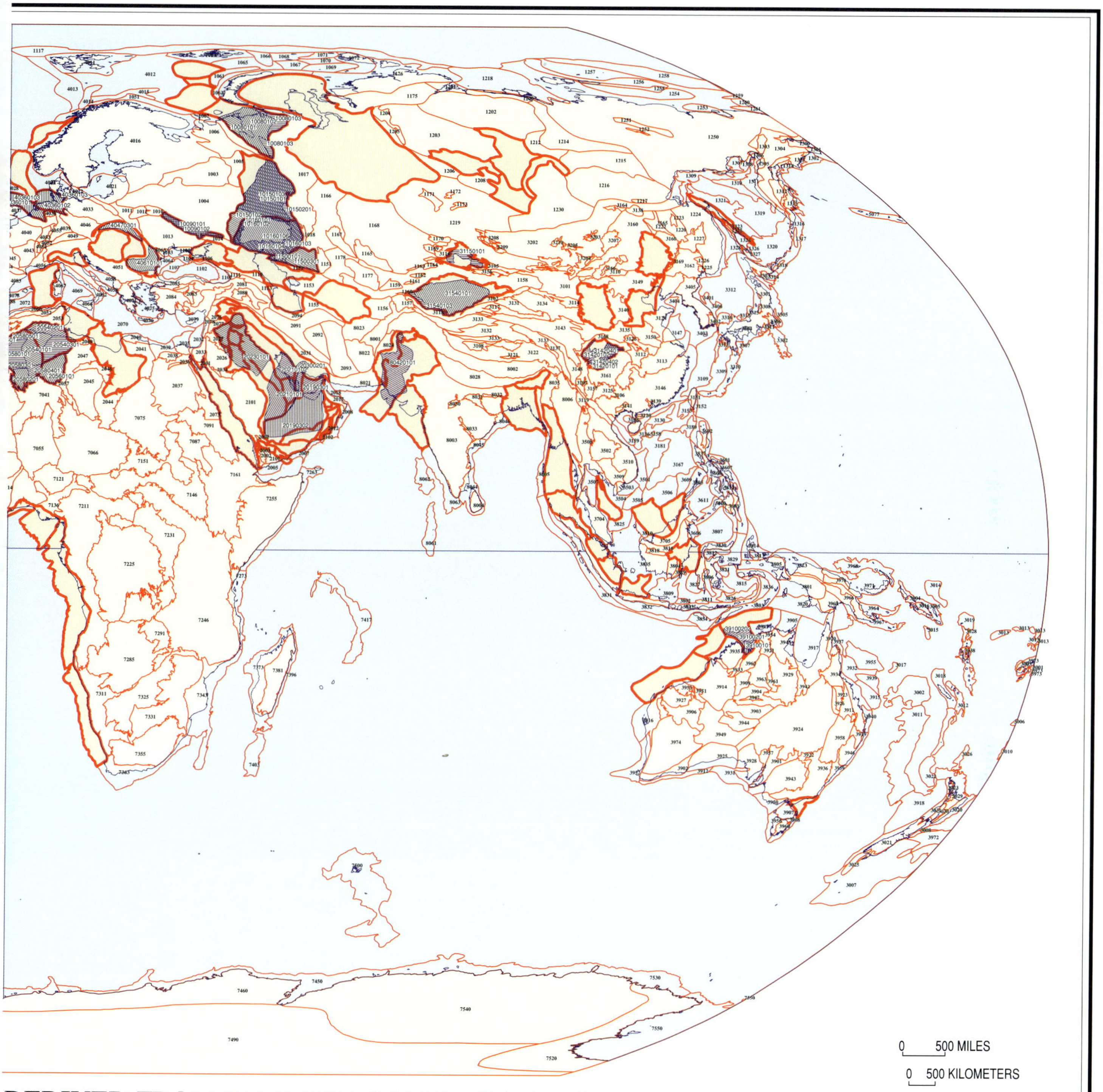

DERIVED FROM PALEOZOIC SOURCE ROCKS

Projection Robinson
Central meridian 0

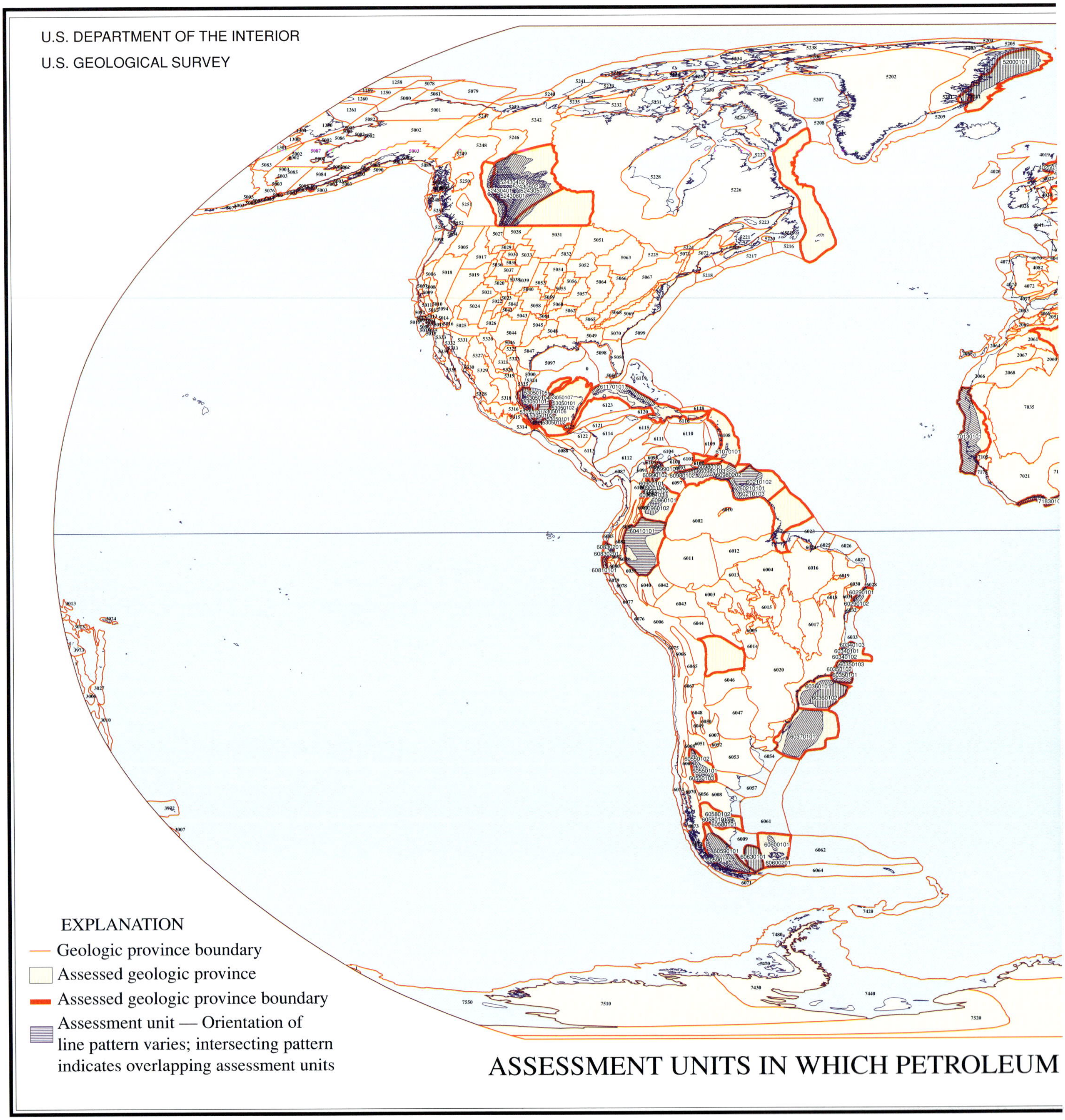

FIGURE 39. Map showing assessment units in which petroleum is derived from Mesozoic source rocks.

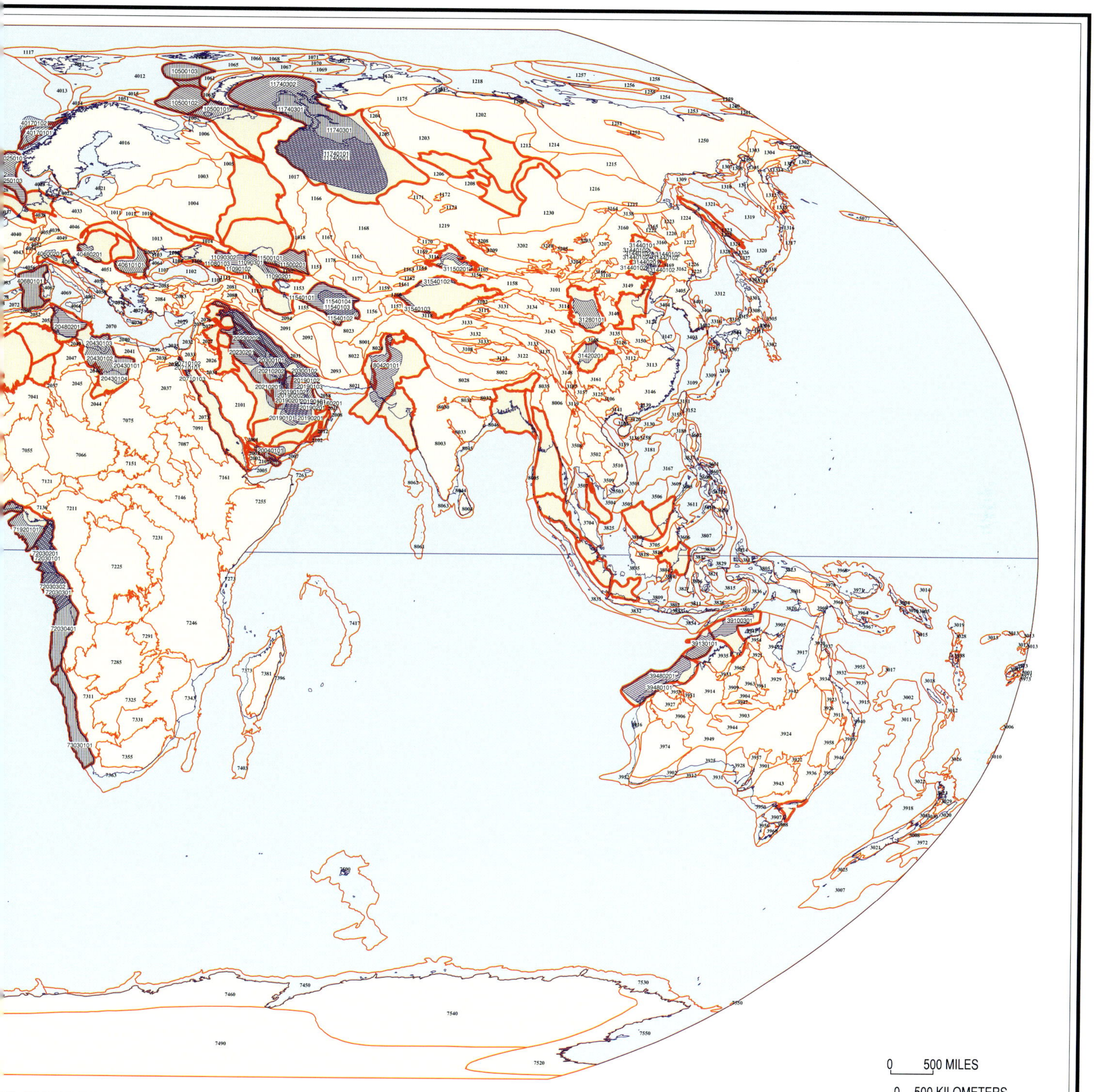

S DERIVED FROM MESOZOIC SOURCE ROCKS

Projection Robinson
Central meridian 0

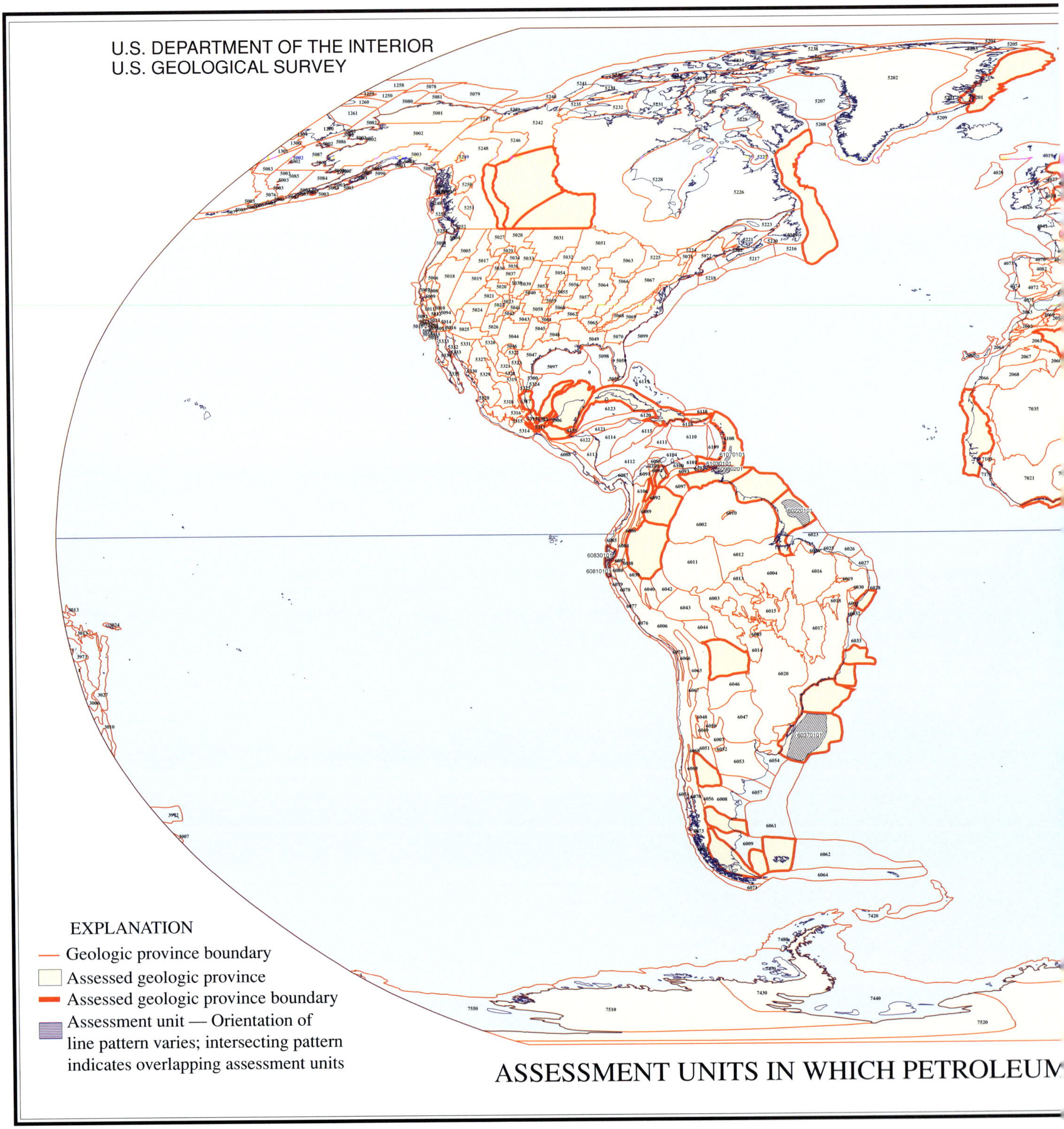

FIGURE 40. Map showing assessment units in which petroleum is derived from Cenozoic source rocks.

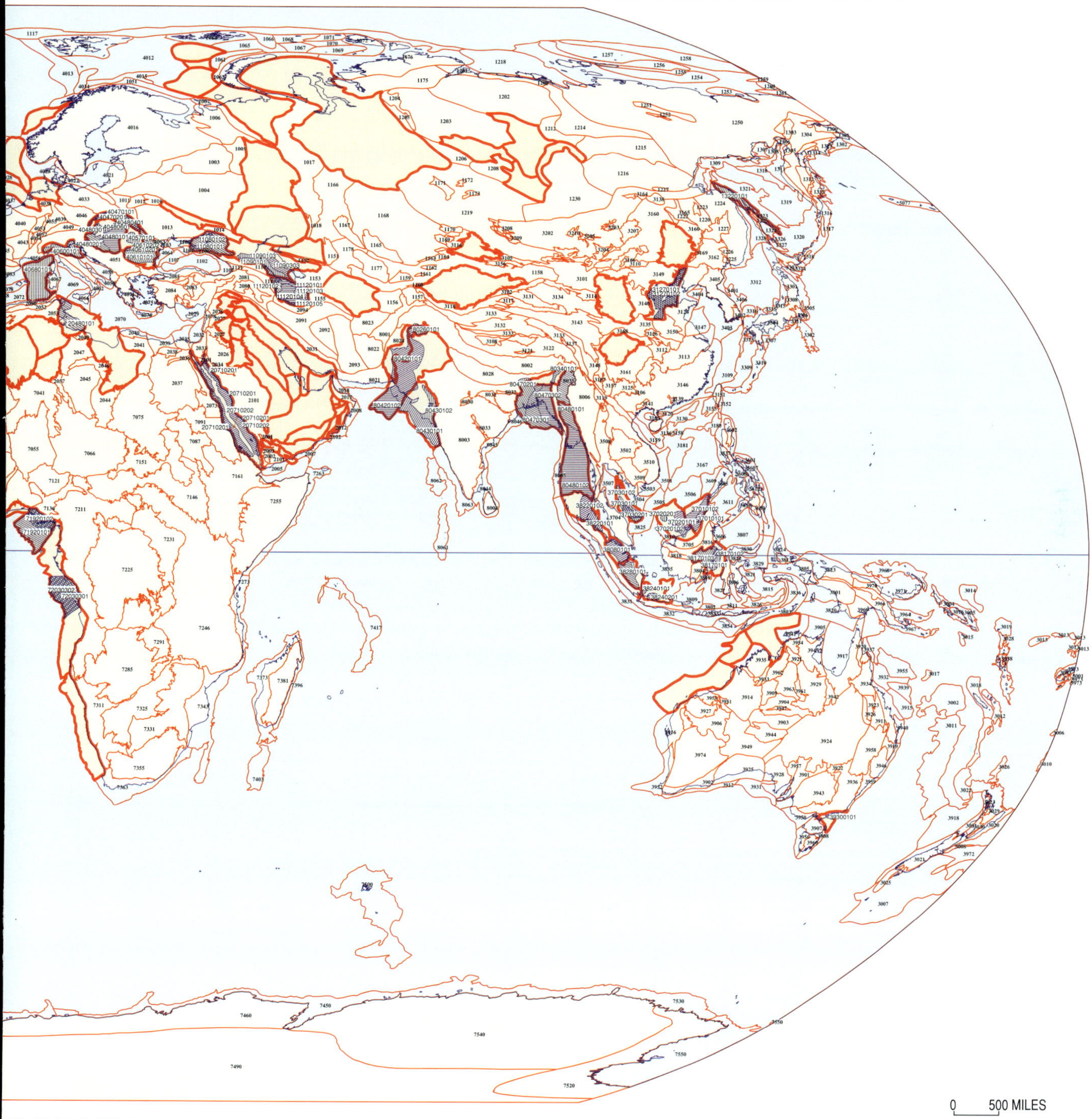

IS DERIVED FROM CENOZOIC SOURCE ROCKS

Projection Robinson
Central meridian 0

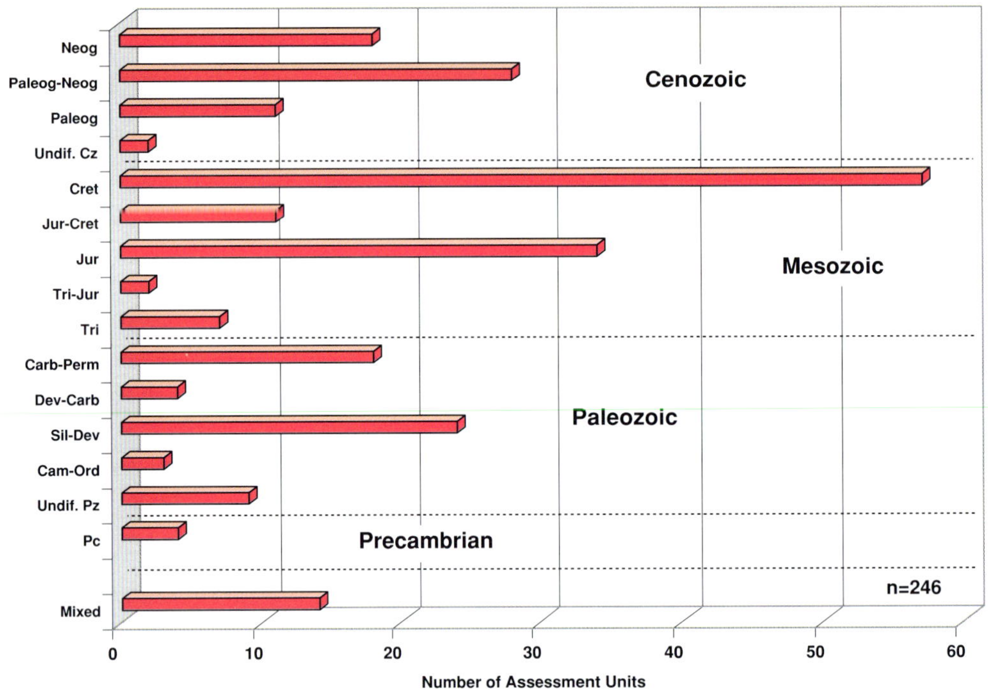

FIGURE 41. Numerical distribution of 246 assessment units by specific geologic age of source rock. Pc = Precambrian; Undif. Pz = Undifferentiated Paleozoic; Cam-Ord = Cambrian–Ordovician; Sil-Dev = Silurian–Devonian; Dev-Carb = Devonian–Carboniferous; Carb-Perm = Carboniferous–Permian; Tri = Triassic; Tri-Jur = Triassic–Jurassic; Jur = Jurassic; Jur-Cret = Jurassic–Cretaceous; Cret = Cretaceous; Undif. Cz = Undifferentiated Cenozoic; Paleog = Paleogene; Paleog-Neog = Paleogene–Neogene; Neog = Neogene; Mixed = source rocks of multiple ages.

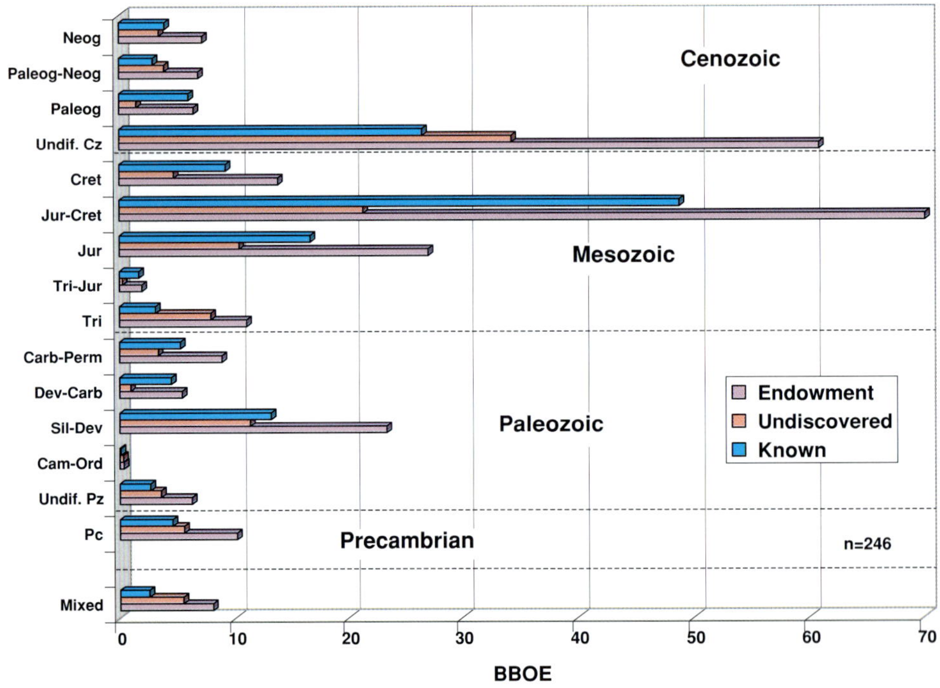

FIGURE 42. Distribution of mean estimated petroleum volumes of 246 assessment units by specific age of source rock. Known volume = cumulative production plus remaining reserves; undiscovered volume = estimated in this study; and conventional endowment = sum of known and undiscovered volumes. Pc = Precambrian; Undif. Pz = Undifferentiated Paleozoic; Cam-Ord = Cambrian–Ordovician; Sil-Dev = Silurian–Devonian; Dev-Carb = Devonian–Carboniferous; Carb-Perm = Carboniferous–Permian; Tri = Triassic; Tri-Jur = Triassic–Jurassic; Jur = Jurassic; Jur-Cret = Jurassic–Cretaceous; Cret = Cretaceous; Undif. Cz = Undifferentiated Cenozoic; Paleog = Paleogene; Paleog-Neog = Paleogene–Neogene; Neog = Neogene; Mixed = source rocks of multiple ages. BBOE = billion barrels of oil equivalent.

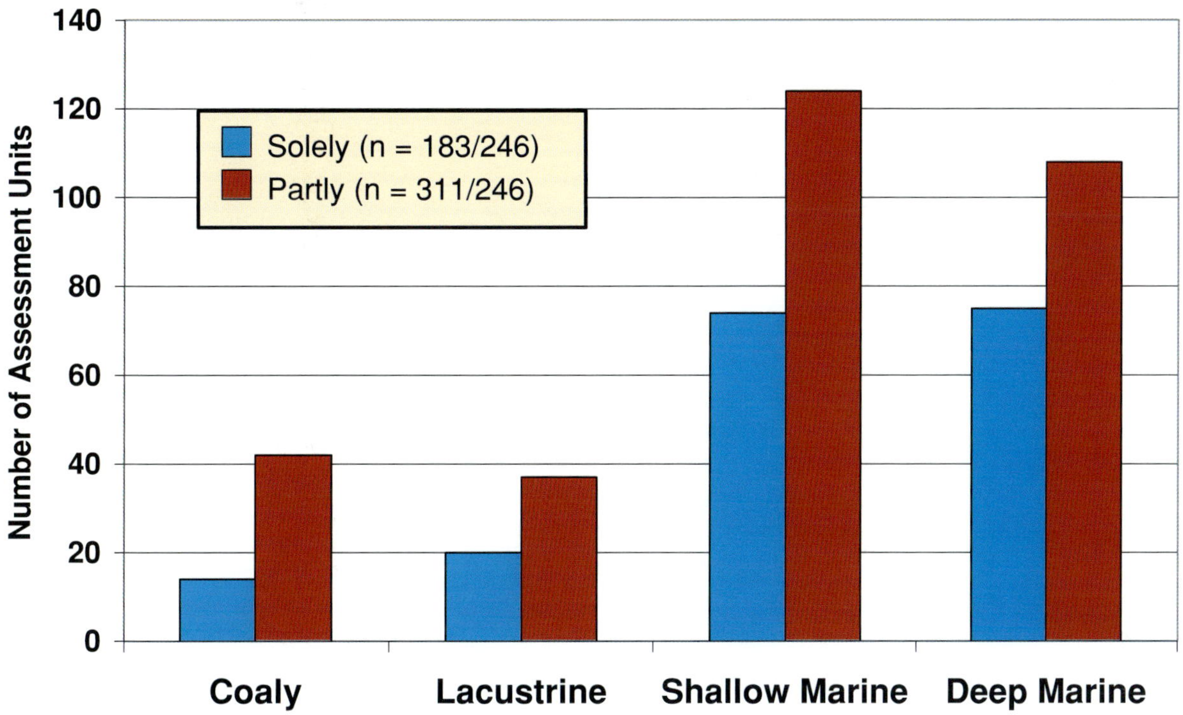

FIGURE 43. Numerical distribution of 246 assessment units by character of source rock. The four categories are coaly, lacustrine, shallow marine, and deep marine as defined for each assessment unit by the assessors. Some source rocks in the assessment units have mixed characters and are referred to as partly (red color) coaly, partly lacustrine, etc. If only one source rock character is determined, the term solely (blue color) is used.

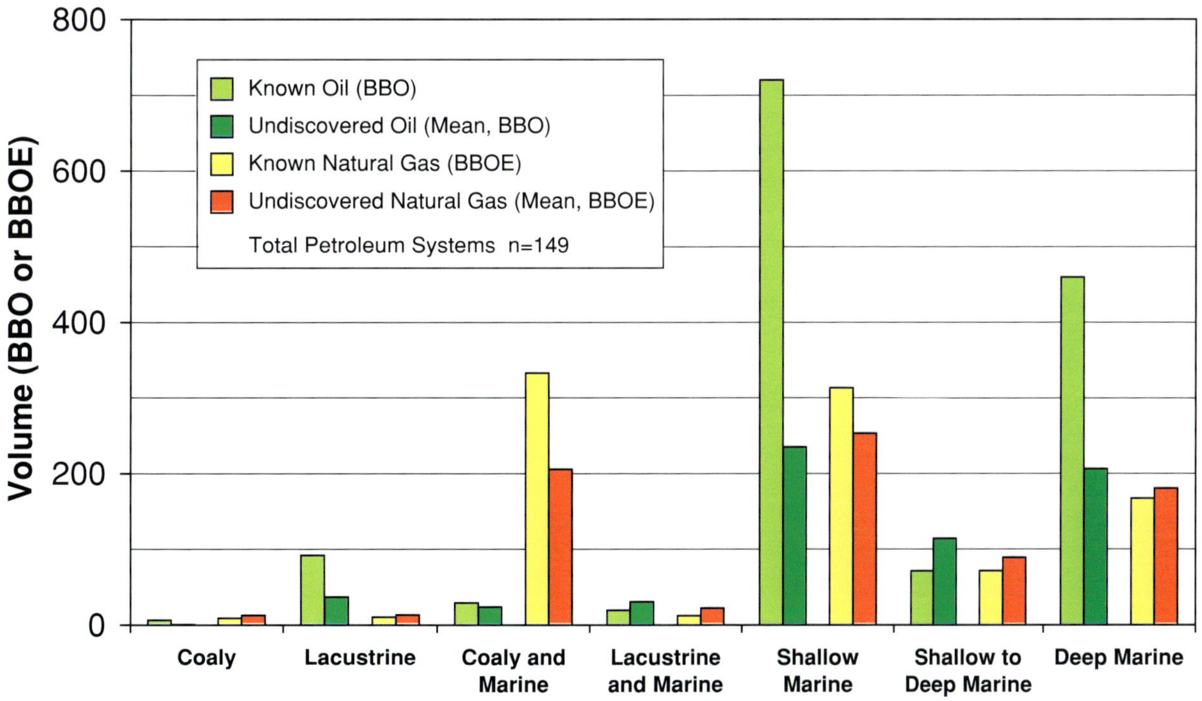

FIGURE 44. Distribution of mean estimated petroleum volumes of total petroleum systems by depositional environment of source rock. Known volume = cumulative production plus remaining reserves; undiscovered volume = estimated in this study. The four distinct categories and three combination categories are coaly, lacustrine, coaly and marine, lacustrine and marine, shallow marine, shallow to deep marine, and deep marine as defined for each total petroleum system by the assessors. BBO = billion barrels of oil; BBOE = billion barrels of oil equivalent.

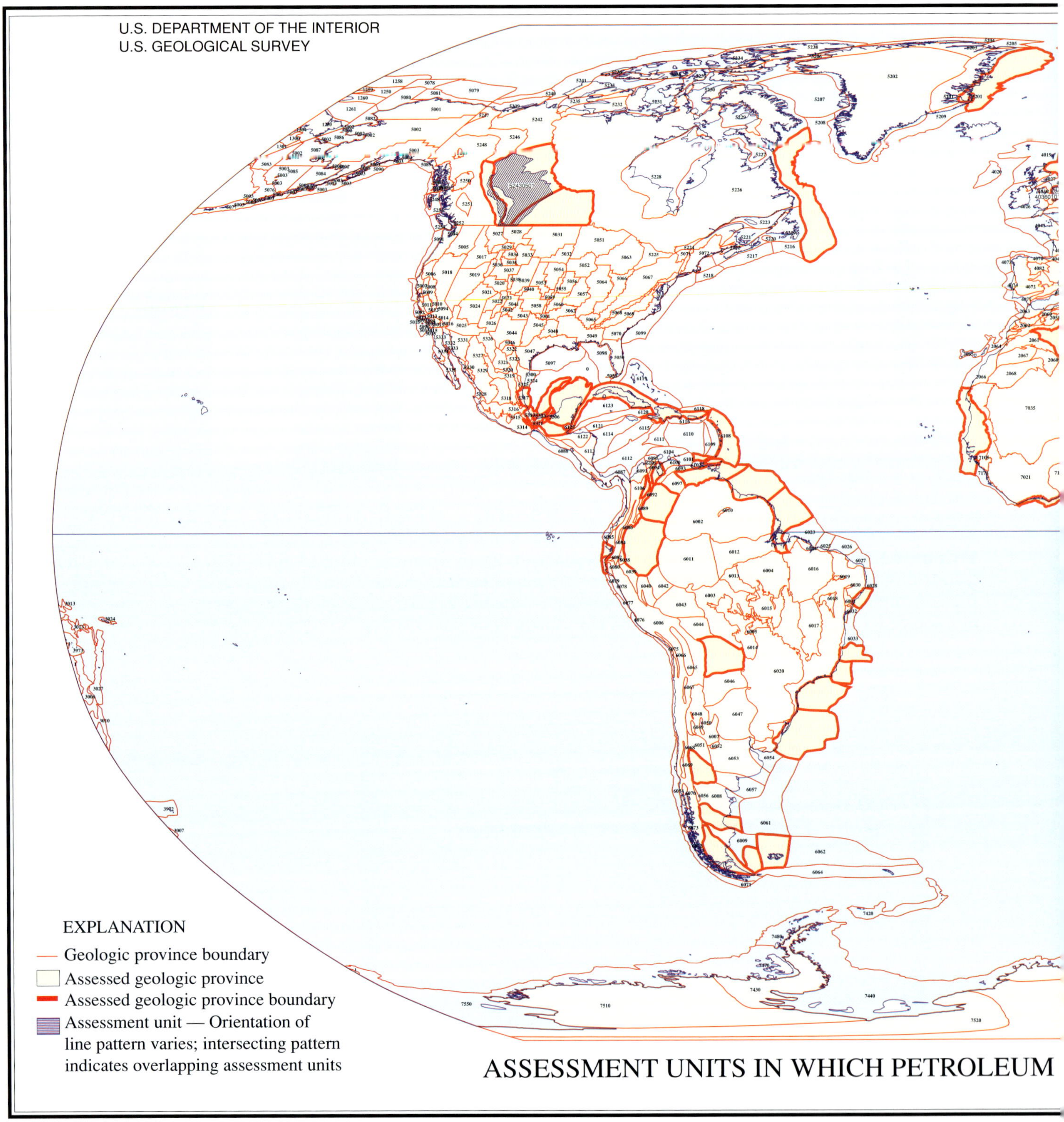

Figure 45. Map showing assessment units in which petroleum is derived from coaly source rocks.

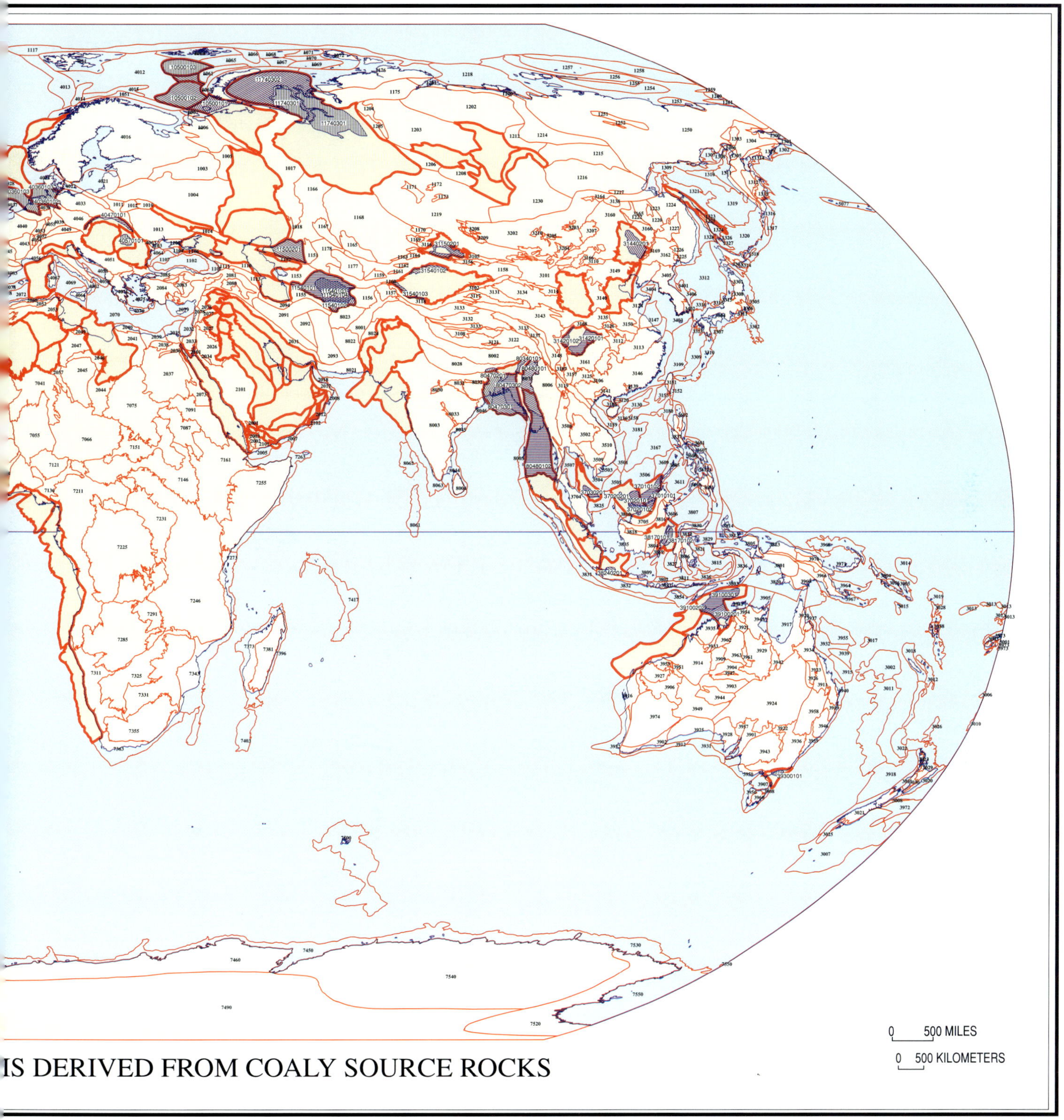

IS DERIVED FROM COALY SOURCE ROCKS

Projection Robinson
Central meridian 0

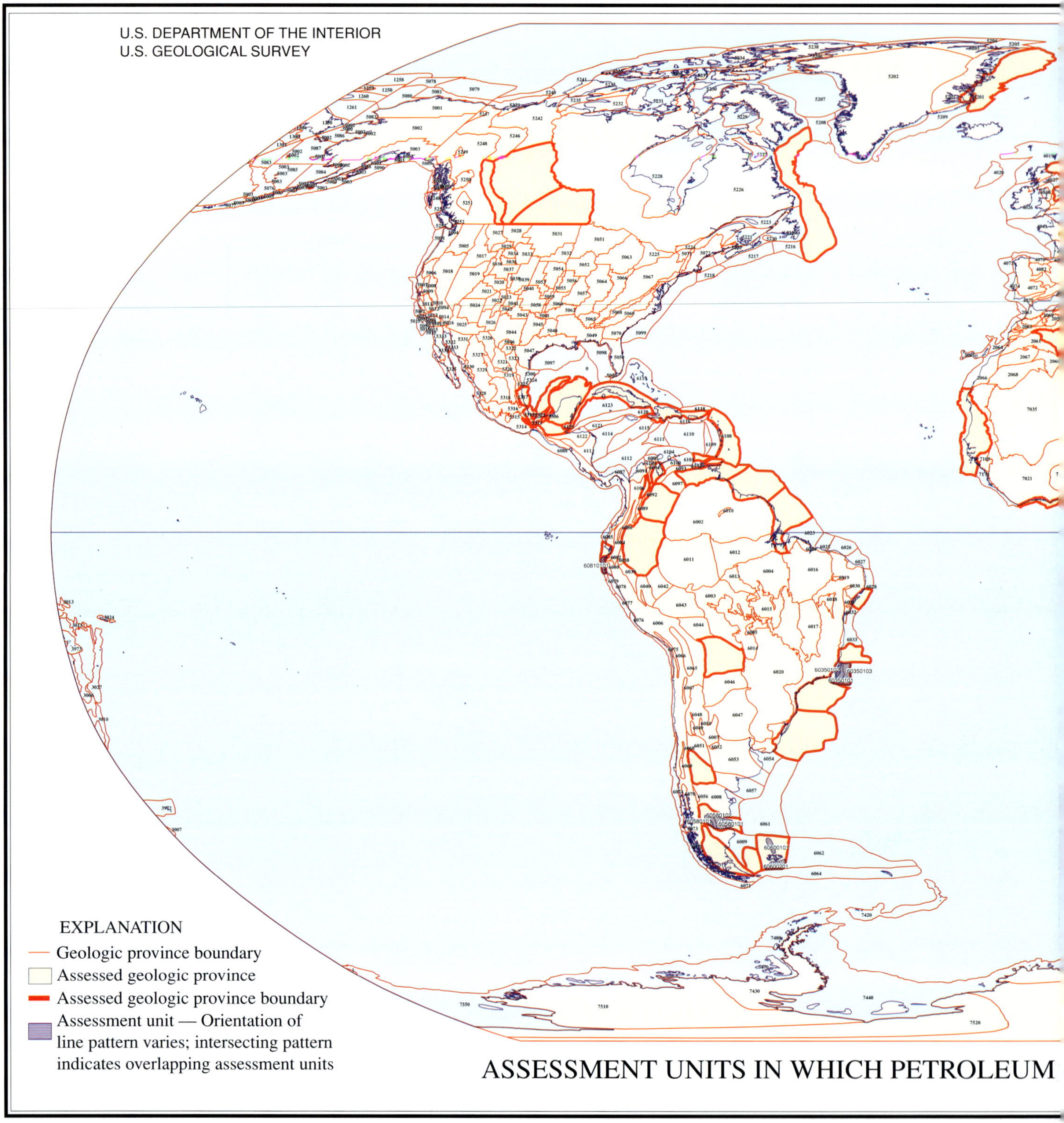

Figure 46. Map showing assessment units in which petroleum is derived from lacustrine source rocks.

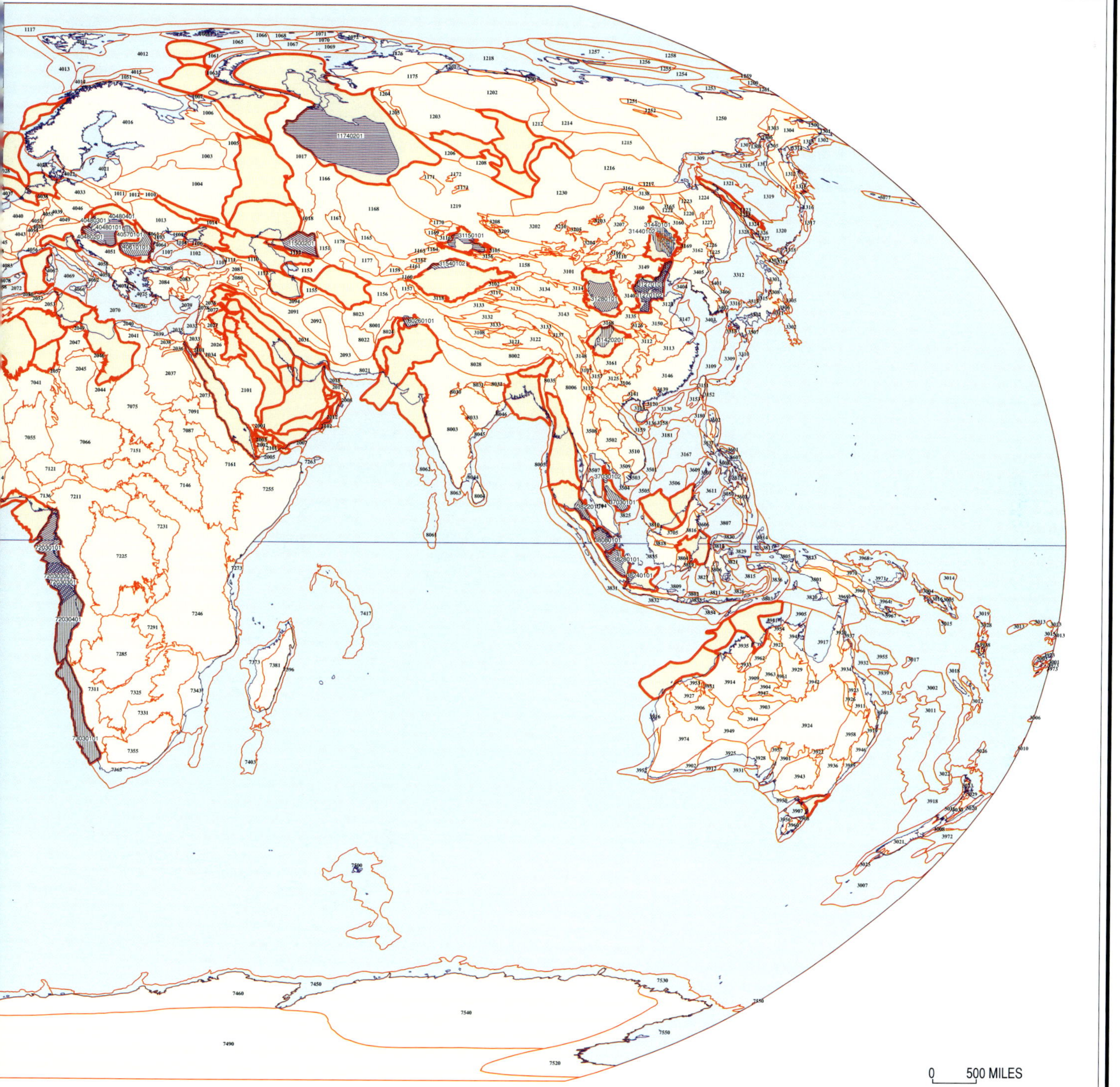

IS DERIVED FROM LACUSTRINE SOURCE ROCKS

Projection Robinson
Central meridian 0

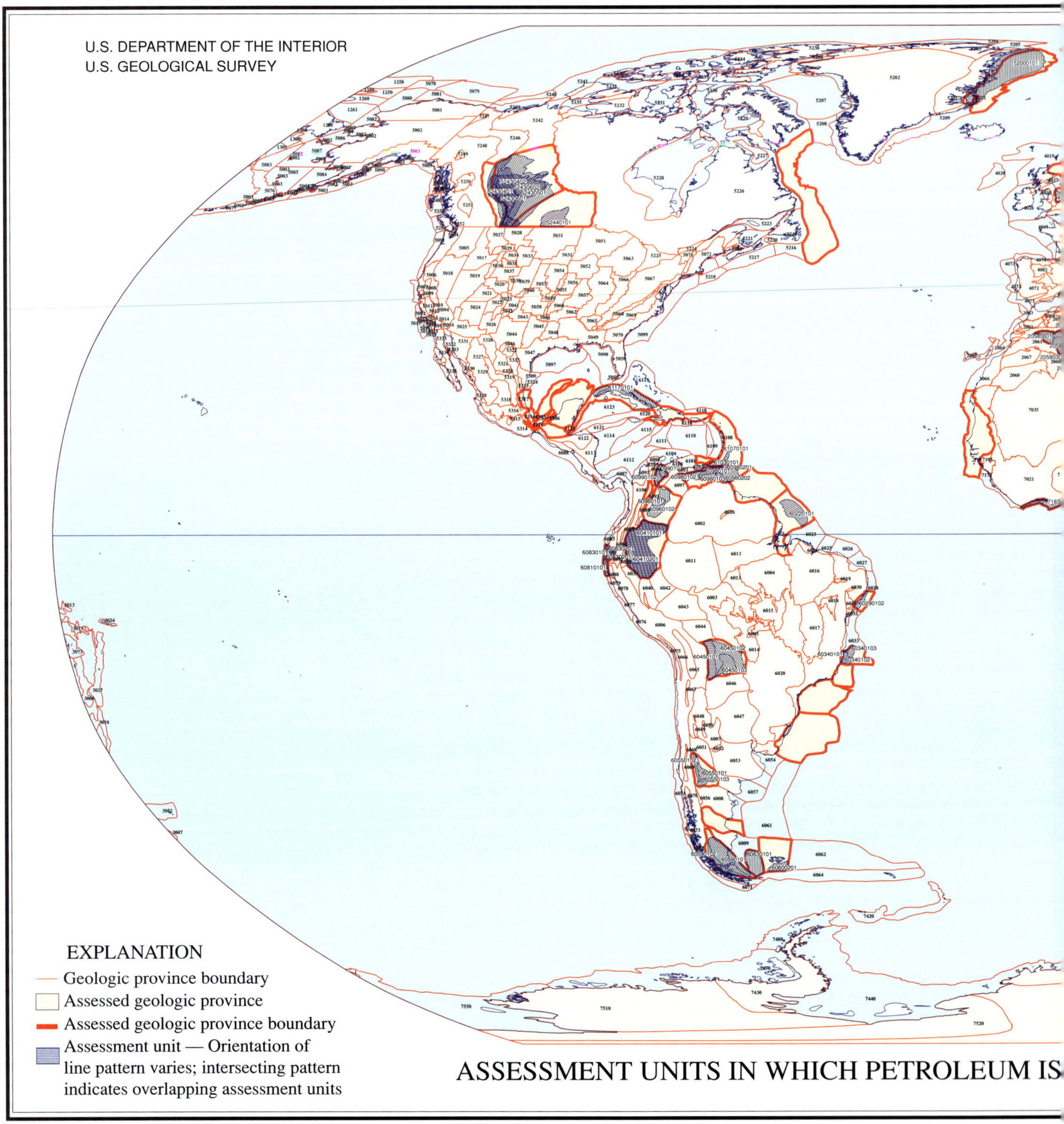

Figure 47. Map showing assessment units in which petroleum is derived from shallow marine source rocks.

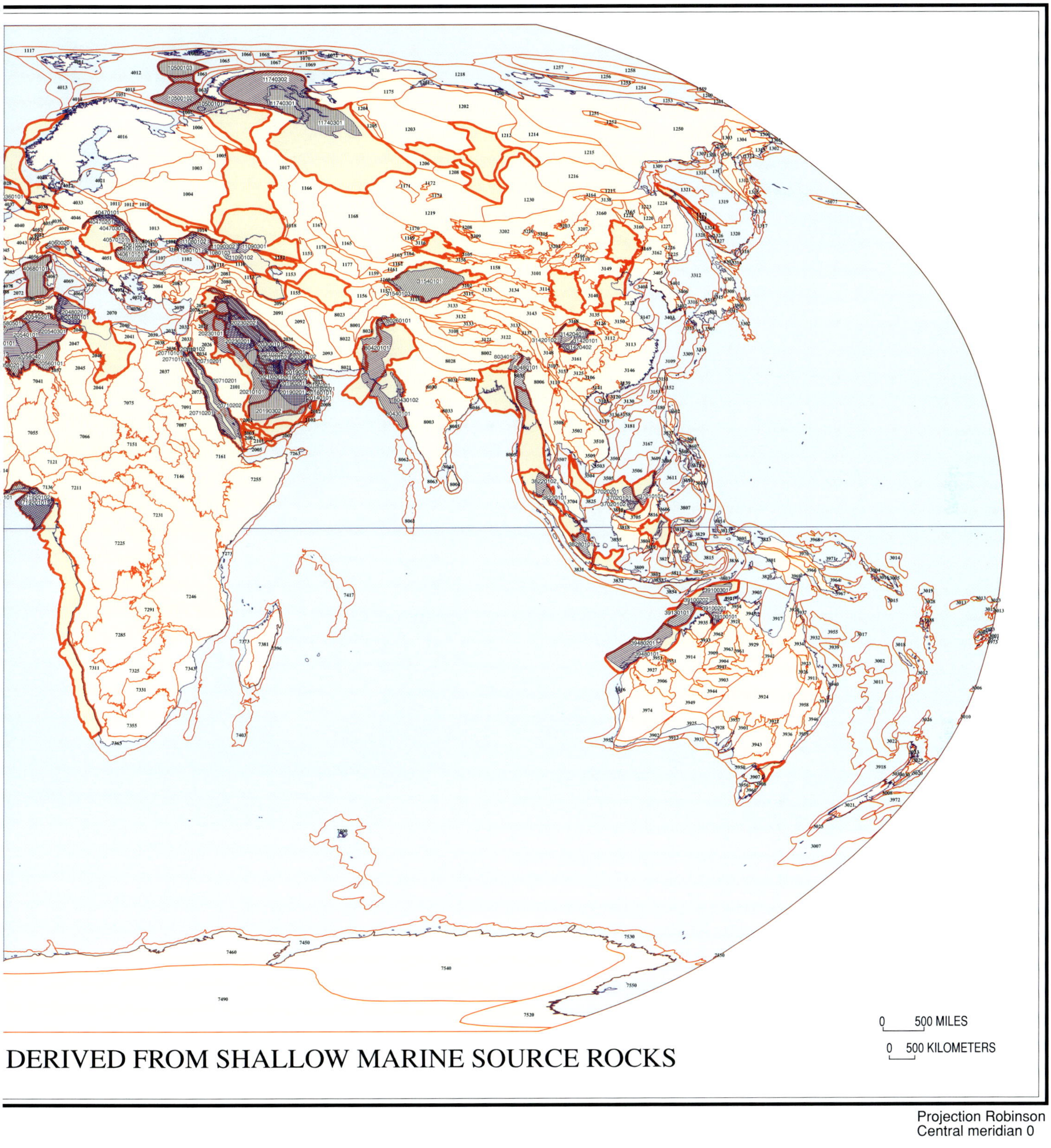
DERIVED FROM SHALLOW MARINE SOURCE ROCKS
0 500 MILES
0 500 KILOMETERS
Projection Robinson
Central meridian 0

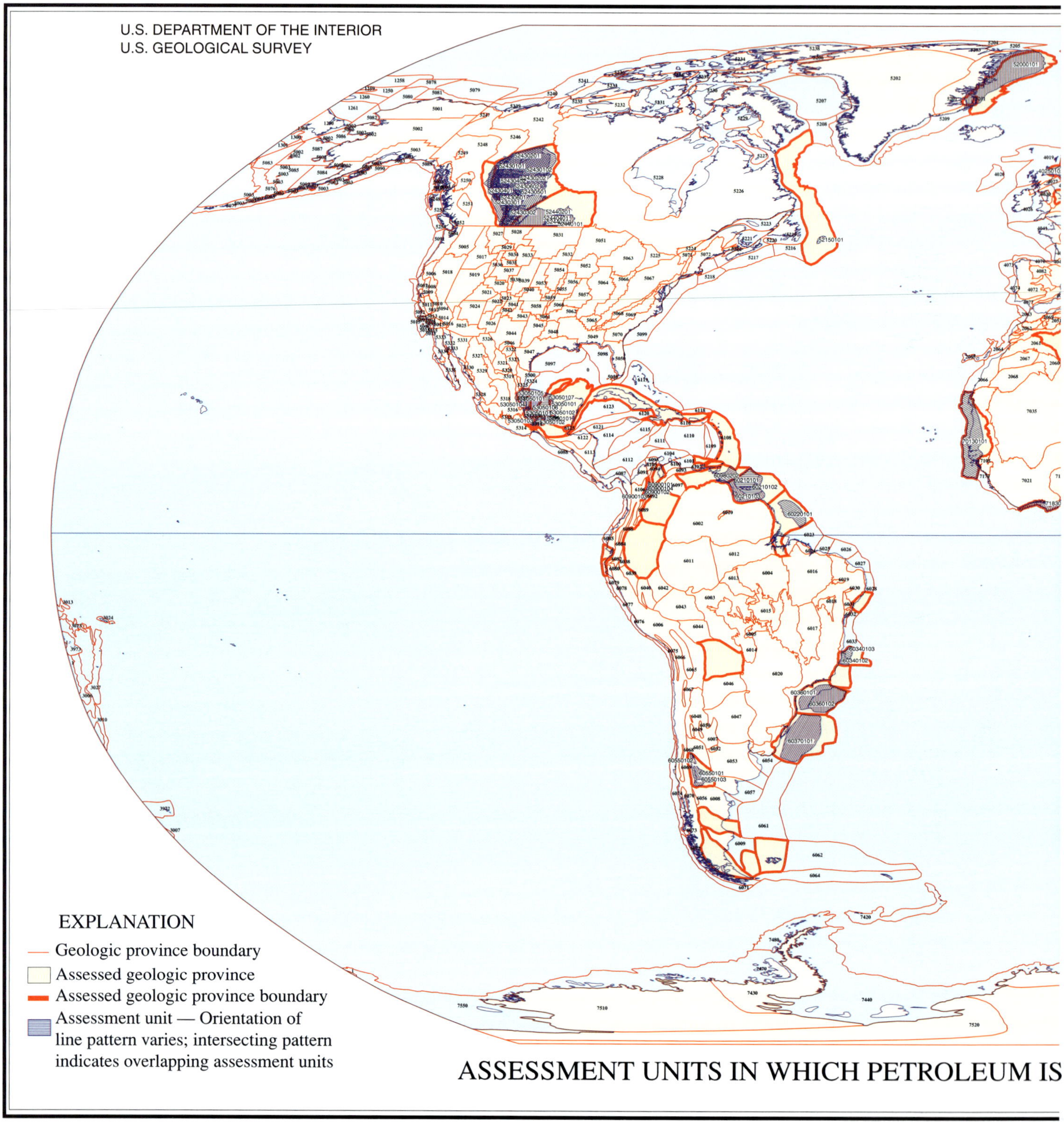

FIGURE 48. Map showing assessment units in which petroleum is derived from deep marine source rocks.

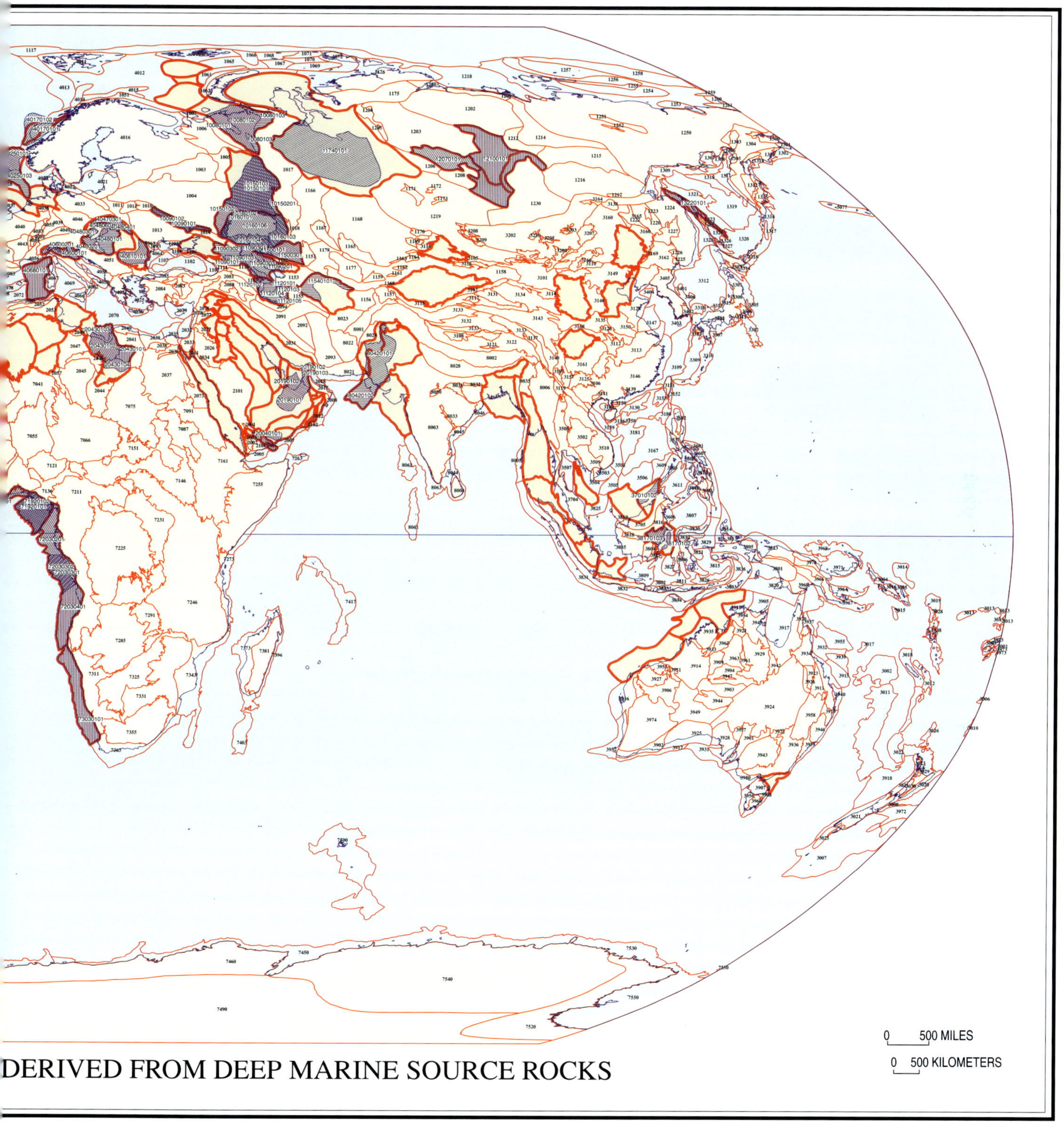

Projection Robinson
Central meridian 0

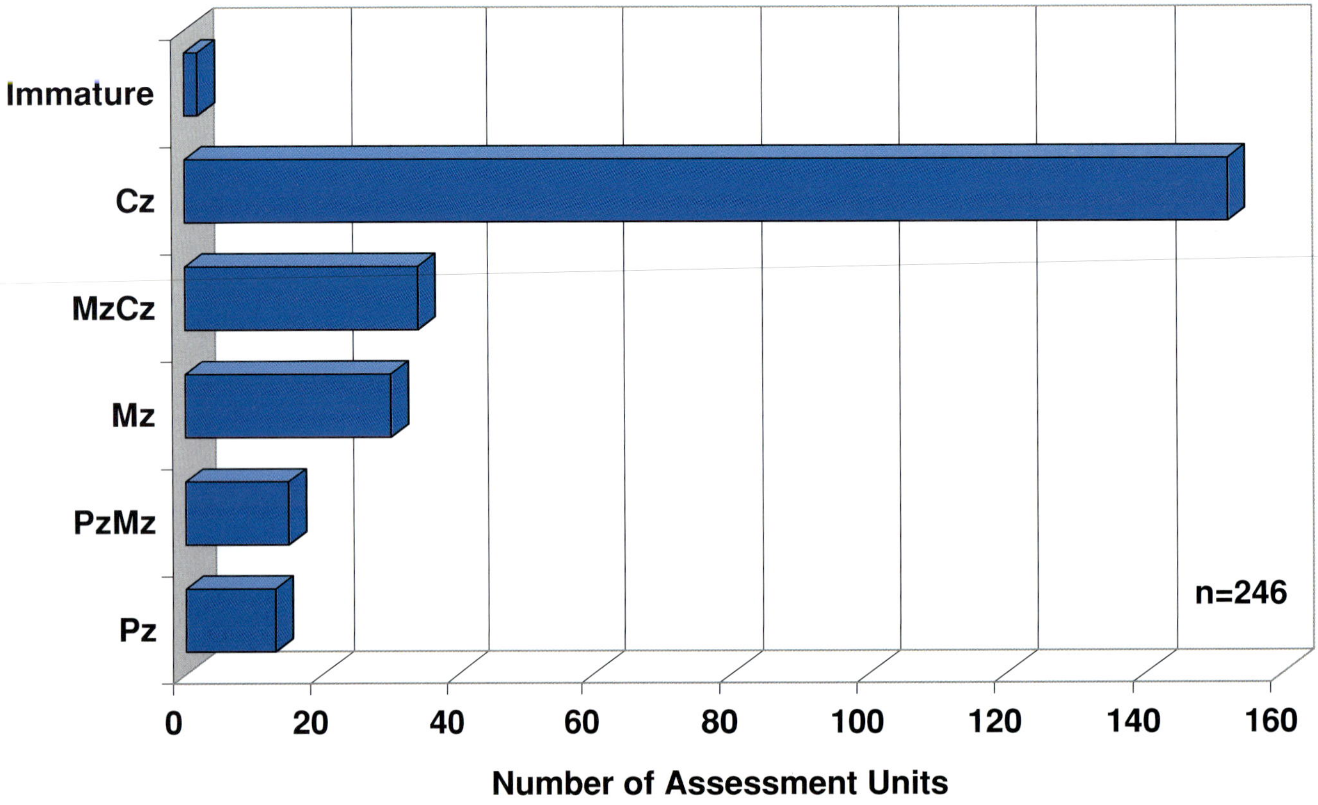

FIGURE 49. Numerical distribution of 246 assessment units by age (era) at which peak maturation occurred. Pz = Paleozoic; PzMz = Paleozoic–Mesozoic; Mz = Mesozoic; MzCz = Mesozoic–Cenozoic; Cz = Cenozoic; Immature = assessment unit in which source rock has not reached peak maturity.

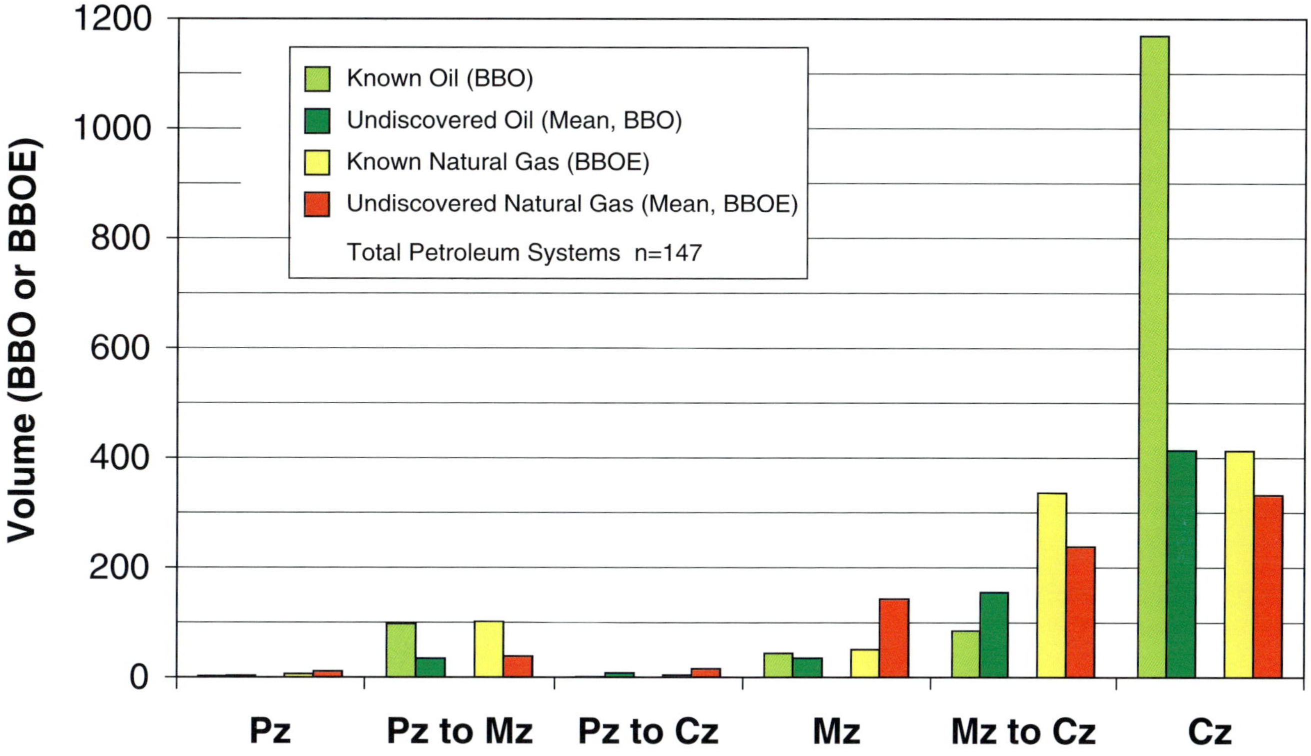

FIGURE 50. Distribution of mean estimated petroleum volumes of total petroleum systems by age at which peak maturation occurred. Known volume = cumulative production plus remaining reserves; undiscovered volume = estimated in this study. Pz = Paleozoic; Pz to Mz = Paleozoic–Mesozoic; Pz to Cz = Paleozoic–Cenozoic; Mz = Mesozoic; Mz to Cz = Mesozoic–Cenozoic; Cz = Cenozoic. BBO = billion barrels of oil; BBOE = billion barrels of oil equivalent.

FIGURE 51. Map showing assessment units that attained peak maturation in the Paleozoic.

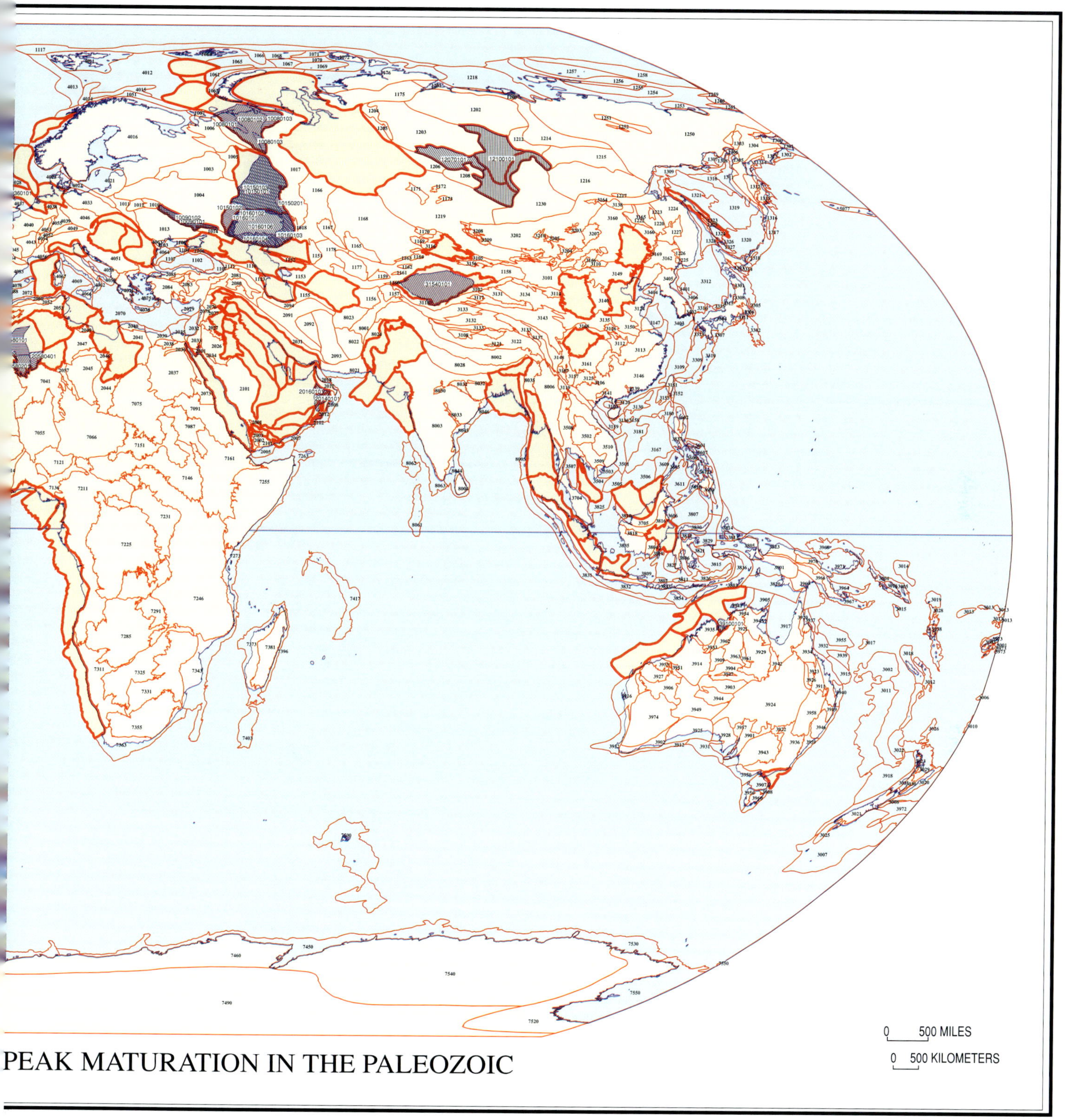

Projection Robinson
Central meridian 0

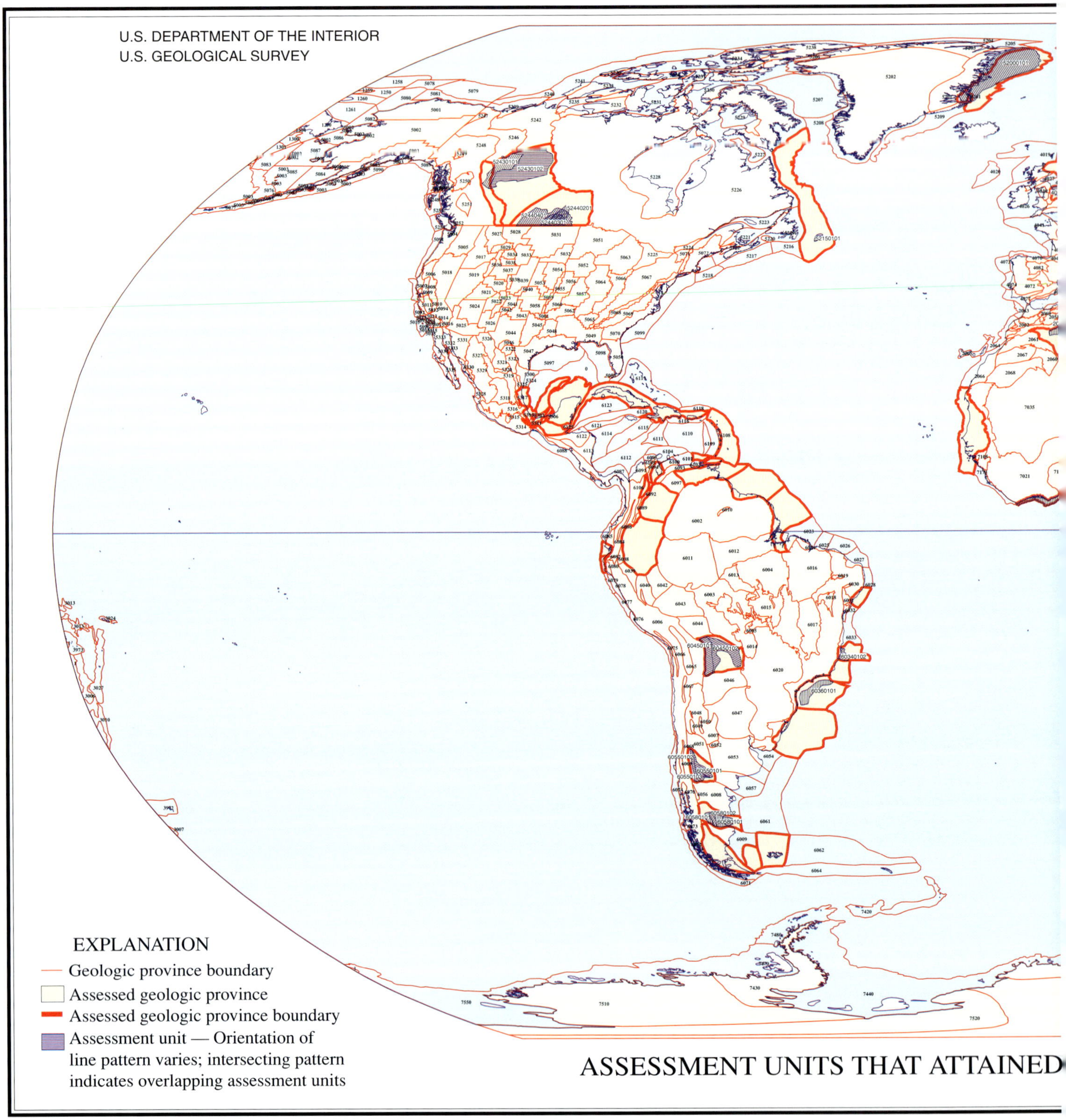

FIGURE 52. Map showing assessment units that attained peak maturation in the Mesozoic.

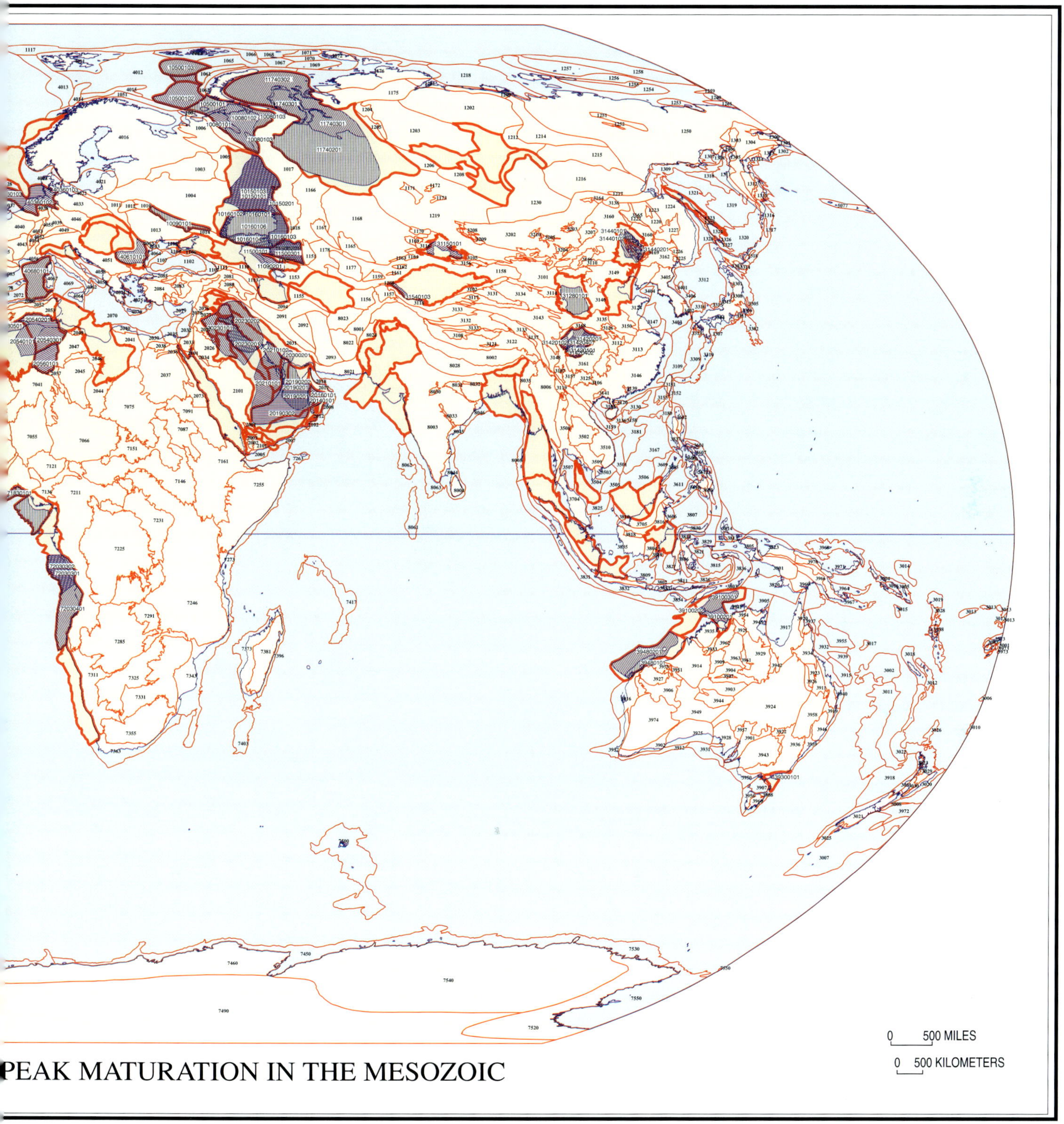

Projection Robinson
Central meridian 0

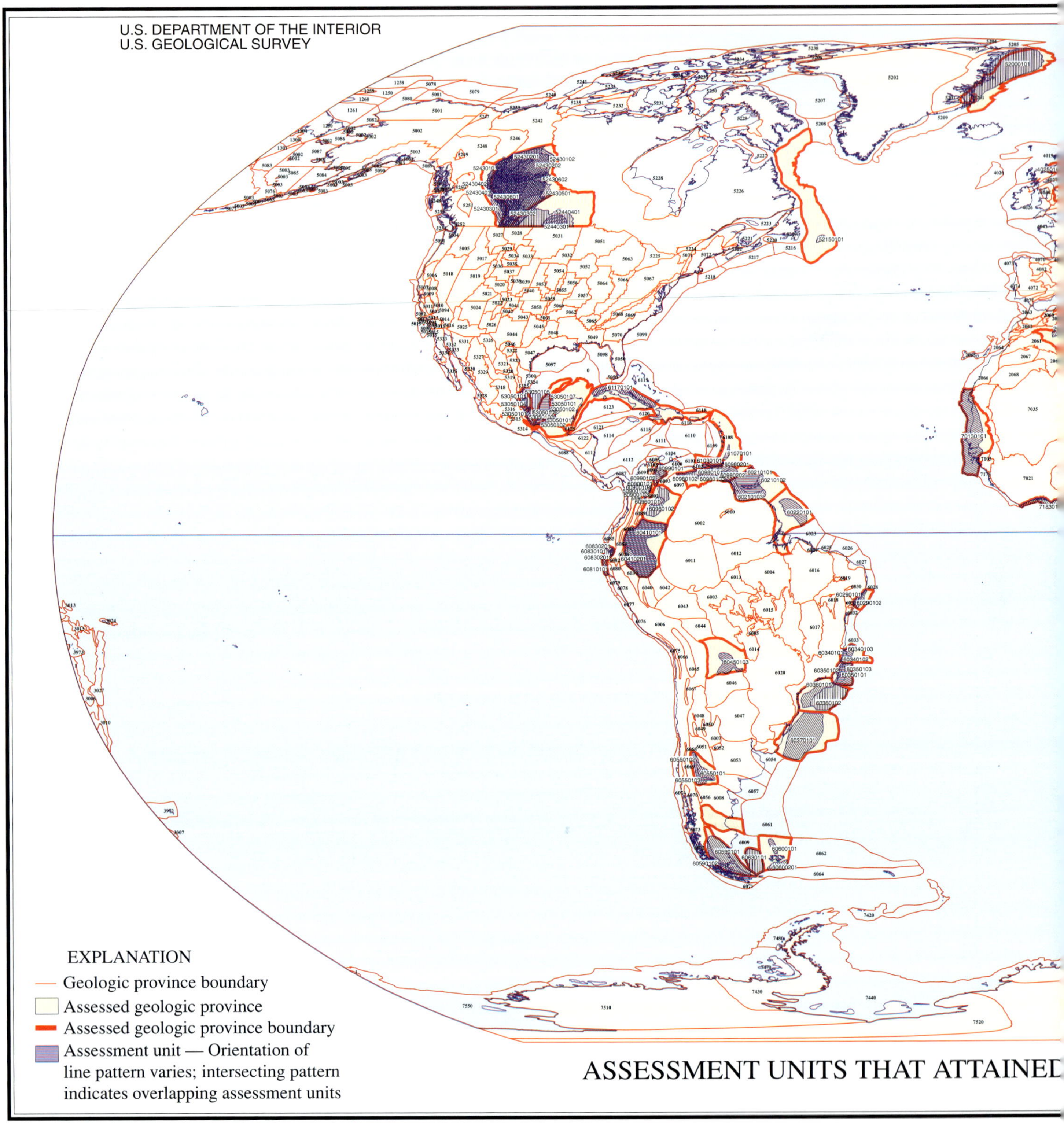

Figure 53. Map showing assessment units that attained peak maturation in the Cenozoic.

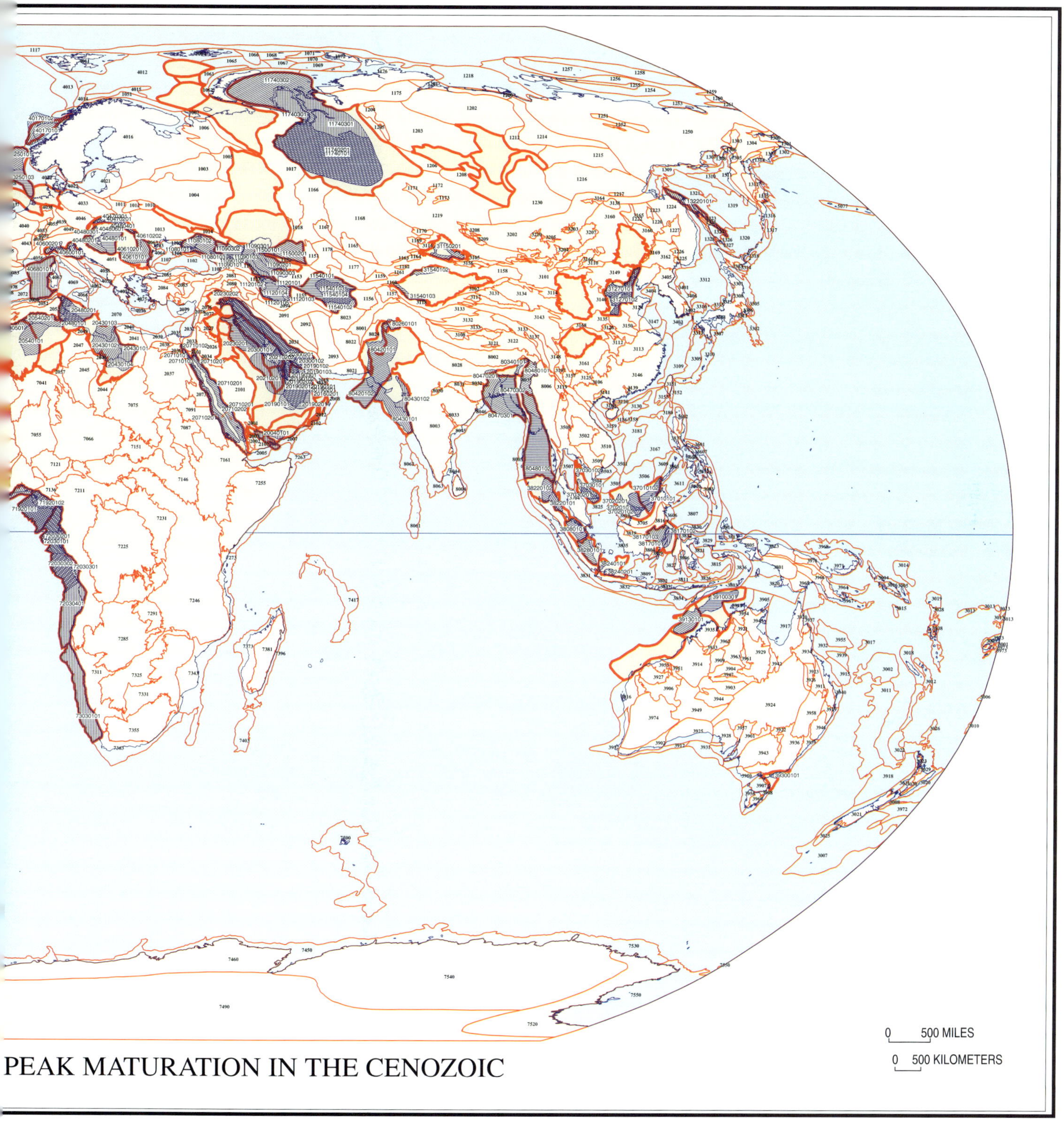

Projection Robinson
Central meridian 0

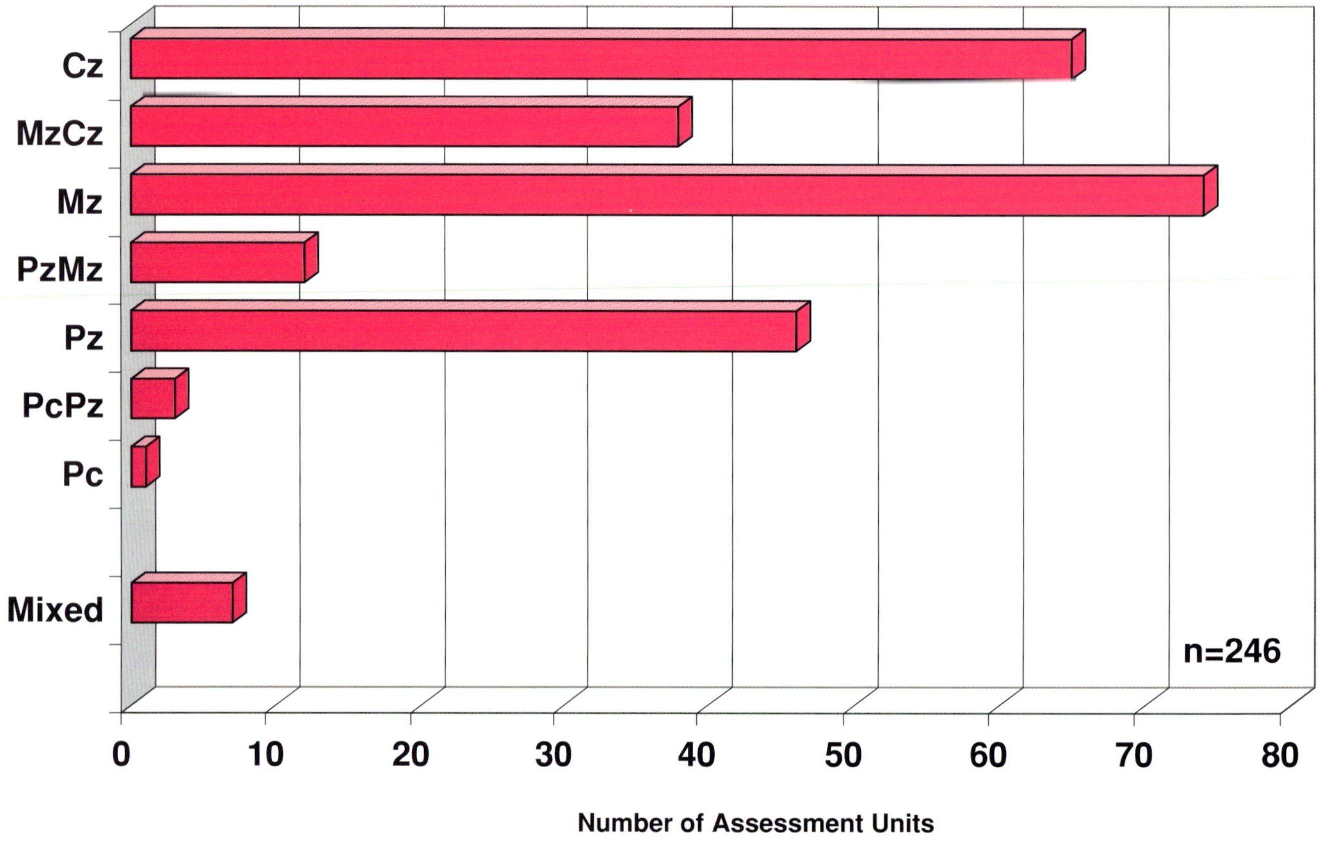

FIGURE 54. Numerical distribution of 246 assessment units by age (era) of reservoir rock. Pc = Precambrian; PcPz = Precambrian–Paleozoic; Pz = Paleozoic; PzMz = Paleozoic–Mesozoic; Mz = Mesozoic; MzCz = Mesozoic–Cenozoic; Cz = Cenozoic; Mixed = reservoir rocks of multiple ages.

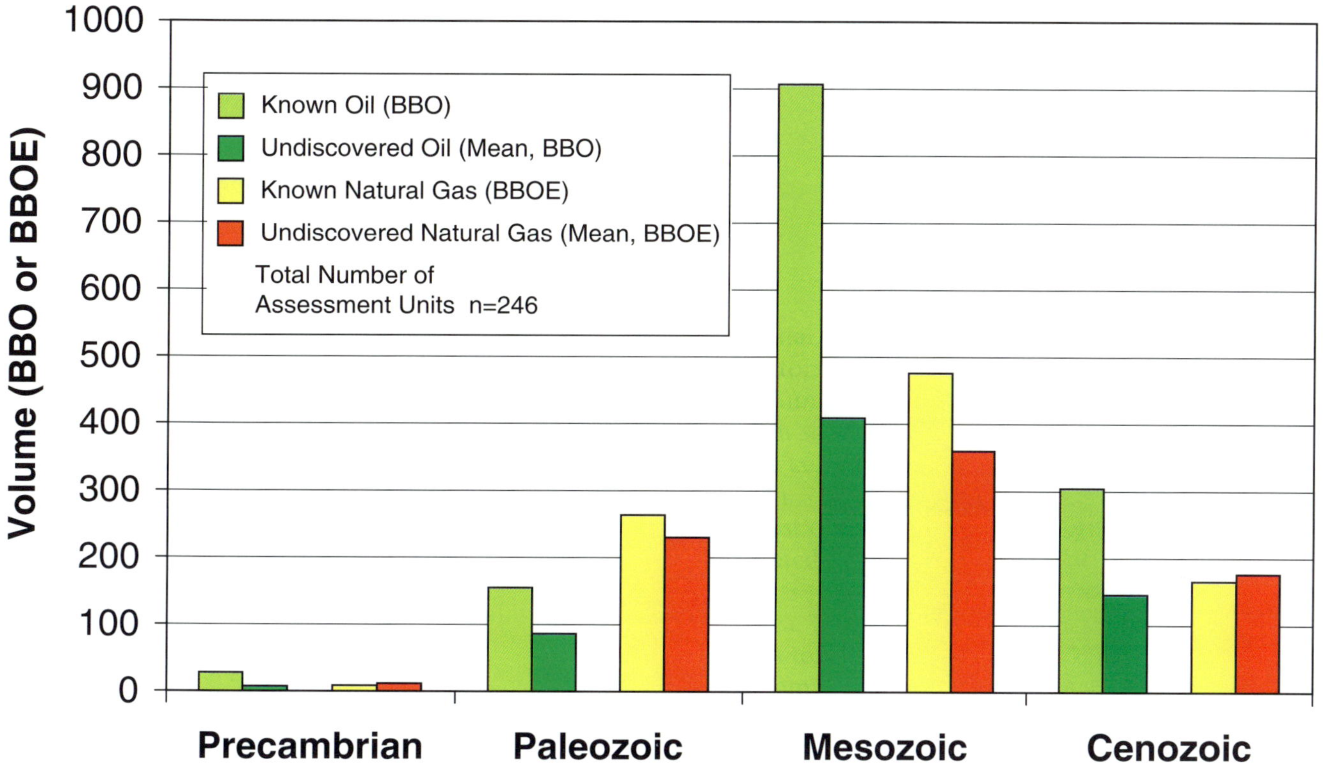

FIGURE 55. Distribution of mean estimated petroleum volumes of assessment units by age (era) of reservoir rock; the oldest reservoir for any multiple reservoir fields is used. Known volume = cumulative production plus remaining reserves; undiscovered volume = estimated in this study. BBO = billion barrels of oil; BBOE = billion barrels of oil equivalent.

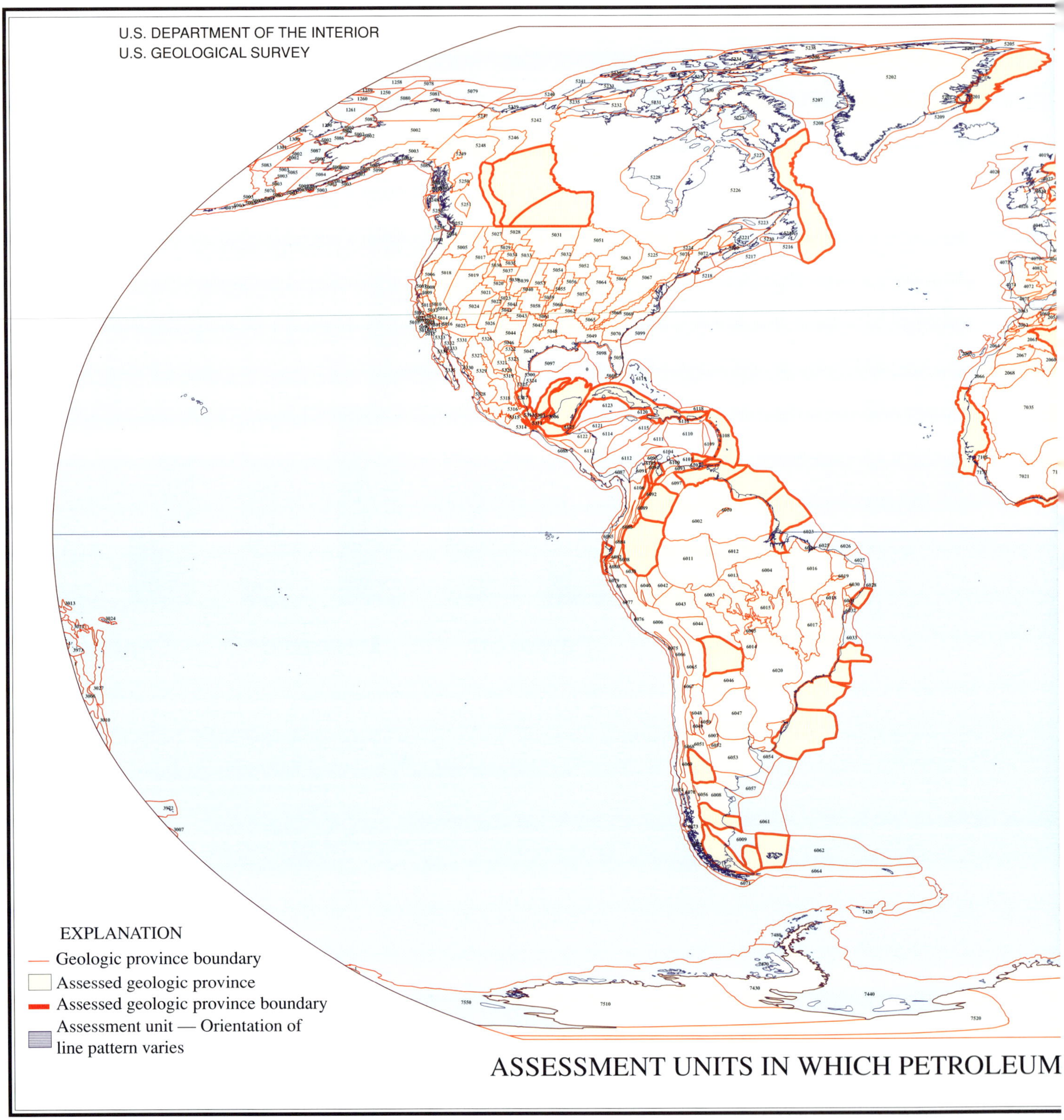

FIGURE 56. Map showing assessment units in which petroleum occurs in Precambrian reservoir rocks.

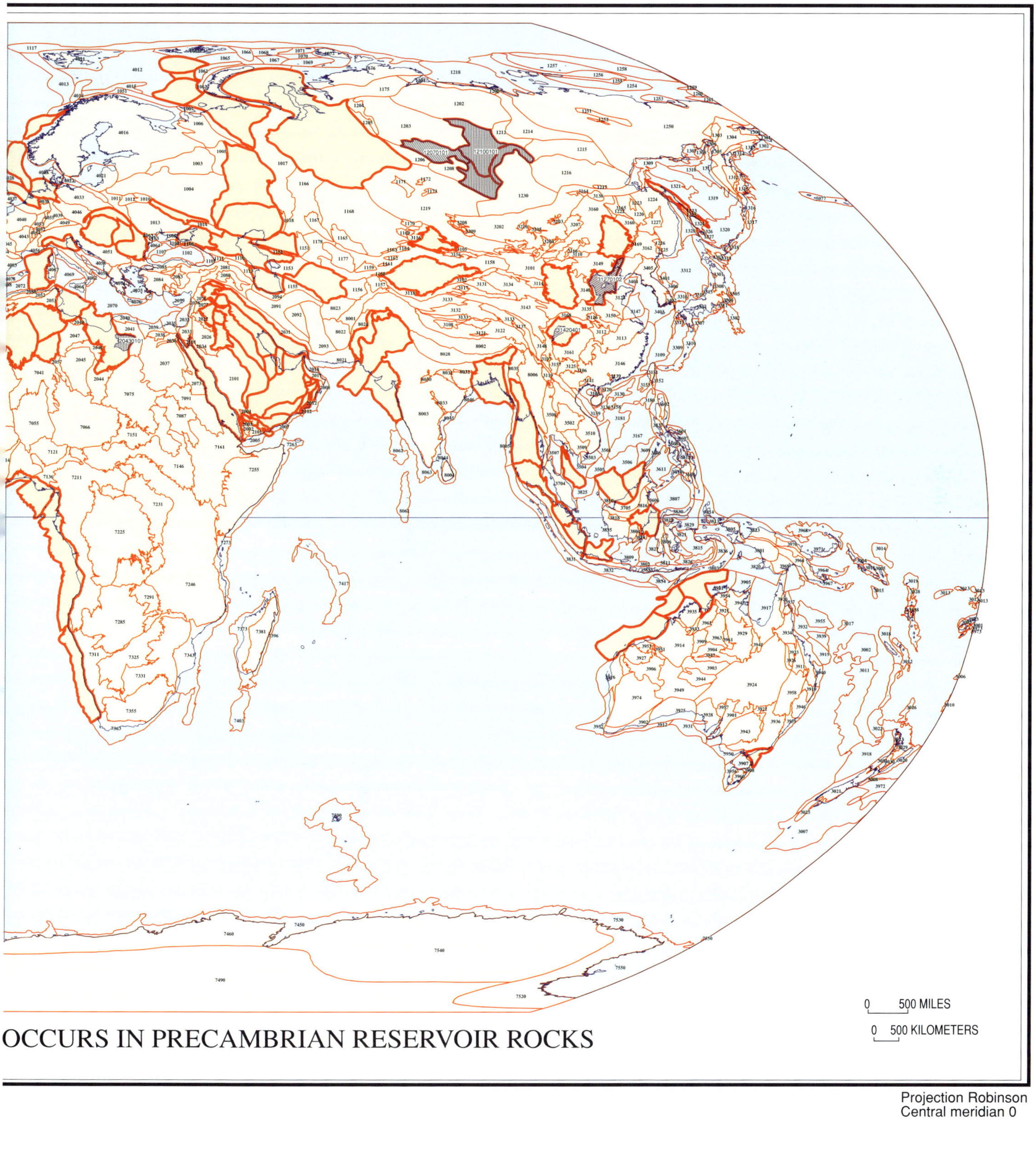

Projection Robinson
Central meridian 0

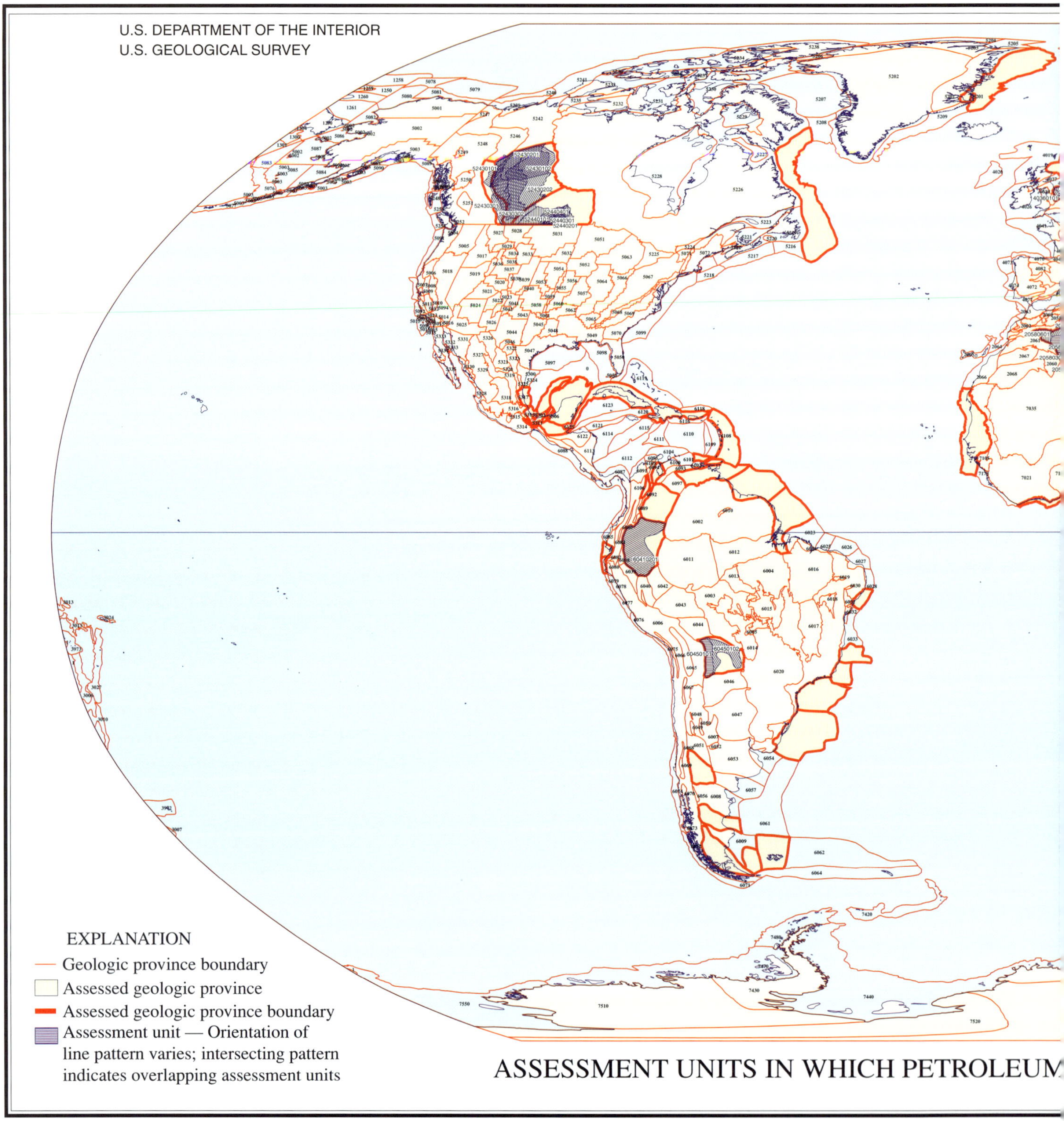

Figure 57. Map showing assessment units in which petroleum occurs in Paleozoic reservoir rocks.

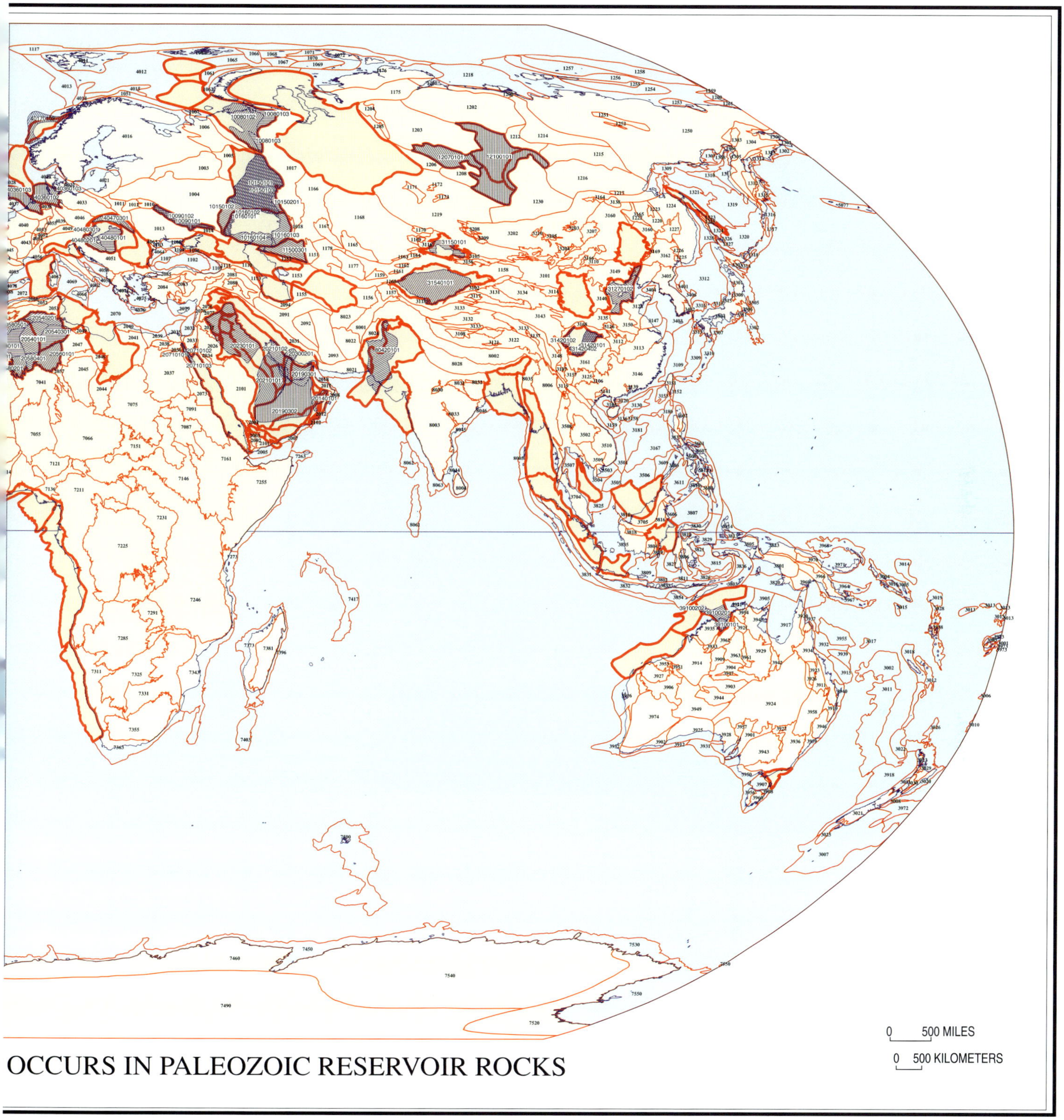

Projection Robinson
Central meridian 0

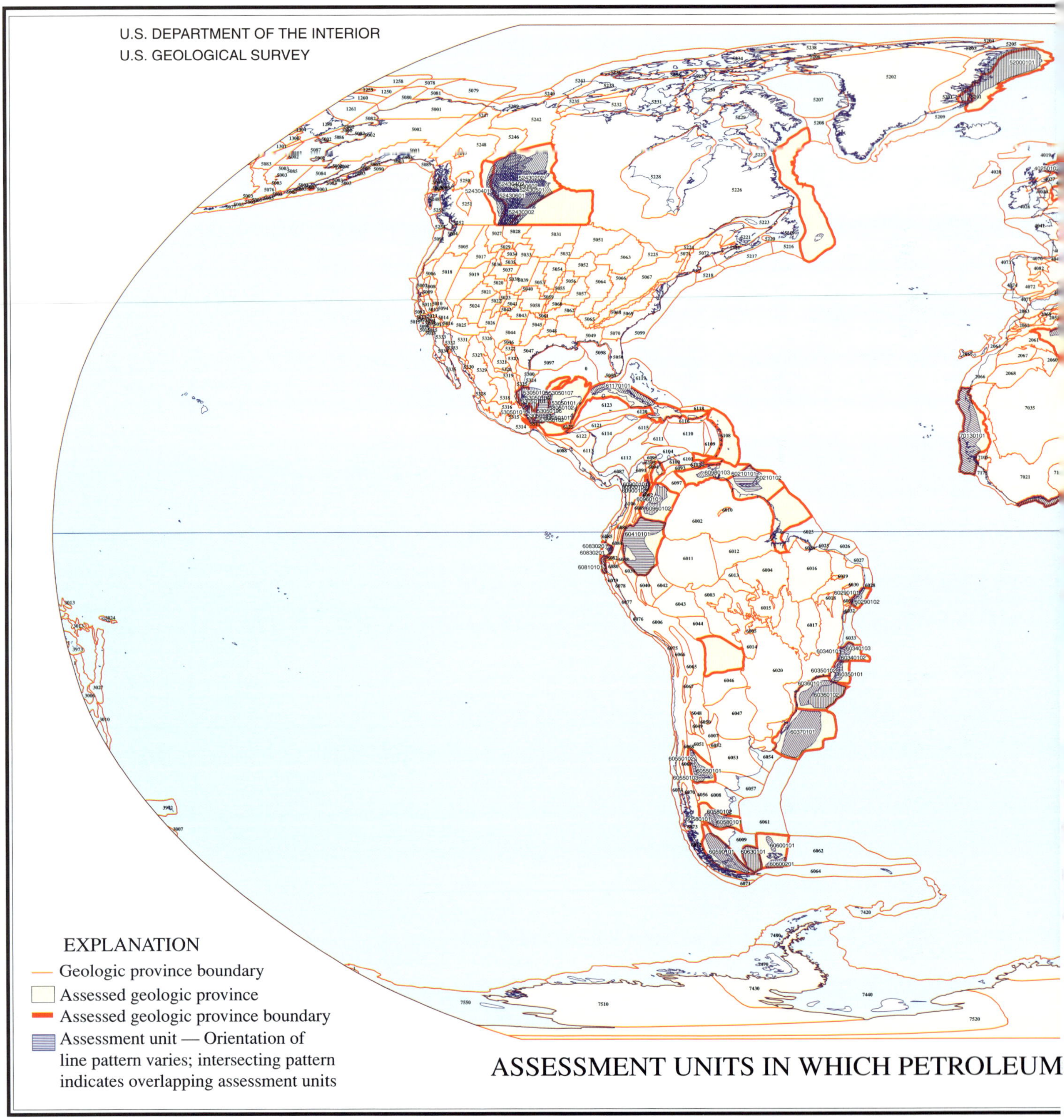

FIGURE 58. Map showing assessment units in which petroleum occurs in Mesozoic reservoir rocks.

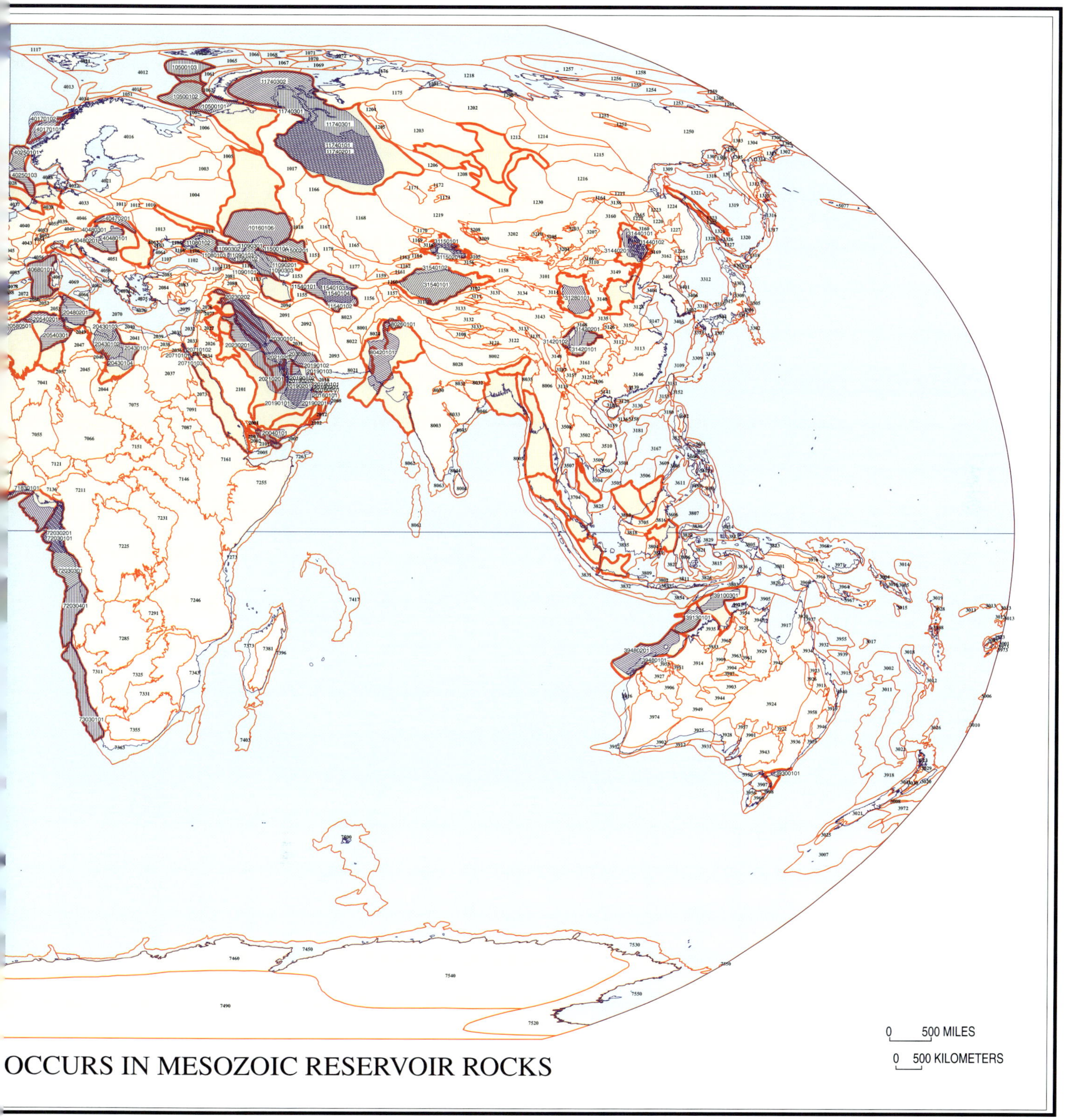

Projection Robinson
Central meridian 0

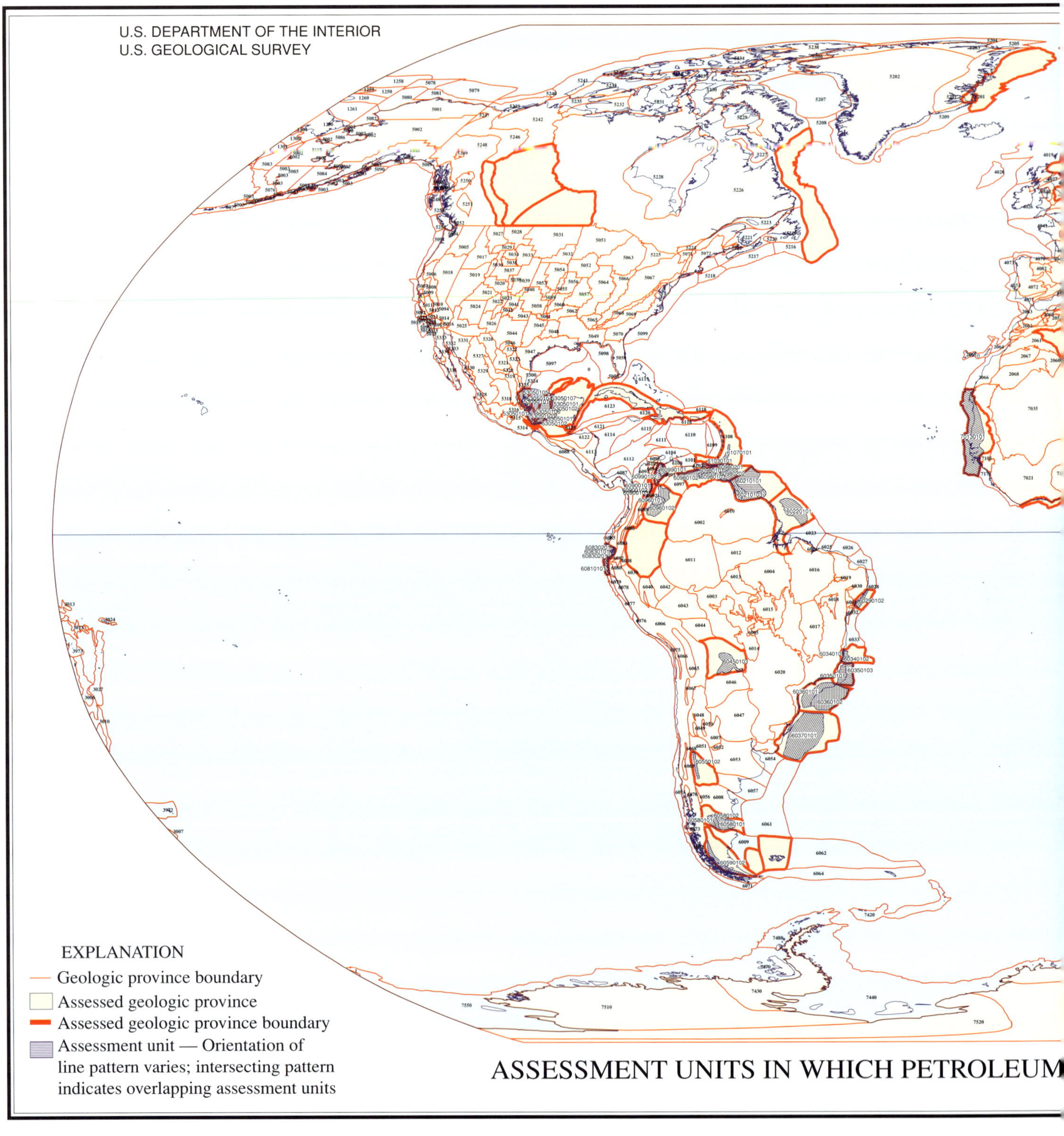

Figure 59. Map showing assessment units in which petroleum occurs in Cenozoic reservoir rocks.

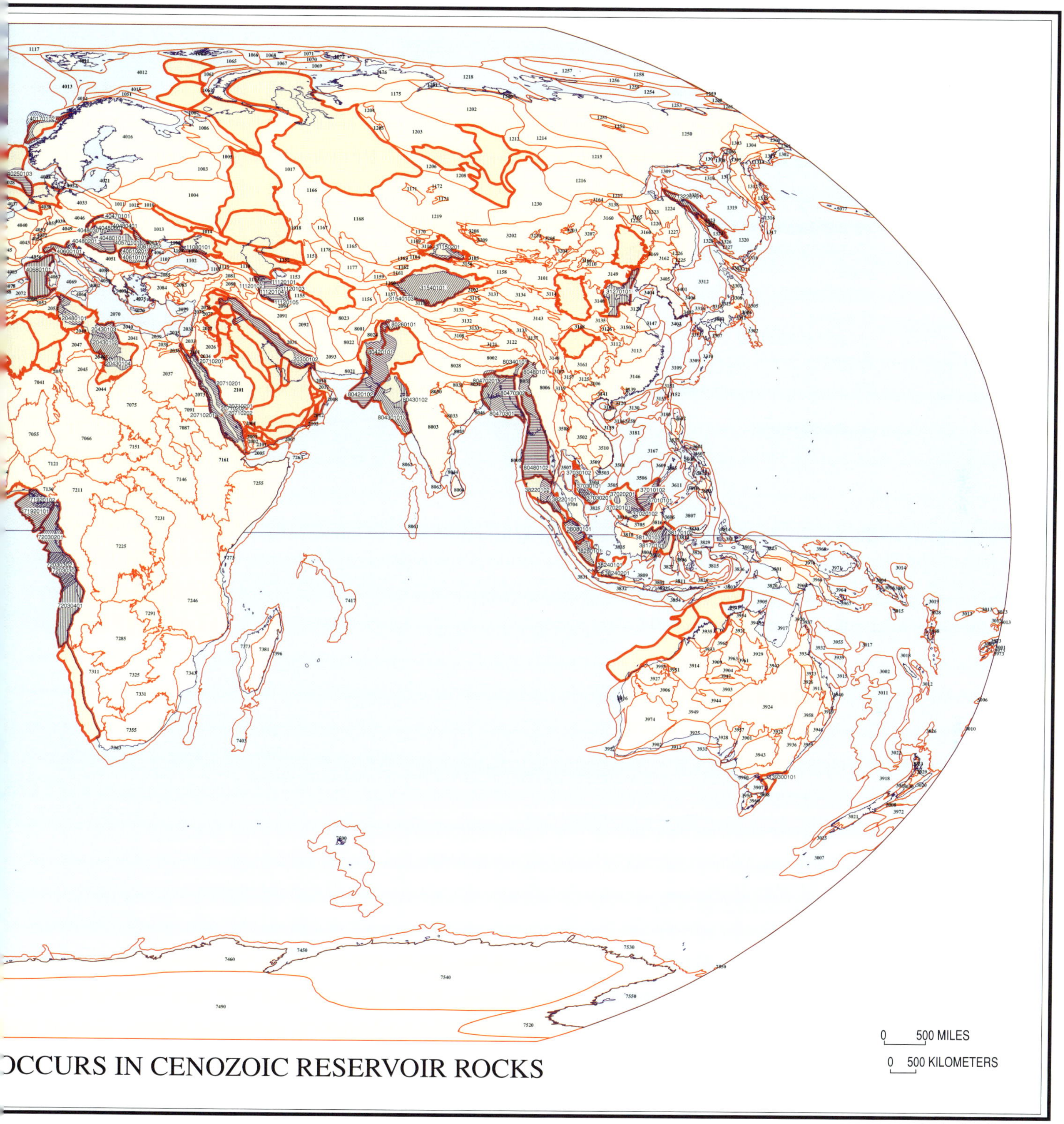

Projection Robinson
Central meridian 0

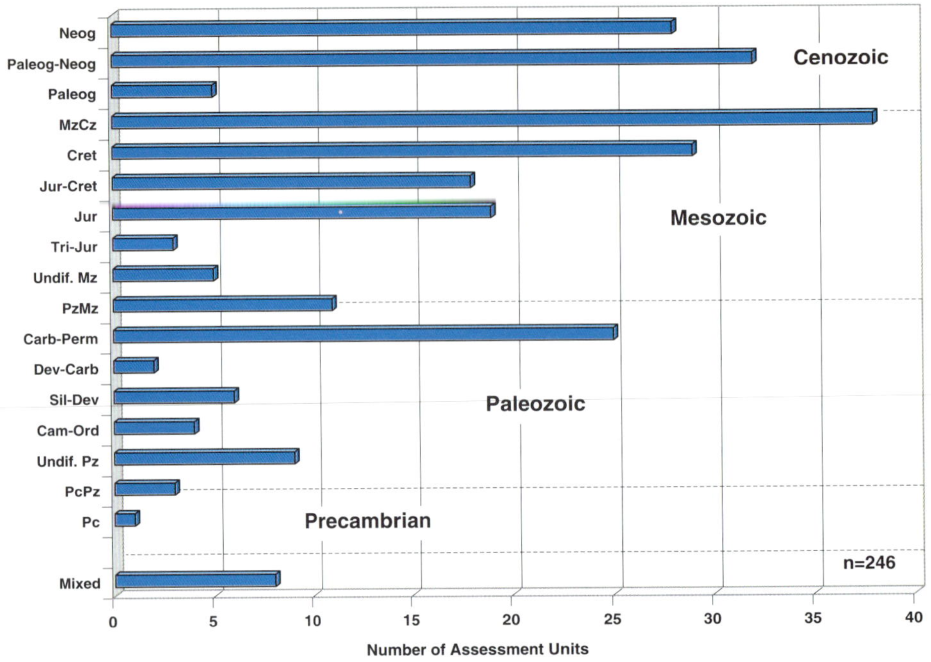

FIGURE 60. Numerical distribution of 246 assessment units by specific age of oldest reservoir rock. Pc = Precambrian; PcPz = Precambrian–Paleozoic; Undif. Pz = Undifferentiated Paleozoic; Cam-Ord = Cambrian–Ordovician; Sil-Dev = Silurian–Devonian; Dev-Carb = Devonian–Carboniferous; Carb-Perm = Carboniferous–Permian; PzMz = Paleozoic–Mesozoic; Undif Mz = Undifferentiated Mesozoic; Tri-Jur = Triassic–Jurassic; Jur = Jurassic; Jur-Cret = Jurassic–Cretaceous; Cret = Cretaceous; MzCz = Mesozoic–Cenozoic; Paleog = Paleogene; Paleog-Neog = Paleogene–Neogene; Neog = Neogene; Mixed = reservoir rocks of multiple ages.

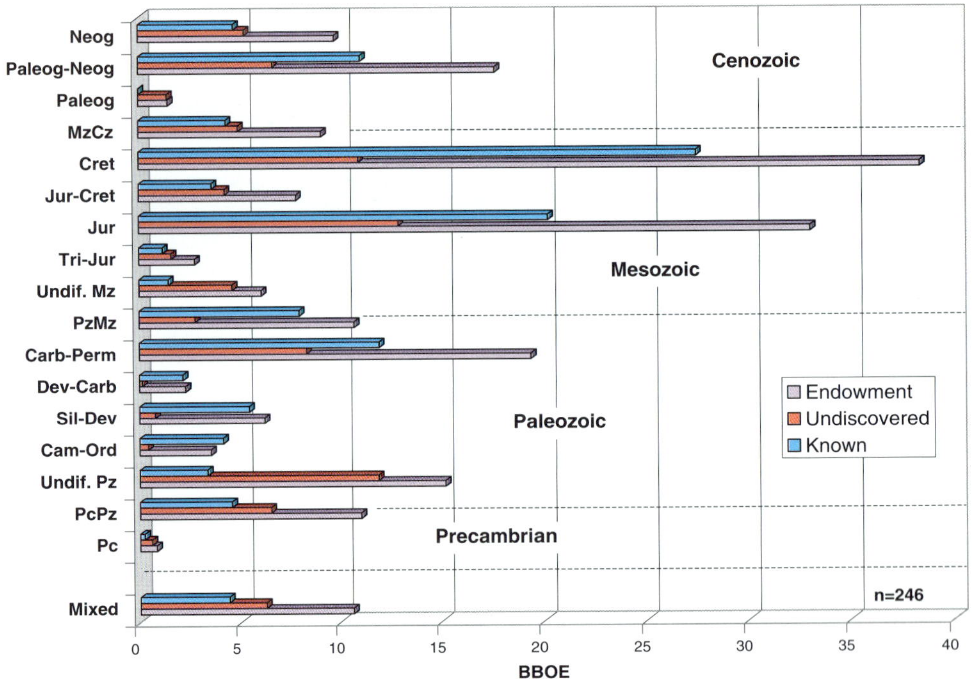

FIGURE 61. Distribution of mean estimated petroleum volumes of 246 assessment units by age of reservoir rock. Known volume = cumulative production plus remaining reserves; undiscovered volume = estimated in this study; and conventional endowment = sum of known and undiscovered volumes. Pc = Precambrian; PcPz = Precambrian–Paleozoic; Undif. Pz = Undifferentiated Paleozoic; Cam-Ord = Cambrian–Ordovician; Sil-Dev = Silurian–Devonian; Dev-Carb = Devonian–Carboniferous; Carb-Perm = Carboniferous–Permian; PzMz = Paleozoic–Mesozoic; Undif. Mz = Undifferentiated Mesozoic; Tri-Jur = Triassic–Jurassic; Jur = Jurassic; Jur-Cret = Jurassic–Cretaceous; Cret = Cretaceous; MzCz = Mesozoic–Cenozoic; Paleog = Paleogene; Paleog-Neog = Paleogene–Neogene; Neog = Neogene; Mixed = reservoir rocks of multiple ages. BBOE = billion barrels of oil equivalent.

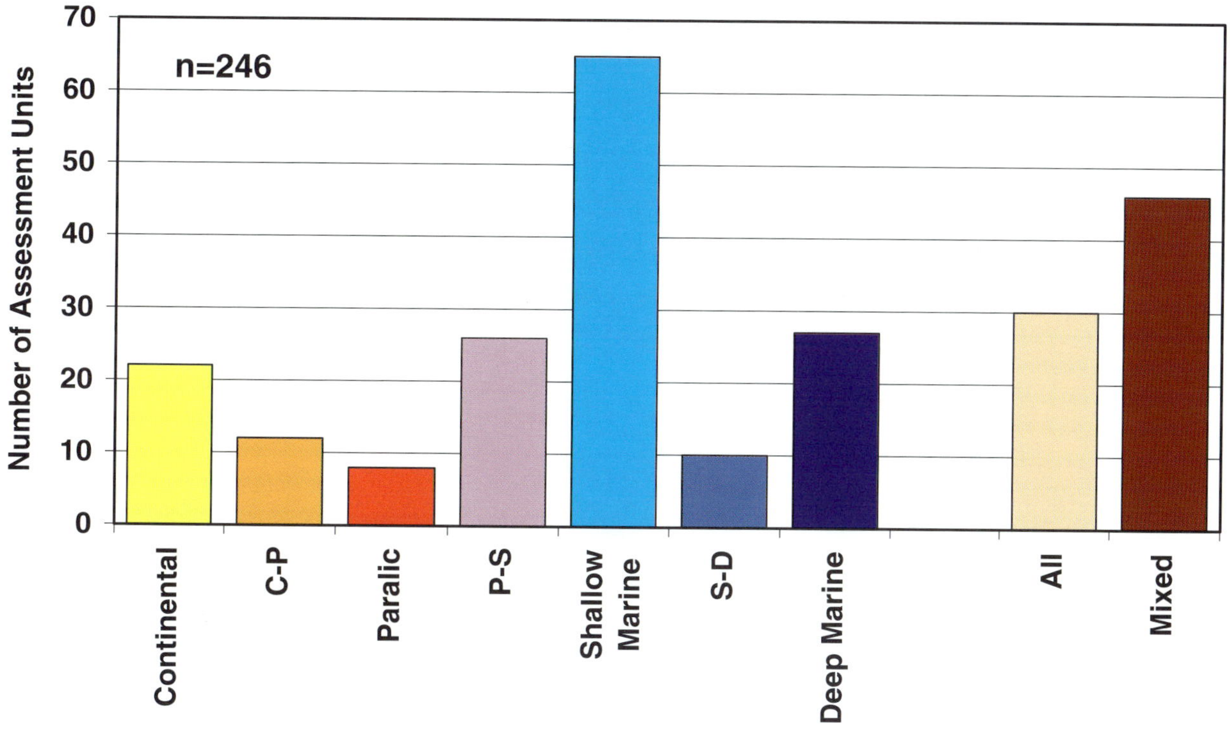

FIGURE 62. Numerical distribution of assessment units by depositional environment of reservoir rock. C-P = continental and paralic mixed; P-S = paralic and shallow marine mixed; S-D = shallow and deep marine mixed; All = all depositional environments represented; Mixed = multiple depositional environments, but not necessarily all.

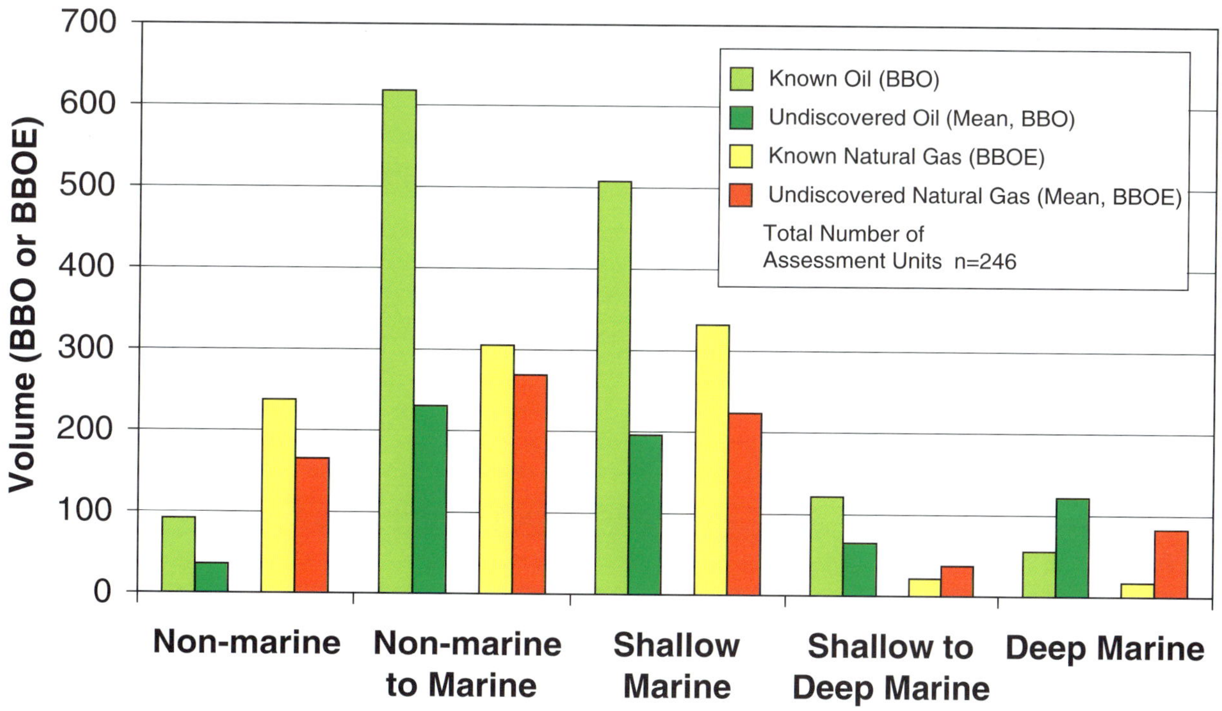

FIGURE 63. Distribution of mean estimated petroleum volumes of 246 assessment units by depositional environment of reservoir rock. Known volume = cumulative production plus remaining reserves; undiscovered volume = estimated in this study. Depositional environments are allocated to non-marine, non-marine to marine, shallow marine, shallow to deep marine, deep marine. BBO = billion barrels of oil; BBOE = billion barrels of oil equivalent.

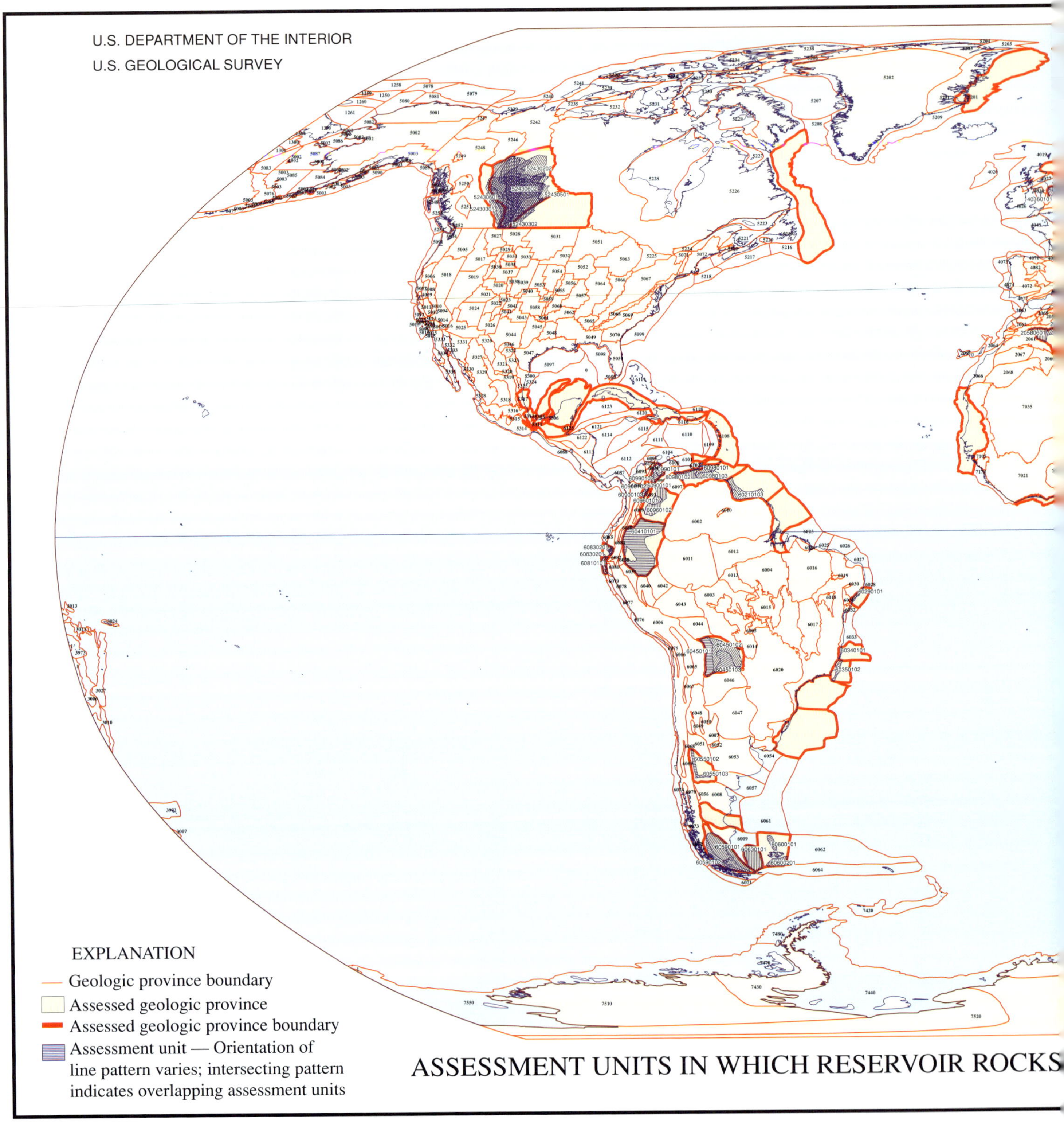

FIGURE 64. Map showing assessment units in which reservoir rocks were deposited in continental environments.

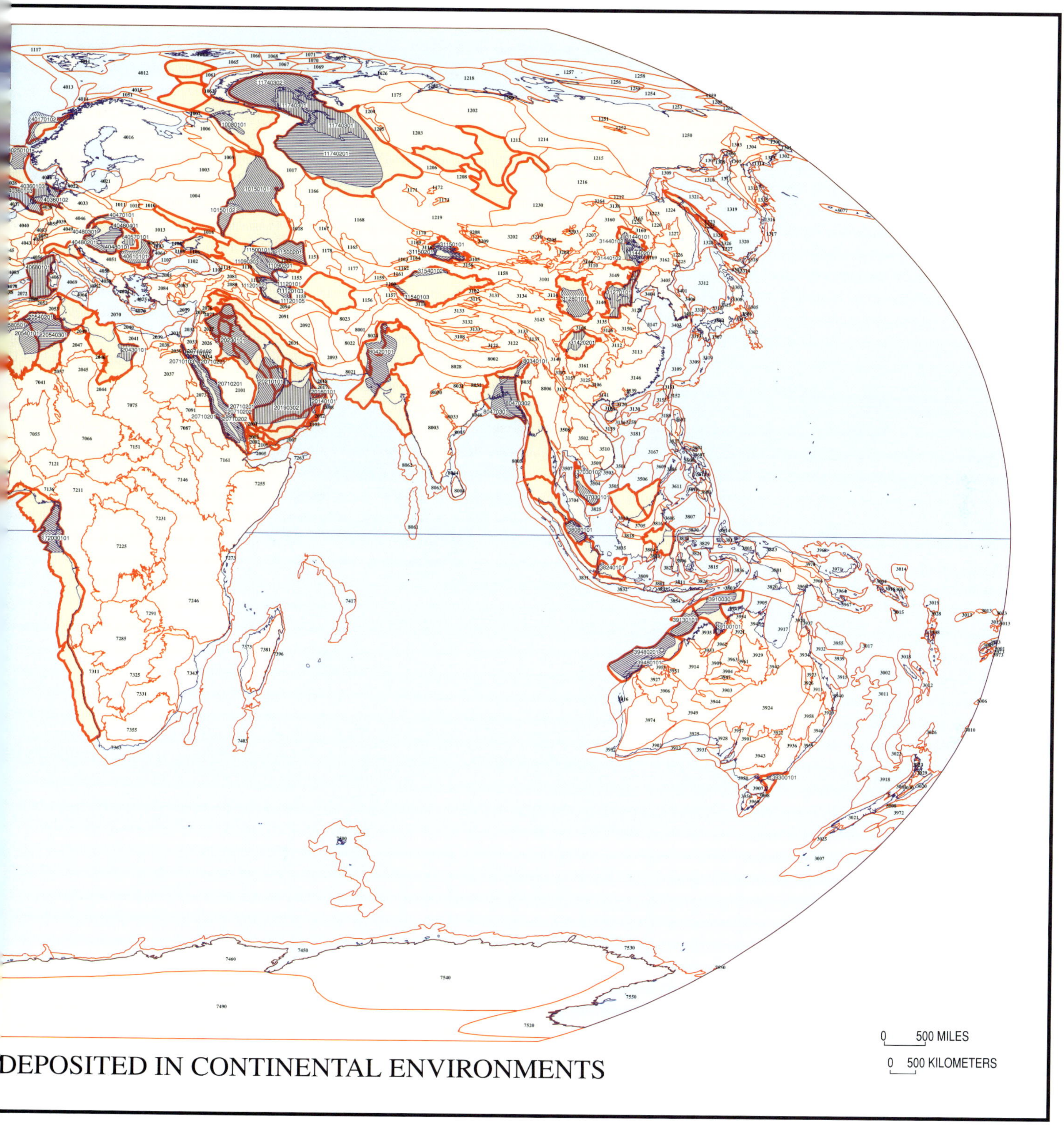

Projection Robinson
Central meridian 0

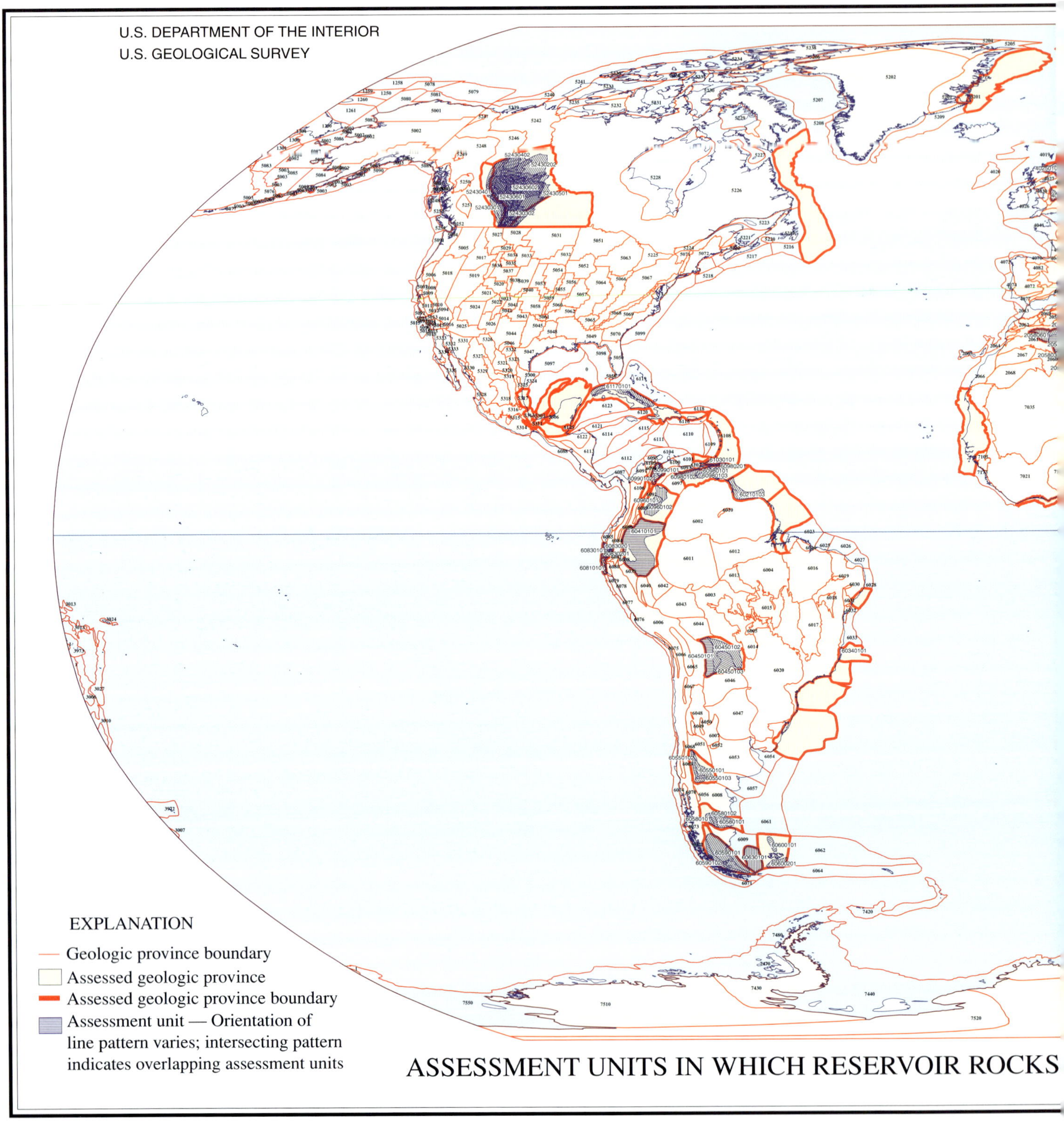

FIGURE 65. Map showing assessment units in which reservoir rocks were deposited in paralic environments.

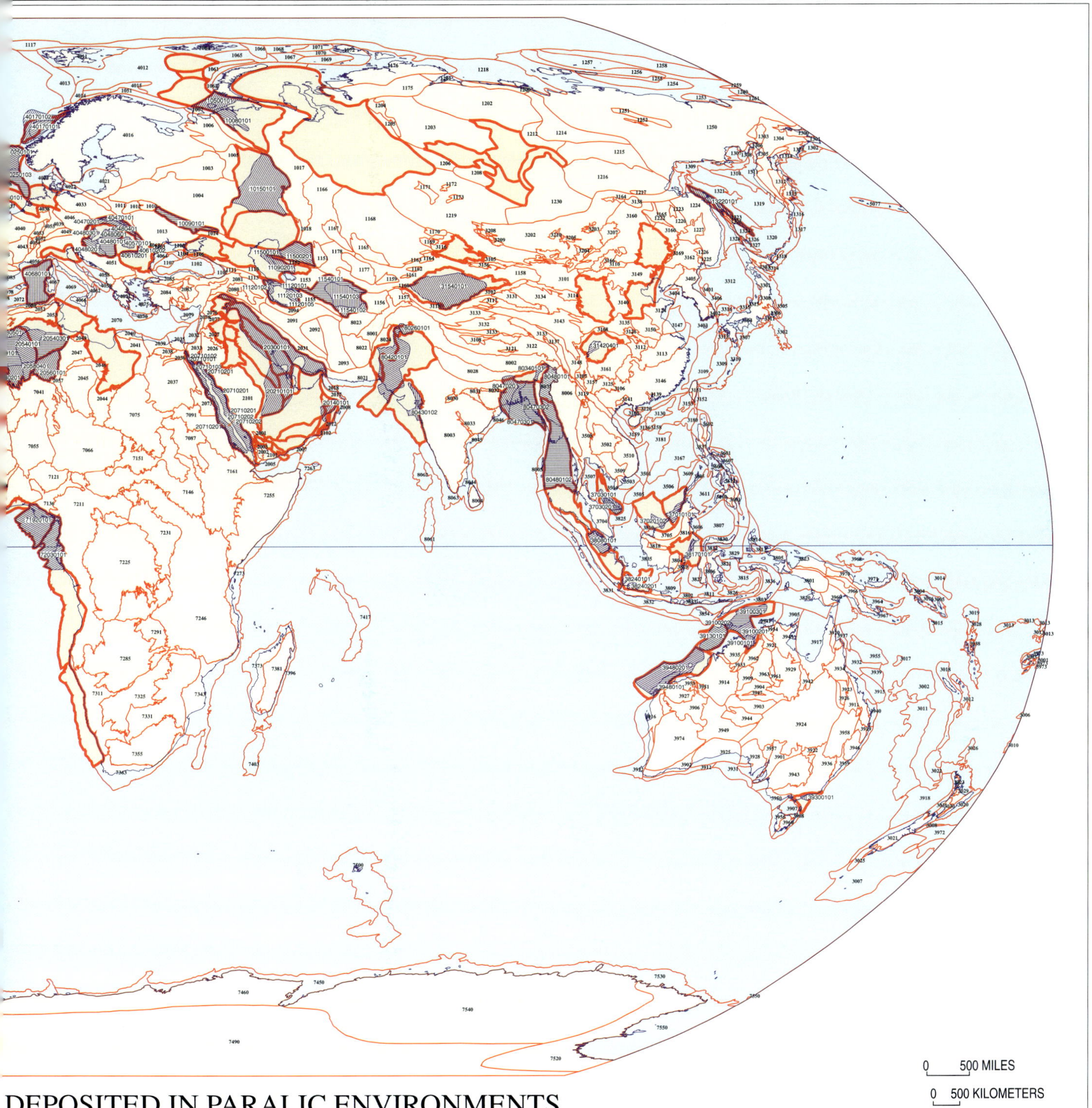

DEPOSITED IN PARALIC ENVIRONMENTS

Projection Robinson
Central meridian 0

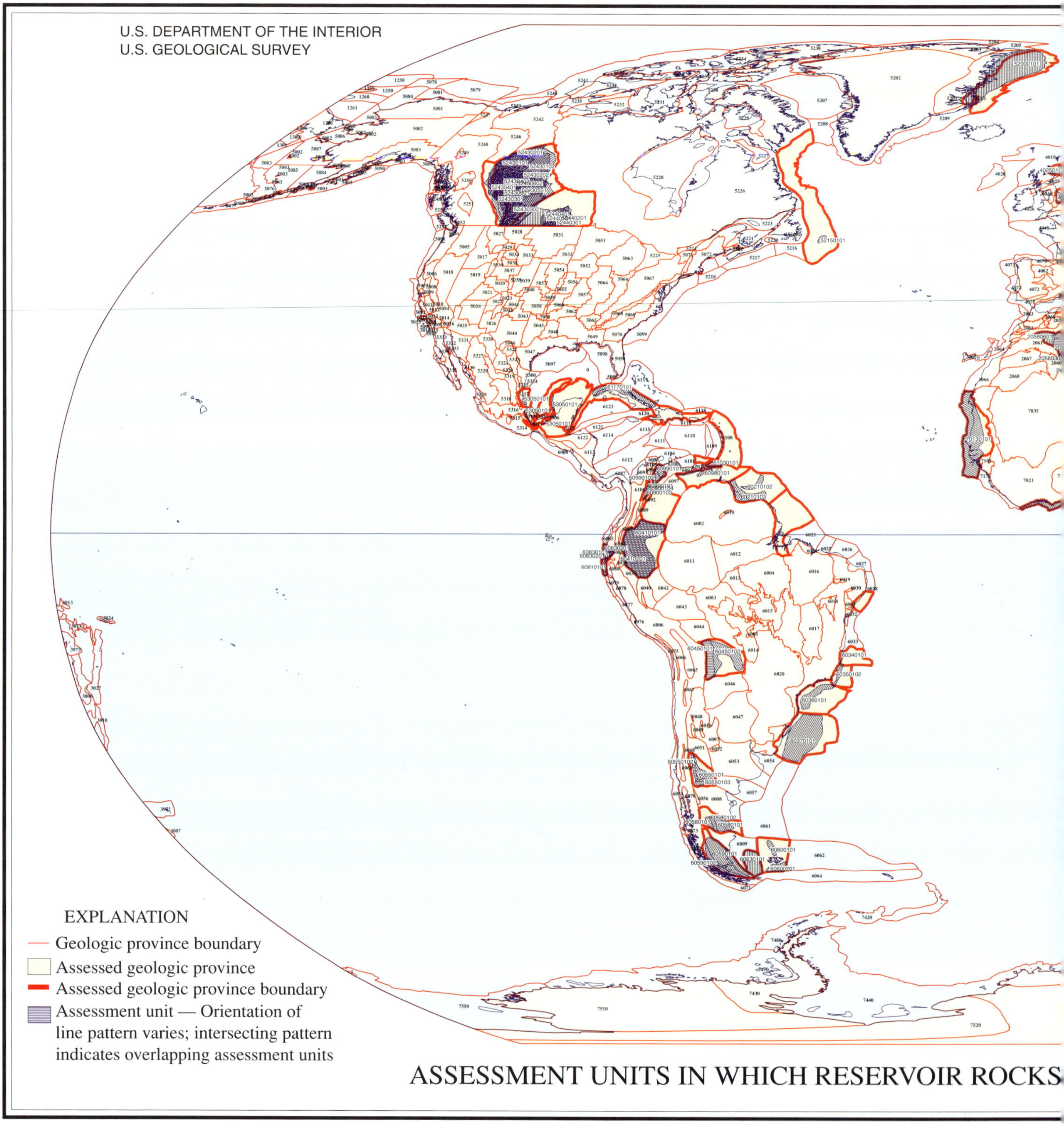

Figure 66. Map showing assessment units in which reservoir rocks were deposited in shallow marine environments.

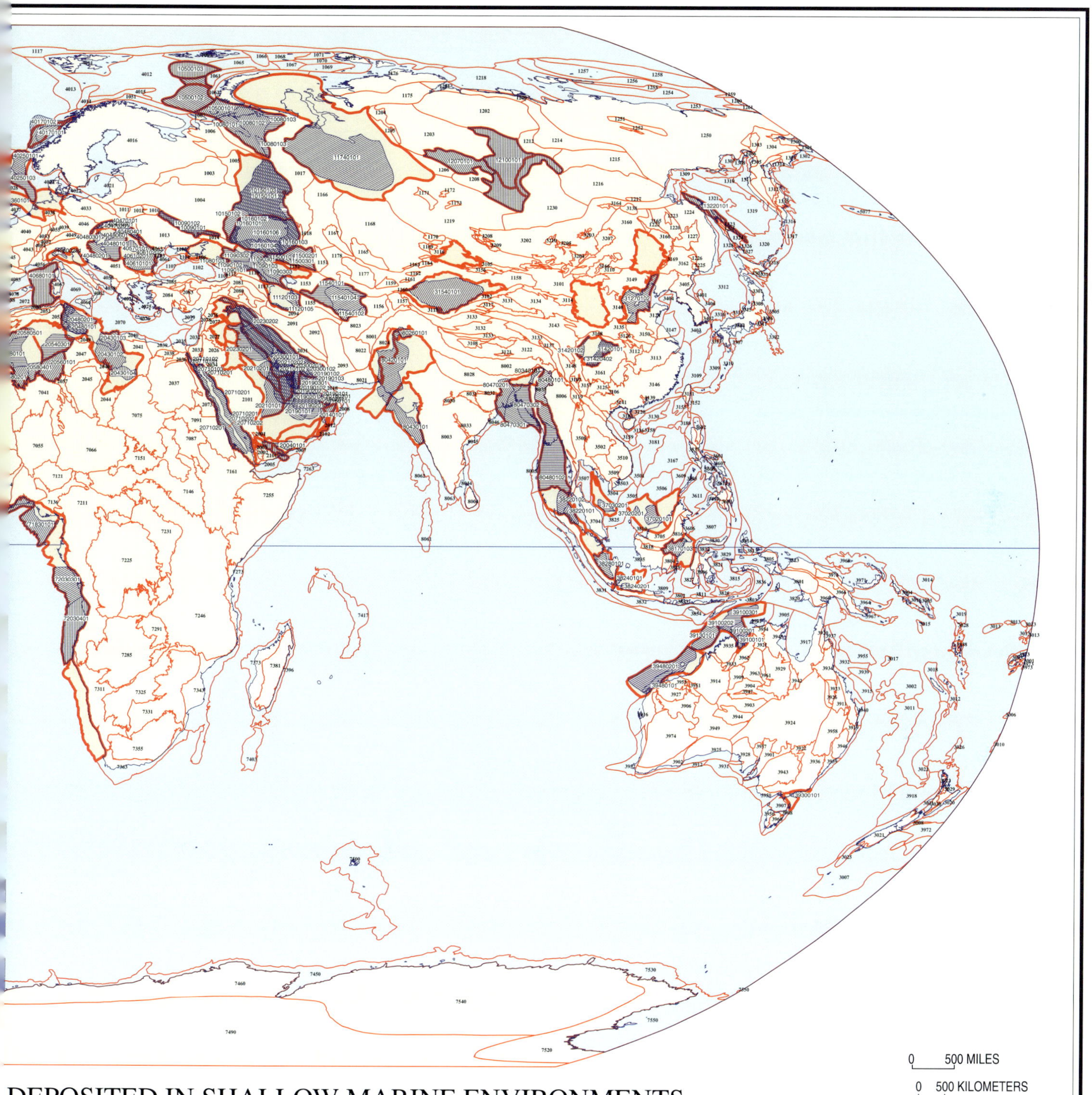

DEPOSITED IN SHALLOW MARINE ENVIRONMENTS

Projection Robinson
Central meridian 0

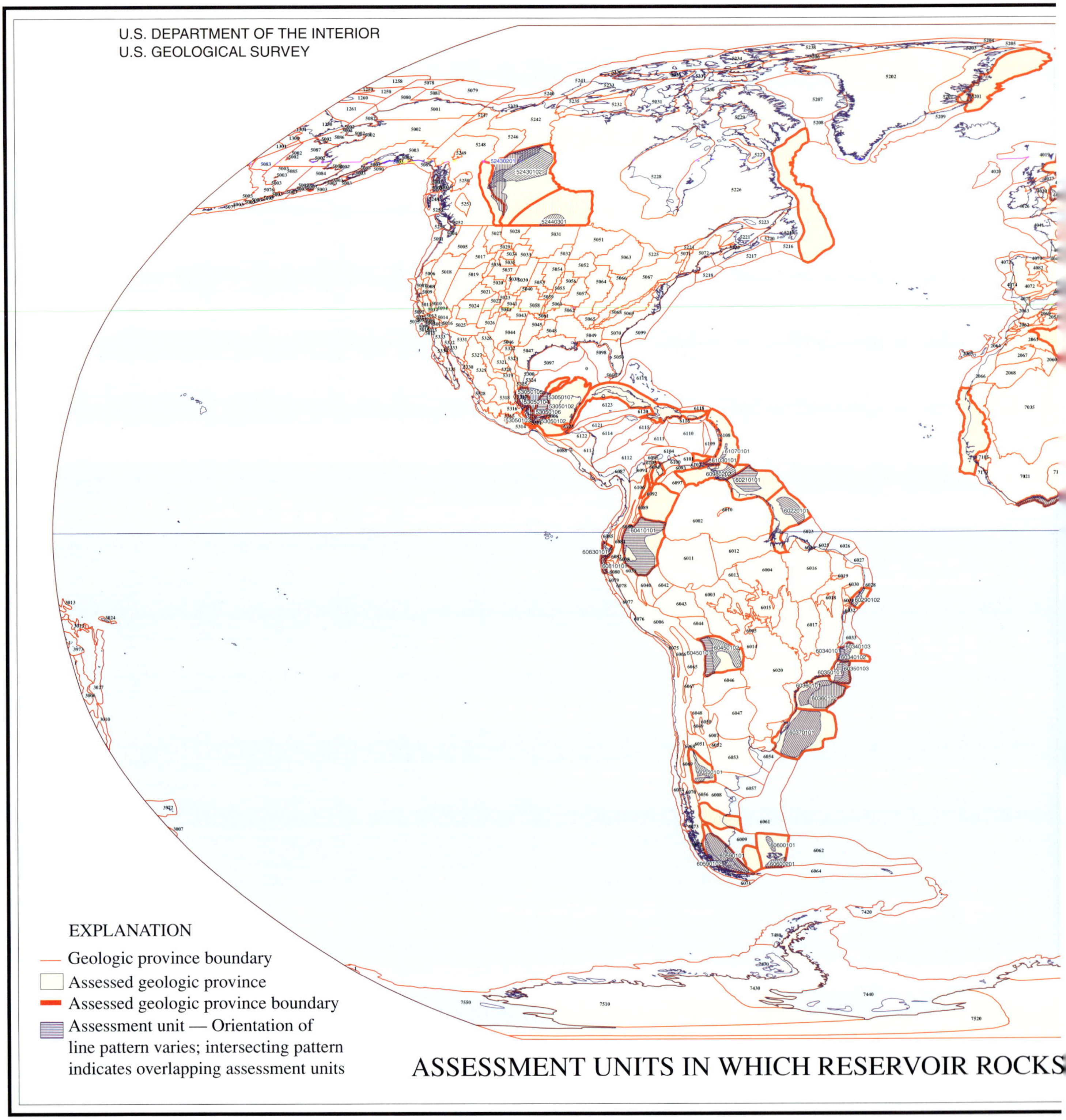

FIGURE 67. Map showing assessment units in which reservoir rocks were deposited in deep marine environments.

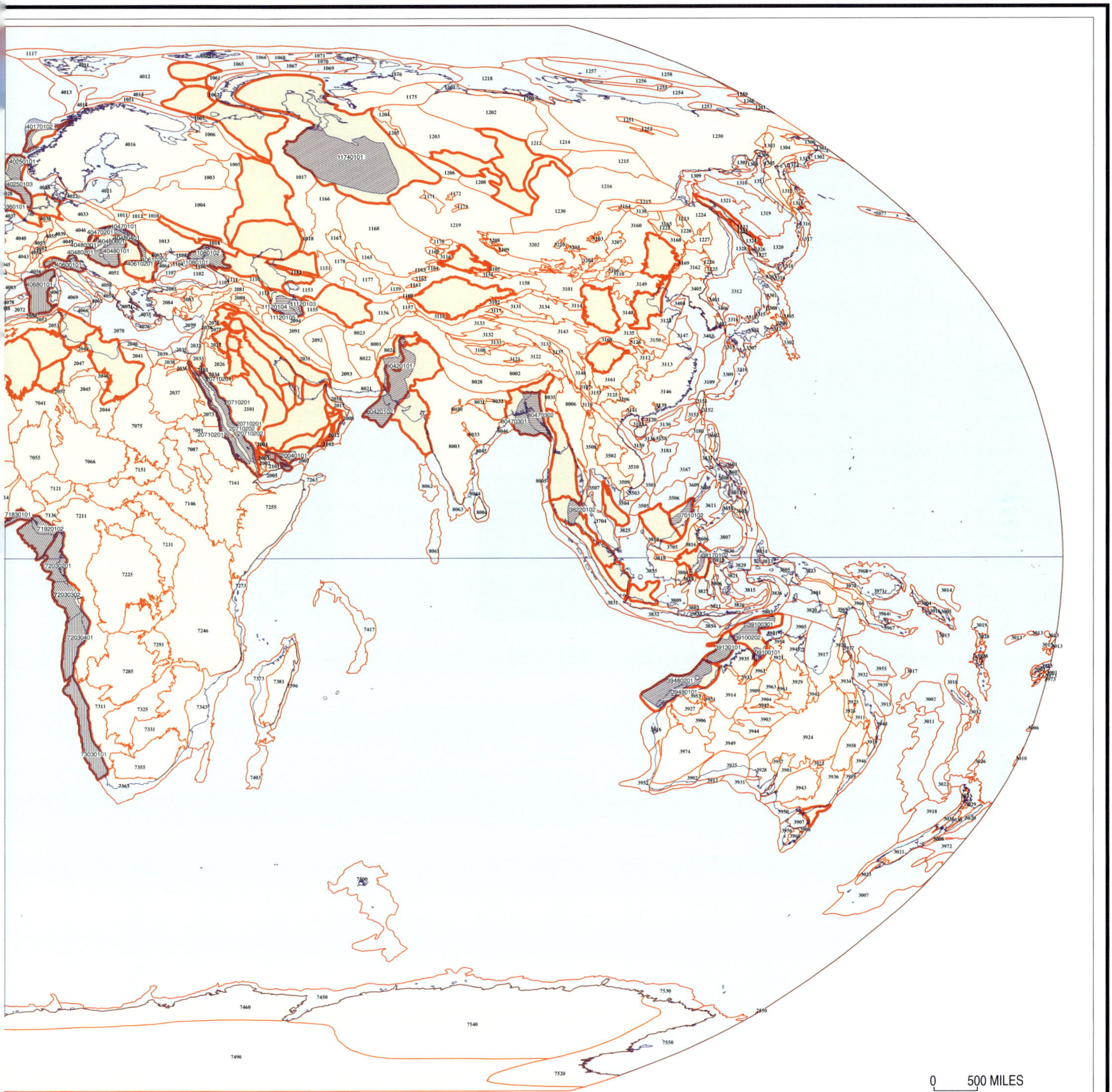

DEPOSITED IN DEEP MARINE ENVIRONMENTS

Projection Robinson
Central meridian 0

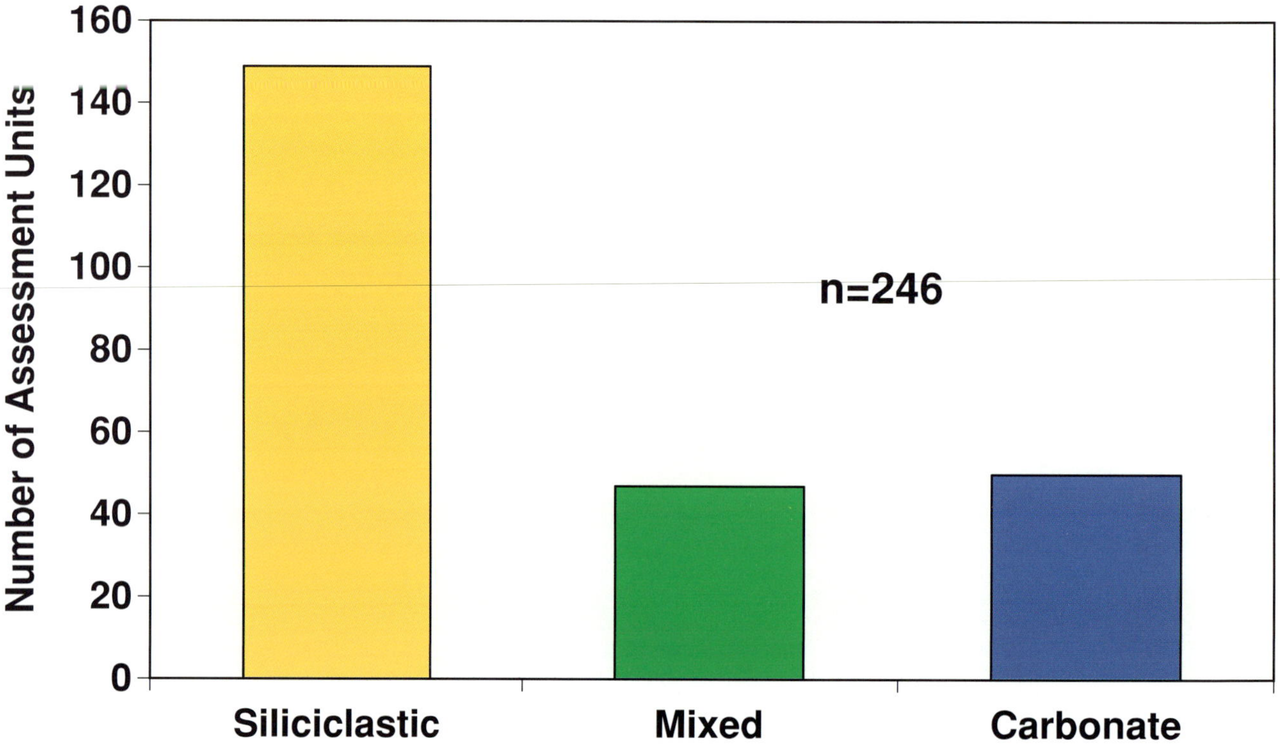

FIGURE 68. Numerical distribution of 246 assessment units by lithology of reservoir rock allocated to siliciclastic, mixed siliciclastic and carbonate, or carbonate lithology.

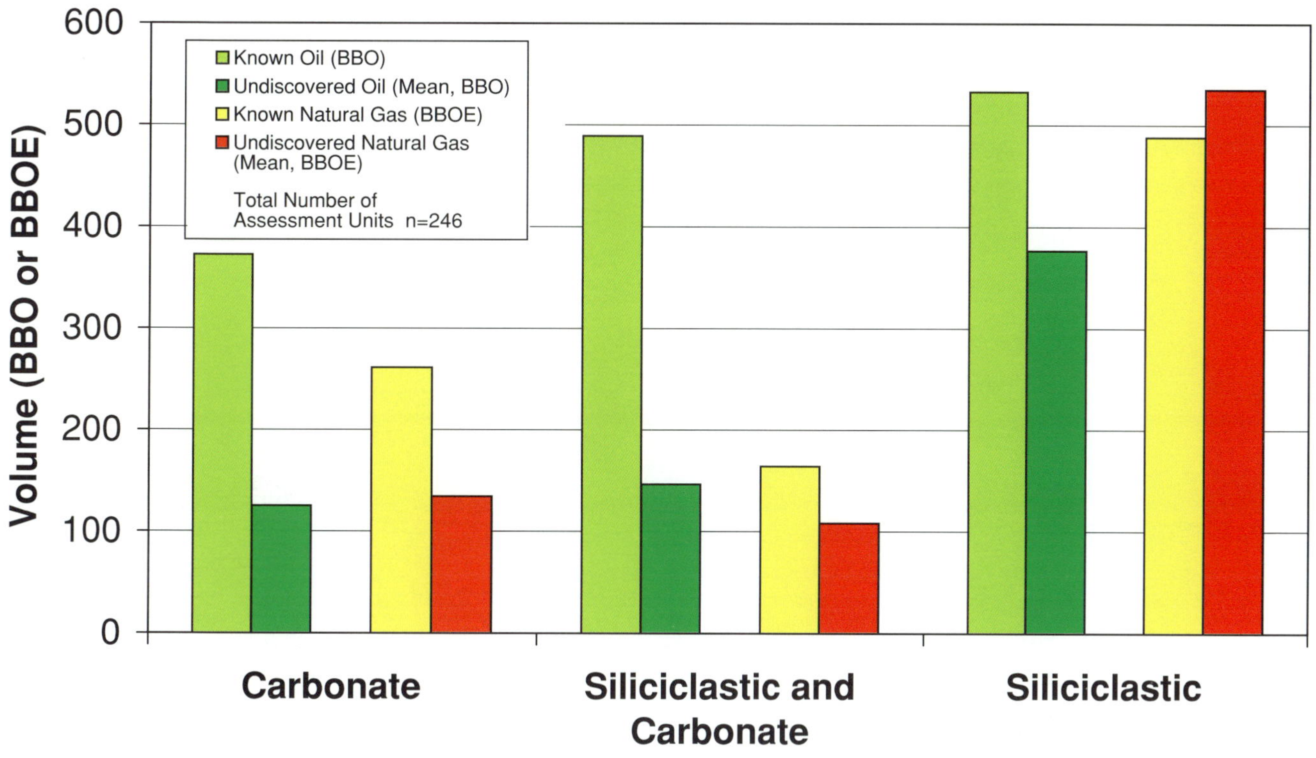

FIGURE 69. Distribution of mean estimated petroleum volumes of assessment units by lithology of reservoir rock allocated to categories of carbonate, mixed siliciclastic and carbonate, or siliciclastic lithology. Known volume = cumulative production plus remaining reserves; undiscovered volume = estimated in this study. BBO = billion barrels of oil; BBOE = billion barrels of oil equivalent.

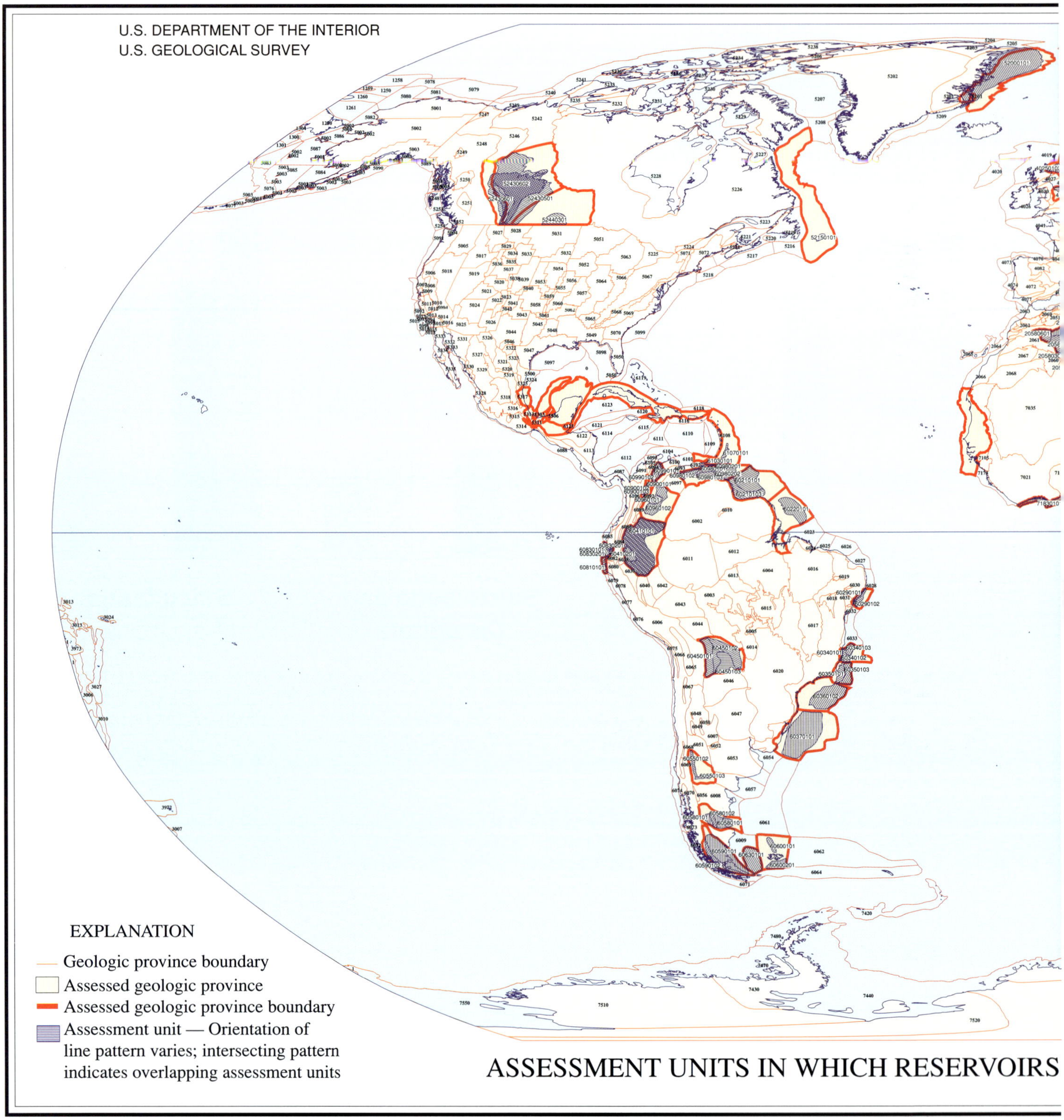

FIGURE 70. Map showing those assessment units and their numbers in which reservoirs are dominantly siliciclastic.

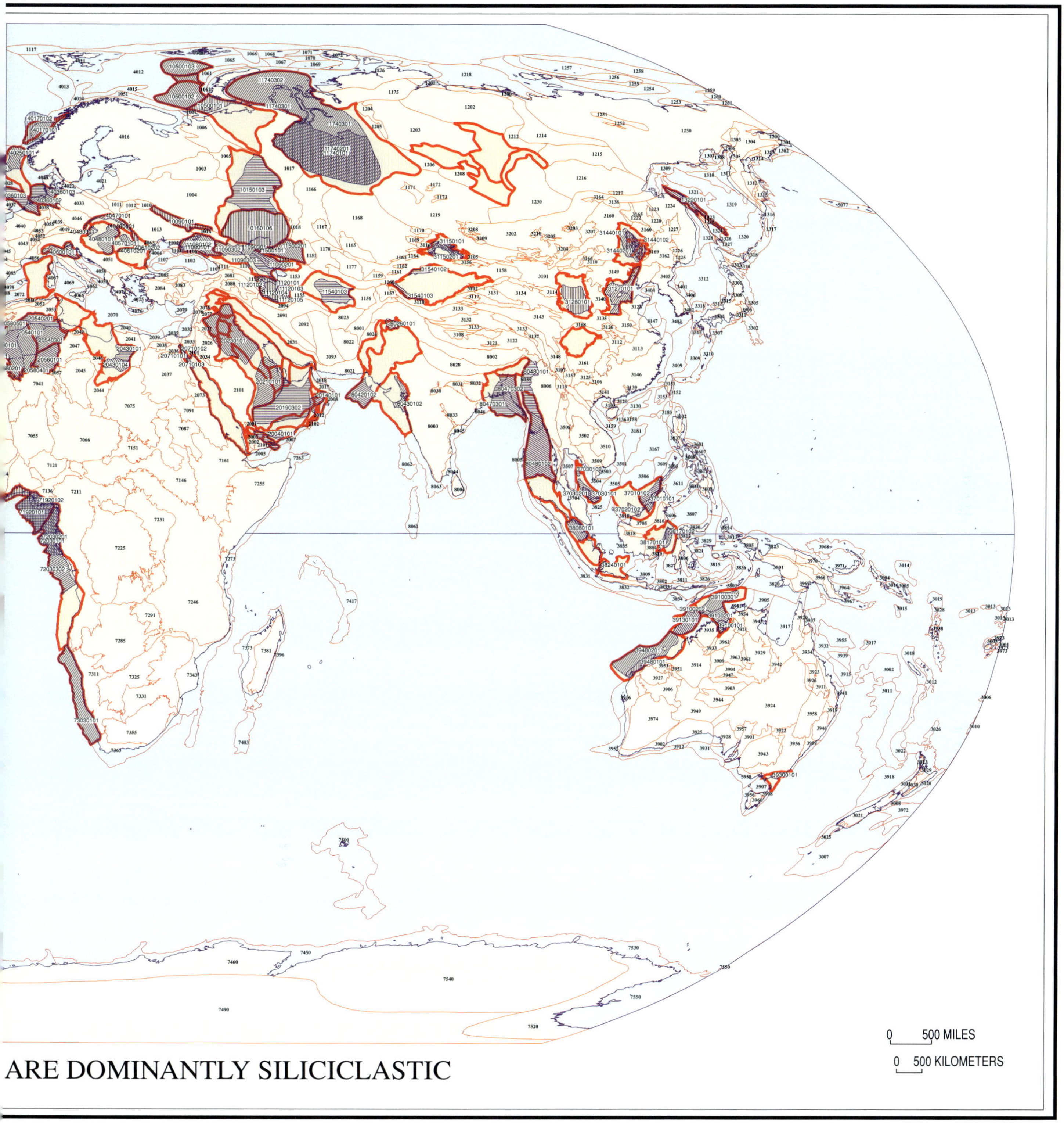

ARE DOMINANTLY SILICICLASTIC

Projection Robinson
Central meridian 0

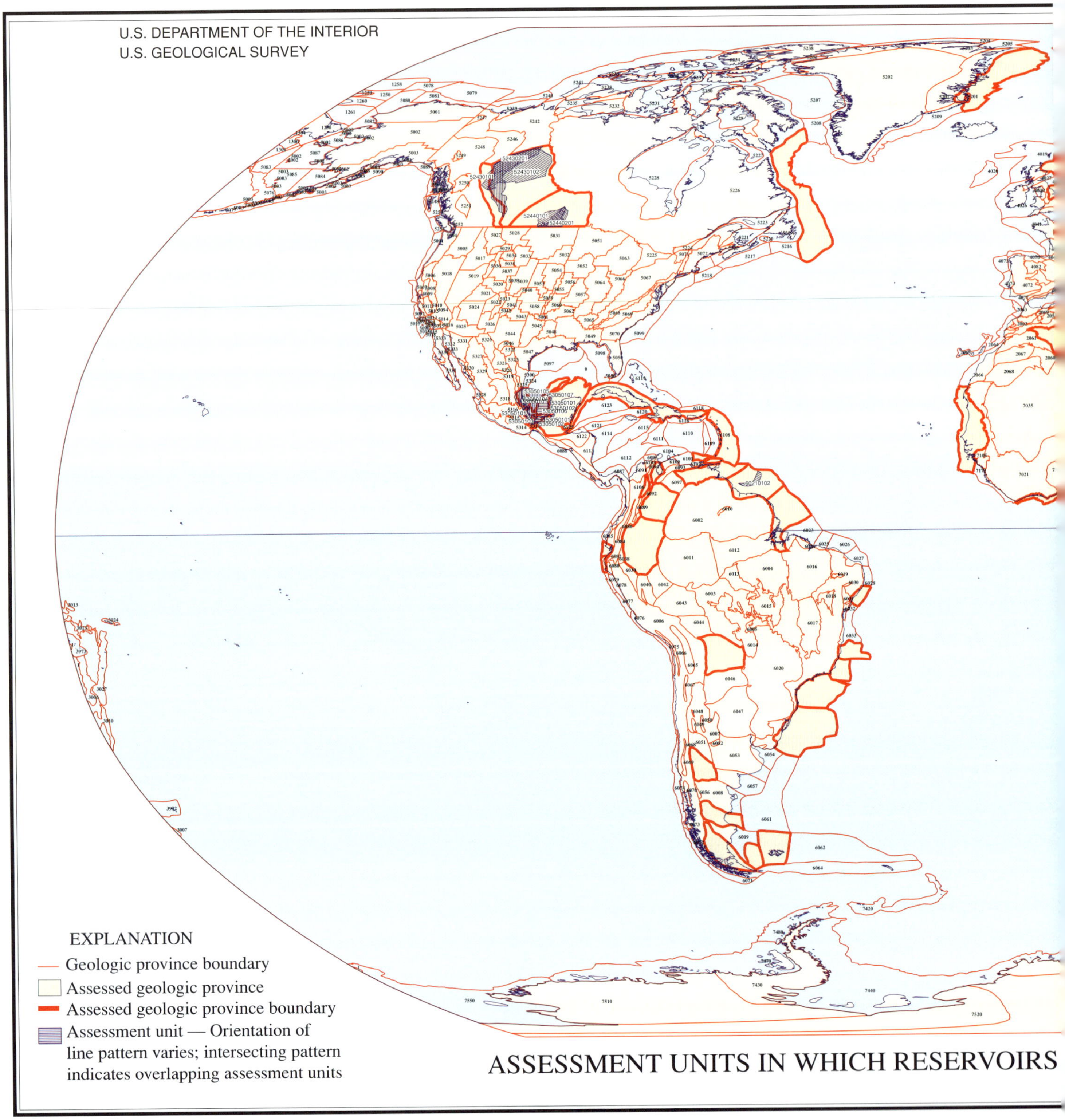

FIGURE 71. Map showing those assessment units and their numbers in which reservoirs are dominantly carbonate.

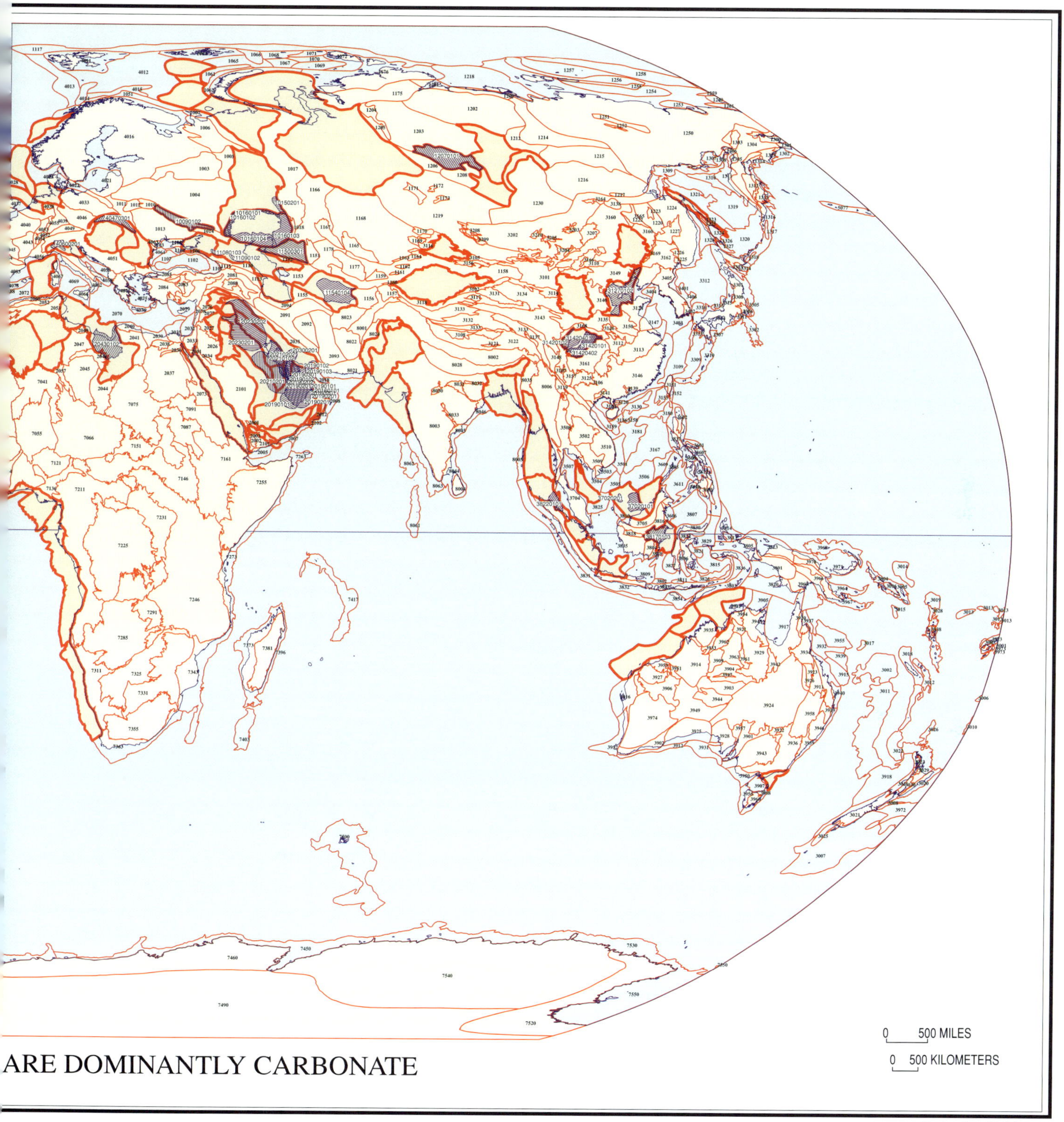

Projection Robinson
Central meridian 0

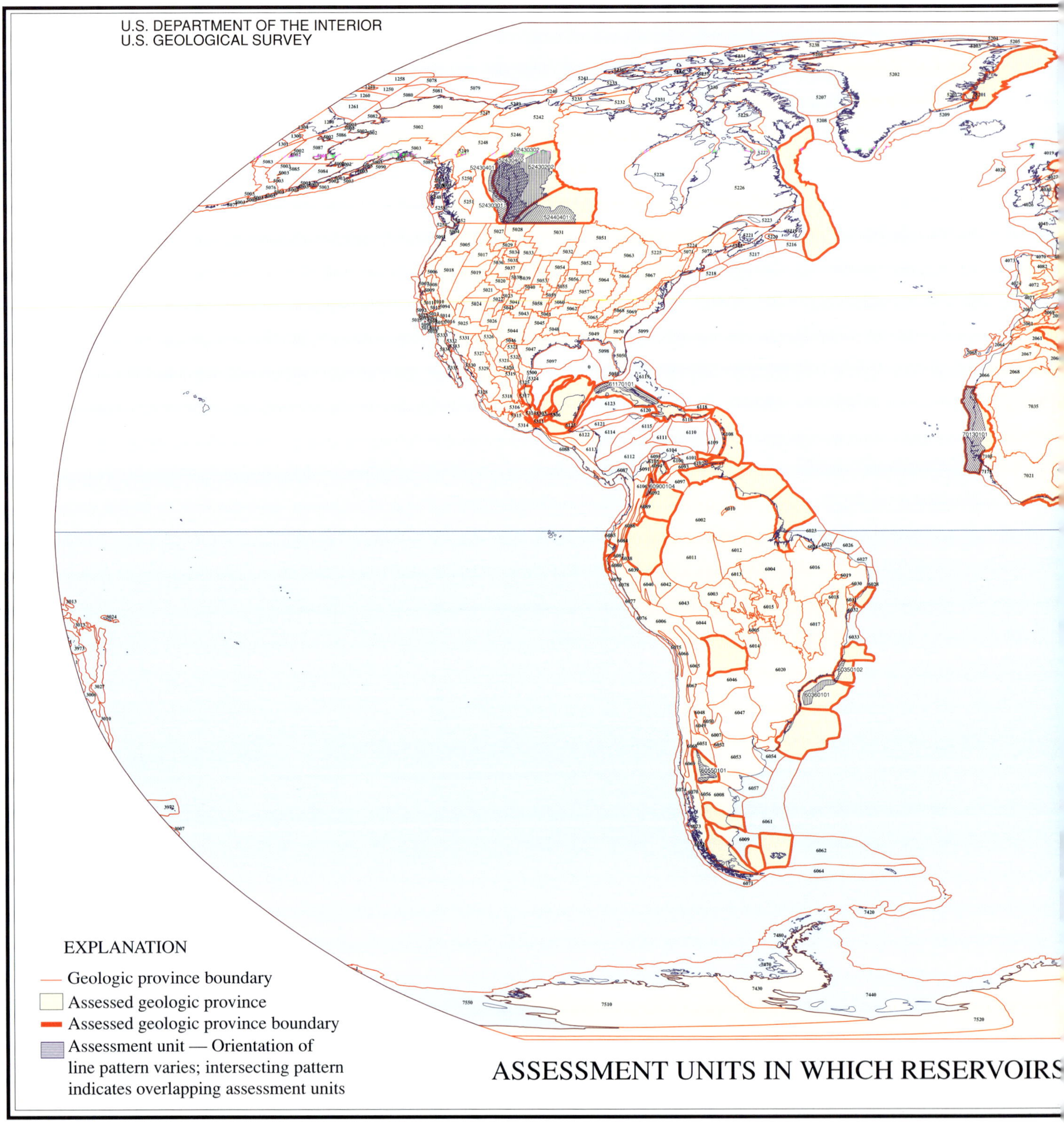

FIGURE 72. Map showing those assessment units and their numbers in which reservoirs are mixed siliciclastic and carbonate.

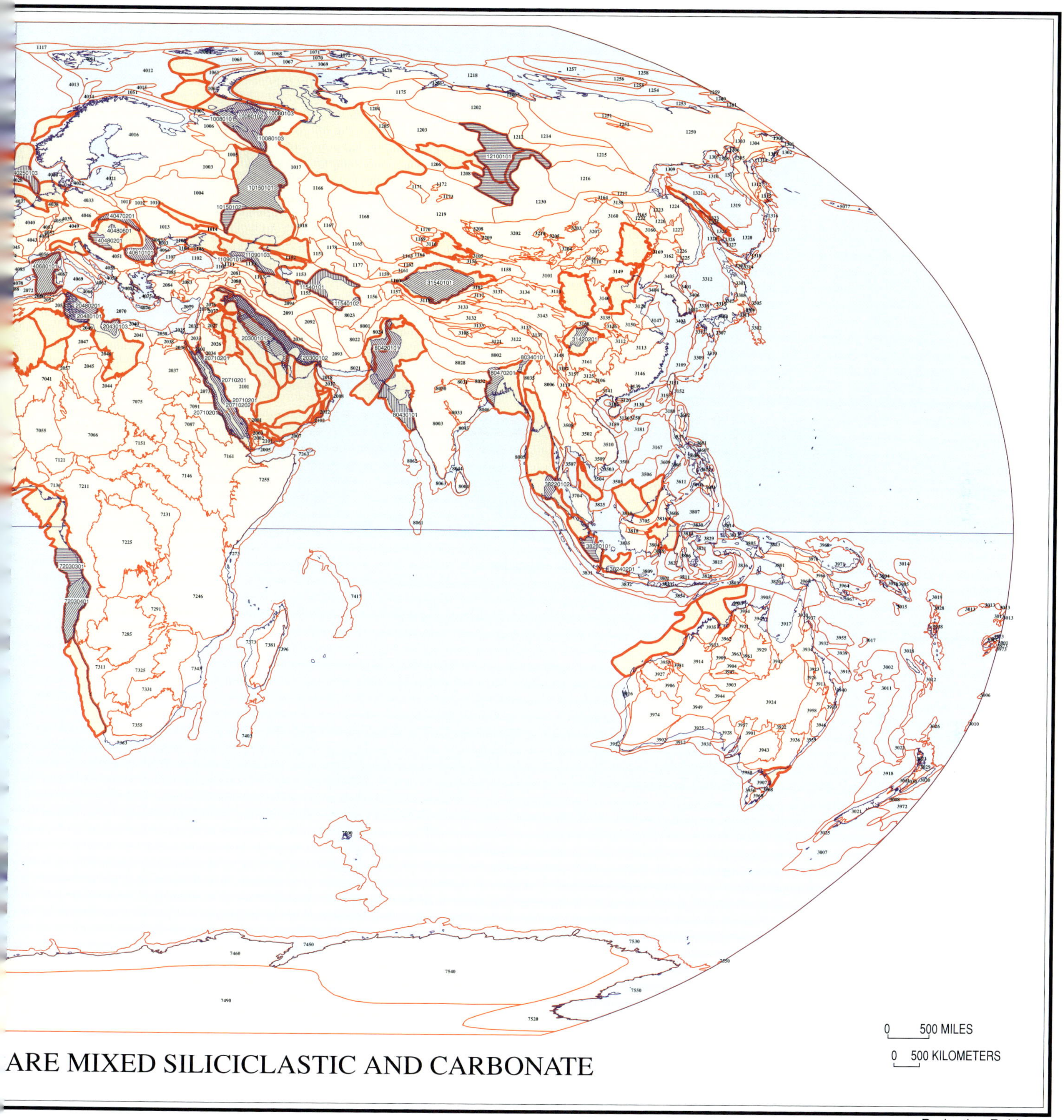

Projection Robinson
Central meridian 0

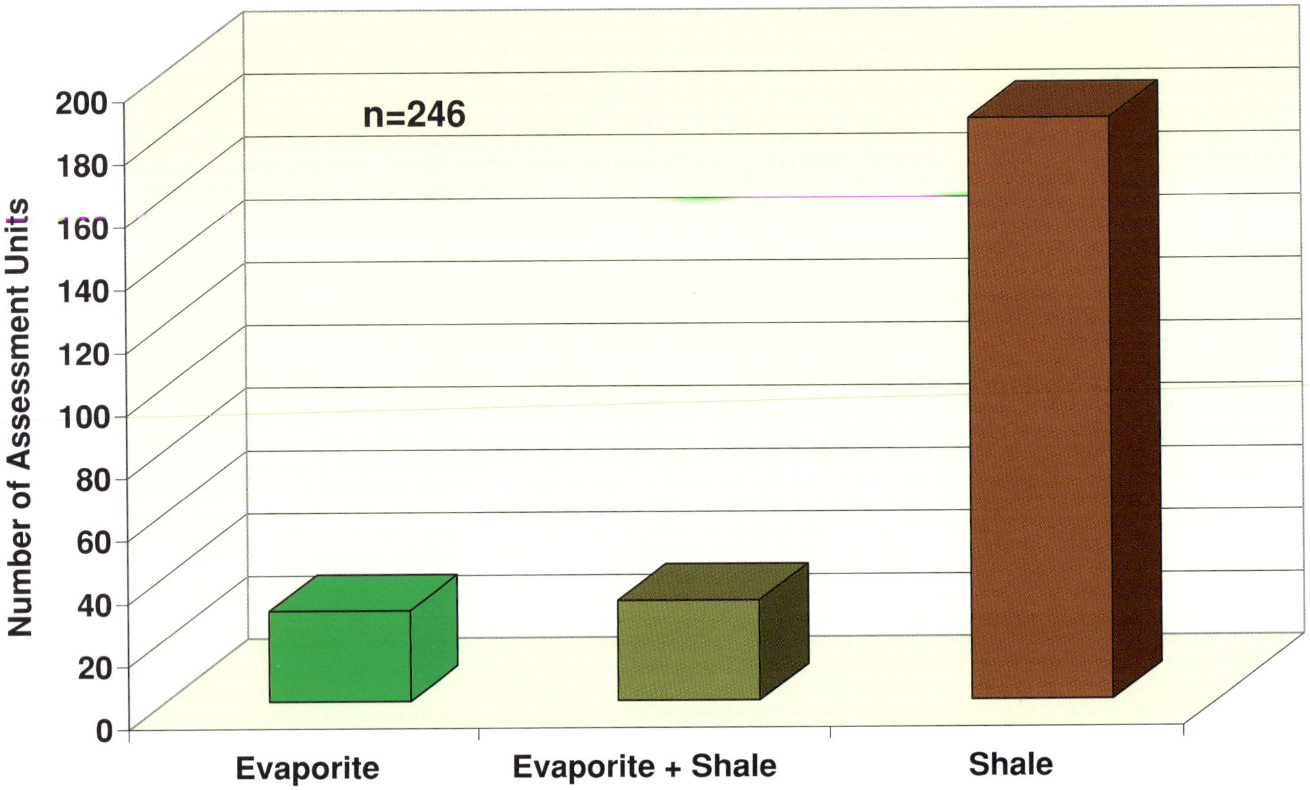

FIGURE 73. Numerical distribution of 246 assessment units by lithology of major seal rock allocated to evaporite, mixed evaporite and shale, and shale categories.

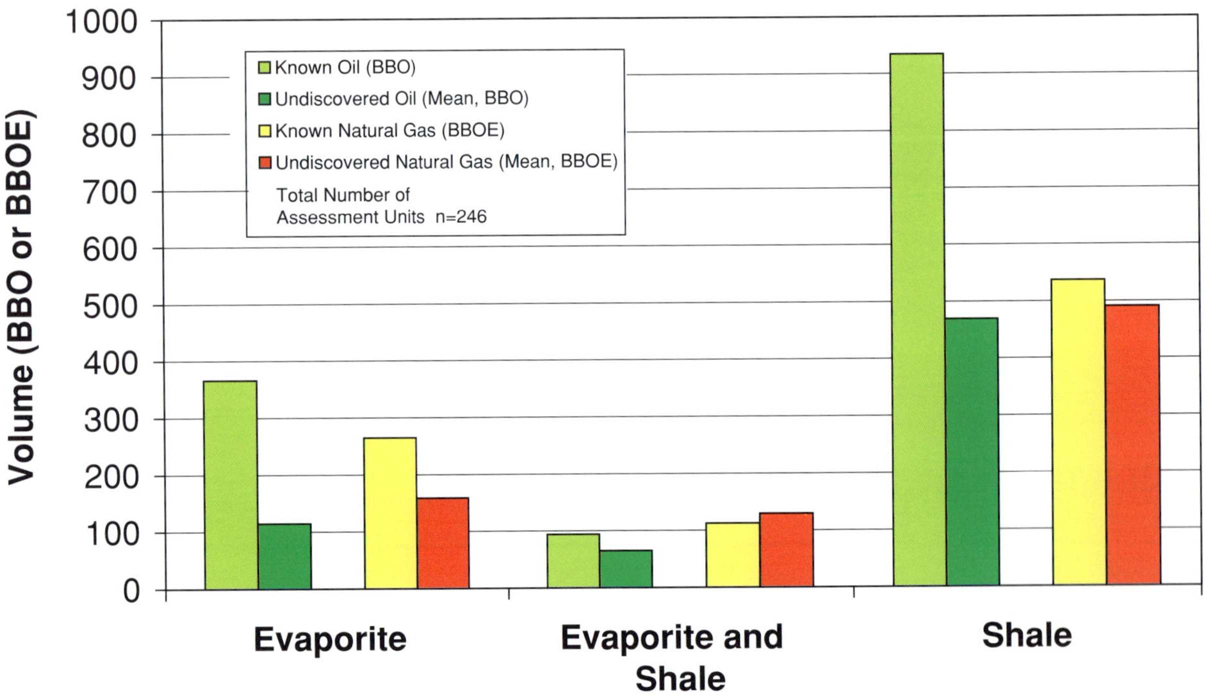

FIGURE 74. Distribution of mean estimated petroleum volumes of assessment units by lithology of major seal rock; i.e. evaporite, mixed evaporite and shale, or shale. Known volume = cumulative production plus remaining reserves; undiscovered volume = estimated in this study. BBO = billion barrels of oil; BBOE = billion barrels of oil equivalent.

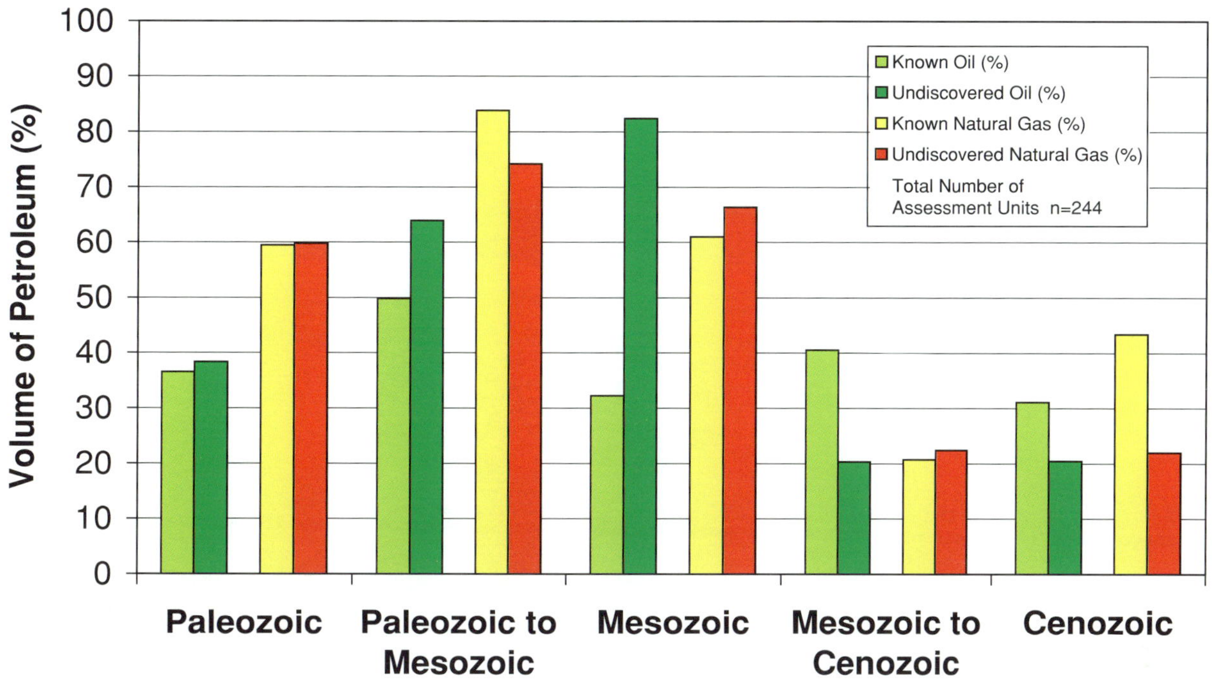

FIGURE 75. Percent of total mean estimated petroleum volumes of assessment units sealed by evaporite and mixed evaporite and shale by age of maturation. Known volume = cumulative production plus remaining reserves; undiscovered volume = estimated in this study. A greater percentage of older AU, allocated to Paleozoic, Mesozoic, or mixed Paleozoic–Mesozoic ages, is sealed by evaporite than younger Mesozoic to Cenozoic or Cenozoic AU. Mixed evaporite and shale lithology of Figure 74 is lumped with evaporite lithology in this calculation.

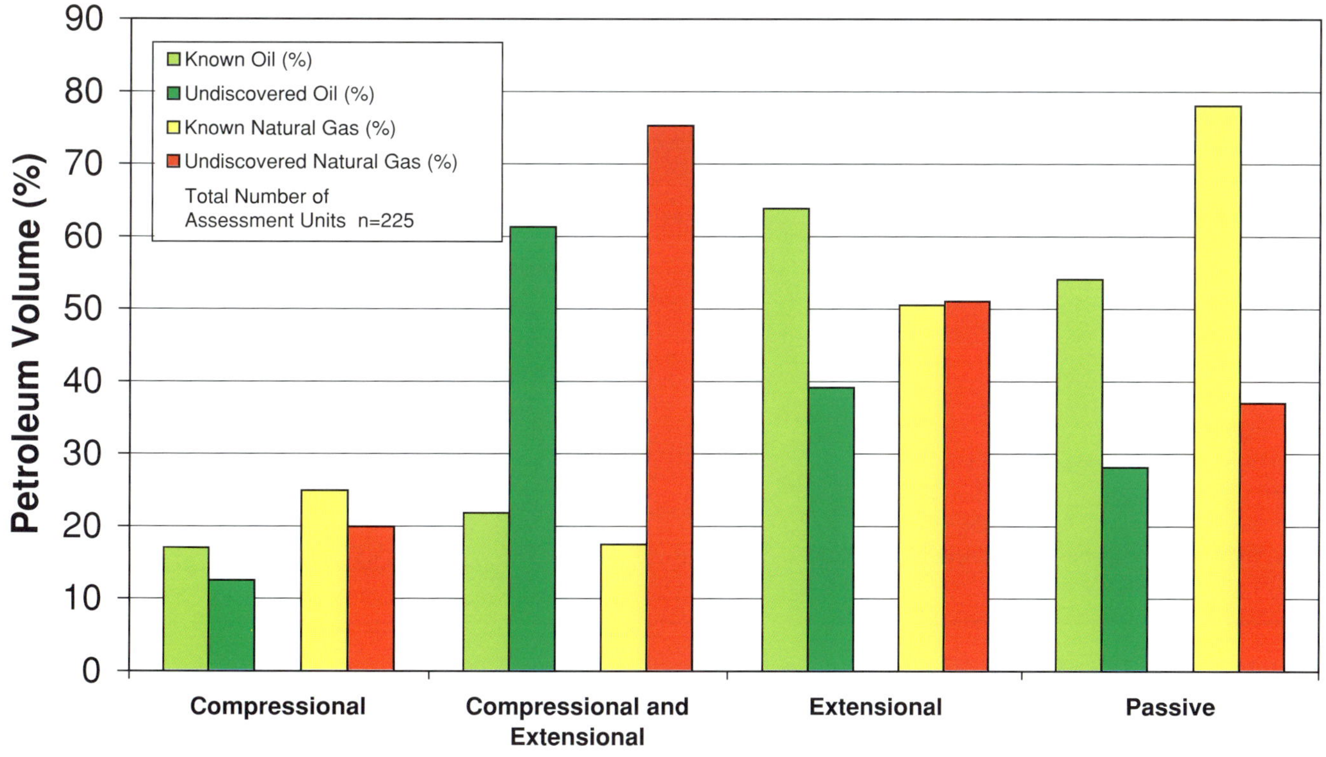

FIGURE 76. Percent of total mean estimated petroleum volume of assessment units sealed by evaporite and their tectonic setting. Known volume = cumulative production plus remaining reserves; undiscovered volume = estimated in this study. Note the greater percentage of AU in extensional and passive settings than in compressional settings. Mixed evaporite and shale lithology of Figure 74 is lumped with evaporite lithology in this calculation.

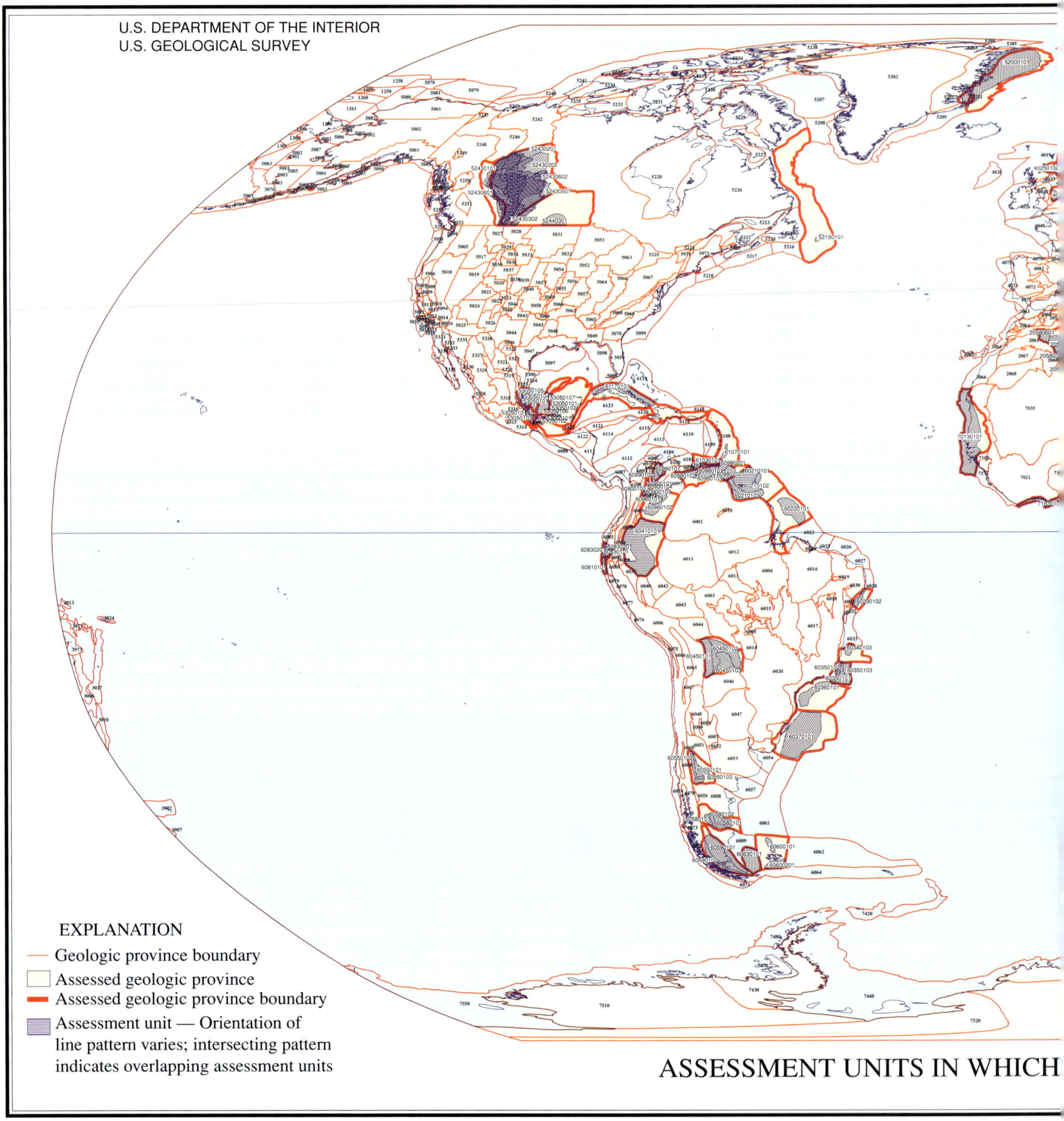

FIGURE 77. Map showing those assessment units and their numbers in which seals are dominantly shale.

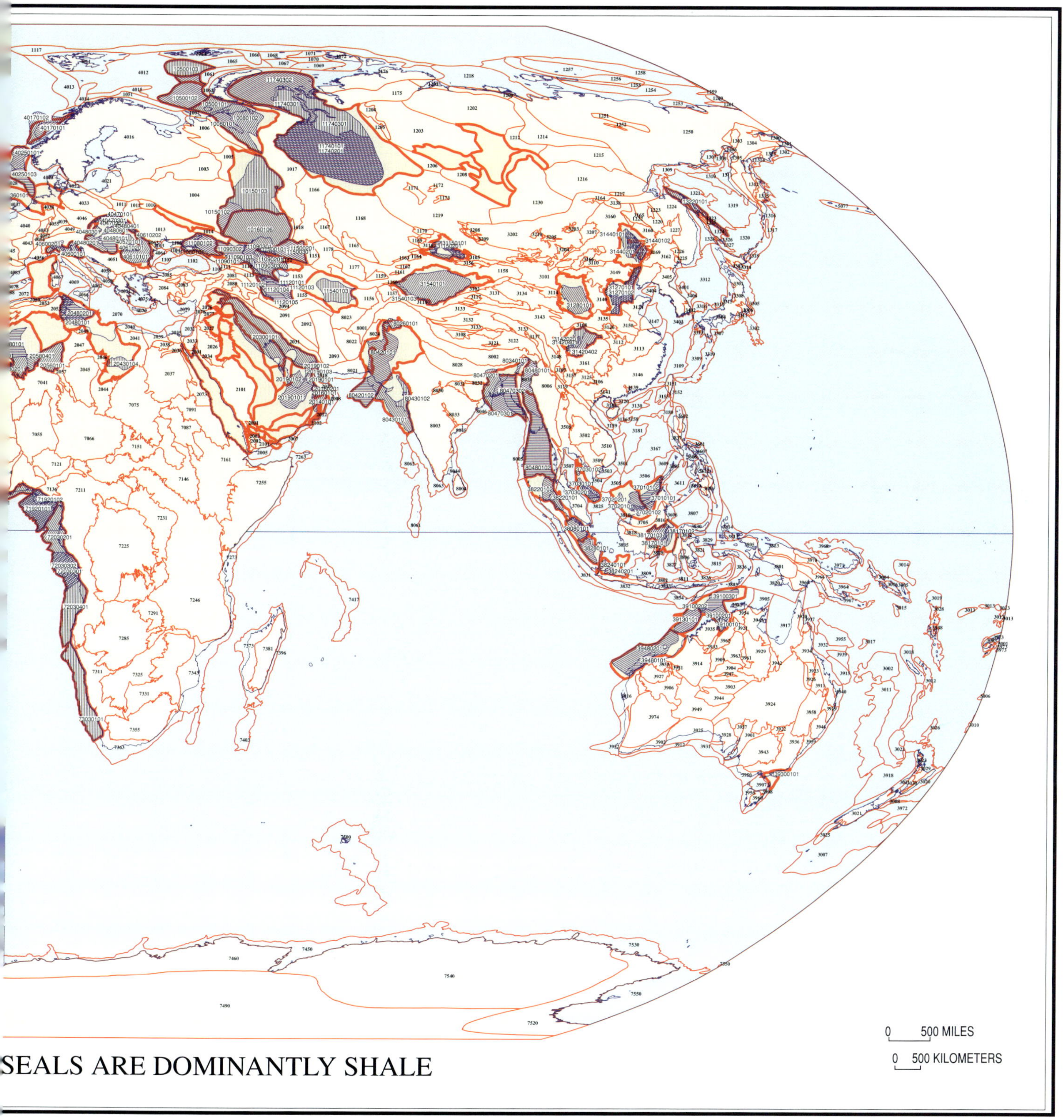

Projection Robinson
Central meridian 0

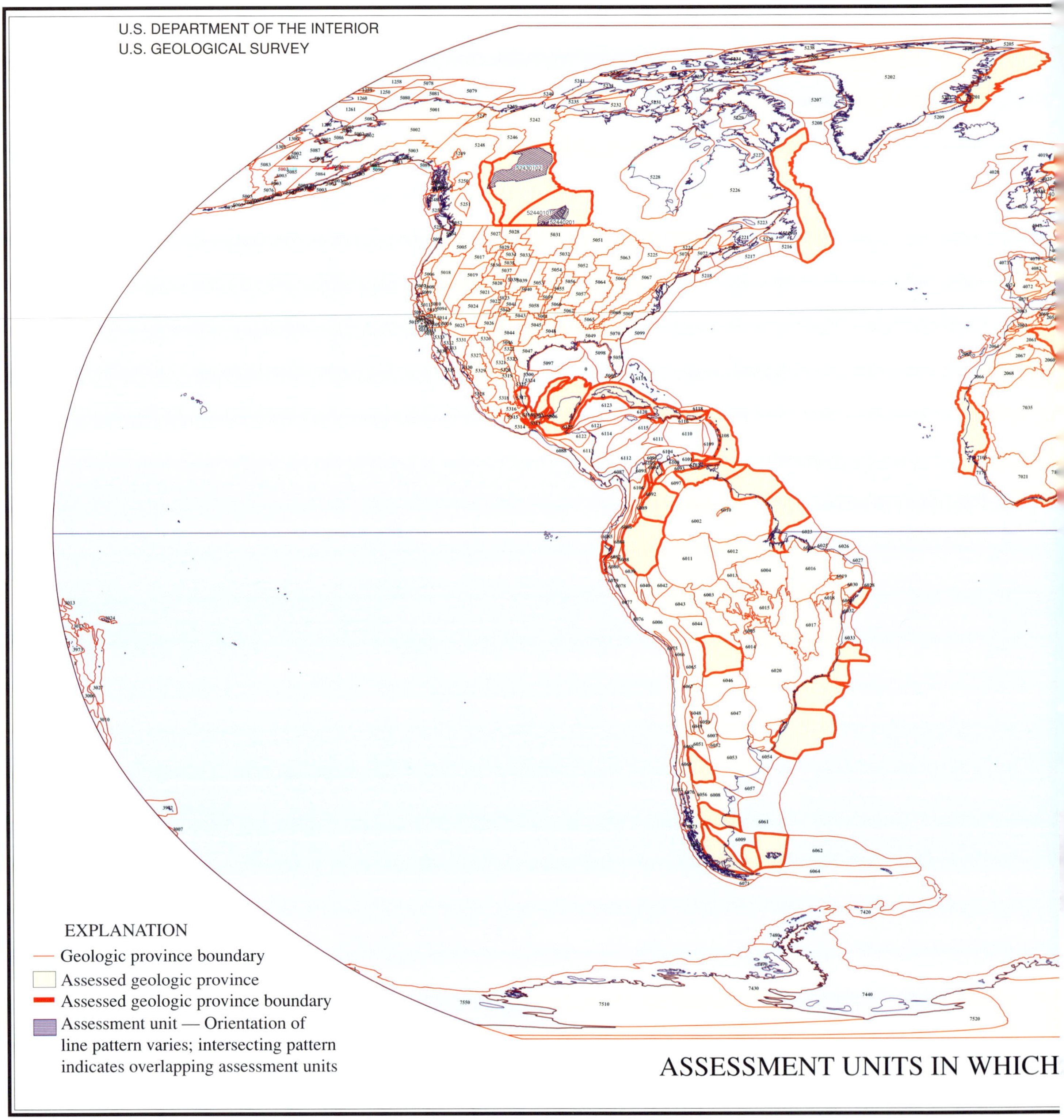

FIGURE 78. Map showing those assessment units and their numbers in which seals are dominantly evaporite.

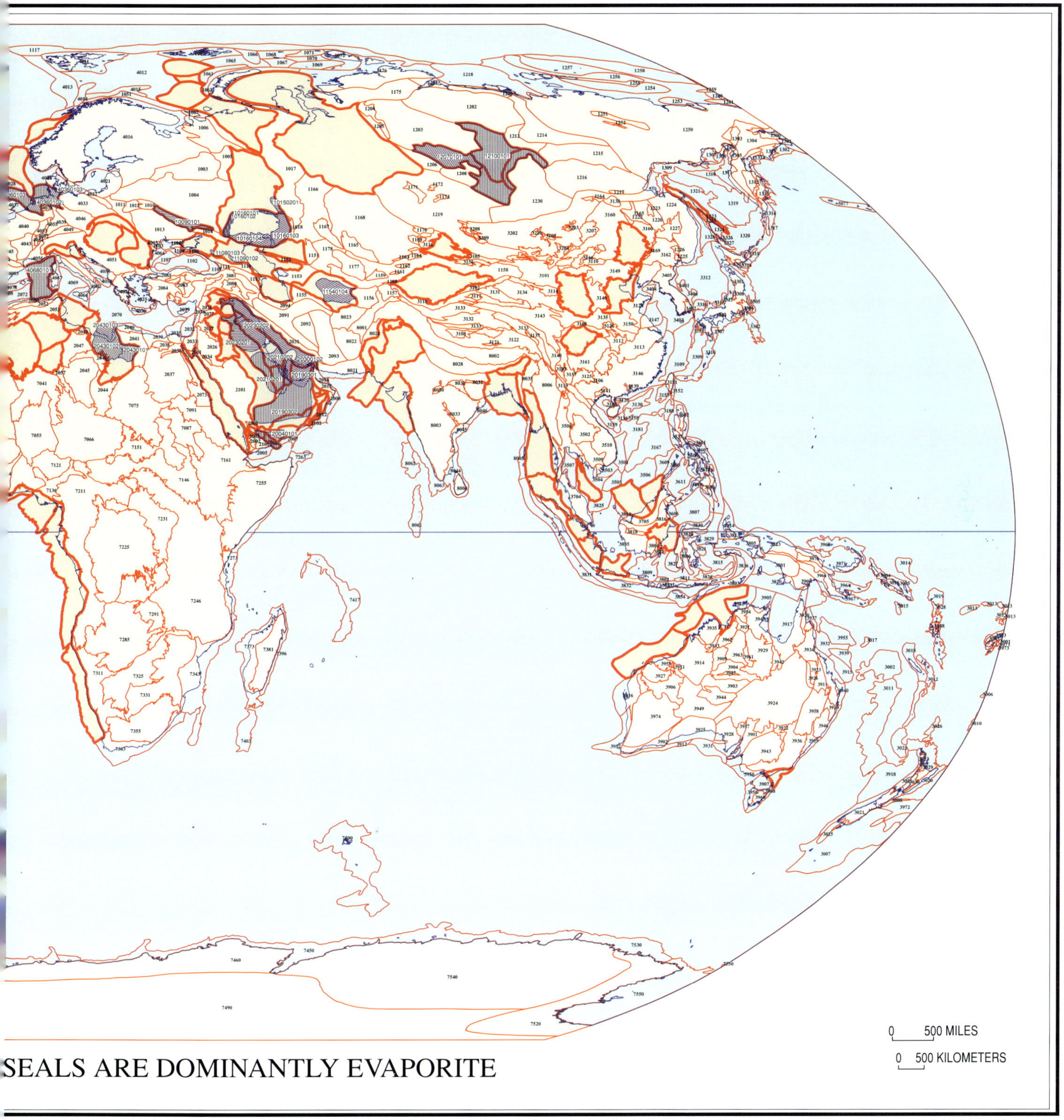

Projection Robinson
Central meridian 0

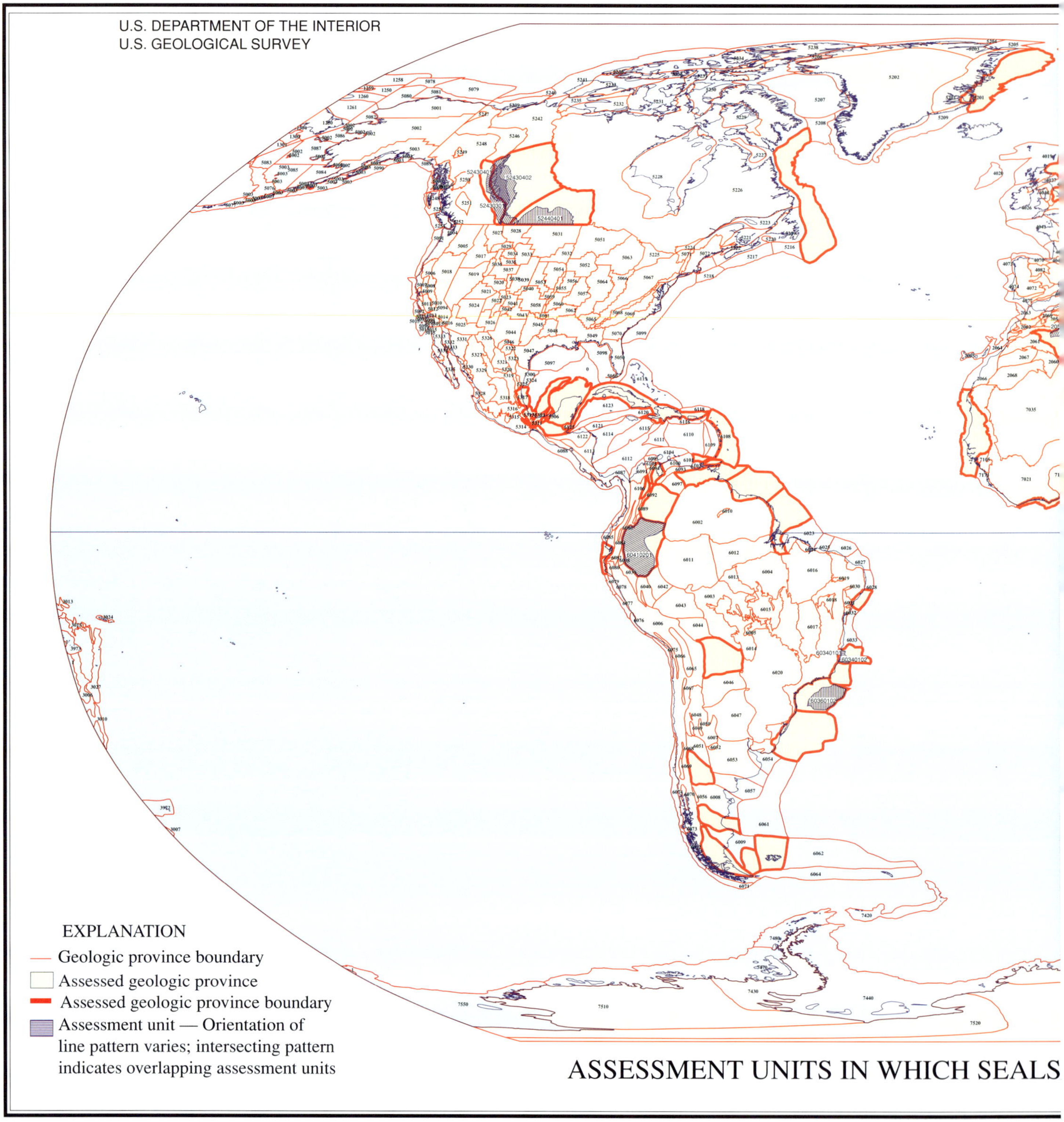

FIGURE 79. Map showing those assessment units and their numbers in which seals are both evaporite and shale.

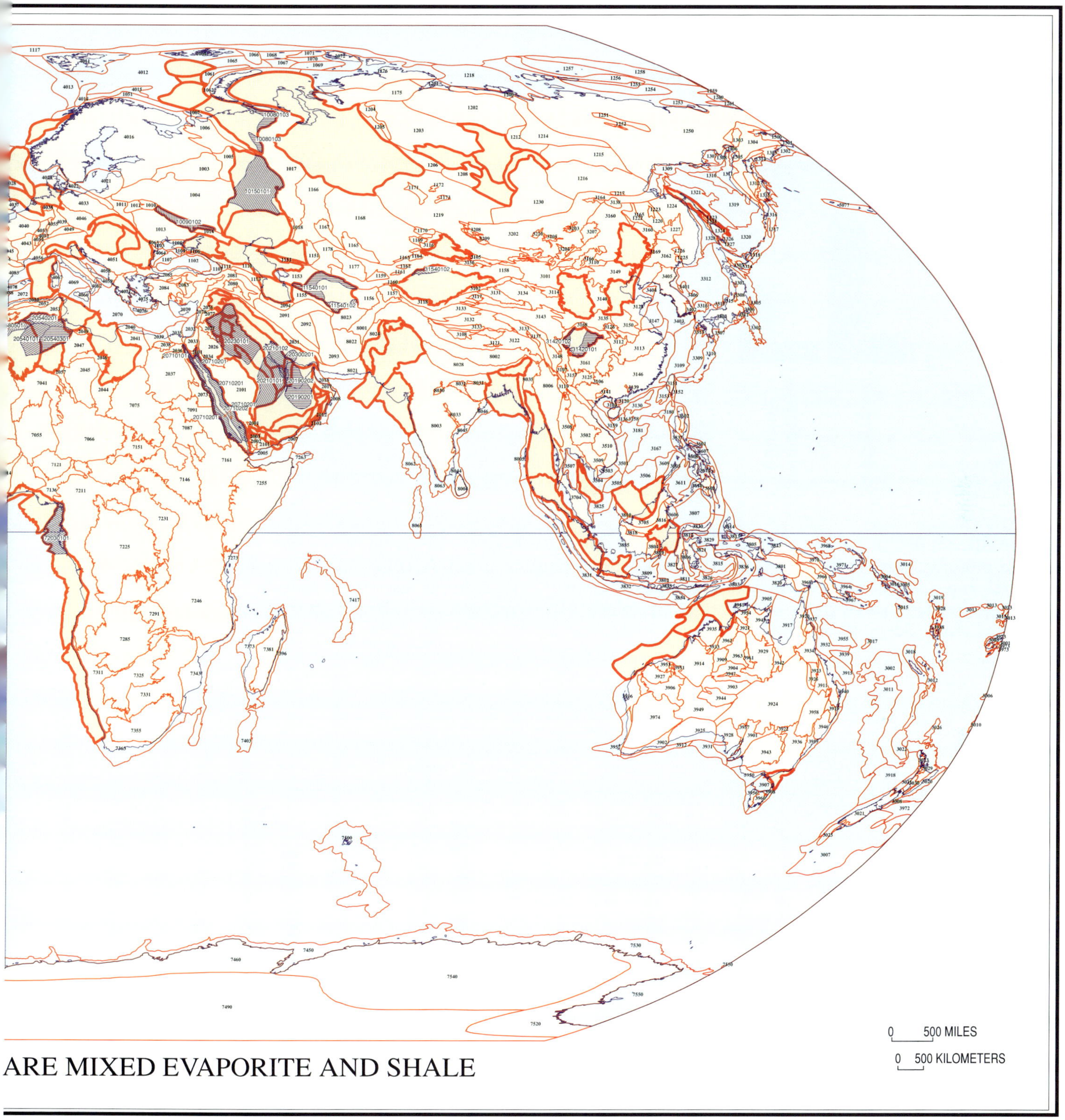

Projection Robinson
Central meridian 0

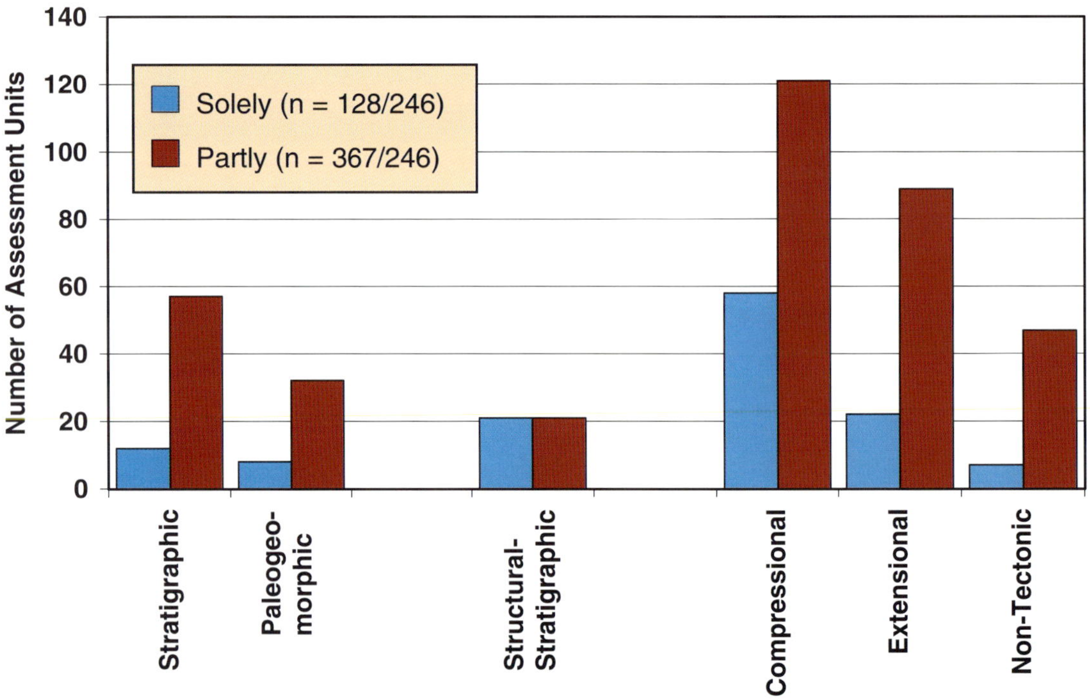

FIGURE 80. Numerical distribution of 246 assessment units by trap type. The six trap categories are stratigraphic, paleogeomorphic (reefs, erosional relief), non-specified structural-stratigraphic, compressional structural, extensional structural, and non-tectonic structural (diapirs, drapes). Some trap types in the assessment units have mixed categories and are referred to as partly (red color) stratigraphic, partly paleogeomorphic, etc. If only one trap category is determined, the term solely (blue color) is used.

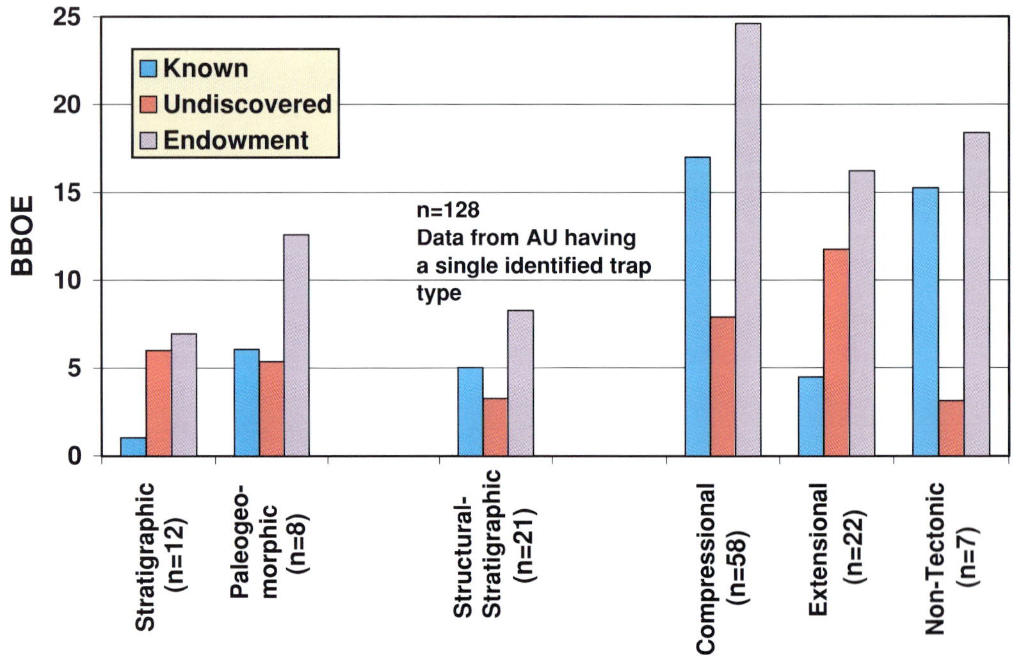

FIGURE 81. Distribution of mean petroleum volumes of 128 assessment units by trap type and tectonic settings. Known volume = cumulative production plus remaining reserves; undiscovered volume = estimated in this study; conventional endowment = sum of known and undiscovered volumes. The six trap categories are stratigraphic, paleogeomorphic (reefs, erosional relief), non-specified structural-stratigraphic, compressional structural, extensional structural and non-tectonic structural, (diapirs, drapes). BBOE = billion barrels of oil equivalent.

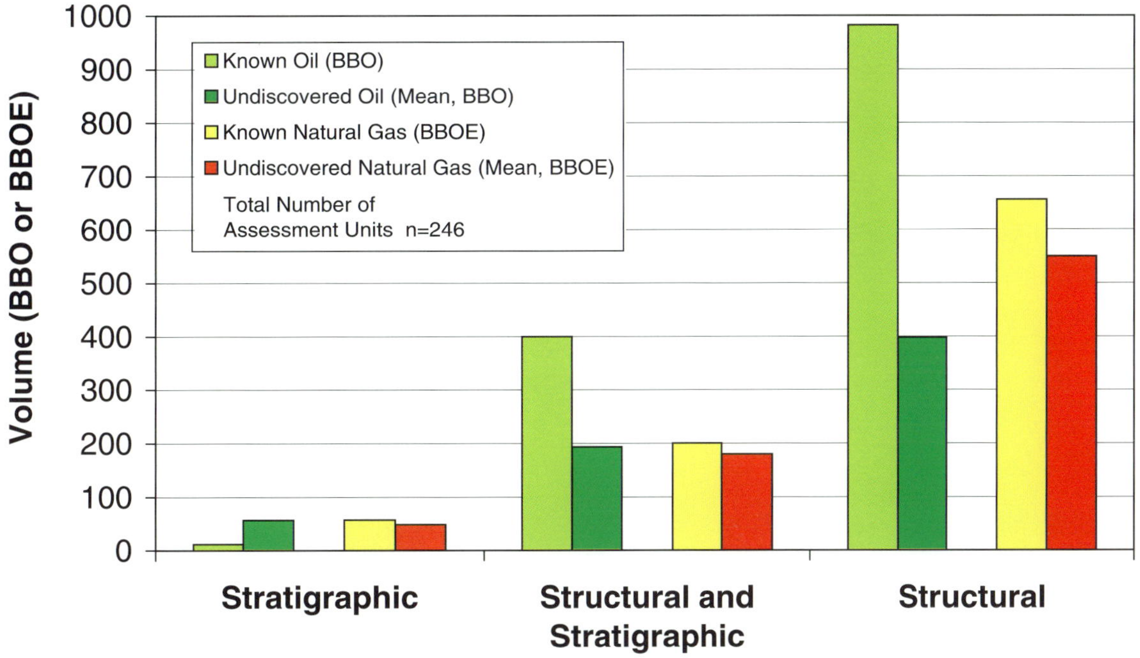

FIGURE 82. Oil and natural gas volumes for known and undiscovered resources by trap type. Known volume = cumulative production plus remaining reserves; undiscovered volume = estimated in this study. The three trap categories, identified as dominant in each assessment unit, are stratigraphic, structural-stratigraphic, and structural. Paleogeomorphic traps are included in stratigraphic category. BBO = billion barrels of oil; BBOE = billion barrels of oil equivalent.

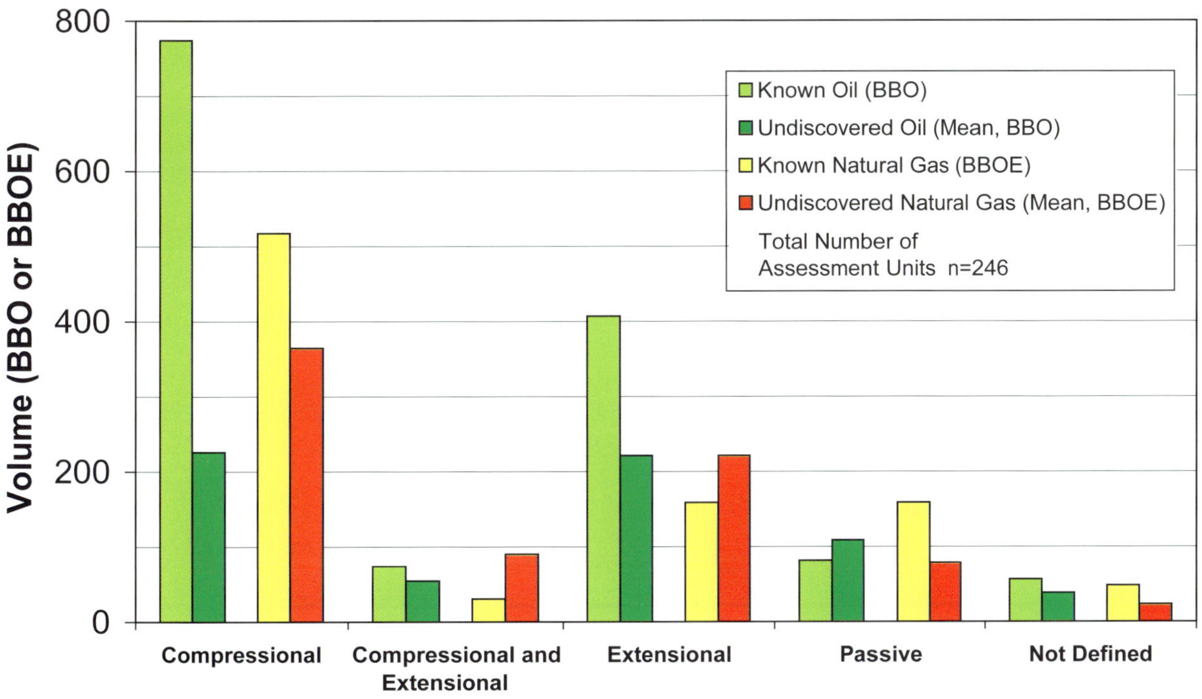

FIGURE 83. Oil and natural gas volumes for known and undiscovered resources by tectonic settings. Known volume = cumulative production plus remaining reserves; undiscovered volume = estimated in this study. The five tectonic categories are compressional, compressional and extensional, extensional, passive, and not defined. BBO = billion barrels of oil; BBOE = billion barrels of oil equivalent.

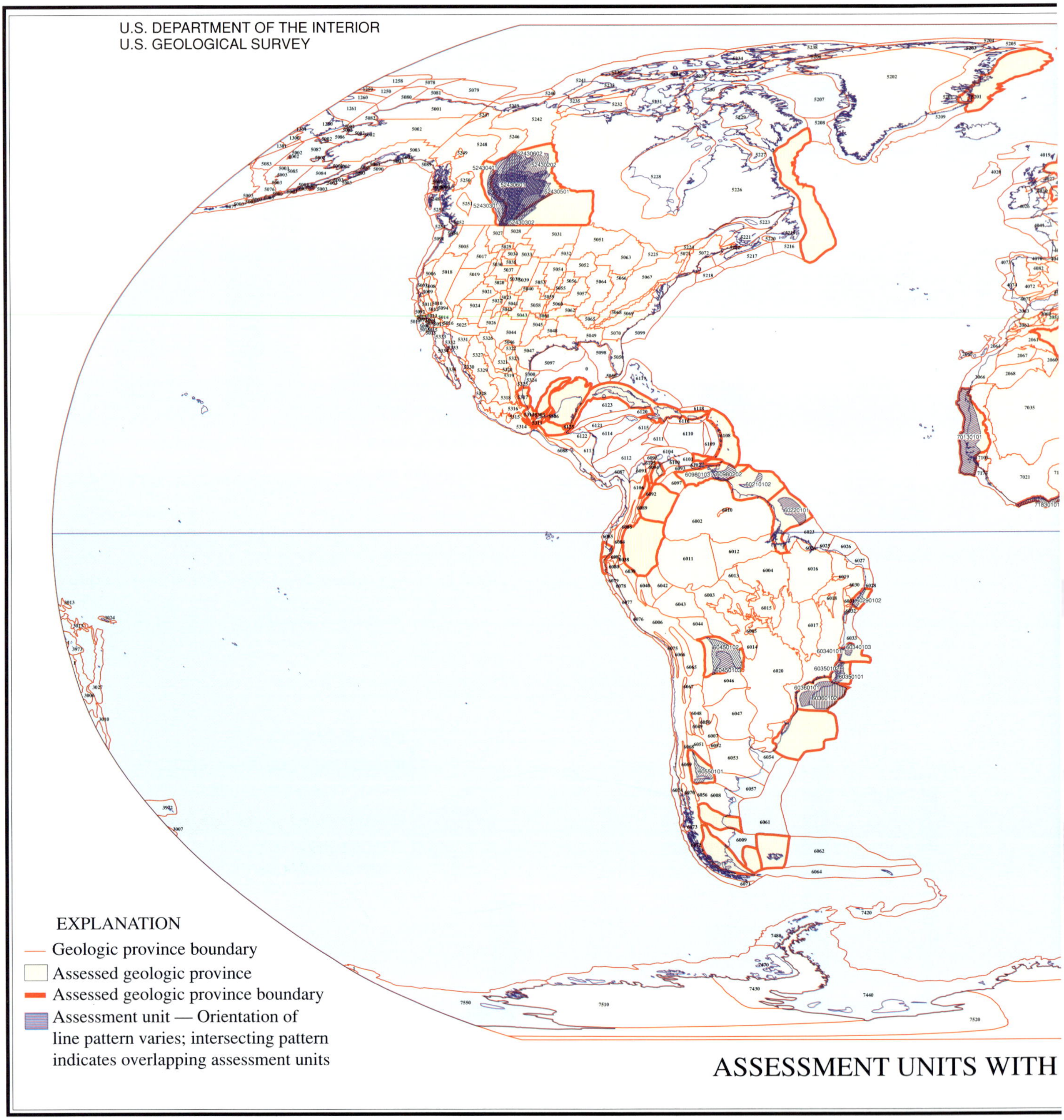

Figure 84. Map showing assessment units with stratigraphic traps.

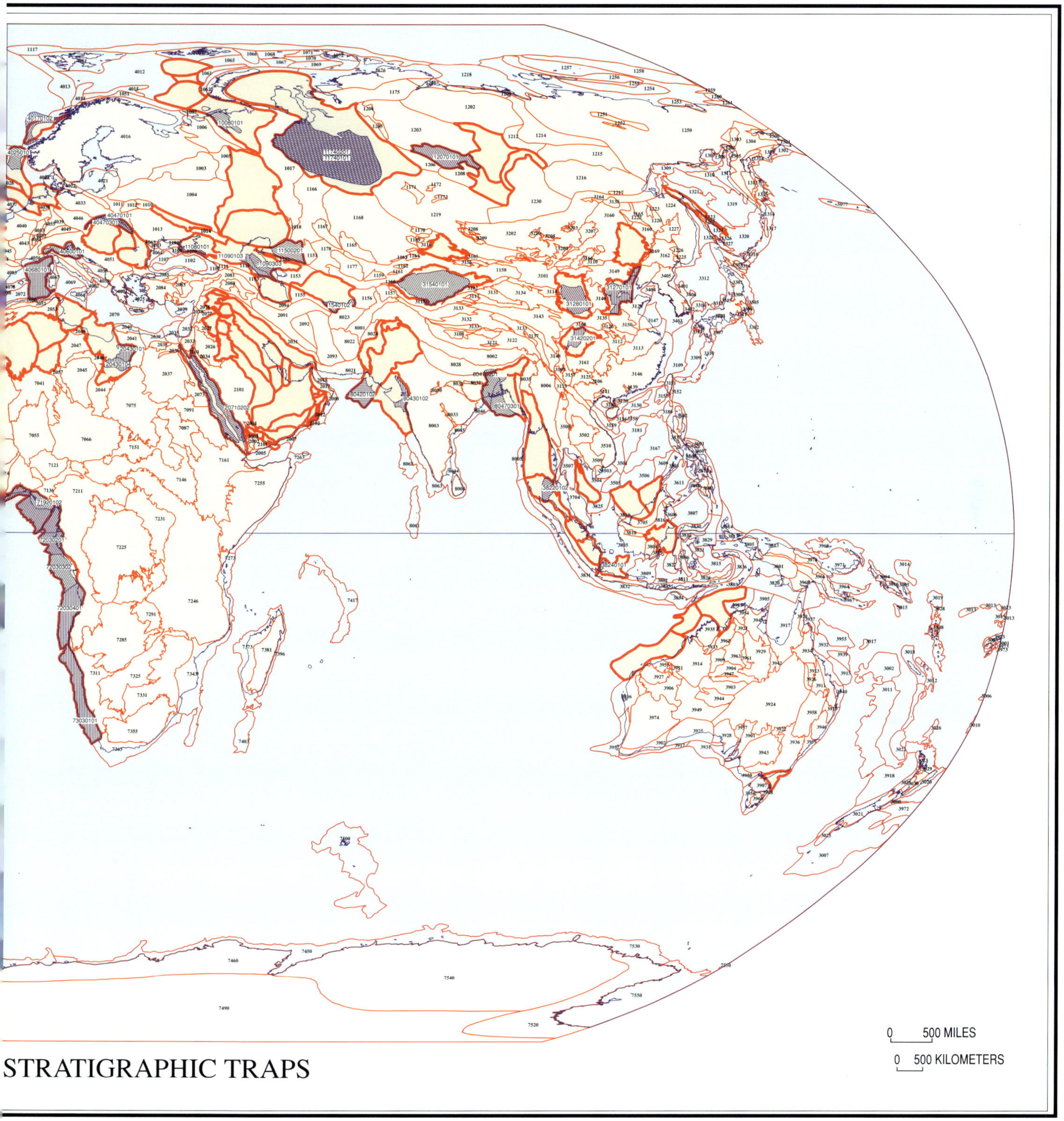

Projection Robinson
Central meridian 0

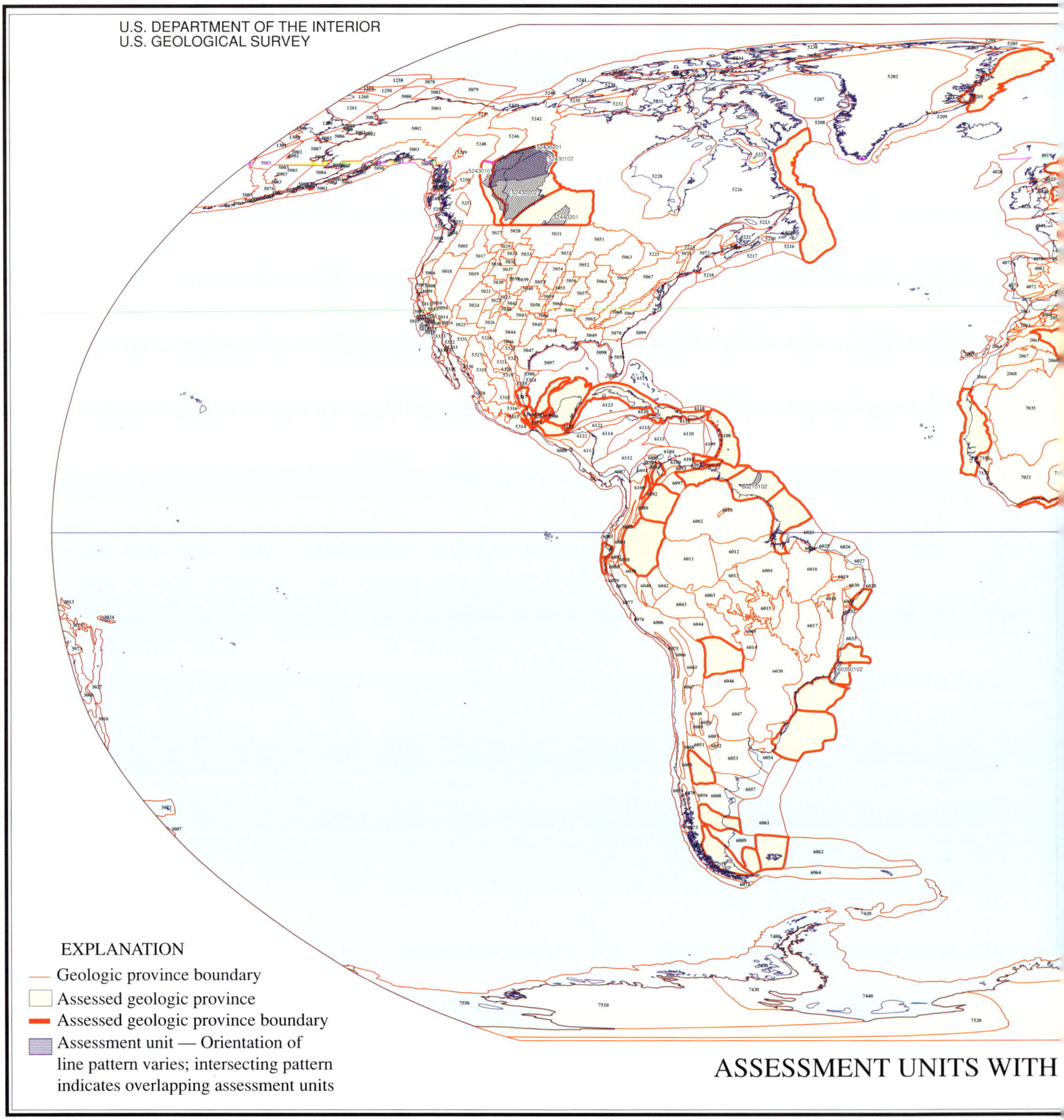

FIGURE 85. Map showing assessment units with paleogeomorphic traps.

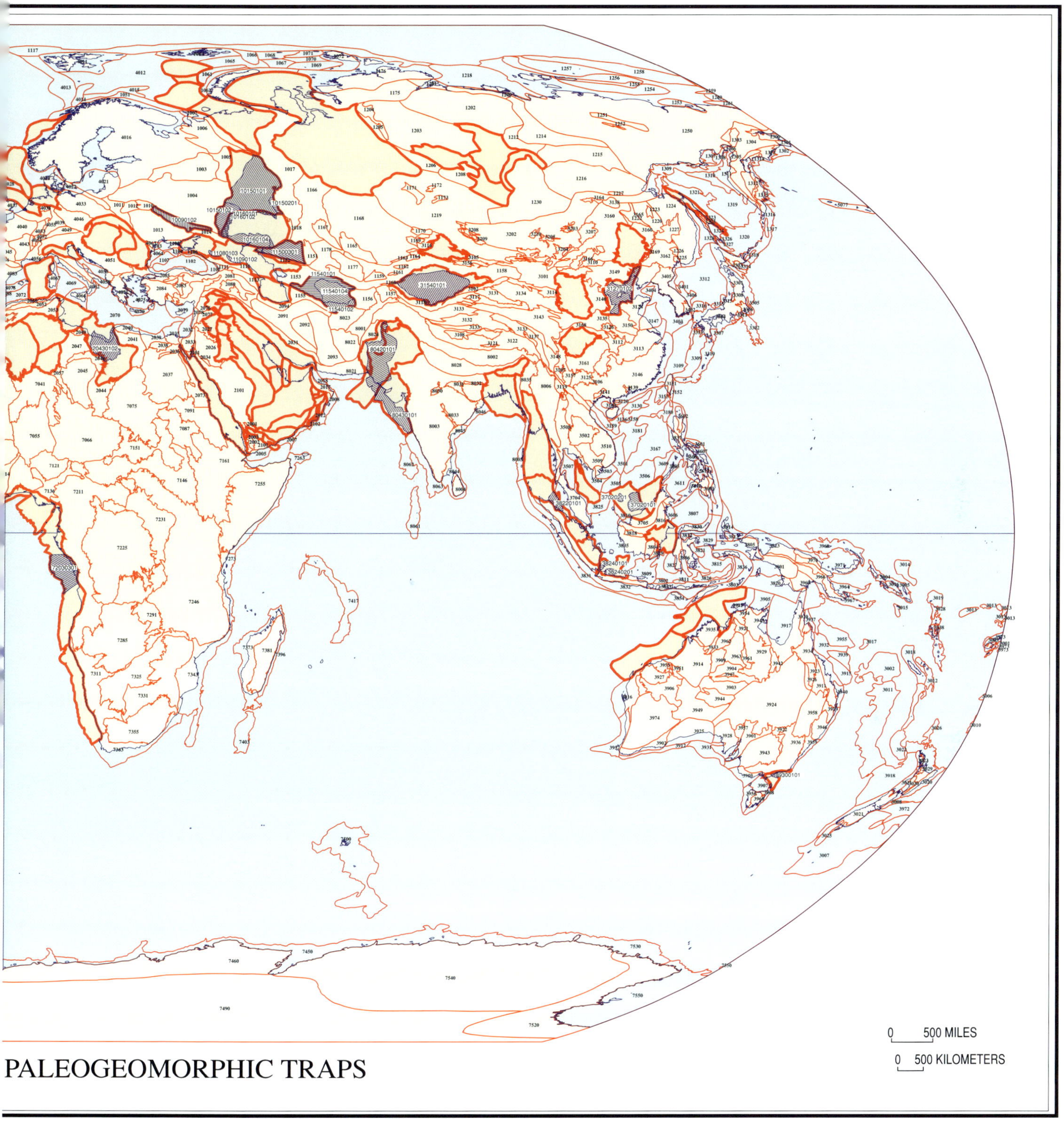

PALEOGEOMORPHIC TRAPS

Projection Robinson
Central meridian 0

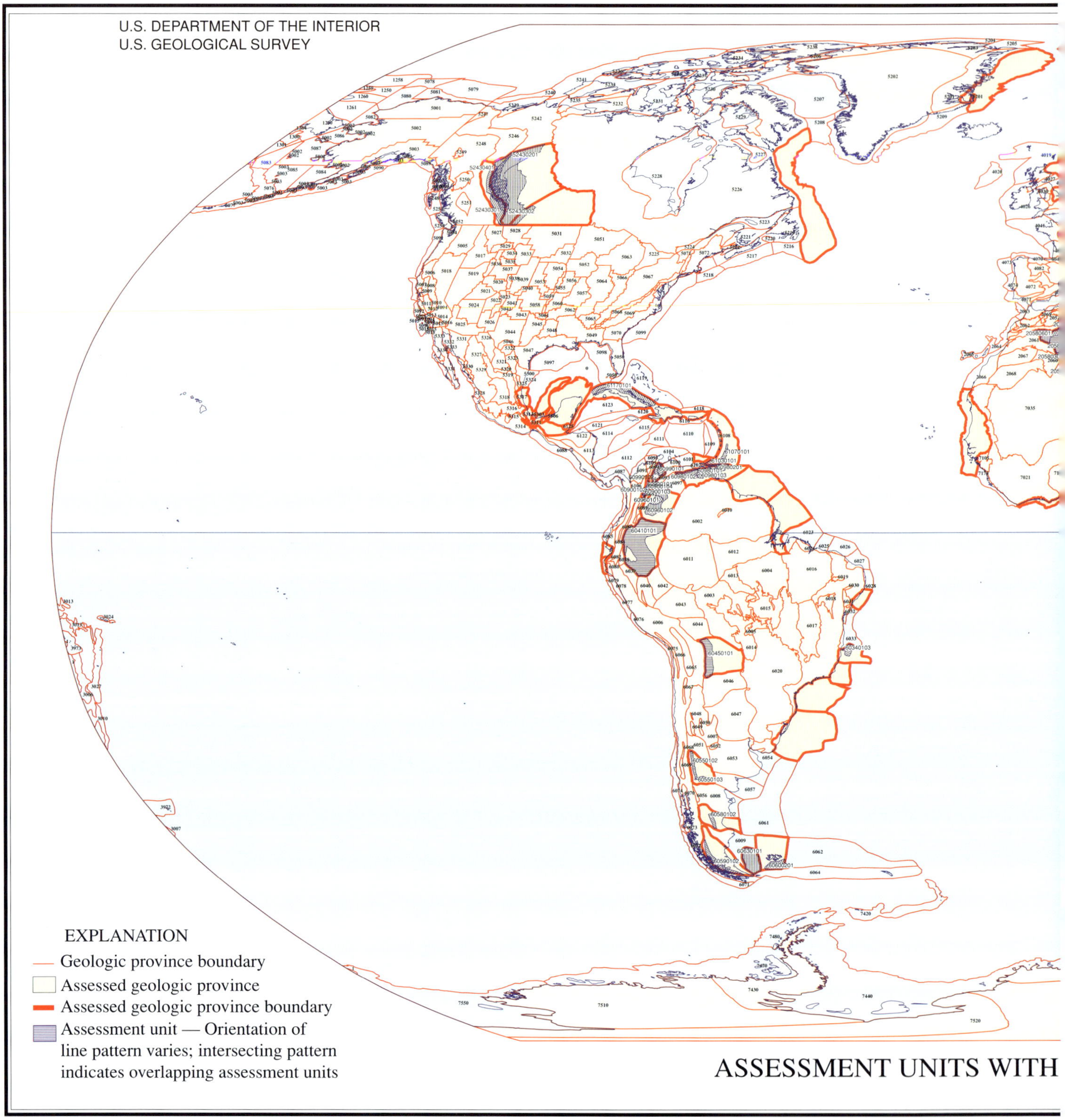

FIGURE 86. Map showing assessment units with compressional structure traps.

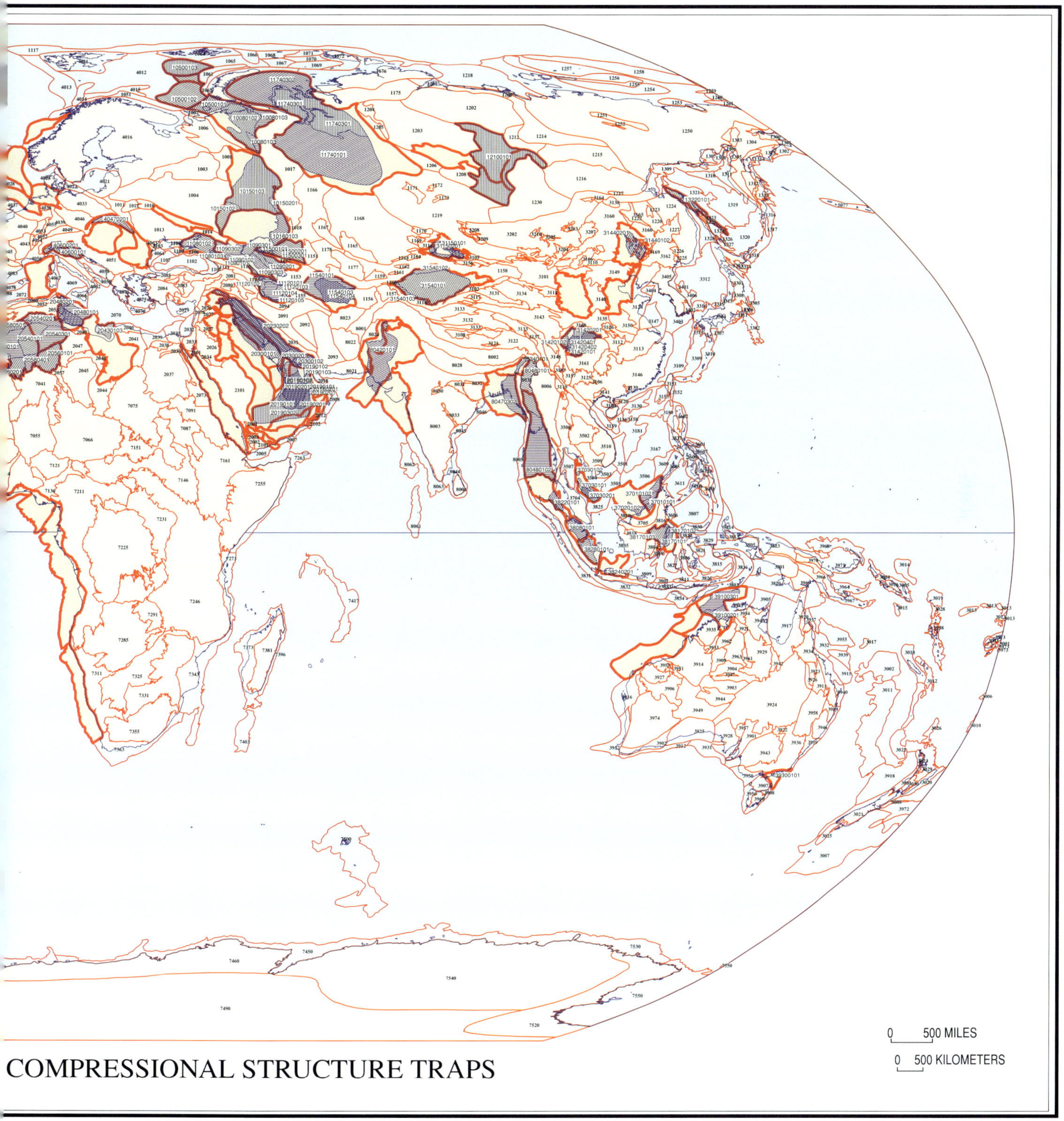

Projection Robinson
Central meridian 0

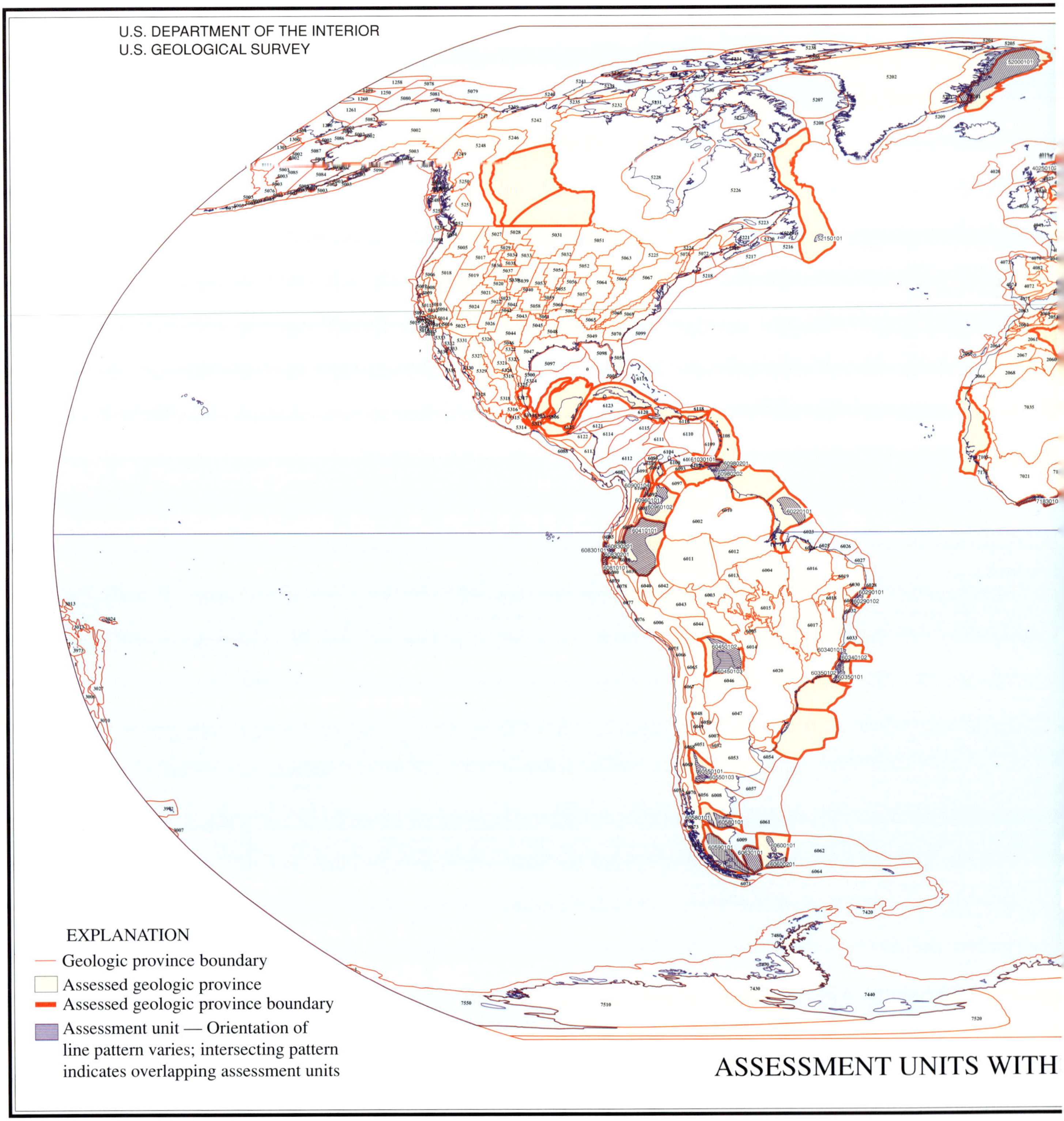

FIGURE 87. Map showing assessment units with extensional structure traps.

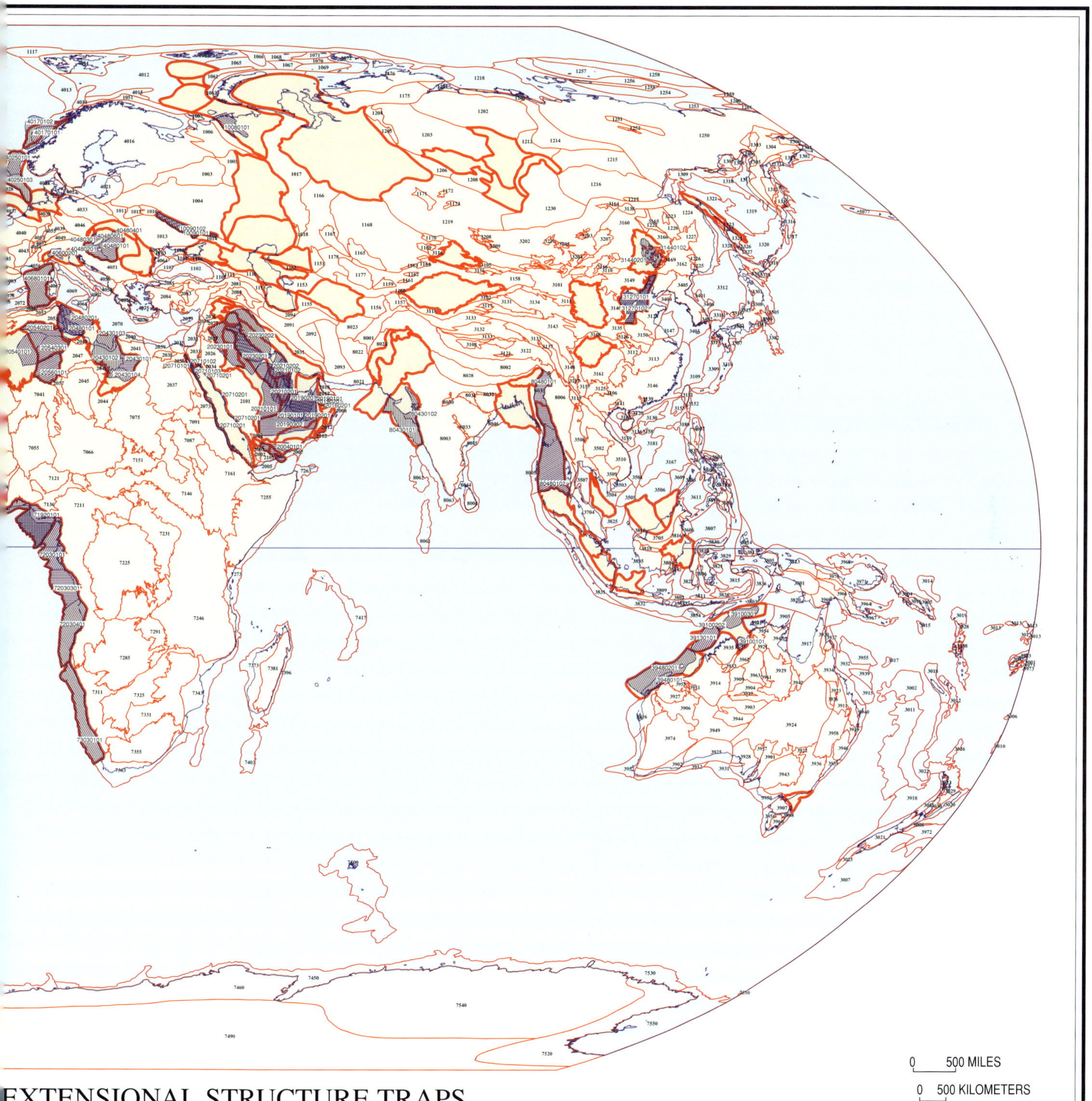

EXTENSIONAL STRUCTURE TRAPS

Projection Robinson
Central meridian 0

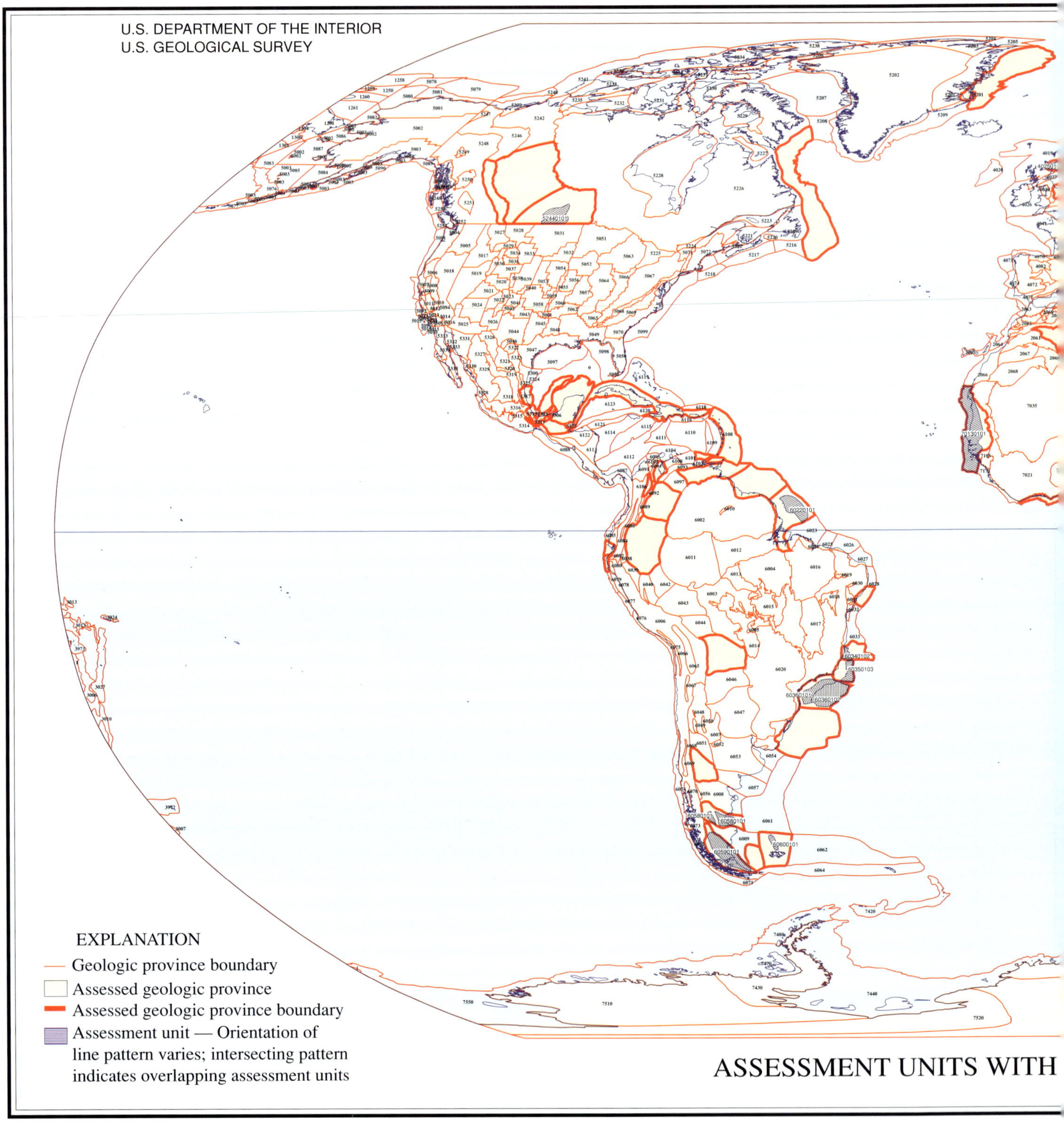

FIGURE 88. Map showing assessment units with non-tectonic-structural traps (e.g., salt, diapirs).

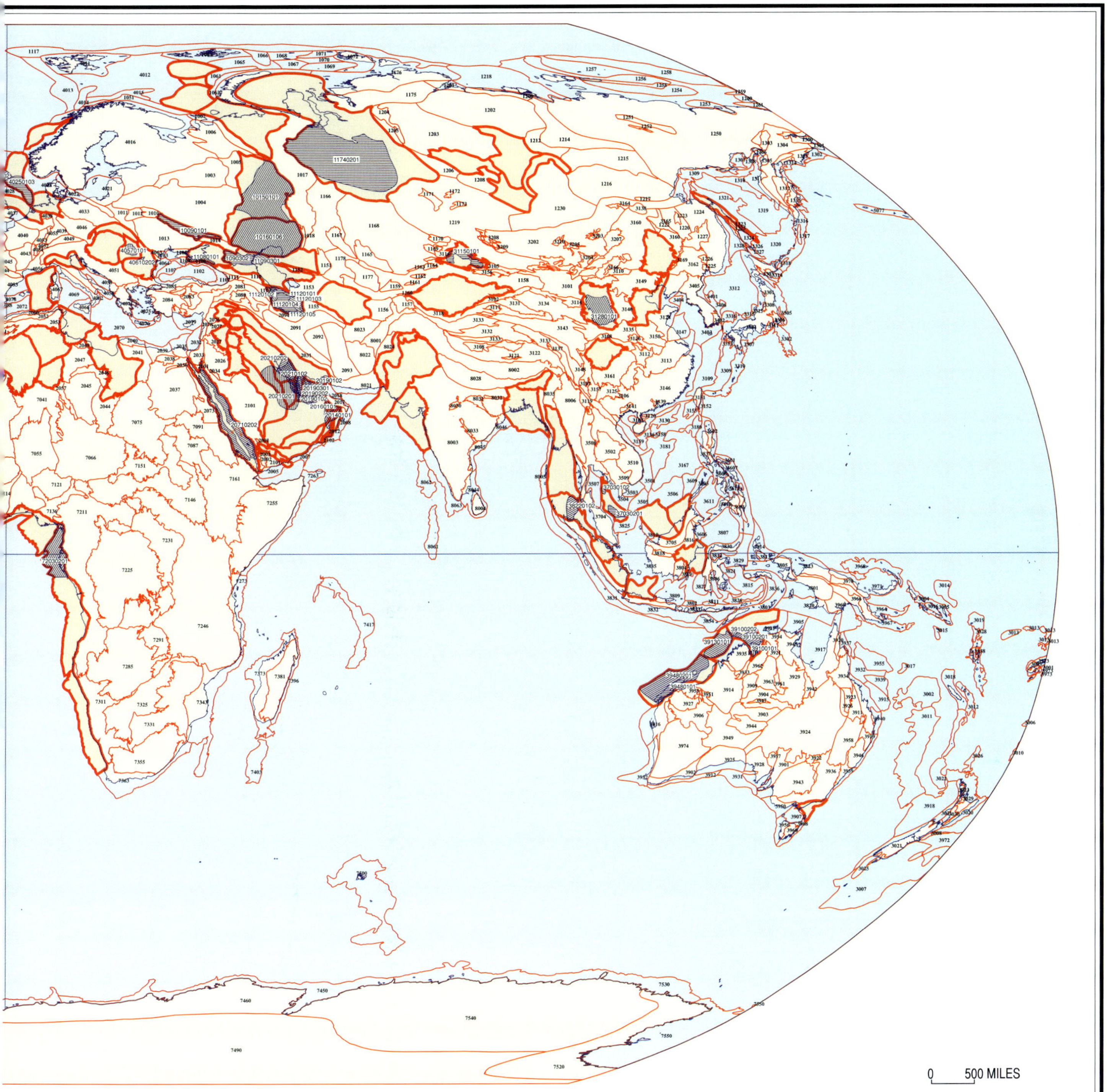

NON-TECTONIC STRUCTURAL TRAPS

0 500 MILES

0 500 KILOMETERS

Projection Robinson
Central meridian 0

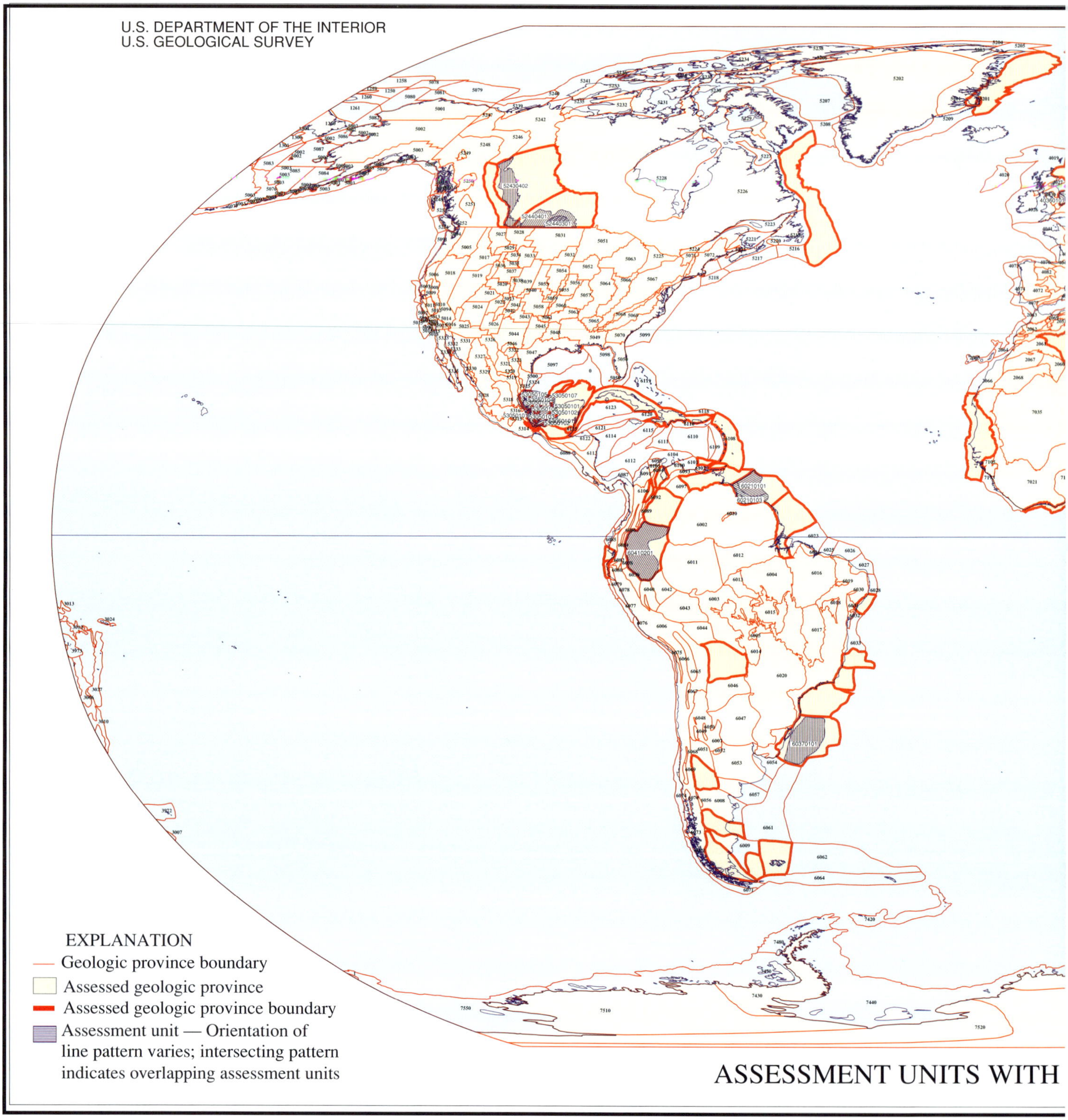

Figure 89. Map showing assessment units with structural-stratigraphic traps.

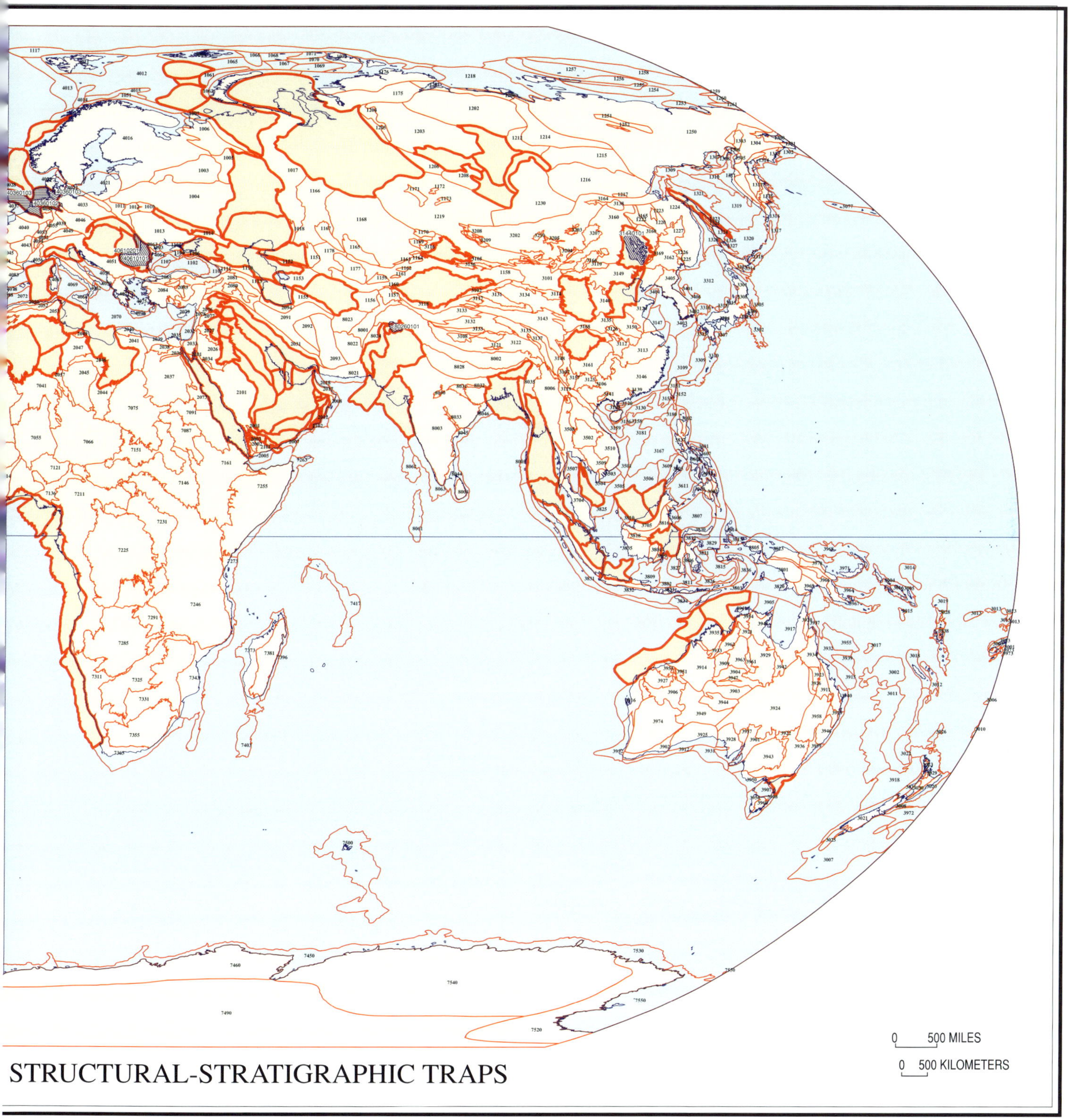

Projection Robinson
Central meridian 0

VI: Field Analyses

A. REMAINING LARGE FIELDS

To understand where the greatest potential for undiscovered resources lies, a summary of the estimated largest undiscovered fields from the USGS (2000) assessment is included here (Appendices 6, 7). Additionally, analyses of the depth, pressure, and temperature attributes of the world's oil and natural gas fields provide insights into the world distribution of these parameters.

Appendices 6, 7, 8, and 9 present information about possible locations of large, undiscovered fields in the world by assessment units (AU). Because of the extreme skewness of field-size distributions, much uncertainty exists as to the size of the largest undiscovered fields. The procedure used provides a distribution that shows the uncertainty for size of the largest undiscovered field for each AU (Charpentier and Klett, 2000). Because of high uncertainty, the statistics in Appendices 6 through 9 should be viewed for their relative, rather than absolute, value.

Appendix 6 presents a ranked list of AU by the estimated size of the largest undiscovered oil field. Appendix 7 shows a similar list for the largest undiscovered natural gas field. Appendix 8 shows the ranking of AU by number of expected oil fields with at least 1024 MMBO. Appendix 9 illustrates AU ranking by number of expected natural gas fields with at least 6144 bcf (1024 million BOE).

All four of these appendices show that many of the largest fields remaining to be discovered are expected in Region 2, the Middle East and North Africa. East Greenland is also high on the oil ranking (Appendix 6) although it is a hypothetical AU (no discovered accumulations). Although the largest expected oil and natural gas fields are of roughly comparable size, they are located geographically in very different areas and in different total petroleum systems (TPS) and AU.

B. FIELD SIZE-DEPTH-PRESSURE-TEMPERATURE RELATIONSHIPS

Analysis of exploration and production data provided by IHS Energy Group (2001) reveals a relation between estimates of recoverable hydrocarbons (cumulative production plus remaining reserves, which is used to express field size) and reservoir depth. Graphs showing this relation, as well as those between field size and reservoir temperature and field size and reservoir pressure, were constructed for fields throughout the world for which data are available (excluding U.S. and Canadian fields).

Figure 90 shows estimated recoverable hydrocarbons, as field size, plotted with respect to the depth to the top of the major producing reservoir in each of the fields. The major producing reservoirs of 98% of the fields with reported values, and all except two of the world's giant fields (those having recoverable hydrocarbon volumes greater than 5 billion BOE (0.8 billion m^3), are at depths shallower than 15,000 ft (4572 m). The major reservoirs of the largest fields (such as Ghawar, Pars South, North Field, Greater Burgan, and Urengoy) are shallower than 10,000 ft (3048 m). Field sizes are significantly smaller below 15,000 ft (4572 m).

When plotted with respect to maximum recorded temperature of the major producing reservoir (Figure 91), most of the fields, and the largest of these fields, have reservoir temperatures less than 266° F (130° C). This population represents 92% of the fields with reported values. When plotted with respect to current pressure of the major producing reservoir (Figure 92), most of the fields, and the largest of these fields, have reservoir pressures less than 12,000 pounds per square inch (psi) (827 megapas-

 DOI:10.1306/1061776M863175

cals). This population represents 99% of the fields with reported values.

The volume-depth relation is most likely attributed to increasing porosity loss by diagenesis within the reservoirs with depth (Bjørkum and Nadeau, 1998). Porosity loss is attributed to increased temperatures and pressures, both of which are depth-related. Bjørkum and Nadeau (1998) suggested that the limit of adequate porosity and the onset of hard overpressure in sandstone reservoirs of the Norwegian Continental Shelf is between 248° F (120° C) and 302° F (150° C), corresponding to depths of between 11,500 ft (3506 m) to 13,600 ft (4491 m). In addition, hard overpressure promotes the development and propagation of fractures resulting in seal rock failure and fluid leakage. Depth-maximum temperature, depth-current pressure, and maximum temperature-current pressure relations of major producing reservoirs in fields throughout the world (excluding U.S. and Canadian fields) are shown in Figures 93, 94, and 95. These graphs may be considered to represent the world's average geothermal and geopressure gradients. Note that two trends with respect to reservoir depth can be observed on the depth-current pressure graph (Figure 94), each trend representing hydrostatic and lithostatic geopressure gradients.

The temperature and depth limits for reservoir quality in sandstones of the Norwegian Continental Shelf are very similar to the field-size limits observed for fields throughout the world, irrespective of reservoir-rock type. The field size-depth relation indicates that the risk for large volumes of undiscovered hydrocarbons greatly increases beyond these limits of depth, temperature and pressure. Exploration for relatively deep accumulations is not common worldwide and the number and sizes of fields producing from deep reservoirs reported in the database used in this study may not be representative of the actual population of deep fields. Therefore, one must be cognizant of this biased distribution of shallower reservoir depths and lower reservoir temperatures and pressures of past discoveries when deriving and applying depth risks to exploration.

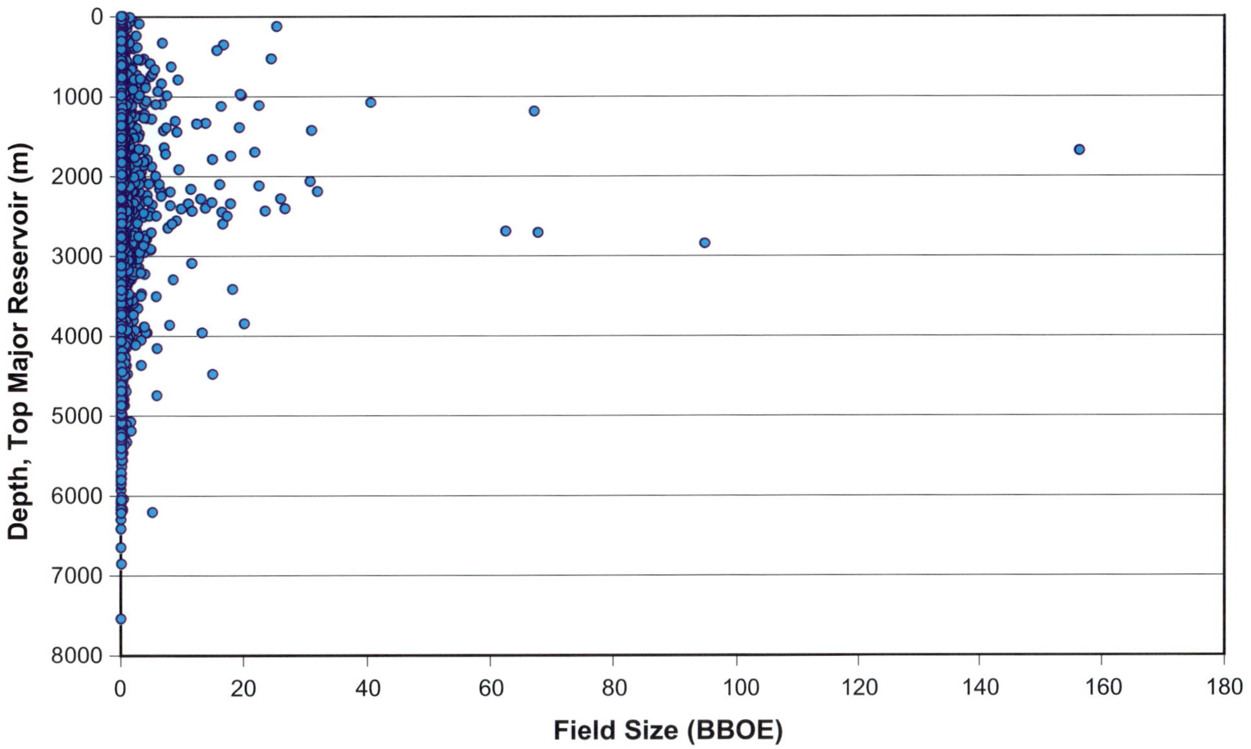

FIGURE 90. Estimated recoverable hydrocarbons, as field size in billion barrels of oil equivalent (BBOE), with respect to the depth to the top of the major producing reservoir in meters (m) in each of the world's fields (excluding U.S. and Canadian fields).

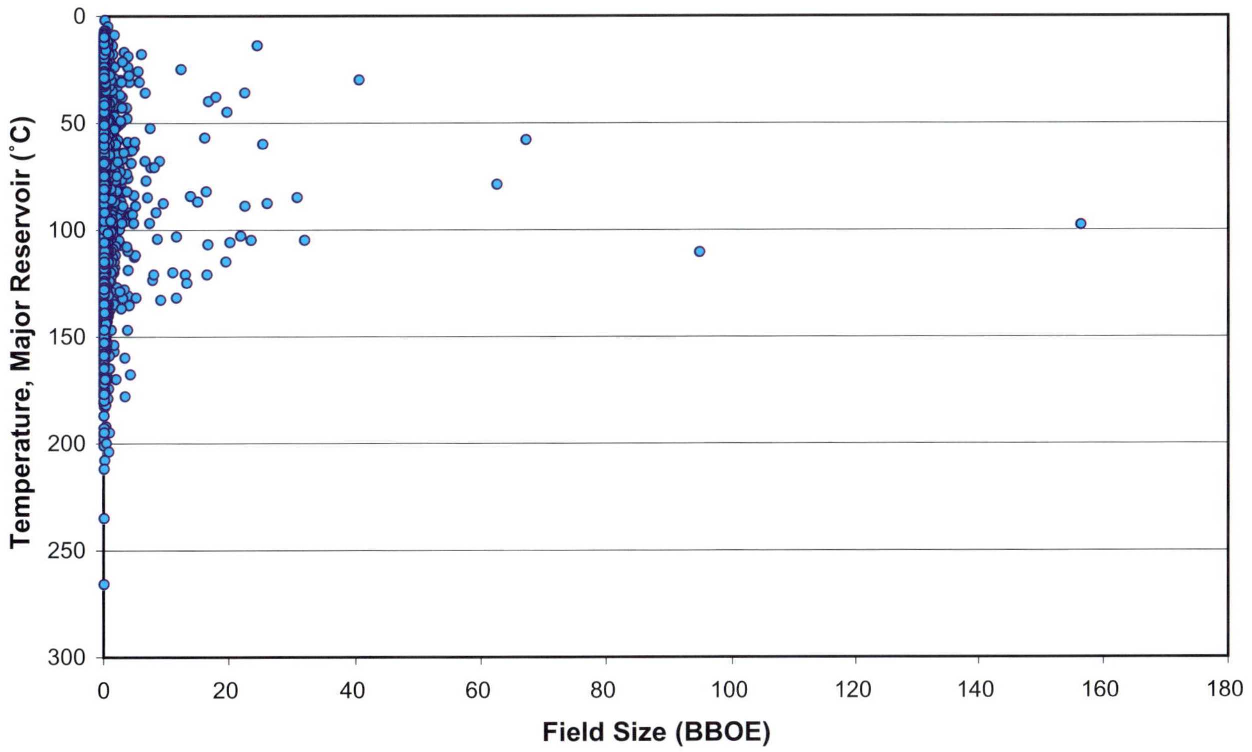

FIGURE 91. Estimated recoverable hydrocarbons, as field size in billion barrels of oil equivalent (BBOE), with respect to the maximum temperature of the major producing reservoir in degrees Centigrade (°C) in each of the world's fields (excluding U.S. and Canadian fields).

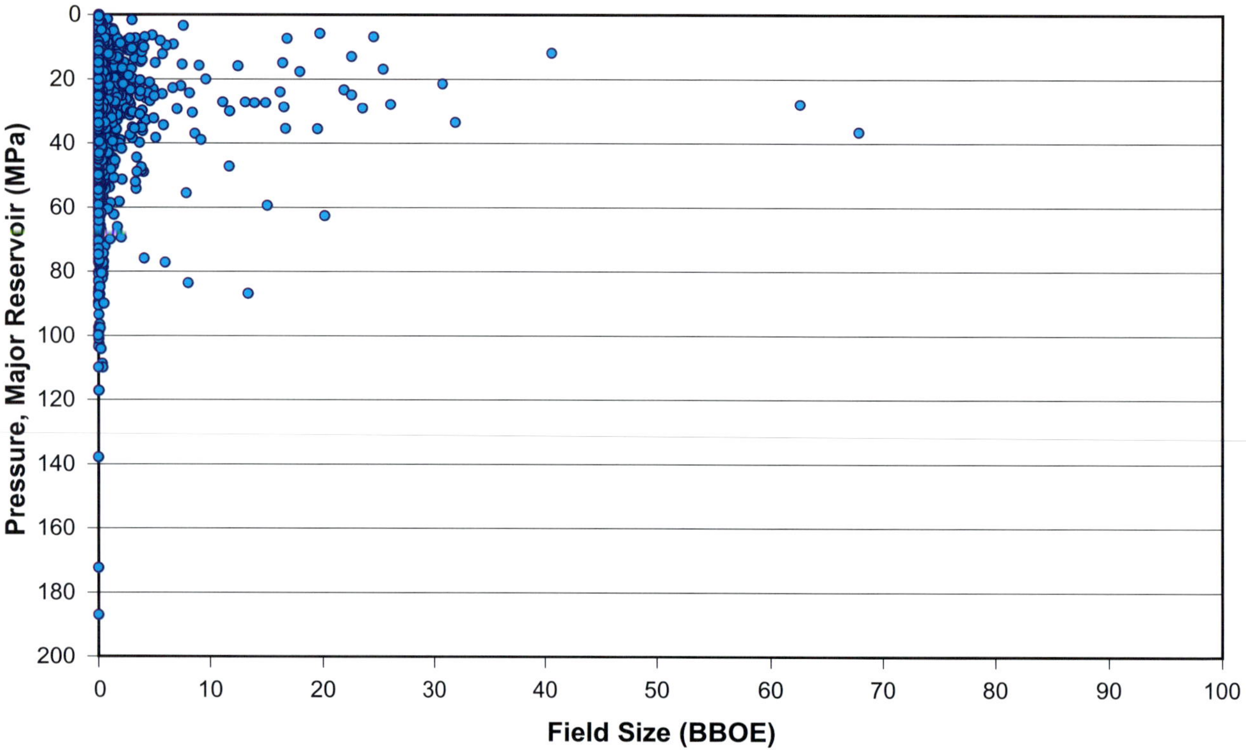

FIGURE 92. Estimated recoverable hydrocarbons, as field size in billion barrels of oil equivalent (BBOE), with respect to the current pressure of the major producing reservoir in megapascals (MPa) in each of the world's fields (excluding U.S. and Canadian fields).

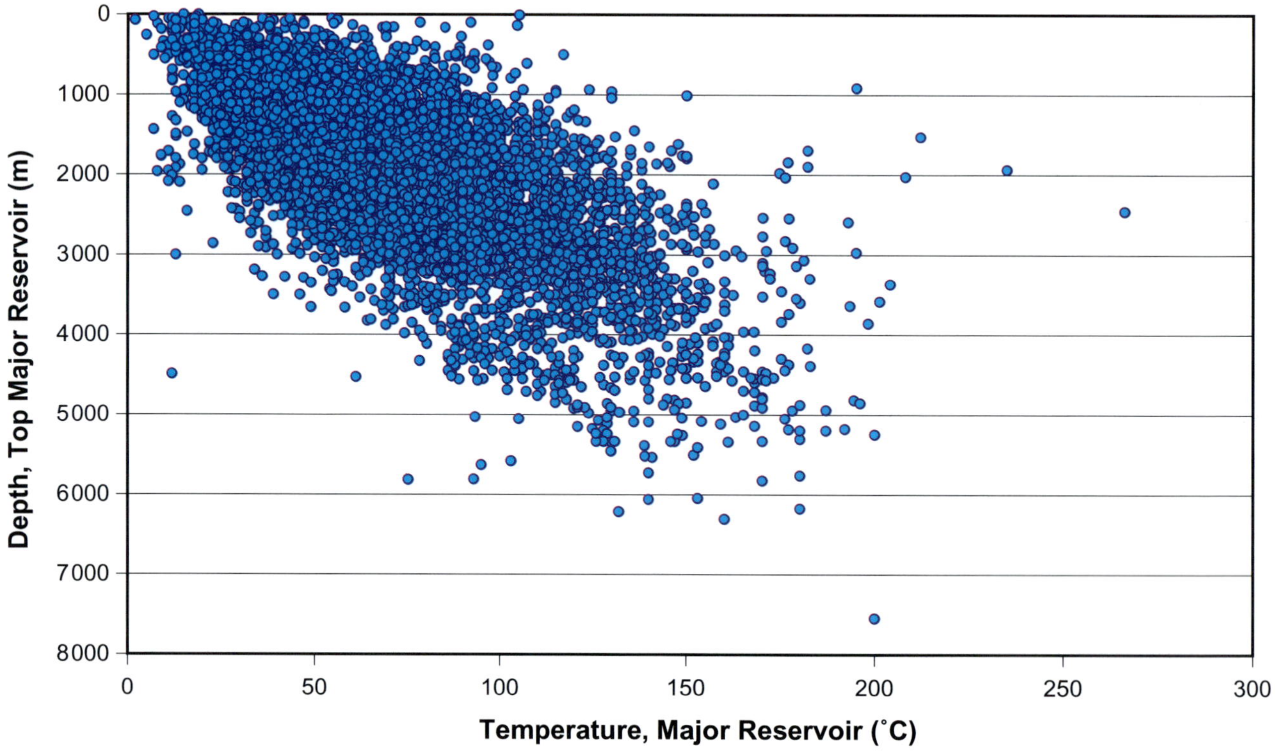

FIGURE 93. Depth-maximum temperature plot for the major producing reservoirs in fields throughout the world (excluding U.S. and Canadian fields). °C = degrees centigrade; m = meters.

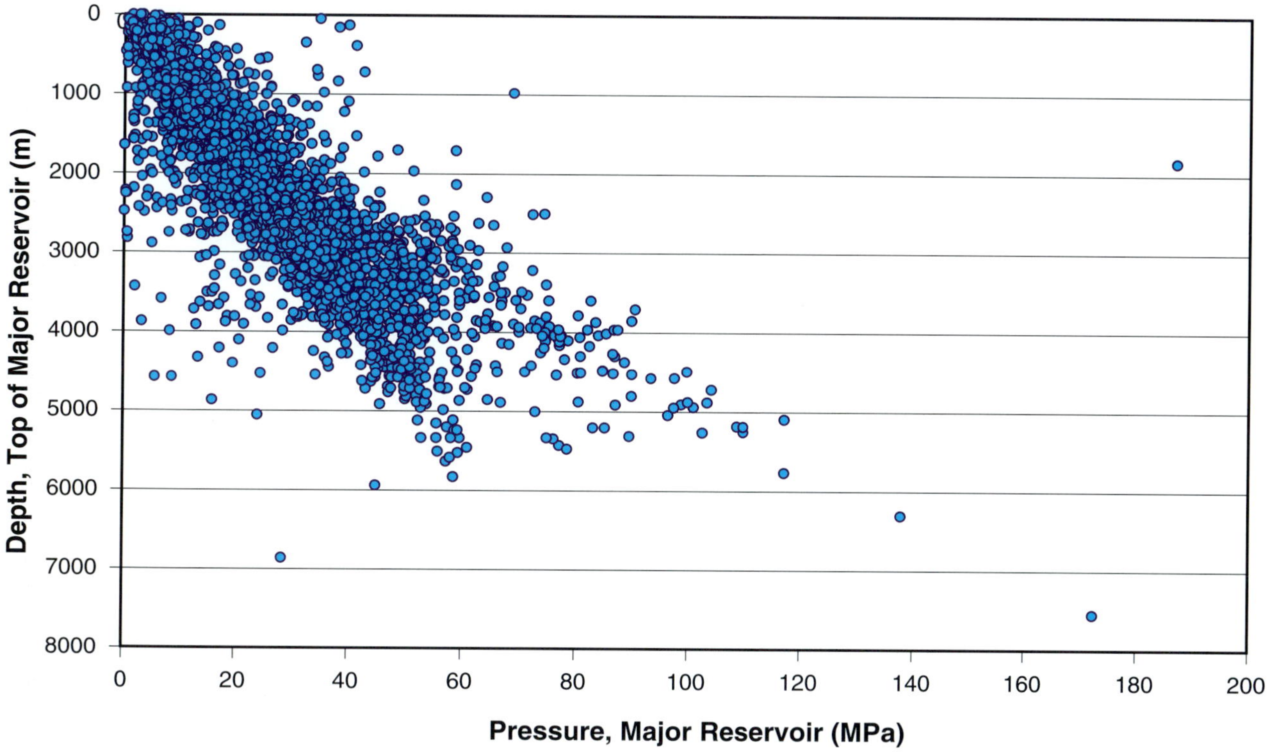

FIGURE 94. Depth-current pressure plot for the major producing reservoirs in fields throughout the world (excluding U.S. and Canadian fields). m = meters; MPa = megapascals.

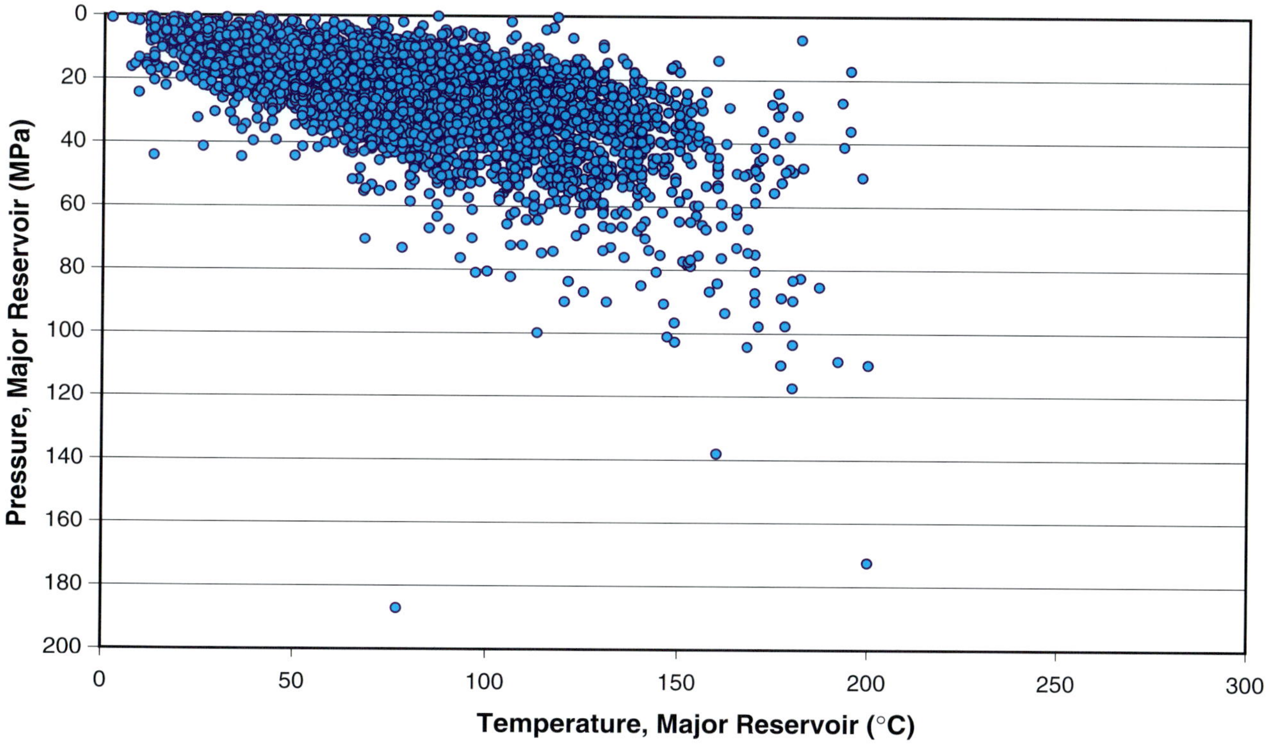

FIGURE 95. Maximum temperature-current pressure plot for the major producing reservoirs in fields throughout the world (excluding U.S. and Canadian fields). °C = degrees centigrade; MPa = megapascals.

VII: Onshore, Offshore, and Country Allocations

A. ONSHORE VERSUS OFFSHORE RESOURCES

Economics of onshore petroleum exploration and production are very different from offshore. In the assessment process, allocations of estimated undiscovered volumes were made to onshore and offshore locations (where applicable) in order to enhance the utility of the results for further economic analysis. Offshore areas were deemed those that required offshore infrastructure, in either fresh or salt water.

Estimated volumes of onshore and offshore undiscovered oil by region (excluding the U.S.) are shown in Figure 96; of undiscovered natural gas, in Figure 97; and of NGL, in Figure 98. Most of the onshore undiscovered resources are expected to be in Regions 1 (Former Soviet Union) and 2 (Middle East and North Africa). Figure 99 shows that, at the world level, approximately equal amounts of each of the three undiscovered commodities are expected to be onshore and offshore. Figure 100 shows that for Regions 3 through 8, about one-fourth of the undiscovered volume of each commodity is expected to be onshore and nearly three-quarters of the potential is offshore. The offshore potential of each assessment unit (AU), total petroleum system (TPS), and province and the water depths associated with the offshore portion for each AU, is documented in USGS (2000). Most offshore provinces extend to water depths of 6562 ft (2000 m) with the exception of areas where exploration has already begun in deeper water depths. For example, the water depth limits of the AU were extended to 13,123 ft (4000 m) in the Campos Basin and other deepwater provinces along the South Atlantic margins.

B. PETROLEUM VOLUMES BY COUNTRY

Undiscovered petroleum volumes allocated to countries are shown in Appendix 10. The appendix is divided into countries belonging to the Organization of Petroleum Exporting Countries (OPEC), the Organisation for Economic Co-operation and Development (OECD), and other nations. Volumes cited in this appendix and used for the figures in this section do not necessarily represent the entire country. Rather, these volumes represent assessed AU in the country. Note that volumes for the entire U.S. (Gautier et al., 1996; MMS, 1996) have been added to OECD volumes in Figures 101–106.

Known (discovered) and estimated undiscovered oil, natural gas, and NGL volumes are presented for OPEC, OECD, and remaining countries of the world in Figures 101, 102 and 103, respectively. Note that these figures do not include estimates for reserve growth. Regional allocations of reserve growth have recently been made as reported here, but this reserve growth allocation was not done for individual countries. The undiscovered resources in OPEC are primarily onshore, whereas the undiscovered resources for OECD and other countries are primarily offshore (Figures 104, 105, 106, respectively).

 DOI:10.1306/1061777M863175

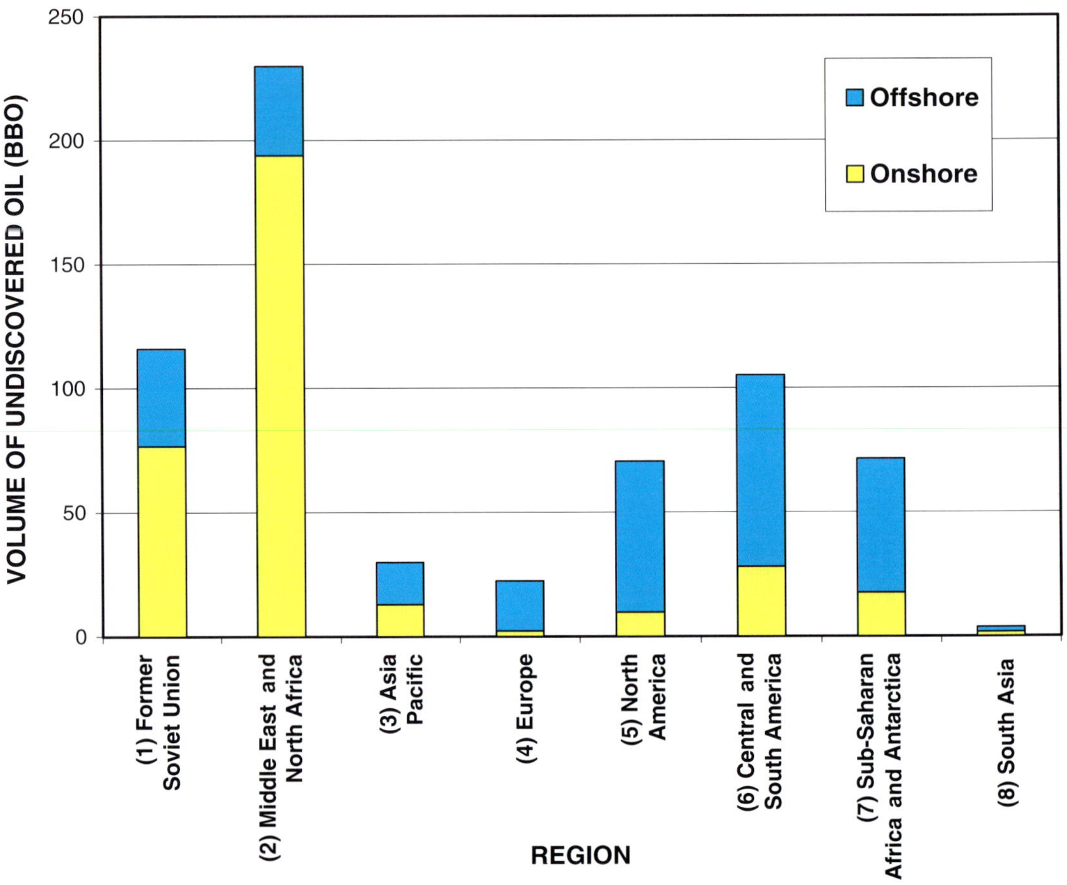

FIGURE 96. Onshore and offshore estimated mean volumes of undiscovered oil by region (excluding the U.S.). BBO = billion barrels of oil.

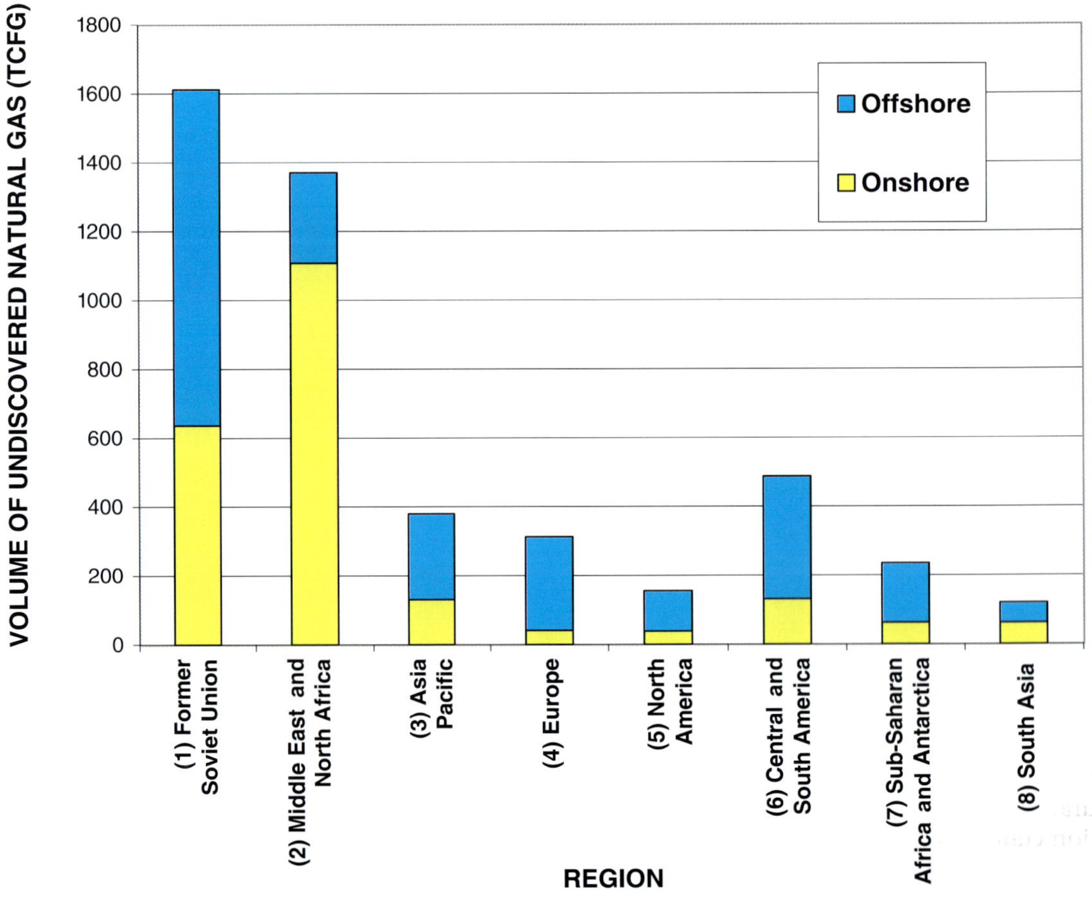

FIGURE 97. Onshore and offshore estimated mean volumes of undiscovered natural gas by region (excluding the U.S.). TCFG = trillion cubic feet of natural gas.

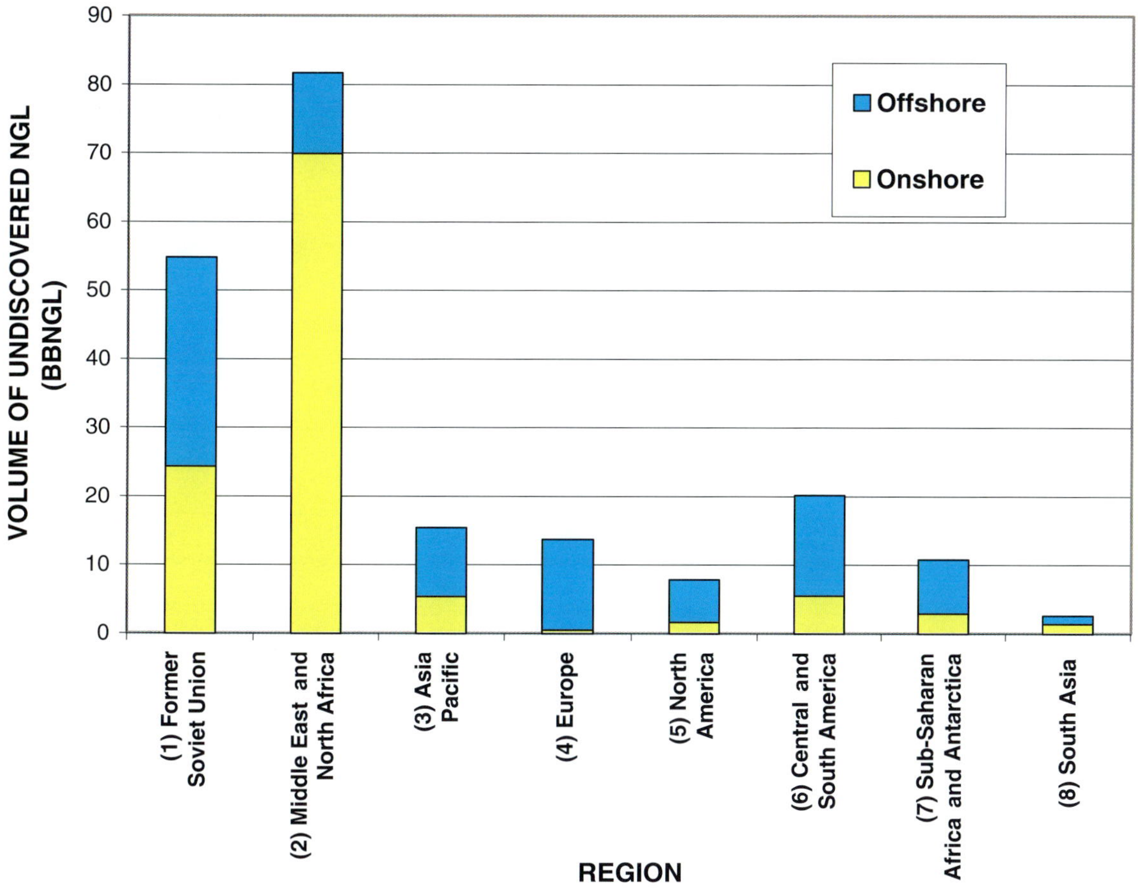

FIGURE 98. Onshore and offshore estimated mean volumes of undiscovered natural gas liquids (NGL) by region (excluding the U.S.). BBNGL = billion barrels of NGL.

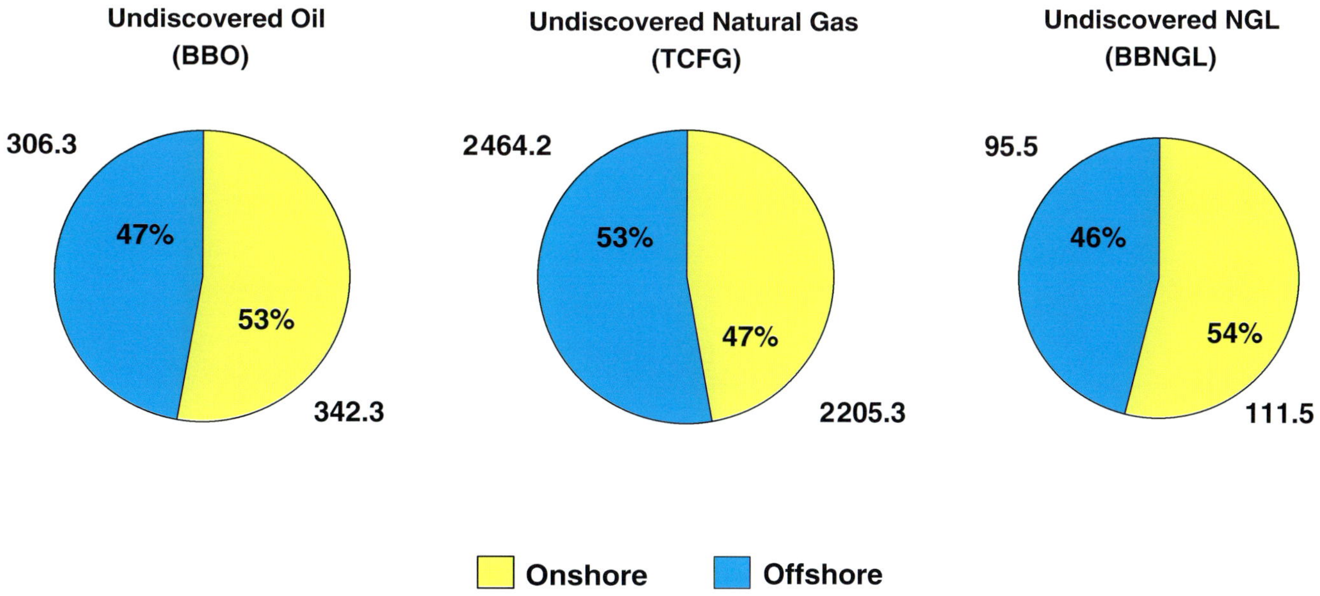

FIGURE 99. Plots comparing the onshore and offshore estimated volumes of undiscovered oil, natural gas, and natural gas liquids (NGL) for the assessed parts of the world (excluding the U.S.). BBO = billion barrels of oil; TCFG = trillion cubic feet of natural gas; BBNGL = billion barrels of NGL.

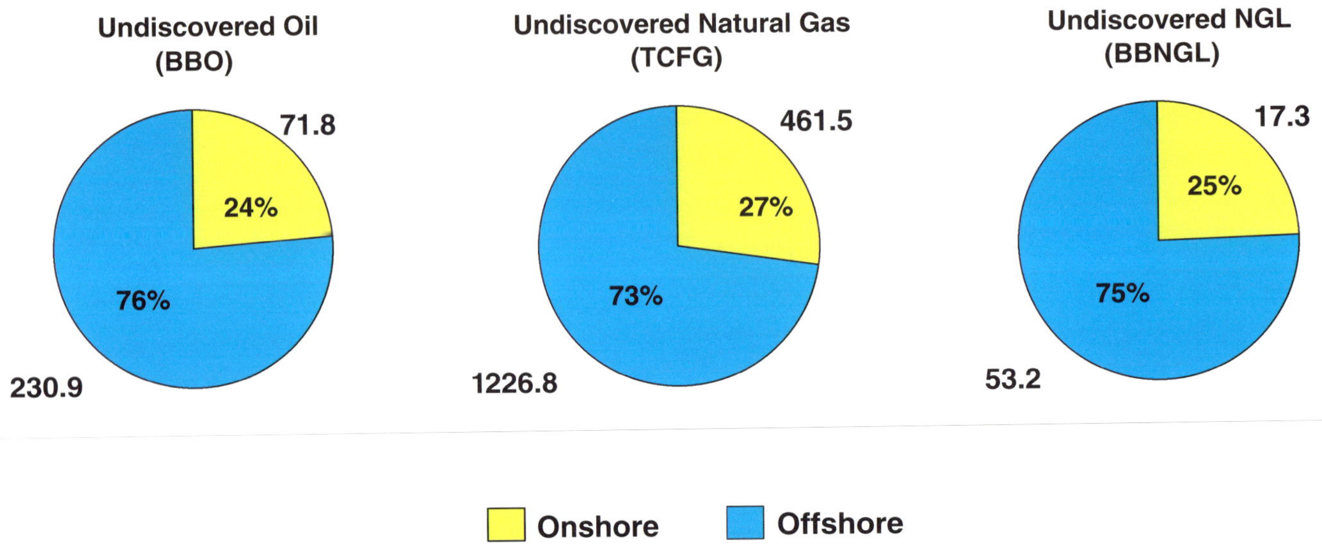

FIGURE 100. Plots comparing the onshore and offshore estimated volumes of undiscovered oil, natural gas, and natural gas liquids (NGL) for the assessed parts of Regions 3 through 8 (excluding the U.S.). BBO = billion barrels of oil; TCFG = trillion cubic feet of natural gas; BBNGL = billion barrels of NGL.

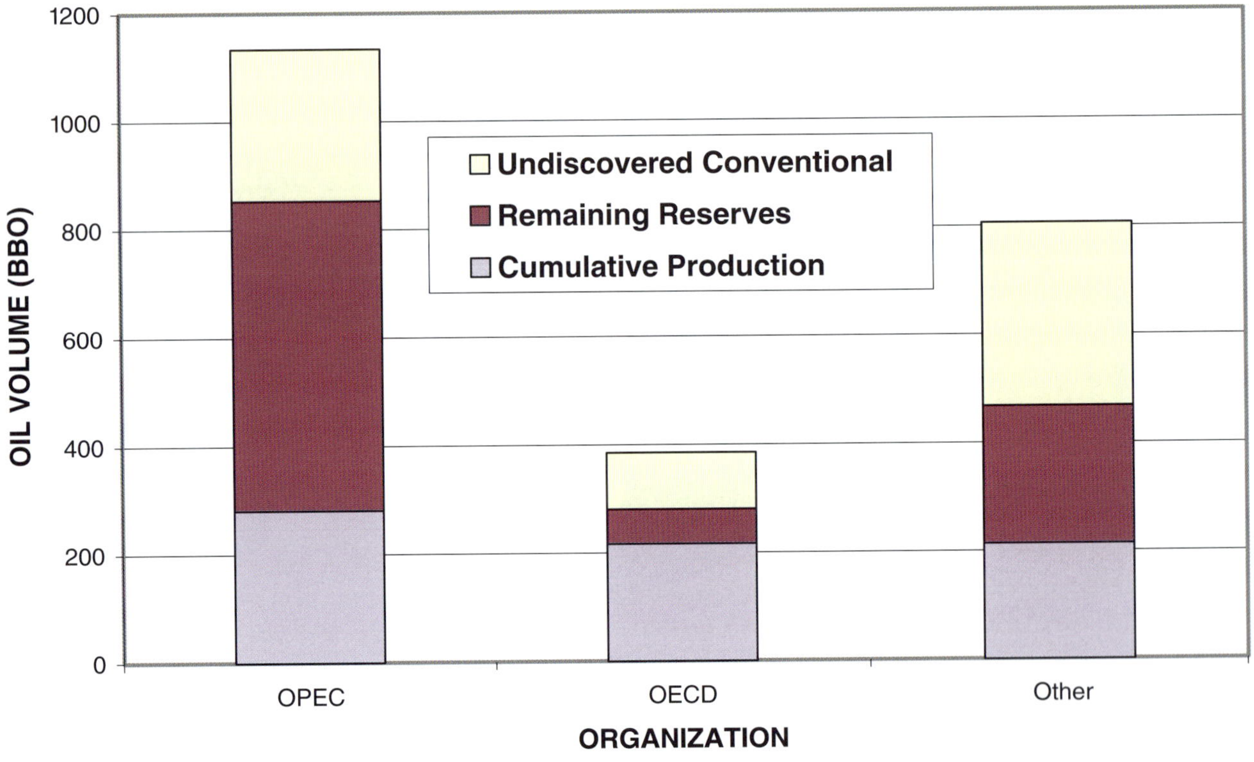

FIGURE 101. Volumes of cumulative production, remaining reserves, and estimated undiscovered oil for the assessed parts of the Organization of the Petroleum Exporting Countries (OPEC); the Organisation for Economic Co-operation and Development (OECD), which includes U.S. volumes; and the remaining countries of the world. BBO = billion barrels of oil.

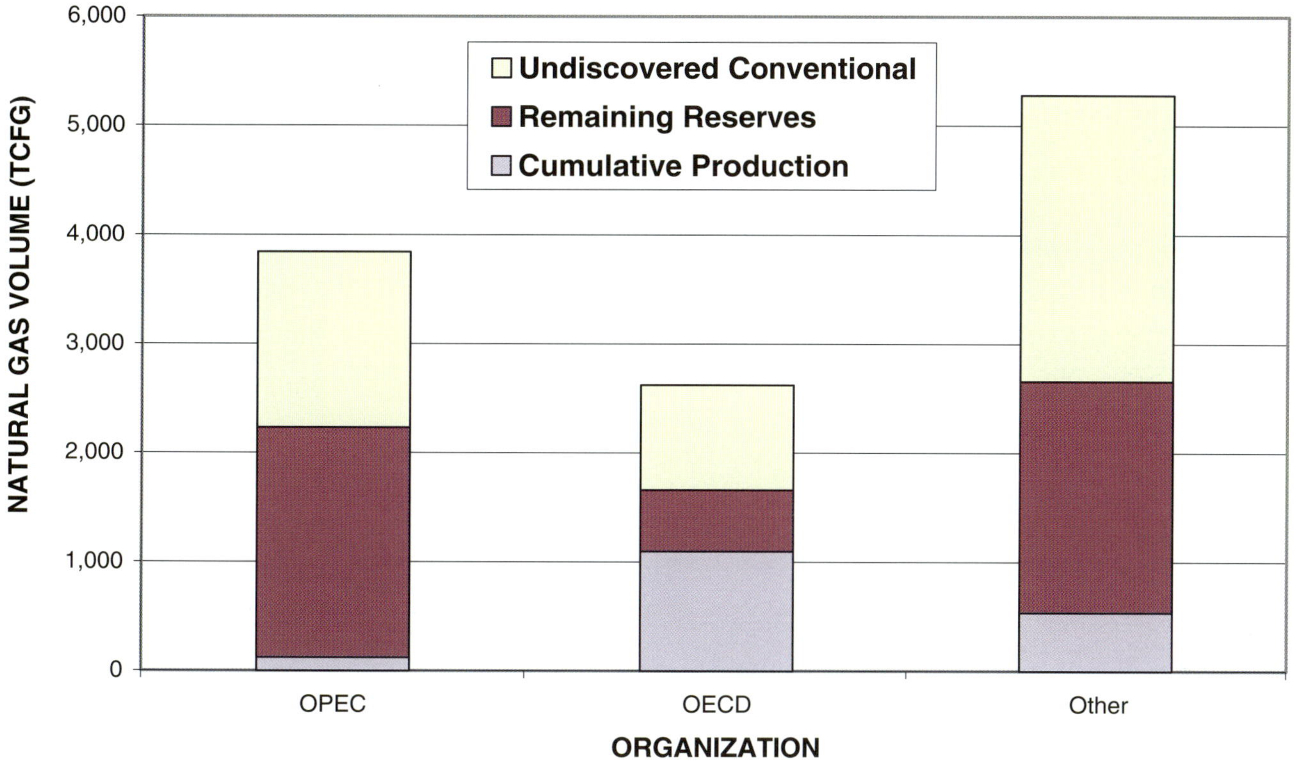

Figure 102. Volumes of cumulative production, remaining reserves, and estimated undiscovered natural gas for the assessed parts of the Organization of the Petroleum Exporting Countries (OPEC); the Organisation for Economic Co-operation and Development (OECD), which includes U.S. volumes; and the remaining countries of the world. TCFG = trillion cubic feet of natural gas.

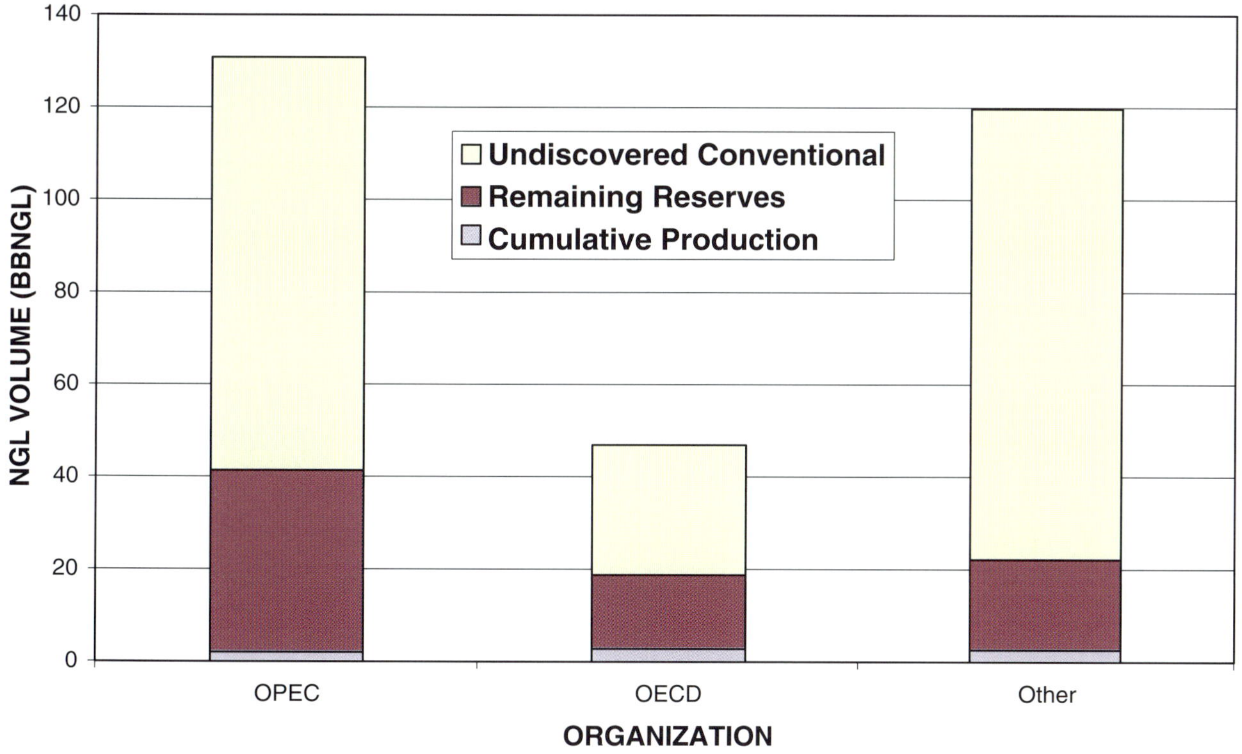

Figure 103. Volumes of cumulative production, remaining reserves, and estimated undiscovered natural gas liquids (NGL) for the assessed parts of the Organization of the Petroleum Exporting Countries (OPEC); the Organisation for Economic Co-operation and Development (OECD), which includes U.S. volumes; and the remaining countries of the world. BBNGL = million barrels of NGL.

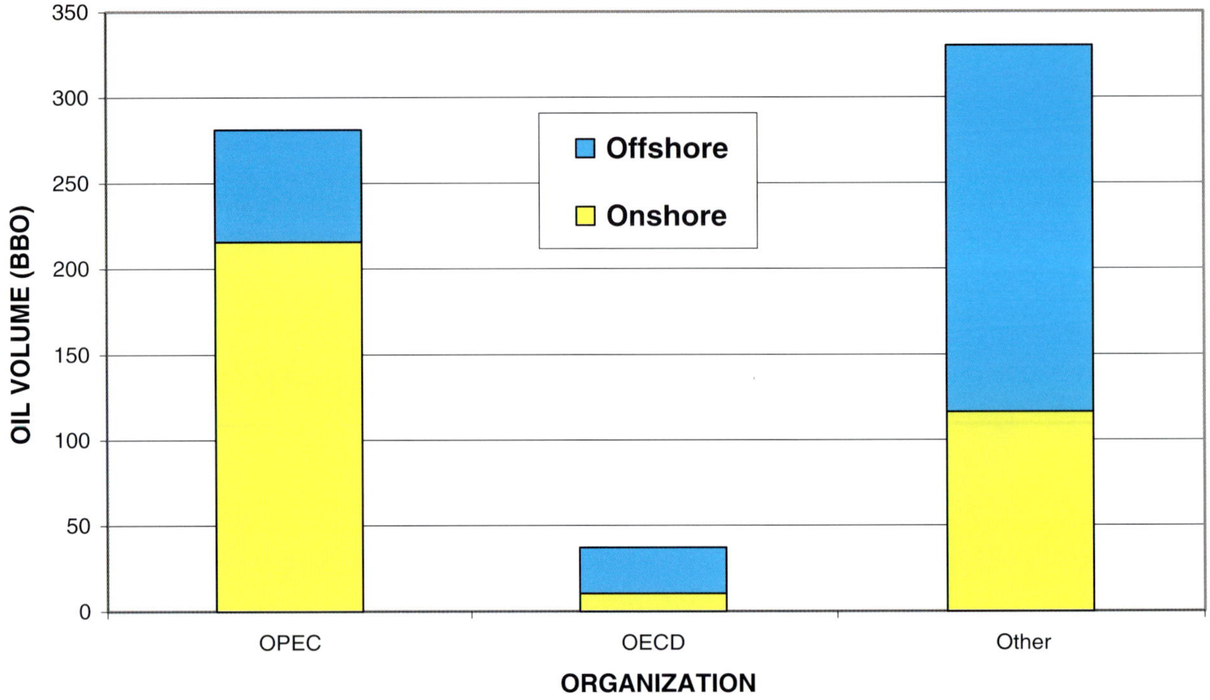

FIGURE 104. Onshore and offshore estimated volumes of undiscovered oil by the assessed parts of the Organization of the Petroleum Exporting Countries (OPEC); the Organisation for Economic Co-operation and Development (OECD), which includes U.S. volumes; and the remaining countries of the world. BBO = million barrels of oil.

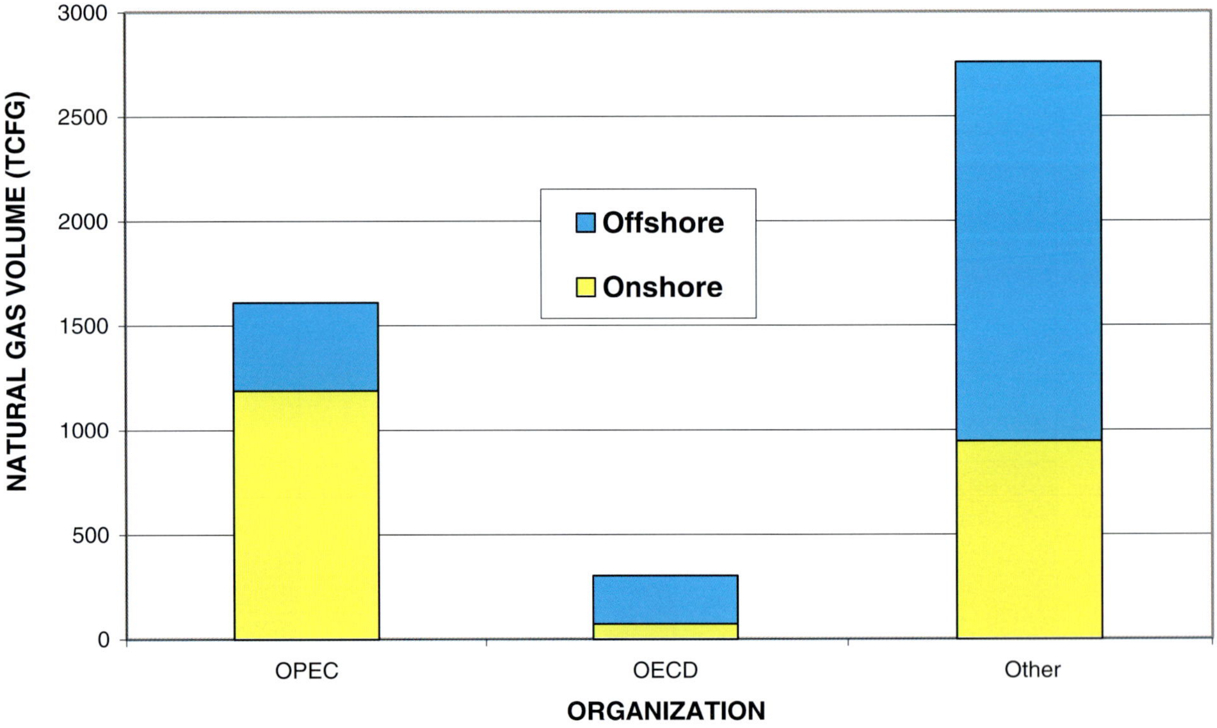

FIGURE 105. Onshore and offshore estimated volumes of undiscovered natural gas by the assessed parts of the Organization of the Petroleum Exporting Countries (OPEC); the Organisation for Economic Co-operation and Development (OECD), which includes U.S. volumes; and the remaining countries of the world. TCFG = trillion cubic feet of natural gas.

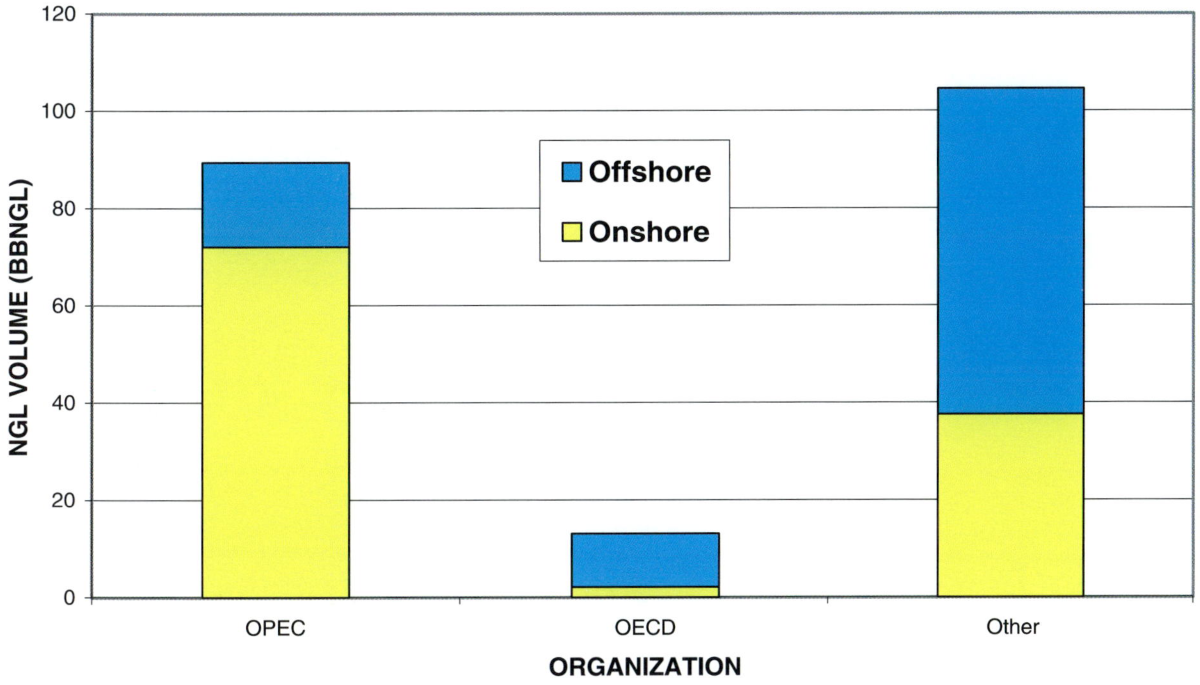

FIGURE 106. Onshore and offshore estimated volumes of undiscovered natural gas liquids (NGL) by the assessed parts of the Organization of the Petroleum Exporting Countries (OPEC); the Organisation for Economic Cooperation and Development (OECD), which includes U.S. volumes; and the remaining countries of the world. BBNGL = million barrels of NGL.

VIII: Discussion

The previous discussion documented the geologic basis and resources volumes estimated in the USGS (2000) assessment. In this section the assessment will be compared within a broader context of other assessments, and its utilization by organizations for energy planning. Uncertainty and related probability are associated with any estimate, including USGS (2000); a range of estimates are given for petroleum resources in Table 1 for those who prefer a conservative (F95), median (F50), or optimistic estimate (F5) instead of the mean estimate. Of this assessed quantity at the mean, 710 billion bbl of oil (23.5%) of oil resource has been consumed, 891 billion bbl of oil (29.5%) are remaining reserves, 688 billion bbl of oil (22.8%) are potential reserve growth, and 732 billion bbl of oil (24.2%) of the oil endowment remains undiscovered as of January 1, 1996 (Figure 107).

Therefore, the grown endowment of recoverable oil is estimated in the USGS (2000) at about 3 trillion bbl of oil (Table 1; Figure 107). The grown natural gas endowment is estimated at about 2.6 trillion BOE (Table 1; Figure 107). Oil reserves as of 12/2001 are 1.1 trillion bbl of oil; world consumption is about 0.028 trillion bbl of oil per year. Natural gas reserves are about 0.8 trillion BOE; world consumption is about 0.014 trillion BOE. Without any additional discoveries of oil, natural gas, or natural gas liquids, there are about 2 trillion BOE of proved petroleum reserves. Of the oil and natural gas grown endowment of about 5.6 trillion BOE, the world has consumed about 1 trillion BOE, or 17% leaving about 83% of the grown endowment to be found and/or utilized (Figures 12, 13, 107; 1 trillion bbl of oil = 1000 billion bbl of oil, 1 trillion BOE = 1000 billion BOE).

The potential additions to reserves from reserve growth are estimated to be nearly as large as the estimated undiscovered resource volumes. The additional remaining reserves and the estimates for reserve growth from these known fields imply that 76% of the world's grown conventional oil endowment (Figures 12, 107) and 66.3% of the world's conventional grown natural gas endowment have already been discovered in the areas assessed (Figures 13, 107). For these areas, 23.5% of the world's total conventional grown oil endowment and 11.4% of the world's total conventional grown natural gas endowment had been produced by the end of 1995 (Figures 12, 13).

For sake of comparison, the overall grown petroleum endowment volume, estimated in the USGS (2000) assessment at the mean, is 5.91 trillion BOE (Figure 107). This estimate is quite comparable to Schollnberger (1998) of 5.79 trillion BOE (produced, proven, field growth, and undiscovered resources) and low compared to the estimates from various sources as shown in Figure 108. For oil volumes alone, the USGS (2000) mean grown endowment estimates are comparable to many estimates of the world as shown in Figure 108. For natural gas volumes, the USGS (2000) natural gas estimates of the world are comparable if lower than other recent estimates (Figure 108).

The USGS (2000) estimates are also used by a variety of groups or individuals estimating the so-called "peak" (EIA, 2001; Shell, 2001; IEA, 2001, 2002; Cavallo, 2002; Greene et al., 2003; Williams, 2003). One example with 12 different scenarios of demand growth and resource estimate scenarios using the USGS (2000) estimates is given by the Energy Information Administration (EIA, 2001b; Williams, 2003) in Figure 109.

These scenarios include only conventional resources; the utilization of unconventional oil and natural gas resources will augment the resource base and likely slow production decline. Unconventional oil resources, such as the heavy oils of Orinoco or tar sands of Canada, are large (some estimates suggest recoverable resources in the hundreds of billions of bbl of oil for each) and increasingly viable sources of petroleum as evidenced by the recent

 DOI:10.1306/1061778M863175

inclusion of 175 billion bbl of oil of Canadian (Athabasca) tar sands in world recoverable reserves. However, these resources will likely be more expensive to develop than conventional accumulations. Dyman et al. (2002) have shown that 17% of the natural gas resource estimated in the USGS (2000) assessment will occur at depths of 15,000 ft (4.5 km) or greater. Supplies of unconventional natural gas from basin-center gas, fractured shale and low permeability reservoirs, coal-bed methane, and gas hydrates are significant and largely undeveloped throughout the world. These resources were identified in the USGS (2000) report but were not formally assessed. They are believed to represent a significant, but yet-to-be developed, source of natural gas (Law and Spencer, 1993; Law et al., 1998).

A. NATURAL GAS

While debates about the relative remaining amounts of petroleum continue, a transition to energy sources other than oil will occur (Nakicenovic, 1993; McCabe, 1998; Odell, 1998; Ausubel, 1998, 2000, 2001; Ausubel and Marchetti, 2001). In contrast to oil (Figures 12, 107), only about 11% of the grown endowment of natural gas of the world has been produced whereas fully one-third of the grown endowment is undiscovered (Figures 13, 107). With the development of technologies to convert natural gas to a liquid, a large petroleum liquid resource will become available to the world, helping to offset crude oil production decline. In the first half of the 21st century, petroleum resources will be increasingly developed offshore, where more than 50% of future undiscovered petroleum is estimated to occur, and then the next great frontier will probably be the Arctic regions (above 65° N latitude, where about 25% of estimated undiscovered petroleum resources occur for those provinces assessed in USGS [2000]). Because both offshore and Arctic areas can be environmentally hostile and sensitive, considerable technologic and international cooperation will be required to access these natural gas resources.

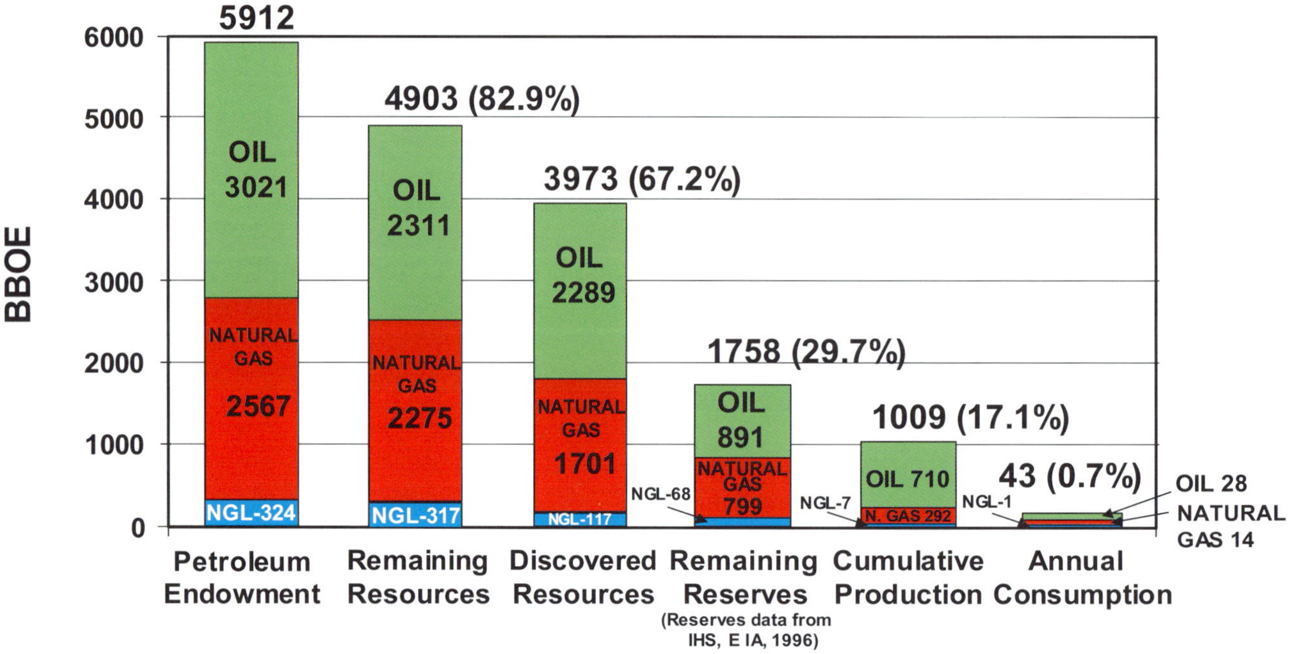

FIGURE 107. World conventional grown petroleum endowment in BBOE as assessed in the USGS (2000). Data represent 128 provinces (exclusive of the United States) as reported in USGS (2000); United States estimates are those of USGS (1995) and MMS (1996). Components of this endowment are shown in the four bars to the right of the grown petroleum endowment, and include cumulative production, remaining reserves, reserve growth, and undiscovered resources at the mean for oil, natural gas, and natural gas liquids (NGL) summed to totals. Remaining resources is the endowment less cumulative production. Discovered resources is the endowment less the undiscovered resources. Remaining reserves is the endowment less cumulative production, reserve growth, and undiscovered resources. Cumulative production reflects production as of January 1, 1996. Annual consumption represents consumption worldwide as reported in EIA (2001b) (from Ahlbrandt, 2002). BBOE = million barrels of oil equivalent.

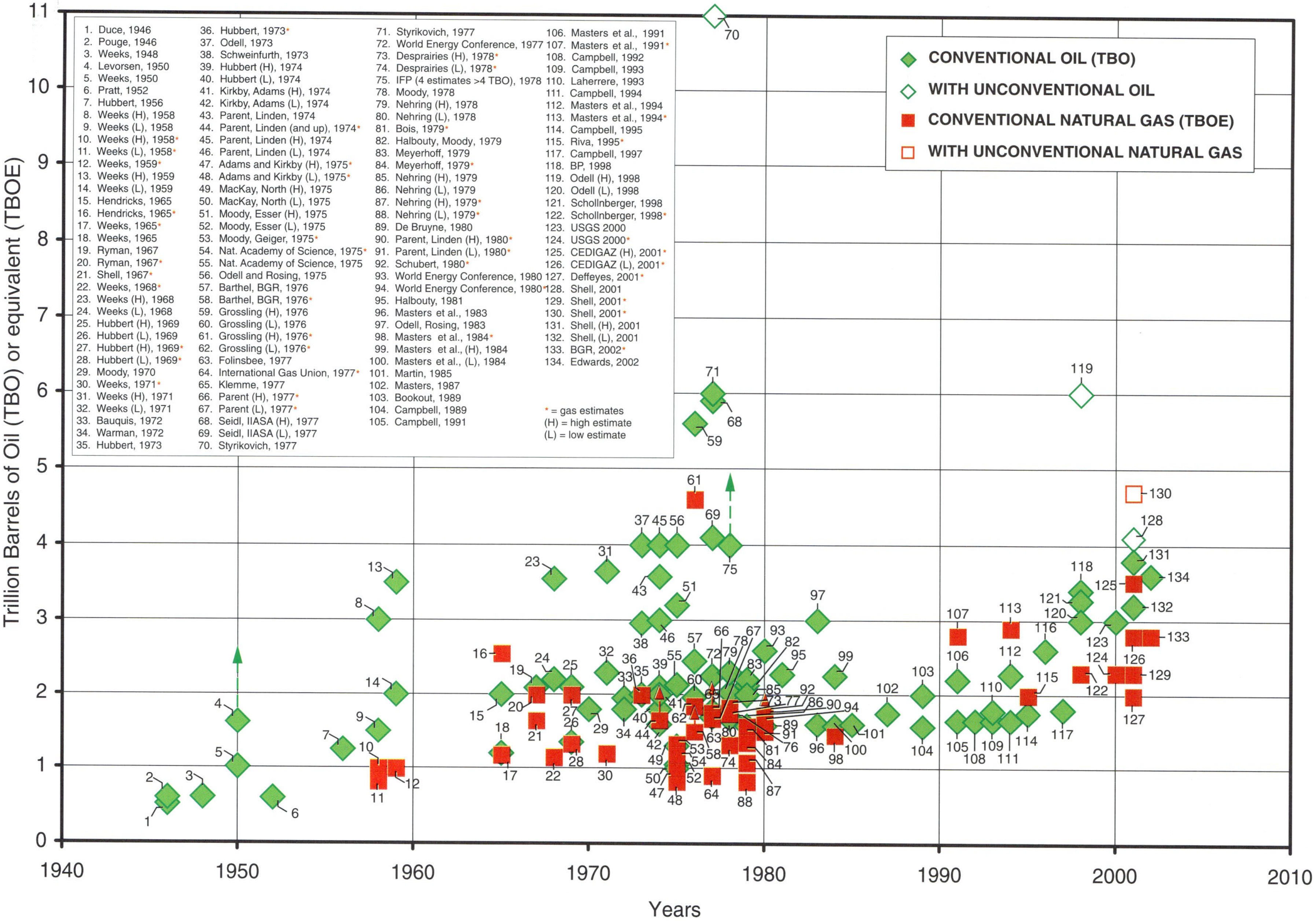

FIGURE 108. Comparison of world oil and natural gas resource endowment estimates. Oil estimates and sources are in trillion barrels of oil (TBO); natural gas estimates are in trillion barrels of oil equivalent (TBOE).

B. NORTH AMERICAN NATURAL GAS SUPPLY CONCERNS

Real resource concerns exist for the United States, particularly relative to the supply of natural gas from Canada and Mexico. Although overall, the U.S. imports much more natural gas than it exports, the United States is increasing natural gas exports to Mexico, now up to 140 bcf per year as well as 158 bcf to Canada (EIA, 2001b). For the U.S., concerns about decreasing production of oil and increasing demands for natural gas coupled with lowered estimates for undiscovered natural gas in the Alberta Basin of Canada as well as for future Mexican oil supplies, are real. For example, the USGS estimates for technically recoverable, conventional natural gas resources in the southern Canadian basins (Alberta Basin and Foothills) are much lower, in some cases by an order of magnitude, than other natural gas assessments of the region (Henry and Charpentier, 2001; National Petroleum Council, 2003). Although many energy scenarios call for increased natural gas consumption in the United States, the U.S. will need considerable additions to imports of both natural gas and liquefied natural gas or extraordinary new domestic natural gas discoveries to provide this new supply.

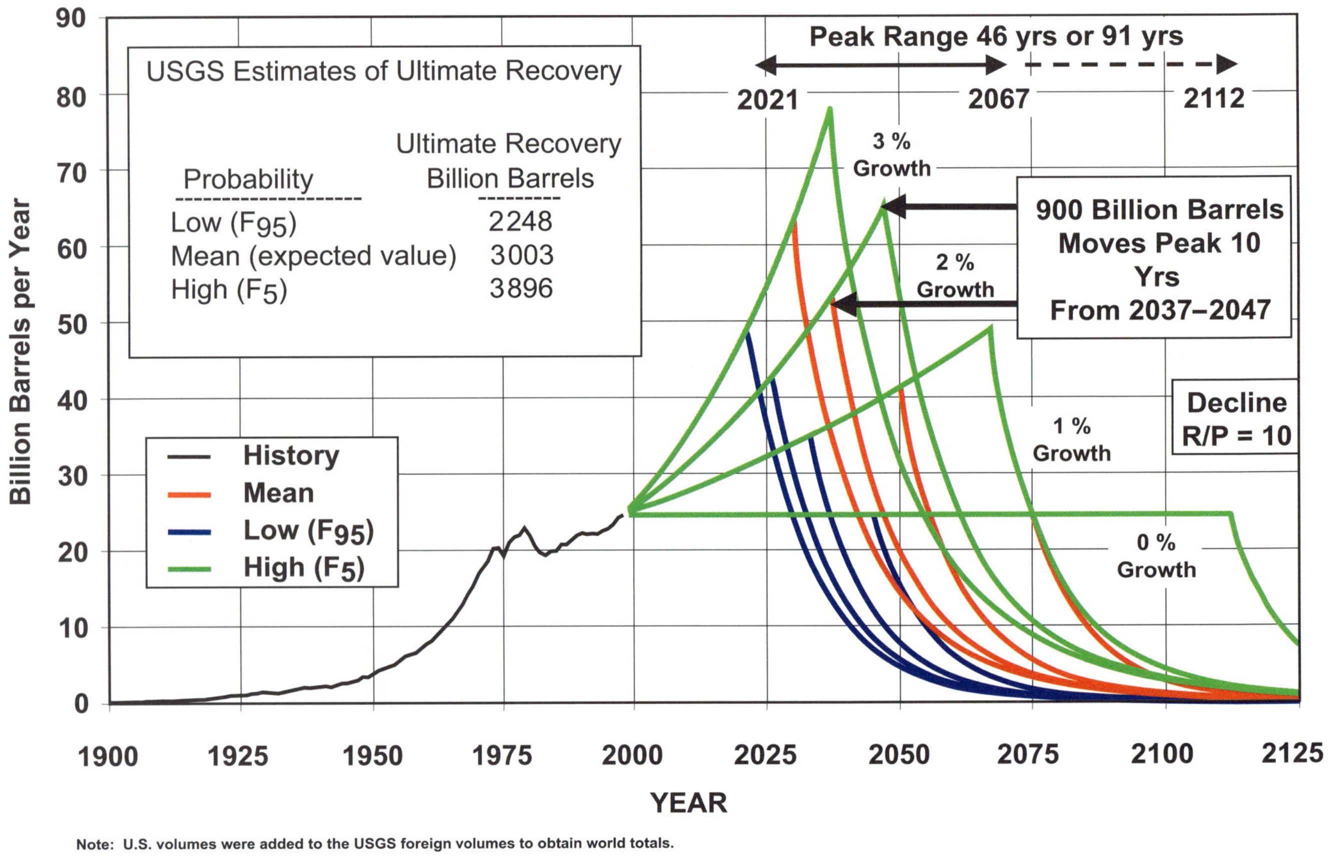

FIGURE 109. Twelve oil production scenarios under various growth rates and USGS (2000) estimate levels (from EIA, 2001b; also shown in Williams, 2003). R/P = reserve to production ratio. F95 (the 95th fractile) represents a 95% chance of the occurrence of at least the amount tabulated. Other fractiles are defined similarly.

IX: Conclusions

The USGS World Petroleum Assessment 2000 (USGS, 2000) provides estimates of the quantities of conventional oil, natural gas, and natural gas liquids outside the United States that have the potential to be added to reserves in the next 30 years (1995 to 2025). Excluding the U.S., the mean estimated volumes of undiscovered resources are 649 billion bbl of oil, 4669 tcf, and 207 billion bbl of NGL. The estimated mean additions to reserves from discovered fields (reserve growth) are 612 billion bbl of oil, 3305 tcf, and 42 billion bbl of NGL.

The potential additions to reserves from reserve growth are nearly as large as the estimated undiscovered resource volumes. These estimates imply that 75% of the world's grown conventional oil endowment and 66% of the world's grown conventional natural gas endowment have already been discovered in the areas assessed (exclusive of the U.S.). Additionally, for these areas, 20% of the world's grown conventional oil endowment and 7% of the world's grown conventional natural gas endowment had been produced by the end of 1995.

By way of calibration, the 8 years since 1995 are less than one-third of the 30-yr forecast span (1995–2025) for the assessment. During this period, 18% of estimated oil resources and 27% of the estimated natural gas resources have been added to reserves from new field discoveries and reserve growth for the provinces assessed in USGS (2000). If the entire world is considered, exclusive of the U.S. and Canada, these percentages increase to 23% of estimated oil and 31% of estimated natural gas in this timeframe. Estimated reserve growth is very significant with 26% of the potential oil volume and 52% of the estimated natural gas volume via reserve growth being realized (Klett et al., in press). Overall, about 10% of the estimated undiscovered oil and natural gas was added to reserves by new discoveries in the USGS (2000) assessed provinces. Results of these analyses suggest that the USGS (2000) estimates are somewhat reasonable assuming linear rates of reserves additions, and moderate relative to recent higher estimates for both oil and natural gas. Additionally, both the geographic and volumetric results of the assessment combined with geologic insights for the significant TPS of the world offer opportunities for many studies beyond the initial analyses in this report.

 DOI:10.1306/1061779M863175

X: Appendices

APPENDIX 1A. Assessment units ranked by mean (expected) undiscovered oil.

[MMBO = million barrels of oil. Major commodity is based on the ratio of gas endowment to oil endowment as ft^3 of gas per bbl of oil (CFG/BO). Major commodity is oil if this ratio is less than 20,000 CFG/BO and gas if this ratio is equal to or greater than 20,000 CFG/BO. Cumulative oil is the reported volume of oil that has been produced. Remaining oil is the calculated remaining reserves; the difference between known and cumulative oil. Known oil is the reported discovered volume (cumulative production plus remaining reserves). Mean undiscovered oil is the mean volume estimated in World Petroleum Assessment 2000 (USGS, 2000). Future oil is the remaining oil plus estimated mean undiscovered oil. Oil endowment is the known oil plus estimated mean undiscovered oil volumes. Discovery maturity is the percentage of oil discovered (known oil) with respect to oil endowment. Calculated values may not be equal to the sums of the component values because numbers have been independently rounded. Negative values result from inconsistent reporting of cumulative and known volumes. Values in parentheses are questionable data. NA means not applicable; it is shown for all values where there is only one field present, and used where the endowment could not be calculated.] Known volumes from Petroconsultants (1996) and NRG Associates (1995). Note that gas = natural gas.

Rank	Assessment Unit Code	Assessment Unit Name	Major Commodity	Cumulative Oil (MMBO)	Remaining Oil (MMBO)	Known Oil (MMBO)	Mean Undiscovered Oil (MMBO)	Future Oil (MMBO)	Oil Endowment (MMBO)	Discovery Maturity, Oil (%)
1	20300101	Cretaceous Reservoirs	Oil	53,182	209,735	262,917	74,911	284,646	337,828	78
2	52000101	Northeast Greenland Shelf Rift Systems	Oil	0	0	0	47,148	47,148	47,148	0
3	11740101	Upper Jurassic-Cretaceous Sandstones	Oil	45,819	72,153	117,972	42,947	115,100	160,919	73
4	20210201	Horst-Block Anticlinal Oil	Oil	53,906	101,239	155,145	26,715	127,954	181,860	85
5	71920101	Agbada Reservoirs	Oil	16,598	17,925	34,522	21,925	39,850	56,447	61
6	20300102	Tertiary Reservoirs	Oil	47,285	62,025	109,310	19,605	81,630	128,914	85
7	20210101	Central Arch Horst-Block Anticlinal Oil and Gas	Oil	84	6142	6226	19,190	25,331	25,416	24
8	71920102	Akata Reservoirs	Oil	0	0	0	18,562	18,562	18,562	0
9	72030302	Central Congo Turbidites	Oil	0	205	205	18,522	18,727	18,727	1
10	10160104	South Margin Subsalt	Oil	0	6245	6245	15,467	21,712	21,712	29
11	20210202	Salt-Involved Structural Oil	Oil	9948	33,902	43,850	14,989	48,891	58,839	75
12	60360101	Santos Shelf	Oil	6	279	285	13,723	14,002	14,008	2
13	20190302	Paleozoic Reservoirs	NA	NA	NA	NA	13,235	NA	NA	NA
14	60210101	Late Cretaceous-Tertiary Turbidites	Oil	0	0	0	12,745	12,745	12,745	0
15	60350101	Late Cretaceous-Tertiary Turbidites	Oil	1508	8122	9630	10,869	18,991	20,499	47
16	20190201	Jurassic Reservoirs in Northwest Desert Anticlines	Oil	0	777	777	10,633	11,410	11,410	7
17	53050102	Tamabra-Like Debris-Flow-Breccia Limestone Overlying Evaporites	Oil	14,366	18,699	33,065	9857	28,556	42,922	77
18	60360102	Salt-Structured Deep-Water Sandstones	Oil	0	0	0	9486	9486	9486	0
19	60990101	Main Maracaibo Basin	Oil	34,861	13,350	48,211	7842	21,192	56,053	86
20	53050106	Tamaulipas-Like Basinal Limestone and Tertiary Strata Overlying Evaporites	Oil	1929	2339	4268	7823	10,163	12,092	35
21	11740201	Pre-Upper Jurassic	Oil	645	11,111	11,756	7474	18,584	19,230	61
22	40250101	Viking Graben	Oil	13,038	13,643	26,681	7412	21,055	34,093	78
23	11120104	Central Offshore	Oil	0	0	0	6410	6410	6410	0
24	40170101	Halten Terrace-Trondelag Platform	Oil	76	2585	2660	6204	8788	8864	30
25	31540101	Tarim Basin Excluding Marginal Foldbelts	Oil	21	634	655	5364	5998	6019	11
26	72030201	Gabon Suprasalt	Oil	1278	836	2114	5031	5867	7145	30

App-1a_Page 1

Rank	Assessment Unit Code	Assessment Unit Name	Major Commodity	Cumulative Oil (MMBO)	Remaining Oil (MMBO)	Known Oil (MMBO)	Mean Undiscovered Oil (MMBO)	Future Oil (MMBO)	Oil Endowment (MMBO)	Discovery Maturity, Oil (%)
27	10080102	Main Basin Platform	Oil	2543	10,442	12,985	4938	15,379	17,922	72
28	60980101	East Venezuela Fold and Thrust Belt	Oil	3253	9978	13,231	4928	14,905	18,159	73
29	11120103	Gograndag-Okarem Zone and Adjacent Shelf	Oil	64	242	306	4764	5005	5070	6
30	40250103	Central Graben	Oil	7181	5588	12,769	4740	10,328	17,509	73
31	20190101	Cretaceous Reservoirs in Northwest Desert Anticlines	Oil	2594	16,386	18,981	4702	21,088	23,683	80
32	20230201	Platform Horst/Graben-Related Oil	Oil	94	379	474	4682	5062	5156	9
33	72030301	Central Congo Delta and Carbonate Platform	Oil	3693	5847	9540	4677	10,523	14,216	67
34	60350103	Salt Dome Province Tertiary Sandstones	Oil	0	0	0	4493	4493	4493	0
35	20540301	Tanezzuft-Ghadames Structural/Stratigraphic	Oil	1629	2909	4538	4461	7370	8999	50
36	60980202	Orinoco Delta and Offshore	Oil	0	0	0	4228	4228	4228	0
37	20190202	Jurassic Reservoirs in South Gulf Suprasalt/Qatar Arch Structural	Oil	5688	12,548	18,236	3845	16,394	22,081	83
38	20430102	Central Sirte Carbonates	Oil	8712	7465	16,177	3840	11,305	20,017	81
39	20190102	Cretaceous Reservoirs in South Gulf Suprasalt Structural	Oil	10,858	41,236	52,094	3764	45,000	55,858	93
40	60600101	North Falklands Basin	Oil	0	0	0	3603	3603	3603	0
41	38170102	Kutei Basin Turbidites	Oil	0	0	0	3569	3569	3569	0
42	11740302	South Kara Sea Offshore	Gas	0	0	0	3510	3510	3510	0
43	10160106	Suprasalt	Oil	690	1140	1830	3497	4637	5327	34
44	11120101	Apsheron-Pribalkhan Zone	Oil	10,581	4512	15,093	3091	7602	18,183	83
45	60410101	Hollin-Napo	Oil	2889	3712	6601	3023	6735	9625	69
46	60370101	Pelotas Platform and Basin	Oil	0	0	0	2938	2938	2938	0
47	20560101	Tanezzuft-Illizi Structural/Stratigraphic	Oil	2286	1384	3670	2814	4198	6484	57
48	20040101	Ma'Rib-Al Jawf/Shabwah/Masila	Oil	649	1284	1933	2740	4025	4673	41
49	20230101	Horst/Graben-Related Oil and Gas	Oil	269	944	1213	2632	3577	3845	32
50	13220101	Onshore and Offshore Northeastern Shelf	Oil	754	1428	2182	2582	4010	4764	46
51	60340102	Late Cretaceous-Tertiary Slide Blocks and Turbidites	NA	NA	NA	NA	2437	NA	NA	NA
52	20430101	Southeast Sirte Clastics	Oil	5274	15,622	20,895	2380	18,002	23,275	90
53	60960102	Peripheral	Oil	1027	1282	2309	2319	3601	4627	50
54	31270101	Tertiary Lacustrine	Oil	8944	10,993	19,937	2268	13,261	22,205	90
55	39480101	Dingo-Mungaroo/Barrow	Gas	489	660	1149	2249	2909	3398	34
56	60600201	South Falklands Basin	Oil	0	0	0	2229	2229	2229	0
57	60210102	Cretaceous Carbonates	Oil	0	0	0	2200	2200	2200	0
58	10160103	East and Southeast Margins Subsalt	Oil	131	1086	1217	2196	3283	3414	36
59	20710101	Gulf of Suez Block-Fault Fairway	Oil	6310	3444	9754	2171	5615	11,925	82
60	31270102	Pre-Tertiary Buried Hills	Oil	1471	3145	4616	2144	5289	6760	68
61	37010101	Brunei-Sabah Deltaics	Oil	3872	2780	6652	2137	4917	8789	76
62	20710201	Red Sea Coastal Block Faults	Gas	22	133	155	2136	2269	2291	7
63	11090101	Foldbelt-Foothills	Oil	752	2738	3490	2094	4832	5584	62
64	60980103	Maturin Sub-Basin	Oil	7468	1978	9446	1919	3897	11,365	83
65	20230202	Basinal Oil and Gas	Oil	0	0	0	1886	1886	1886	0

App-1a_Page 2

Rank	Assessment Unit Code	Assessment Unit Name	Major Commodity	Cumulative Oil (MMBO)	Remaining Oil (MMBO)	Known Oil (MMBO)	Mean Undiscovered Oil (MMBO)	Future Oil (MMBO)	Oil Endowment (MMBO)	Discovery Maturity, Oil (%)
66	20540201	Tanezzuft-Melrhir Structural/Stratigraphic	Oil	0	6	6	1875	1881	1881	0
67	52150101	Jeanne d'Arc	Oil	0	1582	1582	1796	3378	3378	47
68	60810101	Cretaceous-Paleogene Basin	Oil	1346	326	1672	1711	2038	3383	49
69	10150101	Supra-Domanik Carbonates/Clastics	Oil	19,973	14,931	34,904	1676	16,607	36,580	95
70	80430101	Eocene-Miocene Bombay Shelf	Oil	2337	3920	6257	1612	5531	7868	80
71	12100101	Baikal-Patom Foldbelt Riphean-Craton Margin Vendian	Gas	1	2005	2006	1611	3616	3617	55
72	61070101	Inner Forearc Deformation Belt	Oil	0	0	0	1536	1536	1536	0
73	10160101	North and West Margins Subsalt Pinnacle Reefs	NA	NA	NA	NA	1522	NA	NA	NA
74	60290102	Late Cretaceous-Tertiary Deep-Water Sandstones	Oil	55	67	122	1457	1523	1579	8
75	60960101	Central	Oil	152	2942	3094	1312	4254	4406	70
76	11740301	Northern West Siberian Onshore Gas	Gas	66	10,119	10,185	1304	11,424	11,489	89
77	20160101	Fahud-Huqf Combined Structural	Oil	1259	1426	2685	1304	2731	3990	67
78	11120102	Lower Kura Depression and Adjacent Shelf	Oil	1035	1004	2040	1275	2279	3315	62
79	38170101	Kutei Basin Deltaics	Oil	2195	684	2879	1258	1942	4137	70
80	60450102	Foreland Basins	Oil	0	2	2	1254	1256	1256	0
81	37010102	Brunei-Sabah Turbidites	Oil	81	165	246	1252	1417	1498	16
82	12070101	Yenisey Foldbelt Riphean-Craton Margin Riphean	Oil	0	34	34	1213	1246	1246	3
83	39100301	Malita	Oil	123	378	502	1146	1524	1648	30
84	39130101	Late Jurassic/Early Cretaceous-Mesozoic	Gas	19	27	46	1055	1083	1101	4
85	60630101	Malvinas Extensional Structures	Oil	0	10	10	1031	1041	1041	1
86	71830101	Coastal Plain and Offshore	Oil	74	162	236	1004	1166	1240	19
87	37030101	South Malay Lacustrine	Oil	1459	1557	3017	1003	2560	4019	75
88	40250102	Moray Firth	Oil	2248	2196	4444	947	3143	5391	82
89	60350102	Cretaceous Carbonates	Oil	261	166	426	932	1097	1358	31
90	38080101	Pematang/Sihapas Siliciclastics	Oil	8349	4868	13,217	899	5768	14,117	94
91	53050107	Tamabra-Like Debris-Flow-Breccia Limestone North of Campeche	Oil	0	0	0	871	871	871	0
92	60550101	Neuquen Extensional Structures	Oil	1111	605	1716	862	1467	2578	67
93	10090102	Devonian Synrift	Oil	0	0	0	848	848	848	0
94	20540101	Tanezzuft-Oued Mya Structural/Stratigraphic	Oil	5872	4971	10,843	830	5801	11,673	93
95	11500101	Mesozoic Sandstone Reservoirs	Oil	459	1827	2286	814	2641	3100	74
96	53050103	Tamabra-Like Debris-Flow-Breccia Limestone and Overlying Strata	Oil	26	54	80	807	860	886	9
97	72030401	Cuanza-Namibe	Oil	86	20	107	790	810	897	12
98	11090301	Offshore Prikumsk Zone	Oil	0	0	0	789	789	789	0
99	20140101	Ghaba-Makarem Combined Structural	Oil	715	1313	2028	743	2056	2771	73
100	31440101	Subtle Traps	Oil	171	979	1150	735	1714	1885	61
101	72030101	Gabon Subsalt	Oil	708	1044	1752	727	1772	2480	71
102	60450103	Foreland Central Chaco High	Oil	0	0	0	711	711	711	0
103	10150103	Sub-Domanik Devonian Clastics	Oil	20,807	5632	26,439	682	6314	27,121	97
104	20480101	Bou Dabbous-Tertiary Structural/Stratigraphic	Oil	481	1633	2114	667	2300	2782	76
105	11090201	South Mangyshlak (Entire)	Oil	2406	2835	5241	651	3486	5893	89

Rank	Assessment Unit Code	Assessment Unit Name	Major Commodity	Cumulative Oil (MMBO)	Remaining Oil (MMBO)	Known Oil (MMBO)	Mean Undiscovered Oil (MMBO)	Future Oil (MMBO)	Oil Endowment (MMBO)	Discovery Maturity, Oil (%)
106	10160102	North and West Margins Subsalt Barrier Reefs	Oil	0	134	134	637	771	771	17
107	40610201	Romania Flysch Zone	Oil	1039	909	1948	564	1473	2512	78
108	10500101	Kolguyev Terrace	Gas	0	51	51	563	614	614	8
109	53050104	Tamabra-Like Debris-Flow-Breccia Limestone of the Golden Lane	Oil	1917	1392	3308	557	1948	3865	86
110	37020102	Balingian	Oil	168	460	628	529	989	1157	54
111	80480101	Central Burma Basin	Oil	494	142	637	516	658	1152	55
112	20190103	Mesozoic/Tertiary Foredeep Fold and Thrust	Gas	92	64	156	513	577	669	23
113	52430202	Leduc Oil and Gas	Oil	8113	1304	9417	501	1805	9918	95
114	31540103	Southwest Foldbelt	Oil	17	24	41	495	519	536	8
115	60590101	Magallanes Extensional Structures	Gas	595	620	1216	494	1114	1709	71
116	61170101	North Cuba Fold and Thrust Belt	Oil	65	232	297	494	726	790	38
117	80430102	Eocene-Miocene Deltaics	Oil	1038	1146	2183	479	1625	2662	82
118	38280101	South Sumatra	Oil	1842	587	2429	469	1055	2898	84
119	60340103	Abrolhos Sub-Volcanic Structures	Oil	0	0	0	468	468	468	0
120	31150101	Upper Paleozoic/Lower Mesozoic Nonmarine Coarse Clastics	Oil	937	5687	6624	467	6154	7091	93
121	38240101	Sunda/Asri	Oil	828	431	1259	462	894	1722	73
122	11090103	Foreland Slope and Foredeep	Oil	0	0	0	452	452	452	0
123	20710202	Red Sea Salt Basin	Oil	0	0	0	443	443	443	0
124	60980201	Trinidad Basins	Oil	2780	667	3447	416	1083	3863	89
125	20480201	Jurassic-Cretaceous Structural/Stratigraphic	Oil	39	64	103	403	467	506	20
126	80340101	Sylhet-Kopili/Barail-Tipam Composite	Oil	911	1625	2536	399	2024	2935	86
127	40680101	Subsalt	Gas	0	0	0	388	388	388	0
128	60980102	Guarico Sub-Basin	Gas	268	54	322	385	439	707	46
129	20430103	Offshore Sirte Hypothetical	Oil	0	0	0	382	382	382	0
130	20160201	Natih-Fiqa Structural/Stratigraphic	Oil	1299	553	1852	371	925	2224	83
131	53050105	Tamaulipas-Like Basinal Limestone and Tertiary Strata Without Underlying Evaporites	Oil	1354	273	1627	366	638	1993	82
132	40600201	Thermal Triassic	Oil	78	254	332	360	614	692	48
133	60580101	San Jorge Extensional Structures	Oil	2330	776	3106	347	1123	3453	90
134	20710103	Southern Gulf of Suez	Oil	41	15	57	344	359	400	14
135	40470201	Deformed Belt	Oil	640	283	923	344	627	1267	73
136	60990102	Southwest Maracaibo Basin Fold Belt	Oil	720	141	861	341	482	1202	72
137	40170102	Mid-Norway Continental Margin	Gas	0	0	0	334	334	334	0
138	60550102	Neuquen Foothills Structure	Oil	159	138	297	334	472	631	47
139	11090303	Central Caspian Offshore	Oil	0	0	0	331	331	331	0
140	38170103	Kutei Basin Fold and Thrust Belt	NA	NA	NA	NA	318	NA	NA	NA
141	39300101	Gippsland	Oil	3077	784	3860	310	1093	4170	93
142	60210103	Late Cretaceous-Tertiary Nearshore Sandstones	Oil	17	46	63	301	347	364	17
143	31440102	Anticlinal	Oil	12,228	2192	14,420	291	2483	14,711	98
144	53050101	El Abra-Like Reef and Backreef Limestone	Oil	2546	(-483)	2063	289	(-195)	2352	88
145	11090302	Onshore Stavropol-Prikumsk	Oil	1200	(-380)	820	287	(-93)	1107	74

Rank	Assessment Unit Code	Assessment Unit Name	Major Commodity	Cumulative Oil (MMBO)	Remaining Oil (MMBO)	Known Oil (MMBO)	Mean Undiscovered Oil (MMBO)	Future Oil (MMBO)	Oil Endowment (MMBO)	Discovery Maturity, Oil (%)
146	11090102	Terek-Sunzha Subsalt Jurassic	Gas	0	0	0	283	283	283	0
147	11080101	Tertiary Foredeep	Oil	1268	(-770)	498	280	(-490)	778	64
148	40610202	Romania Ploiesti Zone	Oil	1793	374	2167	278	652	2445	89
149	10500102	South Barents and Ludlov Saddle	Gas	0	0	0	257	257	257	0
150	20430104	Southeast Sirte Hypothetical	Oil	0	0	0	253	253	253	0
151	10090101	Carboniferous-Lower Permian Clastics	Gas	110	1501	1611	251	1752	1862	87
152	60900104	La Luna and Older	Oil	11	4	15	231	235	246	6
153	10080103	Foredeep Basins	Gas	803	(-719)	84	214	(-504)	299	28
154	60590102	Andean Fold Belt Structures	Oil	0	0	0	210	211	211	0
155	80480102	Irrawaddy-Andaman	Gas	34	46	79	209	254	288	28
156	60830201	Cretaceous-Paleogene Santa Elena Block	Oil	118	53	171	200	252	370	46
157	40480101	Greater Hungarian Plain Basins	Oil	731	358	1089	196	555	1285	85
158	38240201	Ardjuna	Oil	727	1181	1908	195	1377	2104	91
159	60900101	Northern	Oil	1556	175	1730	191	366	1921	90
160	60900102	Southern	Oil	348	333	681	188	521	869	78
161	11120105	Iran Onshore/Nearshore	Oil	0	0	0	185	185	185	0
162	60450101	Sub-Andean Fold and Thrust Belt	Gas	257	62	318	181	242	499	64
163	40610101	Moesian Platform	Oil	1251	542	1793	171	713	1963	91
164	20580301	Tanezzuft-Sbaa Structural/Stratigraphic	Oil	0	284	284	162	446	446	64
165	70130101	Coastal Plain and Offshore	Oil	0	10	10	157	167	168	6
166	60580102	San Bernardo Fold Belt Structures	Oil	115	124	240	151	275	390	61
167	10150102	Lower Volga	Oil	1803	726	2529	148	874	2677	94
168	80260101	Kohat-Potwar Intrathrust Basin	Oil	204	121	324	145	266	469	69
169	39100101	Barnett	Oil	0	15	15	140	155	155	10
170	11500201	Jurassic-Tertiary	Oil	0	80	80	139	220	220	37
171	31280101	Jurassic/Triassic Fluvial and Lacustrine Sandstone	Oil	81	602	683	138	740	821	83
172	10500103	North Barents	Gas	0	0	0	135	135	135	0
173	39480201	Locker-Mungaroo/Barrow	Gas	0	6	6	132	137	137	4
174	52430302	Exshaw-Rundle Oil and Gas	Oil	1122	415	1537	126	541	1663	92
175	37030102	North Malay Lacustrine	Oil	0	0	0	124	124	124	0
176	80420101	Greater Indus Foreland and Foldbelt	Gas	111	70	180	119	188	299	60
177	73030101	Offshore	Gas	0	0	0	116	116	116	0
178	40480201	Zala-Drava-Sava Basins	Oil	836	265	1101	114	379	1215	91
179	11540104	Murgab Depression Subsalt	Gas	31	485	516	113	597	628	82
180	38220101	North Sumatra	Gas	496	178	674	112	289	786	86
181	52440401	Lodgepole Oil	Oil	2239	539	2778	111	650	2889	96
182	37020201	East Natuna	Gas	0	20	20	102	122	122	16
183	37030201	South Malay Coaly	Gas	58	533	592	101	634	693	85
184	38220102	Mergui	Gas	0	0	0	101	101	101	0
185	60900103	Eastern	Gas	0	0	0	99	99	99	0
186	31150201	Jurassic/Tertiary Fluvial and Lacustrine Sandstone	Oil	16	193	210	96	289	306	69
187	52430402	Combined Triassic/Jurassic Oil and Gas	Oil	965	277	1242	96	373	1338	93
188	20710102	Gulf of Suez Qaa Plain	Oil	0	0	0	93	93	93	0

Rank	Assessment Unit Code	Assessment Unit Name	Major Commodity	Cumulative Oil (MMBO)	Remaining Oil (MMBO)	Known Oil (MMBO)	Mean Undiscovered Oil (MMBO)	Future Oil (MMBO)	Oil Endowment (MMBO)	Discovery Maturity, Oil (%)
189	60550103	Dorsal de Neuquen Structure	Oil	262	95	357	93	189	450	79
190	11080103	Subsalt Jurassic	Gas	4	1	5	90	91	95	5
191	31540102	Kuche (Northern) Foldbelt	Gas	7	0	7	89	89	96	8
192	11540102	Karabil-Badkhyz (Southern Area)	Gas	0	110	111	88	198	198	56
193	52430102	Keg River Oil and Gas	Oil	835	176	1011	84	260	1095	92
194	40360102	Southern Permian Basin-Europe Onshore	Gas	2029	372	2400	74	446	2474	97
195	60410201	Ene	Oil	0	0	0	74	74	74	0
196	20580501	Tanezzuft-Benoud Structural/Stratigraphic	Gas	16	72	88	72	144	160	55
197	11540101	Northern and Western Areas	Gas	48	92	140	70	162	210	67
198	80470201	Western Shelf and Slope	NA	NA	NA	NA	64	NA	NA	NA
199	40360103	Southern Permian Basin-Offshore	Gas	232	176	408	62	238	470	87
200	60340101	Espirito Santo Shelf	Oil	93	37	130	54	91	184	71
201	40480601	Hungarian Paleogene Basin	Oil	10	1	11	49	50	60	18
202	11080102	Foreland Slope	Gas	1	19	20	48	67	68	30
203	40470301	Paleozoic Reservoirs	Oil	0	1	1	47	47	47	1
204	10150201	Permian Reefs/Thrust Folds	Gas	106	(-41)	65	42	(1)	106	61
205	60830101	Neogene Pull-Apart Basin	Oil	4	0	4	37	37	41	11
206	52430602	Second White Specks-Cardium Oil and Gas	Oil	1929	494	2423	36	529	2459	99
207	20580201	Tanezzuft-Ahnet Structural/Stratigraphic	Gas	0	1	1	34	35	35	3
208	20580101	Tanezzuft-Timimoun Structural/Stratigraphic	Gas	0	0	0	31	31	31	0
209	60290101	Western Pre-Aptian Reservoirs	Oil	413	157	570	31	187	601	95
210	31420201	Jurassic Lacustrine	Oil	38	71	109	27	99	136	80
211	80470302	Eastern Fold Belt	Gas	3	2	4	24	26	28	16
212	10080101	Northwest Izhma-Pechora Depression	NA	NA	NA	NA	18	NA	NA	NA
213	20580601	Tanezzuft-Bechar/Abadla Structural/Stratigraphic	Gas	0	0	0	16	16	16	0
214	80470301	Central Basin	Gas	0	0	0	15	15	15	0
215	20580401	Tanezzuft-Mouydir Structural/Stratigraphic	Gas	0	0	0	12	12	12	0
216	52440301	Bakken Sandstone	Oil	5	2	7	11	13	18	40
217	52440201	Brightholme Oil	Oil	11	4	14	8	11	22	66
218	40360101	Southern Permian Basin-U.K. Onshore	Oil	39	25	64	7	32	71	90
219	52440101	Yeoman Oil	Oil	1	0	1	5	6	6	15
220	11500301	Upper Paleozoic Carbonates	NA	NA	NA	NA	0	NA	NA	NA
221	11540103	Murgab Depression Suprasalt	Gas	1	(-1)	0	0	(-1)	0	NA
222	20190301	Khuff Carbonates in Salt Structures	Gas	0	500	500	0	500	500	100
223	20210102	North Gulf Salt Basin Structural Gas	Gas	0	150	150	0	150	150	100
224	20300201	Northern Qatar Arch Extension	Gas	0	0	0	0	0	0	NA
225	31420101	Southeastern Fold Belt	Gas	0	0	0	0	0	0	100
226	31420102	Northwestern Depression/Foldbelt	Gas	0	0	0	0	0	0	NA
227	31420401	Leshan-Longnusi Paleohigh	Gas	9	(-8)	1	0	(-8)	1	100
228	31420402	Lower Paleozoic of Southeastern Fold Belt	Gas	0	0	0	0	0	0	NA
229	31440201	Structural Traps	Gas	0	5	5	0	5	5	100
230	37020101	Central Luconia	Gas	0	170	170	0	170	170	100
231	39100201	Petrel	Gas	0	0	0	0	0	0	NA

Rank	Assessment Unit Code	Assessment Unit Name	Major Commodity	Cumulative Oil (MMBO)	Remaining Oil (MMBO)	Known Oil (MMBO)	Mean Undiscovered Oil (MMBO)	Future Oil (MMBO)	Oil Endowment (MMBO)	Discovery Maturity, Oil (%)
232	39100202	Vulcan Graben	Gas	0	0	0	0	0	0	NA
233	40470101	Foreland Basin	Gas	1	1	1	0	1	1	100
234	40480301	Danube Basin	Gas	0	0	0	0	0	0	NA
235	40480401	Transcarpathian Basin	Gas	0	0	0	0	0	0	NA
236	40570101	Transylvanian Neogene Suprasalt Gas	Gas	0	0	0	0	0	0	NA
237	40600101	Neogene Flysch Gas	Gas	13	0	14	0	0	14	100
238	52430101	Keg River Gas	Gas	0	0	0	0	0	0	NA
239	52430201	Leduc Gas	Gas	39	2	42	0	2	42	100
240	52430301	Exshaw-Rundle Gas	Gas	170	20	191	0	20	191	100
241	52430401	Combined T:iassic/Jurassic Gas	Gas	47	7	54	0	7	54	100
242	52430501	Mannville Gas	Gas	0	0	0	0	0	0	NA
243	52430601	Second White Specks-Cardium Gas	Gas	0	0	0	0	0	0	NA
244	60220101	Amazon Delta and Submarine Fan	NA	NA	NA	NA	0	NA	NA	NA
245	61030101	Carupano Basin Gas	Gas	0	20	20	0	20	20	100
246	80420102	Indus Fan	Gas	0	0	0	0	0	0	NA
NA	10090103	Continuous Basin-Centered Gas Accumulation	NA	0	0	0	NA	NA	NA	NA
NA	10150104	Self-Sourced Domanik	NA	0	0	0	NA	NA	NA	NA
NA	10160105	Basin Center Subsalt	NA	0	0	0	NA	NA	NA	NA
NA	11740102	Self-Sourced Bazhenov Fractured Reservoirs	NA	0	0	0	NA	NA	NA	NA
NA	31150301	Continuous-Type (Basin-Centered) Gas	NA	0	0	0	NA	NA	NA	NA
NA	31270201	Coal-Sourced Gas	NA	0	0	0	NA	NA	NA	NA
NA	31280201	Basin Centered Gas (Continuous) in Ordovician Carbonate and Permian Sandstone	NA	0	0	0	NA	NA	NA	NA
NA	31280202	Carboniferous and Permian Coalbed Methane (Continuous)	NA	0	0	0	NA	NA	NA	NA
NA	31420103	Continuous Gas in Central Uplift/Northwestern Depression	NA	0	0	0	NA	NA	NA	NA
NA	31420301	Continuous Gas in Northwestern Depression/Central Uplift	NA	NA	NA	NA	NA	NA	NA	NA
NA	38240301	Biogenic Gas	NA	0	0	0	NA	NA	NA	NA
NA	38240401	Billiton Basin Hypothetical	NA	0	0	0	NA	NA	NA	NA
NA	40480501	Central Carpathian Paleogene Basin	NA	NA	NA	NA	NA	NA	NA	NA
NA	40570102	Transylvanian Subsalt	NA	0	0	0	NA	NA	NA	NA
NA	52150201	Graben Systems Exclusive of Jeanne d'Arc	NA	0	0	0	NA	NA	NA	NA
NA	52430502	Mannville Continuous (Basin-Center) Gas	NA	0	0	0	NA	NA	NA	NA
NA	52430603	Upper Cretaceous Eastern Shallow Gas	NA	216	49	265	NA	NA	NA	NA
NA	60220102	Offshore Gas Hydrates	NA	0	0	0	NA	NA	NA	NA
NA	60370102	Offshore Gas Hydrates	NA	0	0	0	NA	NA	NA	NA
NA	60580201	Hypothetical Basin-Centered Gas	NA	0	0	0	NA	NA	NA	NA
NA	60980104	Orinoco Heavy Oil and Tar Belt	NA	38	3719	3757	NA	NA	NA	NA
NA	60990103	Basin-Centered Gas	NA	0	0	0	NA	NA	NA	NA
NA	61070102	Barbados Accreationary Prism	NA	8	2	11	NA	NA	NA	NA
NA	80470101	Coalbed Gas	NA	0	0	0	NA	NA	NA	NA

APPENDIX 1B. Assessment units ranked by mean (expected) undiscovered natural gas.

[BCFG = billion ft^3 of gas. Major commodity is based on the ratio of gas endowment to oil endowment as ft^3 of gas per bbl of oil (CFG/BO). Major commodity is oil if this ratio is less than 20,000 CFG/BO and gas if this ratio is equal to or greater than 20,000 CFG/BO. Cumulative gas is the reported volume of gas that has been produced. Remaining gas is the calculated remaining reserves; the difference between known and cumulative gas. Known gas is the reported discovered volume (cumulative production plus remaining reserves). Mean undiscovered gas is the mean volume estimated in World Petroleum Assessment 2000 (USGS, 2000). Future gas is the remaining gas plus estimated mean undiscovered gas. Gas endowment is the known gas plus estimated mean undiscovered gas volumes. Discovery maturity is the percentage of gas discovered (known gas) with respect to gas endowment. Calculated values may not be equal to the sums of the component values because numbers have been independently rounded. NA means not applicable; it is shown for all values where there is only one field present, and used where the endowment could not be calculated.] Known volumes from Petroconsultants (1996) and NRG Associates (1995). Note that gas = natural gas.

Rank	Assessment Unit Code	Assessment Unit Name	Major Commodity	Cumulative Gas (BCFG)	Remaining Gas (BCFG)	Known Gas (BCFG)	Mean Undiscovered Gas (BCFG)	Future Gas (BCFG)	Gas Endowment (BCFG)	Discovery Maturity, Gas (%)
1	11740302	South Kara Sea Offshore	Gas	0	81,700	81,700	403,875	485,575	485,575	17
2	20190302	Paleozoic Reservoirs	NA	NA	NA	NA	323,892	NA	NA	NA
3	20210101	Central Arch Horst-Block Anticlinal Oil and Gas	Oil	0	12,115	12,115	230,387	242,502	242,502	5
4	10500102	South Barents and Ludlov Saddle	Gas	0	78,140	78,140	189,244	267,384	267,384	29
5	20210102	North Gulf Salt Basin Structural Gas	Gas	0	67,000	67,000	152,952	219,952	219,952	30
6	11540104	Murgab Depression Subsalt	Gas	16,456	54,120	70,576	141,762	195,883	212,338	33
7	11740301	Northern West Siberian Onshore Gas	Gas	188,736	896,852	1,085,588	136,608	1,033,460	1,222,196	89
8	40170102	Mid-Norway Continental Margin	Gas	0	0	0	107,832	107,832	107,832	0
9	20300101	Cretaceous Reservoirs	Oil	15,185	218,613	233,798	104,098	322,711	337,896	69
10	11740101	Upper Jurassic-Cretaceous Sandstones	Oil	29,287	68,295	97,582	96,934	165,229	194,516	50
11	20300102	Tertiary Reservoirs	Oil	10,075	249,365	259,440	93,499	342,864	352,939	74
12	71920101	Agbada Reservoirs	Oil	6232	87,579	93,811	85,176	172,755	178,986	52
13	11120104	Central Offshore	Oil	0	0	0	82,836	82,836	82,836	0
14	52000101	Northeast Greenland Shelf Rift Systems	Oil	0	0	0	80,709	80,709	80,709	0
15	20300201	Northern Qatar Arch Extension	Gas	0	131,220	131,220	69,031	200,251	200,251	66
16	10160104	South Margin Subsalt	Oil	428	97,672	98,100	67,521	165,193	165,621	59
17	20230101	Horst/Graben-Related Oil and Gas	Oil	47	1614	1661	64,510	66,124	66,172	3
18	60360101	Santos Shelf	Oil	12	1113	1125	62,874	63,987	63,999	2
19	10500103	North Barents	Gas	0	0	0	59,820	59,820	59,820	0
20	13220101	Onshore and Offshore Northeastern Shelf	Oil	162	22,221	22,383	57,451	79,673	79,835	28
21	40170101	Halten Terrace-Trondelag Platform	Oil	0	15,662	15,662	57,369	73,031	73,031	21
22	39480101	Dingo-Mungaroo/Barrow	Gas	1845	46,400	48,245	56,380	102,780	104,626	46
23	72030302	Central Congo Turbidites	Oil	0	2	2	55,268	55,270	55,270	0
24	40680101	Subsalt	Gas	0	0	0	51,210	51,210	51,210	0
25	71920102	Akata Reservoirs	Oil	0	0	0	47,540	47,540	47,540	0
26	11120103	Gograndag-Okarem Zone and Adjacent Shelf	Oil	173	8868	9042	47,515	56,383	56,557	16
27	20190201	Jurassic Reservoirs in Northwest Desert Anticlines	Oil	0	6500	6500	46,058	52,558	52,558	12
28	12100101	Baikal-Patom Foldbelt Riphean-Craton Margin Vendian	Gas	0	30,210	30,210	46,039	76,249	76,249	40

Rank	Assessment Unit Code	Assessment Unit Name	Major Commodity	Cumulative Gas (BCFG)	Remaining Gas (BCFG)	Known Gas (BCFG)	Mean Undiscovered Gas (BCFG)	Future Gas (BCFG)	Gas Endowment (BCFG)	Discovery Maturity, Gas (%)
29	20710201	Red Sea Coastal Block Faults	Gas	10	3517	3527	43,946	47,463	47,473	7
30	31540101	Tarim Basin Excluding Marginal Foldbelts	Oil	14	5140	5154	41,454	46,594	46,608	11
31	38170101	Kutei Basin Deltaics	Oil	8711	35,762	44,473	35,757	71,519	80,230	55
32	60210101	Late Cretaceous-Tertiary Turbidites	Oil	0	0	0	34,625	34,625	34,625	0
33	11120102	Lower Kura Depression and Adjacent Shelf	Oil	1285	6939	8224	32,091	39,030	40,315	20
34	10160101	North and West Margins Subsalt Pinnacle Reefs	NA	NA	NA	NA	30,544	NA	NA	NA
35	60220101	Amazon Delta and Submarine Fan	NA	NA	NA	NA	29,840	NA	NA	NA
36	60980202	Orinoco Delta and Offshore	Oil	0	0	0	28,049	28,049	28,049	0
37	20560101	Tanezzuft-Illizi Structural/Stratigraphic	Oil	750	44,311	45,061	27,785	72,096	72,846	62
38	80470302	Eastern Fold Belt	Gas	2669	9620	12,289	26,494	36,114	38,783	32
39	38170102	Kutei Basin Turbidites	Oil	0	0	0	25,687	25,687	25,687	0
40	60340102	Late Cretaceous-Tertiary Slide Blocks and Turbidites	NA	NA	NA	NA	25,677	NA	NA	NA
41	80420101	Greater Indus Foreland and Foldbelt	Gas	15,101	20,272	35,373	25,405	45,677	60,778	58
42	80470301	Central Basin	Gas	0	0	0	23,985	23,985	23,985	0
43	60980201	Trinidad Basins	Oil	3156	12,667	15,822	23,960	36,627	39,782	40
44	20190301	Khuff Carbonates in Salt Structures	Gas	1018	450,982	452,000	22,999	473,981	474,999	95
45	60370101	Pelotas Platform and Basin	Oil	0	0	0	22,806	22,806	22,806	0
46	40250101	Viking Graben	Oil	18,474	94,268	112,742	22,192	116,460	134,933	84
47	37030101	South Malay Lacustrine	Oil	454	23,619	24,073	21,935	45,554	46,008	52
48	60980101	East Venezuela Fold and Thrust Belt	Oil	10,673	41,706	52,379	21,298	63,004	73,677	71
49	60450101	Sub-Andean Fold and Thrust Belt	Gas	6813	10,004	16,818	20,521	30,525	37,339	45
50	53050102	Tamabra-Like Debris-Flow-Breccia Limestone Overlying Evaporites	Oil	13,276	22,402	35,678	20,450	42,851	56,127	64
51	61030101	Carupano Basin Gas	Gas	0	22,600	22,600	20,103	42,703	42,703	53
52	39130101	Late Jurassic/Early Cretaceous-Mesozoic	Gas	0	17,960	17,960	20,093	38,053	38,053	47
53	10080103	Foredeep Basins	Gas	11,482	4635	16,117	19,891	24,526	36,008	45
54	53050106	Tamaulipas-Like Basinal Limestone and Tertiary Strata Overlying Evaporites	Oil	1633	6007	7640	18,516	24,522	26,155	29
55	80480102	Irrawaddy-Andaman	Gas	333	8706	9040	18,512	27,219	27,552	33
56	38280101	South Sumatra	Oil	3125	7080	10,204	18,250	25,330	28,455	36
57	20210202	Salt-Involved Structural Oil	Oil	4831	40,849	45,680	17,991	58,840	63,671	72
58	60360102	Salt-Structured Deep-Water Sandstones	Oil	0	0	0	17,673	17,673	17,673	0
59	10500101	Kolguyev Terrace	Gas	0	3	3	17,554	17,556	17,556	0
60	60980102	Guarico Sub-Basin	Gas	2933	44,646	47,579	16,995	61,641	64,574	74
61	12070101	Yenisey Foldbelt Riphean-Craton Margin Riphean	Oil	0	5052	5052	16,983	22,035	22,035	23
62	20040101	Ma'Rib-Al Jawf/Shabwah/Masila	Oil	0	17,120	17,120	16,982	34,102	34,102	50
63	40600101	Neogene Flysch Gas	Gas	9373	9102	18,475	16,267	25,369	34,741	53
64	60990101	Main Maracaibo Basin	Oil	61	24,857	24,918	16,085	40,942	41,002	61
65	20210201	Horst-Block Anticlinal Oil	Oil	19,990	209,625	229,615	16,055	225,680	245,670	93
66	37010101	Brunei-Sabah Deltaics	Oil	9168	24,587	33,755	15,977	40,564	49,732	68
67	20190103	Mesozoic/Tertiary Foredeep Fold and Thrust	Gas	2522	15,470	17,993	15,419	30,889	33,411	54
68	37020101	Central Luconia	Gas	3731	31,304	35,035	15,067	46,370	50,102	70

Rank	Assessment Unit Code	Assessment Unit Name	Major Commodity	Cumulative Gas (BCFG)	Remaining Gas (BCFG)	Known Gas (BCFG)	Mean Undiscovered Gas (BCFG)	Future Gas (BCFG)	Gas Endowment (BCFG)	Discovery Maturity, Gas (%)
69	10090101	Carboniferous-Lower Permian Clastics	Gas	35,453	23,645	59,098	14,735	38,380	73,833	80
70	10080102	Main Basin Platform	Oil	892	19,623	20,515	14,515	34,138	35,030	59
71	11090101	Foldbelt-Foothills	Oil	192	7648	7840	14,512	22,159	22,351	35
72	72030301	Central Congo Delta and Carbonate Platform	Oil	838	8603	9441	14,314	22,916	23,754	40
73	20230202	Basinal Oil and Gas	Oil	0	0	0	14,201	14,201	14,201	0
74	60350101	Late Cretaceous-Tertiary Turbidites	Oil	224	5538	5761	13,917	19,455	19,679	29
75	40250103	Central Graben	Oil	11,950	28,370	40,320	13,801	42,172	54,122	74
76	72030201	Gabon Suprasalt	Oil	203	1632	1835	13,230	14,862	15,065	12
77	40360103	Southern Permian Basin-Offshore	Gas	42,288	31,546	73,834	13,029	44,575	85,862	85
78	40360102	Southern Permian Basin-Europe Onshore	Gas	81,742	66,484	148,227	12,730	79,214	160,957	92
79	20190202	Jurassic Reservoirs in South Gulf Suprasalt/Qatar Arch Structural	Oil	1951	69,061	71,012	12,457	81,518	83,469	85
80	20540301	Tanezzuft-Ghadames Structural/Stratigraphic	Oil	579	15,905	16,484	12,035	27,940	28,519	58
81	61070101	Inner Forearc Deformation Belt	Oil	0	0	0	11,957	11,957	11,957	0
82	11090102	Terek-Sunzha Subsalt Jurassic	Gas	0	0	0	11,464	11,464	11,464	0
83	39100301	Malita	Oil	0	7526	7526	11,346	18,872	18,872	40
84	60630101	Malvinas Extensional Structures	Oil	0	75	75	11,327	11,402	11,402	1
85	31540102	Kuche (Northern) Foldbelt	Gas	2	65	66	10,712	10,777	10,778	1
86	60590101	Magallanes Extensional Structures	Gas	10,042	14,697	24,739	10,386	25,082	35,124	70
87	20140101	Ghaba-Makarem Combined Structural	Oil	149	20,189	20,339	10,112	30,301	30,450	67
88	71830101	Coastal Plain and Offshore	Oil	46	1903	1949	10,071	11,974	12,020	16
89	10160106	Suprasalt	Oil	131	1496	1628	9934	11,430	11,562	14
90	20430102	Central Sirte Carbonates	Oil	4079	24,101	28,180	9918	34,019	38,098	74
91	11540101	Northern and Western Areas	Gas	24,483	27,929	52,412	9547	37,476	61,959	85
92	10090102	Devonian Synrift	Oil	0	0	0	9317	9317	9317	0
93	80430101	Eocene-Miocene Bombay Shelf	Oil	4887	12,641	17,528	9299	21,940	26,827	65
94	38220101	North Sumatra	Gas	13,977	11,582	25,559	9117	20,699	34,677	74
95	52150101	Jeanne d'Arc	Oil	0	2406	2406	8893	11,298	11,298	21
96	60600101	North Falklands Basin	Oil	0	0	0	8871	8871	8871	0
97	20230201	Platform Horst/Graben-Related Oil	Oil	0	100	100	8741	8841	8841	1
98	60340103	Abrolhos Sub-Volcanic Structures	Oil	0	0	0	8545	8545	8545	0
99	11500301	Upper Paleozoic Carbonates	NA	NA	NA	NA	8500	NA	NA	NA
100	39480201	Locker-Mungaroo/Barrow	Gas	27	8075	8102	8331	16,406	16,432	49
101	39100201	Petrel	Gas	0	5730	5730	8328	14,058	14,058	41
102	20160101	Fahud-Huqf Combined Structural	Oil	536	7097	7633	8206	15,303	15,839	48
103	60290102	Late Cretaceous-Tertiary Deep-Water Sandstones	Oil	66	169	236	8166	8336	8402	3
104	11120101	Apsheron-Pribalkhan Zone	Oil	7133	11,596	18,729	7908	19,503	26,636	70
105	31540103	Southwest Foldbelt	Oil	192	368	560	7773	8141	8333	7
106	60550101	Neuquen Extensional Structures	Oil	5293	11,407	16,699	7724	19,131	24,424	68
107	60600201	South Falklands Basin	Oil	0	0	0	7710	7710	7710	0
108	11540102	Karabil-Badkhyz (Southern Area)	Gas	11,053	47,057	58,110	7657	54,714	65,767	88
109	60450102	Foreland Basins	Oil	115	53	168	7375	7429	7543	2
110	37010102	Brunei-Sabah Turbidites	Oil	20	2425	2445	7249	9674	9694	25

Rank	Assessment Unit Code	Assessment Unit Name	Major Commodity	Cumulative Gas (BCFG)	Remaining Gas (BCFG)	Known Gas (BCFG)	Mean Undiscovered Gas (BCFG)	Future Gas (BCFG)	Gas Endowment (BCFG)	Discovery Maturity, Gas (%)
111	11080102	Foreland Slope	Gas	6556	8039	14,595	6894	14,933	21,489	68
112	10160103	East and Southeast Margins Subsalt	Oil	0	5364	5364	6781	12,145	12,145	44
113	60210102	Cretaceous Carbonates	Oil	0	0	0	6595	6595	6595	0
114	38240201	Ardjuna	Oil	794	6529	7322	6579	13,108	13,901	53
115	53050103	Tamabra-Like Debris-Flow-Breccia Limestone and Overlying Strata	Oil	335	195	530	6147	6342	6677	8
116	38220102	Mergui	Gas	0	0	0	6005	6005	6005	0
117	20710103	Southern Gulf of Suez	Oil	4	30	35	5974	6004	6008	1
118	20710202	Red Sea Salt Basin	Oil	0	0	0	5875	5875	5875	0
119	39300101	Gippsland	Oil	3501	6274	9775	5658	11,932	15,433	63
120	11740201	Pre-Upper Jurassic	Oil	443	4738	5180	5516	10,254	10,697	48
121	20540201	Tanezzuft-Melrhir Structural/Stratigraphic	Oil	0	125	125	4887	5012	5012	2
122	60810101	Cretaceous-Paleogene Basin	Oil	1951	941	2892	4795	5736	7687	38
123	11540103	Murgab Depression Suprasalt	Gas	26,549	22,967	49,516	4684	27,651	54,200	91
124	31270101	Tertiary Lacustrine	Oil	2955	8898	11,853	4639	13,537	16,492	72
125	80470201	Western Shelf and Slope	NA	NA	NA	NA	4613	NA	NA	NA
126	52430501	Mannville Gas	Gas	14,229	10,225	24,453	4597	14,821	29,050	84
127	11090303	Central Caspian Offshore	Oil	0	0	0	4572	4572	4572	0
128	31420101	Southeastern Fold Belt	Gas	5368	5313	10,681	4488	9801	15,169	70
129	10160102	North and West Margins Subsalt Barrier Reefs	Oil	8	1262	1270	4271	5533	5540	23
130	20430103	Offshore Sirte Hypothetical	Oil	0	0	0	4199	4199	4199	0
131	20710101	Gulf of Suez Block-Fault Fairway	Oil	835	5125	5960	4172	9296	10,132	59
132	38080101	Pematang/Sihapas Siliciclastics	Oil	910	2956	3866	4078	7034	7944	49
133	20190102	Cretaceous Reservoirs in South Gulf Suprasalt Structural	Oil	2705	65,090	67,795	4073	69,163	71,868	94
134	60960102	Peripheral	Oil	28	214	242	4036	4251	4279	6
135	60350103	Salt Dome Province Tertiary Sandstones	Oil	0	0	0	4031	4031	4031	0
136	80420102	Indus Fan	Gas	0	0	0	3919	3919	3919	0
137	11090301	Offshore Prikumsk Zone	Oil	0	0	0	3910	3910	3910	0
138	31270102	Pre-Tertiary Buried Hills	Oil	32	3787	3819	3875	7662	7693	50
139	37030201	South Malay Coaly	Gas	38	23,863	23,901	3768	27,631	27,669	86
140	60550102	Neuquen Foothills Structure	Oil	159	539	697	3764	4303	4462	16
141	72030101	Gabon Subsalt	Oil	62	780	843	3674	4455	4517	19
142	60590102	Andean Fold Belt Structures	Oil	11	57	69	3655	3712	3723	2
143	80430102	Eocene-Miocene Deltaics	Oil	1342	5324	6665	3645	8969	10,310	65
144	73030101	Offshore	Gas	0	6015	6015	3603	9618	9618	63
145	38170103	Kutei Basin Fold and Thrust Belt	NA	NA	NA	NA	3313	NA	NA	NA
146	60980103	Maturin Sub-Basin	Oil	5288	6791	12,078	3258	10,049	15,337	79
147	31440201	Structural Traps	Gas	0	145	145	3212	3357	3357	4
148	60450103	Foreland Central Chaco High	Oil	0	0	0	3211	3211	3211	0
149	11080101	Tertiary Foredeep	Oil	945	2999	3944	3139	6137	7083	56
150	60580101	San Jorge Extensional Structures	Oil	1435	1534	2968	3038	4572	6006	49
151	20580201	Tanezzuft-Ahnet Structural/Stratigraphic	Gas	0	3117	3117	2973	6090	6090	51

Rank	Assessment Unit Code	Assessment Unit Name	Major Commodity	Cumulative Gas (BCFG)	Remaining Gas (BCFG)	Known Gas (BCFG)	Mean Undiscovered Gas (BCFG)	Future Gas (BCFG)	Gas Endowment (BCFG)	Discovery Maturity, Gas (%)
152	11120105	Iran Onshore/Nearshore	Oil	0	0	0	2960	2960	2960	0
153	20190101	Cretaceous Reservoirs in Northwest Desert Anticlines	Oil	1114	15,198	16,312	2819	18,017	19,131	85
154	31420401	Leshan-Longnusi Paleohigh	Gas	491	859	1350	2763	3622	4113	33
155	20480101	Bou Dabbous-Tertiary Structural/Stratigraphic	Oil	37	15,472	15,509	2746	18,218	18,256	85
156	60900103	Eastern	Gas	0	2000	2000	2735	4735	4735	42
157	39100202	Vulcan Graben	Gas	0	0	0	2668	2668	2668	0
158	20580501	Tanezzuft-Benoud Structural/Stratigraphic	Gas	4980	100,070	105,050	2541	102,611	107,591	98
159	52430201	Leduc Gas	Gas	4964	4257	9221	2442	6699	11,663	79
160	40480101	Greater Hungarian Plain Basins	Oil	5436	4221	9657	2395	6616	12,052	80
161	40470201	Deformed Belt	Oil	3360	2033	5393	2345	4378	7738	70
162	20540101	Tanezzuft-Oued Mya Structural/Stratigraphic	Oil	68	8905	8973	2341	11,246	11,314	79
163	40470101	Foreland Basin	Gas	5455	5205	10,660	2321	7525	12,980	82
164	20480201	Jurassic-Cretaceous Structural/Stratigraphic	Oil	6	1295	1301	2280	3575	3581	36
165	11090201	South Mangyshlak (Entire)	Oil	28	5679	5707	2262	7941	7969	72
166	11080103	Subsalt Jurassic	Gas	84	81	165	2219	2300	2384	7
167	37020102	Balingian	Oil	0	2084	2084	2217	4301	4301	48
168	52430301	Exshaw-Rundle Gas	Gas	10,268	5145	15,413	2197	7342	17,610	88
169	11500201	Jurassic-Tertiary	Oil	0	1450	1450	2180	3630	3630	40
170	60960101	Central	Oil	111	14,961	15,071	2180	17,141	17,251	87
171	40610201	Romania Flysch Zone	Oil	530	2692	3222	2177	4869	5399	60
172	40570101	Transylvanian Neogene Suprasalt Gas	Gas	21,194	9537	30,731	2083	11,620	32,814	94
173	80480101	Central Burma Basin	Oil	389	591	981	2008	2600	2989	33
174	37020201	East Natuna	Gas	0	45,045	45,045	1963	47,008	47,008	96
175	11090103	Foreland Slope and Foredeep	Oil	0	0	0	1804	1804	1804	0
176	31440101	Subtle Traps	Oil	3	909	913	1789	2698	2701	34
177	37030102	North Malay Lacustrine	Oil	0	175	175	1762	1937	1937	9
178	40250102	Moray Firth	Oil	629	5223	5852	1752	6975	7604	77
179	60350102	Cretaceous Carbonates	Oil	79	403	482	1743	2146	2225	22
180	53050107	Tamabra-Like Debris-Flow-Breccia Limestone North of Campeche	Oil	0	0	0	1735	1735	1735	0
181	40600201	Thermal Triassic	Oil	154	257	412	1698	1955	2109	20
182	31420102	Northwestern Depression/Foldbelt	Gas	302	38	340	1650	1687	1990	17
183	72030401	Cuanza-Namibe	Oil	46	13	59	1558	1571	1617	4
184	10150201	Permian Reefs/Thrust Folds	Gas	571	2067	2638	1547	3615	4185	63
185	60990102	Southwest Maracaibo Basin Fold Belt	Oil	918	865	1783	1491	2356	3275	54
186	52430401	Combined Triassic/Jurassic Gas	Gas	1465	2500	3965	1289	3789	5254	75
187	31420402	Lower Paleozoic of Southeastern Fold Belt	Gas	0	0	0	1273	1273	1273	0
188	80340101	Sylhet-Kopili/Barail-Tipam Composite	Oil	2039	4495	6534	1178	5673	7712	85
189	11090302	Onshore Stavropol-Prikumsk	Oil	9670	3829	13,499	1163	4992	14,662	92
190	39100101	Barnett	Oil	0	48	48	1151	1199	1199	4
191	40480201	Zala-Drava-Sava Basins	Oil	1082	2847	3929	1151	3998	5080	77
192	10150101	Supra-Domanik Carbonates/Clastics	Oil	4508	72,570	77,078	1141	73,710	78,218	99

Rank	Assessment Unit Code	Assessment Unit Name	Major Commodity	Cumulative Gas (BCFG)	Remaining Gas (BCFG)	Known Gas (BCFG)	Mean Undiscovered Gas (BCFG)	Future Gas (BCFG)	Gas Endowment (BCFG)	Discovery Maturity, Gas (%)
193	20580101	Tanezzuft-Timimoun Structural/Stratigraphic	Gas	0	4200	4200	1128	5328	5328	79
194	53050104	Tamabra-Like Debris-Flow-Breccia Limestone of the Golden Lane	Oil	1862	1961	3824	1113	3074	4937	77
195	31150201	Jurassic/Tertiary Fluvial and Lacustrine Sandstone	Oil	10	332	342	950	1281	1292	26
196	60550103	Dorsal de Neuquen Structure	Oil	2184	1122	3307	927	2049	4234	78
197	52430102	Keg River Oil and Gas	Oil	570	691	1261	907	1599	2169	58
198	52430402	Combined Triassic/Jurassic Oil and Gas	Oil	3119	3497	6615	877	4374	7493	88
199	38240101	Sunda/Asri	Oil	233	381	614	874	1255	1488	41
200	52430101	Keg River Gas	Gas	4019	1793	5812	868	2661	6680	87
201	70130101	Coastal Plain and Offshore	Oil	4	45	49	856	901	905	5
202	60410201	Ene	Oil	0	0	0	841	841	841	0
203	52430602	Second White Specks-Cardium Oil and Gas	Oil	3298	2151	5449	840	2991	6289	87
204	60210103	Late Cretaceous-Tertiary Nearshore Sandstones	Oil	0	0	0	839	839	839	0
205	20430101	Southeast Sirte Clastics	Oil	1400	8187	9587	808	8995	10,395	92
206	53050105	Tamaulipas-Like Basinal Limestone and Tertiary Strata Without Underlying Evaporites	Oil	293	1443	1736	804	2247	2540	68
207	52430202	Leduc Oil and Gas	Oil	4086	1050	5136	782	1832	5918	87
208	60410101	Hollin-Napo	Oil	661	955	1616	755	1711	2372	68
209	60580102	San Bernardo Fold Belt Structures	Oil	232	365	597	735	1100	1332	45
210	31440102	Anticlinal	Oil	0	685	685	705	1390	1390	49
211	10150103	Sub-Domanik Devonian Clastics	Oil	4720	4728	9448	695	5423	10,144	93
212	20580301	Tanezzuft-Sbaa Structural/Stratigraphic	Oil	0	1490	1490	645	2134	2134	70
213	61170101	North Cuba Fold and Thrust Belt	Oil	20	77	96	592	668	688	14
214	40610101	Moesian Platform	Oil	1546	671	2217	561	1232	2778	80
215	10150102	Lower Volga	Oil	6618	3314	9932	558	3873	10,490	95
216	80260101	Kohat-Potwar Intrathrust Basin	Oil	807	1174	1981	551	1726	2533	78
217	53050101	El Abra-Like Reef and Backreef Limestone	Oil	209	1206	1415	507	1713	1922	74
218	60830101	Neogene Pull-Apart Basin	Oil	0	340	340	471	811	811	42
219	52430302	Exshaw-Rundle Oil and Gas	Oil	4583	1572	6155	466	2038	6621	93
220	20580601	Tanezzuft-Bechar/Abadla Structural/Stratigraphic	Gas	0	0	0	441	441	441	0
221	40610202	Romania Ploiesti Zone	Oil	1050	776	1826	441	1217	2267	81
222	60900104	La Luna and Older	Oil	2	2	5	399	401	403	1
223	40470301	Paleozoic Reservoirs	Oil	0	233	233	383	616	616	38
224	20160201	Natih-Fiqa Structural/Stratigraphic	Oil	616	2485	3100	363	2848	3463	90
225	31150101	Upper Paleozoic/Lower Mesozoic Nonmarine Coarse Clastics	Oil	491	1757	2248	346	2104	2594	87
226	40480601	Hungarian Paleogene Basin	Oil	15	1	16	339	340	355	4
227	20580401	Tanezzuft-Mouydir Structural/Stratigraphic	Gas	0	0	0	292	292	292	0
228	52430601	Second White Specks-Cardium Gas	Gas	974	1097	2071	285	1382	2356	88
229	60830201	Cretaceous-Paleogene Santa Elena Block	Oil	148	10	158	224	233	382	41
230	60290101	Western Pre-Aptian Reservoirs	Oil	282	890	1171	168	1057	1339	87
231	40480301	Danube Basin	Gas	7	240	247	146	386	393	63
232	31420201	Jurassic Lacustrine	Oil	50	90	140	137	227	278	51

Rank	Assessment Unit Code	Assessment Unit Name	Major Commodity	Cumulative Gas (BCFG)	Remaining Gas (BCFG)	Known Gas (BCFG)	Mean Undiscovered Gas (BCFG)	Future Gas (BCFG)	Gas Endowment (BCFG)	Discovery Maturity, Gas (%)
233	10080101	Northwest Izhma-Pechora Depression	NA	NA	NA	NA	120	NA	NA	NA
234	60900101	Northern	Oil	2186	605	2791	119	724	2910	96
235	40480401	Transcarpathian Basin	Gas	0	261	261	112	373	373	70
236	11500101	Mesozoic Sandstone Reservoirs	Oil	0	943	943	89	1033	1033	91
237	60340101	Espirito Santo Shelf	Oil	42	130	172	65	195	237	73
238	52440401	Lodgepole Oil	Oil	728	166	893	63	228	956	93
239	20430104	Southeast Sirte Hypothetical	Oil	0	0	0	60	60	60	0
240	20710102	Gulf of Suez Qaa Plain	Oil	0	0	0	46	46	46	0
241	60900102	Southern	Oil	121	85	206	40	125	246	84
242	40360101	Southern Permian Basin-U.K. Onshore	Oil	18	81	99	31	112	130	76
243	31280101	Jurassic/Triassic Fluvial and Lacustrine Sandstone	Oil	13	7	20	20	27	40	50
244	52440201	Brightholme Oil	Oil	8	3	11	5	8	16	67
245	52440101	Yeoman Oil	Oil	0	0	0	4	4	4	3
246	52440301	Bakken Sandstone	Oil	0	0	0	3	3	4	9
NA	10090103	Continuous Basin-Centered Gas Accumulation	NA	0	0	0	NA	NA	NA	NA
NA	10150104	Self-Sourced Domanik	NA	0	0	0	NA	NA	NA	NA
NA	10160105	Basin Center Subsalt	NA	0	0	0	NA	NA	NA	NA
NA	11740102	Self-Sourced Bazhenov Fractured Reservoirs	NA	0	0	0	NA	NA	NA	NA
NA	31150301	Continuous-Type (Basin-Centered) Gas	NA	0	0	0	NA	NA	NA	NA
NA	31270201	Coal-Sourced Gas	NA	0	0	0	NA	NA	NA	NA
NA	31280201	Basin Centered Gas (Continuous) in Ordovician Carbonate and Permian Sandstone	NA	0	5590	5590	NA	NA	NA	NA
NA	31280202	Carboniferous and Permian Coalbed Methane (Continuous)	NA	0	0	0	NA	NA	NA	NA
NA	31420103	Continuous Gas in Central Uplift/Northwestern Depression	NA	38	13	51	NA	NA	NA	NA
NA	31420301	Continuous Gas in Northwestern Depression/Central Uplift	NA	NA	NA	NA	NA	NA	NA	NA
NA	38240301	Biogenic Gas	NA	0	119	119	NA	NA	NA	NA
NA	38240401	Billiton Basin Hypothetical	NA	0	0	0	NA	NA	NA	NA
NA	40480501	Central Carpathian Paleogene Basin	NA	NA	NA	NA	NA	NA	NA	NA
NA	40570102	Transylvanian Subsalt	NA	0	0	0	NA	NA	NA	NA
NA	52150201	Graben Systems Exclusive of Jeanne d'Arc	NA	0	4224	4224	NA	NA	NA	NA
NA	52430502	Mannville Continuous (Basin-Center) Gas	NA	2942	3335	6278	NA	NA	NA	NA
NA	52430603	Upper Cretaceous Eastern Shallow Gas	NA	11,738	7191	18,929	NA	NA	NA	NA
NA	60220102	Offshore Gas Hydrates	NA	0	0	0	NA	NA	NA	NA
NA	60370102	Offshore Gas Hydrates	NA	0	0	0	NA	NA	NA	NA
NA	60580201	Hypothetical Basin-Centered Gas	NA	0	0	0	NA	NA	NA	NA
NA	60980104	Orinoco Heavy Oil and Tar Belt	NA	3	257	260	NA	NA	NA	NA
NA	60990103	Basin-Centered Gas	NA	0	0	0	NA	NA	NA	NA
NA	61070102	Barbados Accreationary Prism	NA	16	7	22	NA	NA	NA	NA
NA	80470101	Coalbed Gas	NA	0	0	0	NA	NA	NA	NA

App-1b_Page 7

APPENDIX 1C. Assessment units ranked by conventional oil endowment.

[MMBO = million barrels of oil. Major commodity is based on the ratio of gas endowment to oil endowment as ft^3 of gas per bbl of oil (CFG/BO). Major commodity is oil if this ratio is less than 20,000 CFG/BO and gas if this ratio is equal to or greater than 20,000 CFG/BO. Cumulative oil is the reported volume of oil that has been produced. Remaining oil is the calculated remaining reserves; the difference between known and cumulative oil. Known oil is the reported discovered volume (cumulative production plus remaining reserves). Mean undiscovered oil is the mean volume estimated in World Petroleum Assessment 2000 (USGS, 2000). Future oil is the remaining oil plus estimated mean undiscovered oil. Oil endowment is the known oil plus estimated mean undiscovered oil volumes. Discovery maturity is the percentage of oil discovered (known oil) with respect to oil endowment. Calculated values may not be equal to the sums of the component values because numbers have been independently rounded. Negative values result from inconsistent reporting of cumulative and known volumes. Values in parentheses are questionable data. NA means not applicable; it is shown for all values where there is only one field present, and used where the endowment could not be calculated.] Known volumes from Petroconsultants (1996) and NRG Associates (1995). Note that gas = natural gas.

Rank	Assessment Unit Code	Assessment Unit Name	Major Commodity	Cumulative Oil (MMBO)	Remaining Oil (MMBO)	Known Oil (MMBO)	Mean Undiscovered Oil (MMBO)	Future Oil (MMBO)	Oil Endowment (MMBO)	Discovery Maturity, Oil (%)
1	20300101	Cretaceous Reservoirs	Oil	53,182	209,735	262,917	74,911	284,646	337,828	78
2	20210201	Horst-Block Anticlinal Oil	Oil	53,906	101,239	155,145	26,715	127,954	181,860	85
3	11740101	Upper Jurassic-Cretaceous Sandstones	Oil	45,819	72,153	117,972	42,947	115,100	160,919	73
4	20300102	Tertiary Reservoirs	Oil	47,285	62,025	109,310	19,605	81,630	128,914	85
5	20210202	Salt-Involved Structural Oil	Oil	9948	33,902	43,850	14,989	48,891	58,839	75
6	71920101	Agbada Reservoirs	Oil	16,598	17,925	34,522	21,925	39,850	56,447	61
7	60990101	Main Maracaibo Basin	Oil	34,861	13,350	48,211	7842	21,192	56,053	86
8	20190102	Cretaceous Reservoirs in South Gulf Suprasalt Structural	Oil	10,858	41,236	52,094	3764	45,000	55,858	93
9	52000101	Northeast Greenland Shelf Rift Systems	Oil	0	0	0	47,148	47,148	47,148	0
10	53050102	Tamabra-Like Debris-Flow-Breccia Limestone Overlying Evaporites	Oil	14,366	18,699	33,065	9857	28,556	42,922	77
11	10150101	Supra-Domanik Carbonates/Clastics	Oil	19,973	14,931	34,904	1676	16,607	36,580	95
12	40250101	Viking Graben	Oil	13,038	13,643	26,681	7412	21,055	34,093	78
13	10150103	Sub-Domanik Devonian Clastics	Oil	20,807	5632	26,439	682	6314	27,121	97
14	20210101	Central Arch Horst-Block Anticlinal Oil and Gas	Oil	84	6142	6226	19,190	25,331	25,416	24
15	20190101	Cretaceous Reservoirs in Northwest Desert Anticlines	Oil	2594	16,386	18,981	4702	21,088	23,683	80
16	20430101	Southeast Sirte Clastics	Oil	5274	15,622	20,895	2380	18,002	23,275	90
17	31270101	Tertiary Lacustrine	Oil	8944	10,993	19,937	2268	13,261	22,205	90
18	20190202	Jurassic Reservoirs in South Gulf Suprasalt/Qatar Arch Structural	Oil	5688	12,548	18,236	3845	16,394	22,081	83
19	10160104	South Margin Subsalt	Oil	0	6245	6245	15,467	21,712	21,712	29
20	60350101	Late Cretaceous-Tertiary Turbidites	Oil	1508	8122	9630	10,869	18,991	20,499	47
21	20430102	Central Sirte Carbonates	Oil	8712	7465	16,177	3840	11,305	20,017	81
22	11740201	Pre-Upper Jurassic	Oil	645	11,111	11,756	7474	18,584	19,230	61
23	72030302	Central Congo Turbidites	Oil	0	205	205	18,522	18,727	18,727	1
24	71920102	Akata Reservoirs	Oil	0	0	0	18,562	18,562	18,562	0

Rank	Assessment Unit Code	Assessment Unit Name	Major Commodity	Cumulative Oil (MMBO)	Remaining Oil (MMBO)	Known Oil (MMBO)	Mean Undiscovered Oil (MMBO)	Future Oil (MMBO)	Oil Endowment (MMBO)	Discovery Maturity, Oil (%)
25	11120101	Apsheron-Pribalkhan Zone	Oil	10,581	4512	15,093	3091	7602	18,183	83
26	60980101	East Venezuela Fold and Thrust Belt	Oil	3253	9978	13,231	4928	14,905	18,159	73
27	10080102	Main Basin Platform	Oil	2543	10,442	12,985	4938	15,379	17,922	72
28	40250103	Central Graben	Oil	7181	5588	12,769	4740	10,328	17,509	73
29	31440102	Anticlinal	Oil	12,228	2192	14,420	291	2483	14,711	98
30	72030301	Central Congo Delta and Carbonate Platform	Oil	3693	5847	9540	4677	10,523	14,216	67
31	38080101	Pematang/Sihapas Siliciclastics	Oil	8349	4868	13,217	899	5768	14,117	94
32	60360101	Santos Shelf	Oil	6	279	285	13,723	14,002	14,008	2
33	60210101	Late Cretaceous-Tertiary Turbidites	Oil	0	0	0	12,745	12,745	12,745	0
34	53050106	Tamaulipas-Like Basinal Limestone and Tertiary Strata Overlying Evaporites	Oil	1929	2339	4268	7823	10,163	12,092	35
35	20710101	Gulf of Suez Block-Fault Fairway	Oil	6310	3444	9754	2171	5615	11,925	82
36	20540101	Tanezzuft-Oued Mya Structural/Stratigraphic	Oil	5872	4971	10,843	830	5801	11,673	93
37	11740301	Northern West Siberian Onshore Gas	Gas	66	10,119	10,185	1304	11,424	11,489	89
38	20190201	Jurassic Reservoirs in Northwest Desert Anticlines	Oil	0	777	777	10,633	11,410	11,410	7
39	60980103	Maturin Sub-Basin	Oil	7468	1978	9446	1919	3897	11,365	83
40	52430202	Leduc Oil and Gas	Oil	8113	1304	9417	501	1805	9918	95
41	60410101	Hollin-Napo	Oil	2889	3712	6601	3023	6735	9625	69
42	60360102	Salt-Structured Deep-Water Sandstones	Oil	0	0	0	9486	9486	9486	0
43	20540301	Tanezzuft-Ghadames Structural/Stratigraphic	Oil	1629	2909	4538	4461	7370	8999	50
44	40170101	Halten Terrace-Trondelag Platform	Oil	76	2585	2660	6204	8788	8864	30
45	37010101	Brunei-Sabah Deltaics	Oil	3872	2780	6652	2137	4917	8789	76
46	80430101	Eocene-Miocene Bombay Shelf	Oil	2337	3920	6257	1612	5531	7868	80
47	72030201	Gabon Suprasalt	Oil	1278	836	2114	5031	5867	7145	30
48	31150101	Upper Paleozoic/Lower Mesozoic Nonmarine Coarse Clastics	Oil	937	5687	6624	467	6154	7091	93
49	31270102	Pre-Tertiary Buried Hills	Oil	1471	3145	4616	2144	5289	6760	68
50	20560101	Tanezzuft-Illizi Structural/Stratigraphic	Oil	2286	1384	3670	2814	4198	6484	57
51	11120104	Central Offshore	Oil	0	0	0	6410	6410	6410	0
52	31540101	Tarim Basin Excluding Marginal Foldbelts	Oil	21	634	655	5364	5998	6019	11
53	11090201	South Mangyshlak (Entire)	Oil	2406	2835	5241	651	3486	5893	89
54	11090101	Foldbelt-Foothills	Oil	752	2738	3490	2094	4832	5584	62
55	40250102	Moray Firth	Oil	2248	2196	4444	947	3143	5391	82
56	10160106	Suprasalt	Oil	690	1140	1830	3497	4637	5327	34
57	20230201	Platform Horst/Graben-Related Oil	Oil	94	379	474	4682	5062	5156	9
58	11120103	Gograndag-Okarem Zone and Adjacent Shelf	Oil	64	242	306	4764	5005	5070	6
59	13220101	Onshore and Offshore Northeastern Shelf	Oil	754	1428	2182	2582	4010	4764	46
60	20040101	Ma'Rib-Al Jawf/Shabwah/Masila	Oil	649	1284	1933	2740	4025	4673	41
61	60960102	Peripheral	Oil	1027	1282	2309	2319	3601	4627	50
62	60350103	Salt Dome Province Tertiary Sandstones	Oil	0	0	0	4493	4493	4493	0
63	60960101	Central	Oil	152	2942	3094	1312	4254	4406	70
64	60980202	Orinoco Delta and Offshore	Oil	0	0	0	4228	4228	4228	0
65	39300101	Gippsland	Oil	3077	784	3860	310	1093	4170	93

App-1c_Page 2

Rank	Assessment Unit Code	Assessment Unit Name	Major Commodity	Cumulative Oil (MMBO)	Remaining Oil (MMBO)	Known Oil (MMBO)	Mean Undiscovered Oil (MMBO)	Future Oil (MMBO)	Oil Endowment (MMBO)	Discovery Maturity, Oil (%)
66	38170101	Kutei Basin Deltaics	Oil	2195	684	2879	1258	1942	4137	70
67	37030101	South Malay Lacustrine	Oil	1459	1557	3017	1003	2560	4019	75
68	20160101	Fahud-Huqf Combined Structural	Oil	1259	1426	2685	1304	2731	3990	67
69	53050104	Tamabra-Like Debris-Flow-Breccia Limestone of the Golden Lane	Oil	1917	1392	3308	557	1948	3865	86
70	60980201	Trinidad Basins	Oil	2780	667	3447	416	1083	3863	89
71	20230101	Horst/Graben-Related Oil and Gas	Oil	269	944	1213	2632	3577	3845	32
72	12100101	Baikal-Patom Foldbelt Riphean-Craton Margin Vendian	Gas	1	2005	2006	1611	3616	3617	55
73	60600101	North Falklands Basin	Oil	0	0	0	3603	3603	3603	0
74	38170102	Kutei Basin Turbidites	Oil	0	0	0	3569	3569	3569	0
75	11740302	South Kara Sea Offshore	Gas	0	0	0	3510	3510	3510	0
76	60580101	San Jorge Extensional Structures	Oil	2330	776	3106	347	1123	3453	90
77	10160103	East and Southeast Margins Subsalt	Oil	131	1086	1217	2196	3283	3414	36
78	39480101	Dingo-Mungaroo/Barrow	Gas	489	660	1149	2249	2909	3398	34
79	60810101	Cretaceous-Paleogene Basin	Oil	1346	326	1672	1711	2038	3383	49
80	52150101	Jeanne d'Arc	Oil	0	1582	1582	1796	3378	3378	47
81	11120102	Lower Kura Depression and Adjacent Shelf	Oil	1035	1004	2040	1275	2279	3315	62
82	11500101	Mesozoic Sandstone Reservoirs	Oil	459	1827	2286	814	2641	3100	74
83	60370101	Pelotas Platform and Basin	Oil	0	0	0	2938	2938	2938	0
84	80340101	Sylhet-Kopili/Barail-Tipam Composite	Oil	911	1625	2536	399	2024	2935	86
85	38280101	South Sumatra	Oil	1842	587	2429	469	1055	2898	84
86	52440401	Lodgepole Oil	Oil	2239	539	2778	111	650	2889	96
87	20480101	Bou Dabbous-Tertiary Structural/Stratigraphic	Oil	481	1633	2114	667	2300	2782	76
88	20140101	Ghaba-Makarem Combined Structural	Oil	715	1313	2028	743	2056	2771	73
89	10150102	Lower Volga	Oil	1803	726	2529	148	874	2677	94
90	80430102	Eocene-Miocene Deltaics	Oil	1038	1146	2183	479	1625	2662	82
91	60550101	Neuquen Extensional Structures	Oil	1111	605	1716	862	1467	2578	67
92	40610201	Romania Flysch Zone	Oil	1039	909	1948	564	1473	2512	78
93	72030101	Gabon Subsalt	Oil	708	1044	1752	727	1772	2480	71
94	40360102	Southern Permian Basin-Europe Onshore	Gas	2029	372	2400	74	446	2474	97
95	52430602	Second White Specks-Cardium Oil and Gas	Oil	1929	494	2423	36	529	2459	99
96	40610202	Romania Ploiesti Zone	Oil	1793	374	2167	278	652	2445	89
97	53050101	El Abra-Like Reef and Backreef Limestone	Oil	2546	(-483)	2063	289	(-195)	2352	88
98	20710201	Red Sea Coastal Block Faults	Gas	22	133	155	2136	2269	2291	7
99	60600201	South Falklands Basin	Oil	0	0	0	2229	2229	2229	0
100	20160201	Natih-Fiqa Structural/Stratigraphic	Oil	1299	553	1852	371	925	2224	83
101	60210102	Cretaceous Carbonates	Oil	0	0	0	2200	2200	2200	0
102	38240201	Ardjuna	Oil	727	1181	1908	195	1377	2104	91
103	53050105	Tamaulipas-Like Basinal Limestone and Tertiary Strata Without Underlying Evaporites	Oil	1354	273	1627	366	638	1993	82
104	40610101	Moesian Platform	Oil	1251	542	1793	171	713	1963	91
105	60900101	Northern	Oil	1556	175	1730	191	366	1921	90

Rank	Assessment Unit Code	Assessment Unit Name	Major Commodity	Cumulative Oil (MMBO)	Remaining Oil (MMBO)	Known Oil (MMBO)	Mean Undiscovered Oil (MMBO)	Future Oil (MMBO)	Oil Endowment (MMBO)	Discovery Maturity, Oil (%)
106	20230202	Basinal Oil and Gas	Oil	0	0	0	1886	1886	1886	0
107	31440101	Subtle Traps	Oil	171	979	1150	735	1714	1885	61
108	20540201	Tanezzuft-Melrhir Structural/Stratigraphic	Oil	0	6	6	1875	1881	1881	0
109	10090101	Carboniferous-Lower Permian Clastics	Gas	110	1501	1611	251	1752	1862	87
110	38240101	Sunda/Asri	Oil	828	431	1259	462	894	1722	73
111	60590101	Magallanes Extensional Structures	Gas	595	620	1216	494	1114	1709	71
112	52430302	Exshaw-Rundle Oil and Gas	Oil	1122	415	1537	126	541	1663	92
113	39100301	Malita	Oil	123	378	502	1146	1524	1648	30
114	60290102	Late Cretaceous-Tertiary Deep-Water Sandstones	Oil	55	67	122	1457	1523	1579	8
115	61070101	Inner Forearc Deformation Belt	Oil	0	0	0	1536	1536	1536	0
116	37010102	Brunei-Sabah Turbidites	Oil	81	165	246	1252	1417	1498	16
117	60350102	Cretaceous Carbonates	Oil	261	166	426	932	1097	1358	31
118	52430402	Combined Triassic/Jurassic Oil and Gas	Oil	965	277	1242	96	373	1338	93
119	40480101	Greater Hungarian Plain Basins	Oil	731	358	1089	196	555	1285	85
120	40470201	Deformed Belt	Oil	640	283	923	344	627	1267	73
121	60450102	Foreland Basins	Oil	0	2	2	1254	1256	1256	0
122	12070101	Yenisey Foldbelt Riphean-Craton Margin Riphean	Oil	0	34	34	1213	1246	1246	3
123	71830101	Coastal Plain and Offshore	Oil	74	162	236	1004	1166	1240	19
124	40480201	Zala-Drava-Sava Basins	Oil	836	265	1101	114	379	1215	91
125	60990102	Southwest Maracaibo Basin Fold Belt	Oil	720	141	861	341	482	1202	72
126	37020102	Balingian	Oil	168	460	628	529	989	1157	54
127	80480101	Central Burma Basin	Oil	494	142	637	516	658	1152	55
128	11090302	Onshore Stavropol-Prikumsk	Oil	1200	(-380)	820	287	(-93)	1107	74
129	39130101	Late Jurassic/Early Cretaceous-Mesozoic	Gas	19	27	46	1055	1083	1101	4
130	52430102	Keg River Oil and Gas	Oil	835	176	1011	84	260	1095	92
131	60630101	Malvinas Extensional Structures	Oil	0	10	10	1031	1041	1041	1
132	72030401	Cuanza-Namibe	Oil	86	20	107	790	810	897	12
133	53050103	Tamabra-Like Debris-Flow-Breccia Limestone and Overlying Strata	Oil	26	54	80	807	860	886	9
134	53050107	Tamabra-Like Debris-Flow-Breccia Limestone North of Campeche	Oil	0	0	0	871	871	871	0
135	60900102	Southern	Oil	348	333	681	188	521	869	78
136	10090102	Devonian Synrift	Oil	0	0	0	848	848	848	0
137	31280101	Jurassic/Triassic Fluvial and Lacustrine Sandstone	Oil	81	602	683	138	740	821	83
138	61170101	North Cuba Fold and Thrust Belt	Oil	65	232	297	494	726	790	38
139	11090301	Offshore Prikumsk Zone	Oil	0	0	0	789	789	789	0
140	38220101	North Sumatra	Gas	496	178	674	112	289	786	86
141	11080101	Tertiary Foredeep	Oil	1268	(-770)	498	280	(-490)	778	64
142	10160102	North and West Margins Subsalt Barrier Reefs	Oil	0	134	134	637	771	771	17
143	60450103	Foreland Central Chaco High	Oil	0	0	0	711	711	711	0
144	60980102	Guarico Sub-Basin	Gas	268	54	322	385	439	707	46
145	37030201	South Malay Coaly	Gas	58	533	592	101	634	693	85
146	40600201	Thermal Triassic	Oil	78	254	332	360	614	692	48

Rank	Assessment Unit Code	Assessment Unit Name	Major Commodity	Cumulative Oil (MMBO)	Remaining Oil (MMBO)	Known Oil (MMBO)	Mean Undiscovered Oil (MMBO)	Future Oil (MMBO)	Oil Endowment (MMBO)	Discovery Maturity, Oil (%)
147	20190103	Mesozoic/Tertiary Foredeep Fold and Thrust	Gas	92	64	156	513	577	669	23
148	60550102	Neuquen Foothills Structure	Oil	159	138	297	334	472	631	47
149	11540104	Murgab Depression Subsalt	Gas	31	485	516	113	597	628	82
150	10500101	Kolguyev Terrace	Gas	0	51	51	563	614	614	8
151	60290101	Western Pre-Aptian Reservoirs	Oil	413	157	570	31	187	601	95
152	31540103	Southwest Foldbelt	Oil	17	24	41	495	519	536	8
153	20480201	Jurassic-Cretaceous Structural/Stratigraphic	Oil	39	64	103	403	467	506	20
154	20190301	Khuff Carbonates in Salt Structures	Gas	0	500	500	0	500	500	100
155	60450101	Sub-Andean Fold and Thrust Belt	Gas	257	62	318	181	242	499	64
156	40360103	Southern Permian Basin-Offshore	Gas	232	176	408	62	238	470	87
157	80260101	Kohat-Potwar Intrathrust Basin	Oil	204	121	324	145	266	469	69
158	60340103	Abrolhos Sub-Volcanic Structures	Oil	0	0	0	468	468	468	0
159	11090103	Foreland Slope and Foredeep	Oil	0	0	0	452	452	452	0
160	60550103	Dorsal de Neuquen Structure	Oil	262	95	357	93	189	450	79
161	20580301	Tanezzuft-Sbaa Structural/Stratigraphic	Oil	0	284	284	162	446	446	64
162	20710202	Red Sea Salt Basin	Oil	0	0	0	443	443	443	0
163	20710103	Southern Gulf of Suez	Oil	41	15	57	344	359	400	14
164	60580102	San Bernardo Fold Belt Structures	Oil	115	124	240	151	275	390	61
165	40680101	Subsalt	Gas	0	0	0	388	388	388	0
166	20430103	Offshore Sirte Hypothetical	Oil	0	0	0	382	382	382	0
167	60830201	Cretaceous-Paleogene Santa Elena Block	Oil	118	53	171	200	252	370	46
168	60210103	Late Cretaceous-Tertiary Nearshore Sandstones	Oil	17	46	63	301	347	364	17
169	40170102	Mid-Norway Continental Margin	Gas	0	0	0	334	334	334	0
170	11090303	Central Caspian Offshore	Oil	0	0	0	331	331	331	0
171	31150201	Jurassic/Tertiary Fluvial and Lacustrine Sandstone	Oil	16	193	210	96	289	306	69
172	10080103	Foredeep Basins	Gas	803	(-719)	84	214	(-504)	299	28
173	80420101	Greater Indus Foreland and Foldbelt	Gas	111	70	180	119	188	299	60
174	80480102	Irrawaddy-Andaman	Gas	34	46	79	209	254	288	28
175	11090102	Terek-Sunzha Subsalt Jurassic	Gas	0	0	0	283	283	283	0
176	10500102	South Barents and Ludlov Saddle	Gas	0	0	0	257	257	257	0
177	20430104	Southeast Sirte Hypothetical	Oil	0	0	0	253	253	253	0
178	60900104	La Luna and Older	Oil	11	4	15	231	235	246	6
179	11500201	Jurassic-Tertiary	Oil	0	80	80	139	220	220	37
180	60590102	Andean Fold Belt Structures	Oil	0	0	0	210	211	211	0
181	11540101	Northern and Western Areas	Gas	48	92	140	70	162	210	67
182	11540102	Karabil-Badkhyz (Southern Area)	Gas	0	110	111	88	198	198	56
183	52430301	Exshaw-Rundle Gas	Gas	170	20	191	0	20	191	100
184	11120105	Iran Onshore/Nearshore	Oil	0	0	0	185	185	185	0
185	60340101	Espirito Santo Shelf	Oil	93	37	130	54	91	184	71
186	37020101	Central Luconia	Gas	0	170	170	0	170	170	100
187	70130101	Coastal Plain and Offshore	Oil	0	10	10	157	167	168	6
188	20580501	Tanezzuft-Benoud Structural/Stratigraphic	Gas	16	72	88	72	144	160	55
189	39100101	Barnett	Oil	0	15	15	140	155	155	10

App-1c_Page 5

Rank	Assessment Unit Code	Assessment Unit Name	Major Commodity	Cumulative Oil (MMBO)	Remaining Oil (MMBO)	Known Oil (MMBO)	Mean Undiscovered Oil (MMBO)	Future Oil (MMBO)	Oil Endowment (MMBO)	Discovery Maturity, Oil (%)
190	20210102	North Gulf Salt Basin Structural Gas	Gas	0	150	150	0	150	150	100
191	39480201	Locker-Mungaroo/Barrow	Gas	0	6	6	132	137	137	4
192	31420201	Jurassic Lacustrine	Oil	38	71	109	27	99	136	80
193	10500103	North Barents	Gas	0	0	0	135	135	135	0
194	37030102	North Malay Lacustrine	Oil	0	0	0	124	124	124	0
195	37020201	East Natuna	Gas	0	20	20	102	122	122	16
196	73030101	Offshore	Gas	0	0	0	116	116	116	0
197	10150201	Permian Reefs/Thrust Folds	Gas	106	(-41)	65	42	(1)	106	61
198	38220102	Mergui	Gas	0	0	0	101	101	101	0
199	60900103	Eastern	Gas	0	0	0	99	99	99	0
200	31540102	Kuche (Northern) Foldbelt	Gas	7	0	7	89	89	96	8
201	11080103	Subsalt Jurassic	Gas	4	1	5	90	91	95	5
202	20710102	Gulf of Suez Qaa Plain	Oil	0	0	0	93	93	93	0
203	60410201	Ene	Oil	0	0	0	74	74	74	0
204	40360101	Southern Permian Basin-U.K. Onshore	Oil	39	25	64	7	32	71	90
205	11080102	Foreland Slope	Gas	1	19	20	48	67	68	30
206	40480601	Hungarian Paleogene Basin	Oil	10	1	11	49	50	60	18
207	52430401	Combined Triassic/Jurassic Gas	Gas	47	7	54	0	7	54	100
208	40470301	Paleozoic Reservoirs	Oil	0	1	1	47	47	47	1
209	52430201	Leduc Gas	Gas	39	2	42	0	2	42	100
210	60830101	Neogene Pull-Apart Basin	Oil	4	0	4	37	37	41	11
211	20580201	Tanezzuft-Ahnet Structural/Stratigraphic	Gas	0	1	1	34	35	35	3
212	20580101	Tanezzuft-Timimoun Structural/Stratigraphic	Gas	0	0	0	31	31	31	0
213	80470302	Eastern Fold Belt	Gas	3	2	4	24	26	28	16
214	52440201	Brightholme Oil	Oil	11	4	14	8	11	22	66
215	61030101	Carupano Basin Gas	Gas	0	20	20	0	20	20	100
216	52440301	Bakken Sandstone	Oil	5	2	7	11	13	18	40
217	20580601	Tanezzuft-Bechar/Abadla Structural/Stratigraphic	Gas	0	0	0	16	16	16	0
218	80470301	Central Basin	Gas	0	0	0	15	15	15	0
219	40600101	Neogene Flysch Gas	Gas	13	0	14	0	0	14	100
220	20580401	Tanezzuft-Mouydir Structural/Stratigraphic	Gas	0	0	0	12	12	12	0
221	52440101	Yeoman Oil	Oil	1	0	1	5	6	6	15
222	31440201	Structural Traps	Gas	0	5	5	0	5	5	100
223	31420401	Leshan-Longnusi Paleohigh	Gas	9	(-8)	1	0	(-8)	1	100
224	40470101	Foreland Basin	Gas	1	1	1	0	1	1	100
225	11540103	Murgab Depression Suprasalt	Gas	1	(-1)	0	0	(-1)	0	NA
226	20300201	Northern Qatar Arch Extension	Gas	0	0	0	0	0	0	NA
227	31420101	Southeastern Fold Belt	Gas	0	0	0	0	0	0	100
228	31420102	Northwestern Depression/Foldbelt	Gas	0	0	0	0	0	0	NA
229	31420402	Lower Paleozoic of Southeastern Fold Belt	Gas	0	0	0	0	0	0	NA
230	39100201	Petrel	Gas	0	0	0	0	0	0	NA
231	39100202	Vulcan Graben	Gas	0	0	0	0	0	0	NA
232	40480301	Danube Basin	Gas	0	0	0	0	0	0	NA

Rank	Assessment Unit Code	Assessment Unit Name	Major Commodity	Cumulative Oil (MMBO)	Remaining Oil (MMBO)	Known Oil (MMBO)	Mean Undiscovered Oil (MMBO)	Future Oil (MMBO)	Oil Endowment (MMBO)	Discovery Maturity, Oil (%)
233	40480401	Transcarpathian Basin	Gas	0	0	0	0	0	0	NA
234	40570101	Transylvanian Neogene Suprasalt Gas	Gas	0	0	0	0	0	0	NA
235	52430101	Keg River Gas	Gas	0	0	0	0	0	0	NA
236	52430501	Mannville Gas	Gas	0	0	0	0	0	0	NA
237	52430601	Second White Specks-Cardium Gas	Gas	0	0	0	0	0	0	NA
238	80420102	Indus Fan	Gas	0	0	0	0	0	0	NA
NA	10080101	Northwest Izhma-Pechora Depression	NA	NA	NA	NA	18	NA	NA	NA
NA	10090103	Continuous Basin-Centered Gas Accumulation	NA	0	0	0	NA	NA	NA	NA
NA	10150104	Self-Sourced Domanik	NA	0	0	0	NA	NA	NA	NA
NA	10160101	North and West Margins Subsalt Pinnacle Reefs	NA	NA	NA	NA	1522	NA	NA	NA
NA	10160105	Basin Center Subsalt	NA	0	0	0	NA	NA	NA	NA
NA	11500301	Upper Paleozoic Carbonates	NA	NA	NA	NA	0	NA	NA	NA
NA	11740102	Self-Sourced Bazhenov Fractured Reservoirs	NA	0	0	0	NA	NA	NA	NA
NA	20190302	Paleozoic Reservoirs	NA	NA	NA	NA	13,235	NA	NA	NA
NA	31150301	Continuous-Type (Basin-Centered) Gas	NA	0	0	0	NA	NA	NA	NA
NA	31270201	Coal-Sourced Gas	NA	0	0	0	NA	NA	NA	NA
NA	31280201	Basin Centered Gas (Continuous) in Ordovician Carbonate and Permian Sandstone	NA	0	0	0	NA	NA	NA	NA
NA	31280202	Carboniferous and Permian Coalbed Methane (Continuous)	NA	0	0	0	NA	NA	NA	NA
NA	31420103	Continuous Gas in Central Uplift/Northwestern Depression	NA	0	0	0	NA	NA	NA	NA
NA	31420301	Continuous Gas in Northwestern Depression/Central Uplift	NA	NA	NA	NA	NA	NA	NA	NA
NA	38170103	Kutei Basin Fold and Thrust Belt	NA	NA	NA	NA	318	NA	NA	NA
NA	38240301	Biogenic Gas	NA	0	0	0	NA	NA	NA	NA
NA	38240401	Billiton Basin Hypothetical	NA	0	0	0	NA	NA	NA	NA
NA	40480501	Central Carpathian Paleogene Basin	NA	NA	NA	NA	NA	NA	NA	NA
NA	40570102	Transylvanian Subsalt	NA	0	0	0	NA	NA	NA	NA
NA	52150201	Graben Systems Exclusive of Jeanne d'Arc	NA	0	0	0	NA	NA	NA	NA
NA	52430502	Mannville Continuous (Basin-Center) Gas	NA	0	0	0	NA	NA	NA	NA
NA	52430603	Upper Cretaceous Eastern Shallow Gas	NA	216	49	265	NA	NA	NA	NA
NA	60220101	Amazon Delta and Submarine Fan	NA	NA	NA	NA	0	NA	NA	NA
NA	60220102	Offshore Gas Hydrates	NA	0	0	0	NA	NA	NA	NA
NA	60340102	Late Cretaceous-Tertiary Slide Blocks and Turbidites	NA	NA	NA	NA	2437	NA	NA	NA
NA	60370102	Offshore Gas Hydrates	NA	0	0	0	NA	NA	NA	NA
NA	60580201	Hypothetical Basin-Centered Gas	NA	0	0	0	NA	NA	NA	NA
NA	60980104	Orinoco Heavy Oil and Tar Belt	NA	38	3719	3757	NA	NA	NA	NA
NA	60990103	Basin-Centered Gas	NA	0	0	0	NA	NA	NA	NA
NA	61070102	Barbados Accreationary Prism	NA	8	2	11	NA	NA	NA	NA
NA	80470101	Coalbed Gas	NA	0	0	0	NA	NA	NA	NA
NA	80470201	Western Shelf and Slope	NA	NA	NA	NA	64	NA	NA	NA

Appendix 1D. Assessment units ranked by conventional natural gas endowment.

[BCFG = billion cubic feet of gas. Major commodity is based on the ratio of gas endowment to oil endowment as ft^3 of gas per bbl of oil (CFG/BO). Major commodity is oil if this ratio is less than 20,000 CFG/BO and gas if this ratio is equal to or greater than 20,000 CFG/BO. Cumulative gas is the reported volume of gas that has been produced. Remaining gas is the calculated remaining reserves; the difference between known and cumulative gas. Known gas is the reported discovered volume (cumulative production plus remaining reserves). Mean undiscovered gas is the mean volume estimated in World Petroleum Assessment 2000 (USGS, 2000). Future gas is the remaining gas plus estimated mean undiscovered gas. Gas endowment is the known gas plus estimated mean undiscovered gas volumes. Discovery maturity is the percentage of gas discovered (known gas) with respect to gas endowment. Calculated values may not be equal to the sums of the component values because numbers have been independently rounded. NA means not applicable; it is shown for all values where there is only one field present, and used where the endowment could not be calculated.] Known volumes from Petroconsultants (1996) and NRG Associates (1995). Note that gas = natural gas.

Rank	Assessment Unit Code	Assessment Unit Name	Major Commodity	Cumulative Gas (BCFG)	Remaining Gas (BCFG)	Known Gas (BCFG)	Mean Undiscovered Gas (BCFG)	Future Gas (BCFG)	Gas Endowment (BCFG)	Discovery Maturity, Gas (%)
1	11740301	Northern West Siberian Onshore Gas	Gas	188,736	896,852	1,085,588	136,608	1,033,460	1,222,196	89
2	11740302	South Kara Sea Offshore	Gas	0	81,700	81,700	403,875	485,575	485,575	17
3	20190301	Khuff Carbonates in Salt Structures	Gas	1018	450,982	452,000	22,999	473,981	474,999	95
4	20300102	Tertiary Reservoirs	Oil	10,075	249,365	259,440	93,499	342,864	352,939	74
5	20300101	Cretaceous Reservoirs	Oil	15,185	218,613	233,798	104,098	322,711	337,896	69
6	10500102	South Barents and Ludlov Saddle	Gas	0	78,140	78,140	189,244	267,384	267,384	29
7	20210201	Horst-Block Anticlinal Oil	Oil	19,990	209,625	229,615	16,055	225,680	245,670	93
8	20210101	Central Arch Horst-Block Anticlinal Oil and Gas	Oil	0	12,115	12,115	230,387	242,502	242,502	5
9	20210102	North Gulf Salt Basin Structural Gas	Gas	0	67,000	67,000	152,952	219,952	219,952	30
10	11540104	Murgab Depression Subsalt	Gas	16,456	54,120	70,576	141,762	195,883	212,338	33
11	20300201	Northern Qatar Arch Extension	Gas	0	131,220	131,220	69,031	200,251	200,251	66
12	11740101	Upper Jurassic-Cretaceous Sandstones	Oil	29,287	68,295	97,582	96,934	165,229	194,516	50
13	71920101	Agbada Reservoirs	Oil	6232	87,579	93,811	85,176	172,755	178,986	52
14	10160104	South Margin Subsalt	Oil	428	97,672	98,100	67,521	165,193	165,621	59
15	40360102	Southern Permian Basin-Europe Onshore	Gas	81,742	66,484	148,227	12,730	79,214	160,957	92
16	40250101	Viking Graben	Oil	18,474	94,268	112,742	22,192	116,460	134,933	84
17	40170102	Mid-Norway Continental Margin	Gas	0	0	0	107,832	107,832	107,832	0
18	20580501	Tanezzuft-Benoud Structural/Stratigraphic	Gas	4980	100,070	105,050	2541	102,611	107,591	98
19	39480101	Dingo-Mungaroo/Barrow	Gas	1845	46,400	48,245	56,380	102,780	104,626	46
20	40360103	Southern Permian Basin-Offshore	Gas	42,288	31,546	73,834	13,029	44,575	86,862	85
21	20190202	Jurassic Reservoirs in South Gulf Suprasalt/Qatar Arch Structural	Oil	1951	69,061	71,012	12,457	81,518	83,469	85
22	11120104	Central Offshore	Oil	0	0	0	82,836	82,836	82,836	0
23	52000101	Northeast Greenland Shelf Rift Systems	Oil	0	0	0	80,709	80,709	80,709	0
24	38170101	Kutei Basin Deltaics	Oil	8711	35,762	44,473	35,757	71,519	80,230	55
25	13220101	Onshore and Offshore Northeastern Shelf	Oil	162	22,221	22,383	57,451	79,673	79 835	28
26	10150101	Supra-Domanik Carbonates/Clastics	Oil	4508	72,570	77,078	1141	73,710	78 218	99
27	12100101	Baikal-Patom Foldbelt Riphean-Craton Margin Vendian	Gas	0	30,210	30,210	46,039	76,249	76 249	40

Rank	Assessment Unit Code	Assessment Unit Name	Major Commodity	Cumulative Gas (BCFG)	Remaining Gas (BCFG)	Known Gas (BCFG)	Mean Undiscovered Gas (BCFG)	Future Gas (BCFG)	Gas Endowment (BCFG)	Discovery Maturity, Gas (%)
28	10090101	Carboniferous-Lower Permian Clastics	Gas	35,453	23,645	59,098	14,735	38,380	73,833	80
29	60980101	East Venezuela Fold and Thrust Belt	Oil	10,673	41,706	52,379	21,298	63,004	73,677	71
30	40170101	Halten Terrace-Trondelag Platform	Oil	0	15,662	15,662	57,369	73,031	73,031	21
31	20560101	Tanezzuft-Illizi Structural/Stratigraphic	Oil	750	44,311	45,061	27,785	72,096	72,846	62
32	20190102	Cretaceous Reservoirs in South Gulf Suprasalt Structural	Oil	2705	65,090	67,795	4073	69,163	71,868	94
33	20230101	Horst/Graben-Related Oil and Gas	Oil	47	1614	1661	64,510	66,124	66,172	3
34	11540102	Karabil-Badkhyz (Southern Area)	Gas	11,053	47,057	58,110	7657	54,714	65,767	88
35	60980102	Guarico Sub-Basin	Gas	2933	44,646	47,579	16,995	61,641	64,574	74
36	60360101	Santos Shelf	Oil	12	1113	1125	62,874	63,987	63,999	2
37	20210202	Salt-Involved Structural Oil	Oil	4831	40,849	45,680	17,991	58,840	63,671	72
38	11540101	Northern and Western Areas	Gas	24,483	27,929	52,412	9547	37,476	61,959	85
39	80420101	Greater Indus Foreland and Foldbelt	Gas	15,101	20,272	35,373	25,405	45,677	60,778	58
40	10500103	North Barents	Gas	0	0	0	59,820	59,820	59,820	0
41	11120103	Gograndag-Okarem Zone and Adjacent Shelf	Oil	173	8868	9042	47,515	56,383	56,557	16
42	53050102	Tamabra-Like Debris-Flow-Breccia Limestone Overlying Evaporites	Oil	13,276	22,402	35,678	20,450	42,851	56,127	64
43	72030302	Central Congo Turbidites	Oil	0	2	2	55,268	55,270	55,270	0
44	11540103	Murgab Depression Suprasalt	Gas	26,549	22,967	49,516	4684	27,651	54,200	91
45	40250103	Central Graben	Oil	11,950	28,370	40,320	13,801	42,172	54,122	74
46	20190201	Jurassic Reservoirs in Northwest Desert Anticlines	Oil	0	6500	6500	46,058	52,558	52,558	12
47	40680101	Subsalt	Gas	0	0	0	51,210	51,210	51,210	0
48	37020101	Central Luconia	Gas	3731	31,304	35,035	15,067	46,370	50,102	70
49	37010101	Brunei-Sabah Deltaics	Oil	9168	24,587	33,755	15,977	40,564	49,732	68
50	71920102	Akata Reservoirs	Oil	0	0	0	47,540	47,540	47,540	0
51	20710201	Red Sea Coastal Block Faults	Gas	10	3517	3527	43,946	47,463	47,473	7
52	37020201	East Natuna	Gas	0	45,045	45,045	1963	47,008	47,008	96
53	31540101	Tarim Basin Excluding Marginal Foldbelts	Oil	14	5140	5154	41,454	46,594	46,608	11
54	37030101	South Malay Lacustrine	Oil	454	23,619	24,073	21,935	45,554	46,008	52
55	61030101	Carupano Basin Gas	Gas	0	22,600	22,600	20,103	42,703	42,703	53
56	60990101	Main Maracaibo Basin	Oil	61	24,857	24,918	16,085	40,942	41,002	61
57	11120102	Lower Kura Depression and Adjacent Shelf	Oil	1285	6939	8224	32,091	39,030	40,315	20
58	60980201	Trinidad Basins	Oil	3156	12,667	15,822	23,960	36,627	39,782	40
59	80470302	Eastern Fold Belt	Gas	2669	9620	12,289	26,494	36,114	38,783	32
60	20430102	Central Sirte Carbonates	Oil	4079	24,101	28,180	9918	34,019	38,098	74
61	39130101	Late Jurassic/Early Cretaceous-Mesozoic	Gas	0	17,960	17,960	20,093	38,053	38,053	47
62	60450101	Sub-Andean Fold and Thrust Belt	Gas	6813	10,004	16,818	20,521	30,525	37,339	45
63	10080103	Foredeep Basins	Gas	11,482	4635	16,117	19,891	24,526	36,008	45
64	60590101	Magallanes Extensional Structures	Gas	10,042	14,697	24,739	10,386	25,082	35,124	70
65	10080102	Main Basin Platform	Oil	892	19,623	20,515	14,515	34,138	35,030	59
66	40600101	Neogene Flysch Gas	Gas	9373	9102	18,475	16,267	25,369	34,741	53
67	38220101	North Sumatra	Gas	13,977	11,582	25,559	9117	20,699	34,677	74
68	60210101	Late Cretaceous-Tertiary Turbidites	Oil	0	0	0	34,625	34,625	34,625	0

App-1d_Page 2

Rank	Assessment Unit Code	Assessment Unit Name	Major Commodity	Cumulative Gas (BCFG)	Remaining Gas (BCFG)	Known Gas (BCFG)	Mean Undiscovered Gas (BCFG)	Future Gas (BCFG)	Gas Endowment (ECFG)	Discovery Maturity, Gas (%)
69	20040101	Ma'Rib-Al Jawf/Shabwah/Masila	Oil	0	17,120	17,120	16,982	34,102	34,102	50
70	20190103	Mesozoic/Tertiary Foredeep Fold and Thrust	Gas	2522	15,470	17,993	15,419	30,889	33,411	54
71	40570101	Transylvanian Neogene Suprasalt Gas	Gas	21,194	9537	30,731	2083	11,620	32,814	94
72	20140101	Ghaba-Makarem Combined Structural	Oil	149	20,189	20,339	10,112	30,301	30,450	67
73	52430501	Mannville Gas	Gas	14,229	10,225	24,453	4597	14,821	29,050	84
74	20540301	Tanezzuft-Ghadames Structural/Stratigraphic	Oil	579	15,905	16,484	12,035	27,940	28,519	58
75	38280101	South Sumatra	Oil	3125	7080	10,204	18,250	25,330	28,455	36
76	60980202	Orinoco Delta and Offshore	Oil	0	0	0	28,049	28,049	28,049	0
77	37030201	South Malay Coaly	Gas	38	23,863	23,901	3768	27,631	27,669	86
78	80480102	Irrawaddy-Andaman	Gas	333	8706	9040	18,512	27,219	27,552	33
79	80430101	Eocene-Miocene Bombay Shelf	Oil	4887	12,641	17,528	9299	21,940	26,827	65
80	11120101	Apsheron-Pribalkhan Zone	Oil	7133	11,596	18,729	7908	19,503	26,636	70
81	53050106	Tamaulipas-Like Basinal Limestone and Tertiary Strata Overlying Evaporites	Oil	1633	6007	7640	18,516	24,522	26,155	29
82	38170102	Kutei Basin Turbidites	Oil	0	0	0	25,687	25,687	25,687	0
83	60550101	Neuquen Extensional Structures	Oil	5293	11,407	16,699	7724	19,131	24,424	68
84	80470301	Central Basin	Gas	0	0	0	23,985	23,985	23,985	0
85	72030301	Central Congo Delta and Carbonate Platform	Oil	838	8603	9441	14,314	22,916	23,754	40
86	60370101	Pelotas Platform and Basin	Oil	0	0	0	22,806	22,806	22,806	0
87	11090101	Foldbelt-Foothills	Oil	192	7648	7840	14,512	22,159	22,351	35
88	12070101	Yenisey Foldbelt Riphean-Craton Margin Riphean	Oil	0	5052	5052	16,983	22,035	22,035	23
89	11080102	Foreland Slope	Gas	6556	8039	14,595	6894	14,933	21,489	68
90	60350101	Late Cretaceous-Tertiary Turbidites	Oil	224	5538	5761	13,917	19,455	19,679	29
91	20190101	Cretaceous Reservoirs in Northwest Desert Anticlines	Oil	1114	15,198	16,312	2819	18,017	19,131	85
92	39100301	Malita	Oil	0	7526	7526	11,346	18,872	18,872	40
93	20480101	Bou Dabbous-Tertiary Structural/Stratigraphic	Oil	37	15,472	15,509	2746	18,218	18,256	85
94	60360102	Salt-Structured Deep-Water Sandstones	Oil	0	0	0	17,673	17,673	17,673	0
95	52430301	Exshaw-Rundle Gas	Gas	10,268	5145	15,413	2197	7342	17,610	88
96	10500101	Kolguyev Terrace	Gas	0	3	3	17,554	17,556	17,556	0
97	60960101	Central	Oil	111	14,961	15,071	2180	17,141	17,251	87
98	31270101	Tertiary Lacustrine	Oil	2955	8898	11,853	4639	13,537	16,492	72
99	39480201	Locker-Mungaroo/Barrow	Gas	27	8075	8102	8331	16,406	16,432	49
100	20160101	Fahud-Huqf Combined Structural	Oil	536	7097	7633	8206	15,303	15,339	48
101	39300101	Gippsland	Oil	3501	6274	9775	5658	11,932	15,433	63
102	60980103	Maturin Sub-Basin	Oil	5288	6791	12,078	3258	10,049	15,337	79
103	31420101	Southeastern Fold Belt	Gas	5368	5313	10,681	4488	9801	15,169	70
104	72030201	Gabon Suprasalt	Oil	203	1632	1835	13,230	14,862	15,065	12
105	11090302	Onshore Stavropol-Prikumsk	Oil	9670	3829	13,499	1163	4992	14,662	92
106	20230202	Basinal Oil and Gas	Oil	0	0	0	14,201	14,201	14,201	0
107	39100201	Petrel	Gas	0	5730	5730	8328	14,058	14,058	41
108	38240201	Ardjuna	Oil	794	6529	7322	6579	13,108	13,901	53
109	40470101	Foreland Basin	Gas	5455	5205	10,660	2321	7525	12,980	82

Rank	Assessment Unit Code	Assessment Unit Name	Major Commodity	Cumulative Gas (BCFG)	Remaining Gas (BCFG)	Known Gas (BCFG)	Mean Undiscovered Gas (BCFG)	Future Gas (BCFG)	Gas Endowment (BCFG)	Discovery Maturity, Gas (%)
110	10160103	East and Southeast Margins Subsalt	Oil	0	5364	5364	6781	12,145	12,145	44
111	40480101	Greater Hungarian Plain Basins	Oil	5436	4221	9657	2395	6616	12,052	80
112	71830101	Coastal Plain and Offshore	Oil	46	1903	1949	10,071	11,974	12,020	16
113	61070101	Inner Forearc Deformation Belt	Oil	0	0	0	11,957	11,957	11,957	0
114	52430201	Leduc Gas	Gas	4964	4257	9221	2442	6699	11,663	79
115	10160106	Suprasalt	Oil	131	1496	1628	9934	11,430	11,562	14
116	11090102	Terek-Sunzha Subsalt Jurassic	Gas	0	0	0	11,464	11,464	11,464	0
117	60630101	Malvinas Extensional Structures	Oil	0	75	75	11,327	11,402	11,402	1
118	20540101	Tanezzuft-Oued Mya Structural/Stratigraphic	Oil	68	8905	8973	2341	11,246	11,314	79
119	52150101	Jeanne d'Arc	Oil	0	2406	2406	8893	11,298	11,298	21
120	31540102	Kuche (Northern) Foldbelt	Gas	2	65	66	10,712	10,777	10,778	1
121	11740201	Pre-Upper Jurassic	Oil	443	4738	5180	5516	10,254	10,697	48
122	10150102	Lower Volga	Oil	6618	3314	9932	558	3873	10,490	95
123	20430101	Southeast Sirte Clastics	Oil	1400	8187	9587	808	8995	10,395	92
124	80430102	Eocene-Miocene Deltaics	Oil	1342	5324	6665	3645	8969	10,310	65
125	10150103	Sub-Domanik Devonian Clastics	Oil	4720	4728	9448	695	5423	10,144	93
126	20710101	Gulf of Suez Block-Fault Fairway	Oil	835	5125	5960	4172	9296	10,132	59
127	37010102	Brunei-Sabah Turbidites	Oil	20	2425	2445	7249	9674	9694	25
128	73030101	Offshore	Gas	0	6015	6015	3603	9618	9618	63
129	10090102	Devonian Synrift	Oil	0	0	0	9317	9317	9317	0
130	60600101	North Falklands Basin	Oil	0	0	0	8871	8871	8871	0
131	20230201	Platform Horst/Graben-Related Oil	Oil	0	100	100	8741	8841	8841	1
132	60340103	Abrolhos Sub-Volcanic Structures	Oil	0	0	0	8545	8545	8545	0
133	60290102	Late Cretaceous-Tertiary Deep-Water Sandstones	Oil	66	169	236	8166	8336	8402	3
134	31540103	Southwest Foldbelt	Oil	192	368	560	7773	8141	8333	7
135	11090201	South Mangyshlak (Entire)	Oil	28	5679	5707	2262	7941	7969	72
136	38080101	Pematang/Sihapas Siliciclastics	Oil	910	2956	3866	4078	7034	7944	49
137	40470201	Deformed Belt	Oil	3360	2033	5393	2345	4378	7738	70
138	80340101	Sylhet-Kopili/Barail-Tipam Composite	Oil	2039	4495	6534	1178	5673	7712	85
139	60600201	South Falklands Basin	Oil	0	0	0	7710	7710	7710	0
140	31270102	Pre-Tertiary Buried Hills	Oil	32	3787	3819	3875	7662	7693	50
141	60810101	Cretaceous-Paleogene Basin	Oil	1951	941	2892	4795	5736	7687	38
142	40250102	Moray Firth	Oil	629	5223	5852	1752	6975	7604	77
143	60450102	Foreland Basins	Oil	115	53	168	7375	7429	7543	2
144	52430402	Combined Triassic/Jurassic Oil and Gas	Oil	3119	3497	6615	877	4374	7493	88
145	11080101	Tertiary Foredeep	Oil	945	2999	3944	3139	6137	7083	56
146	52430101	Keg River Gas	Gas	4019	1793	5812	868	2661	6680	87
147	53050103	Tamabra-Like Debris-Flow-Breccia Limestone and Overlying Strata	Oil	335	195	530	6147	6342	6677	8
148	52430302	Exshaw-Rundle Oil and Gas	Oil	4583	1572	6155	466	2038	6621	93
149	60210102	Cretaceous Carbonates	Oil	0	0	0	6595	6595	6595	0
150	52430602	Second White Specks-Cardium Oil and Gas	Oil	3298	2151	5449	840	2991	6289	87
151	20580201	Tanezzuft-Ahnet Structural/Stratigraphic	Gas	0	3117	3117	2973	6090	6090	51

Rank	Assessment Unit Code	Assessment Unit Name	Major Commodity	Cumulative Gas (BCFG)	Remaining Gas (BCFG)	Known Gas (BCFG)	Mean Undiscovered Gas (BCFG)	Future Gas (BCFG)	Gas Endowment (BCFG)	Discovery Maturity, Gas (%)
152	20710103	Southern Gulf of Suez	Oil	4	30	35	5974	6004	6008	1
153	60580101	San Jorge Extensional Structures	Oil	1435	1534	2968	3038	4572	6006	49
154	38220102	Mergui	Gas	0	0	0	6005	6005	6005	0
155	52430202	Leduc Oil and Gas	Oil	4086	1050	5136	782	1832	5918	87
156	20710202	Red Sea Salt Basin	Oil	0	0	0	5875	5875	5875	0
157	10160102	North and West Margins Subsalt Barrier Reefs	Oil	8	1262	1270	4271	5533	5540	23
158	40610201	Romania Flysch Zone	Oil	530	2692	3222	2177	4869	5399	60
159	20580101	Tanezzuft-Timimoun Structural/Stratigraphic	Gas	0	4200	4200	1128	5328	5328	79
160	52430401	Combined Triassic/Jurassic Gas	Gas	1465	2500	3965	1289	3789	5254	75
161	40480201	Zala-Drava-Sava Basins	Oil	1082	2847	3929	1151	3998	5080	77
162	20540201	Tanezzuft-Melrhir Structural/Stratigraphic	Oil	0	125	125	4887	5012	5012	2
163	53050104	Tamabra-Like Debris-Flow-Breccia Limestone of the Golden Lane	Oil	1862	1961	3824	1113	3074	4937	77
164	60900103	Eastern	Gas	0	2000	2000	2735	4735	4735	42
165	11090303	Central Caspian Offshore	Oil	0	0	0	4572	4572	4572	0
166	72030101	Gabon Subsalt	Oil	62	780	843	3674	4455	4517	19
167	60550102	Neuquen Foothills Structure	Oil	159	539	697	3764	4303	4462	16
168	37020102	Balingian	Oil	0	2084	2084	2217	4301	4301	48
169	60960102	Peripheral	Oil	28	214	242	4036	4251	4279	6
170	60550103	Dorsal de Neuquen Structure	Oil	2184	1122	3307	927	2049	4234	78
171	20430103	Offshore Sirte Hypothetical	Oil	0	0	0	4199	4199	4199	0
172	10150201	Permian Reefs/Thrust Folds	Gas	571	2067	2638	1547	3615	4185	63
173	31420401	Leshan-Longnusi Paleohigh	Gas	491	859	1350	2763	3622	4113	33
174	60350103	Salt Dome Province Tertiary Sandstones	Oil	0	0	0	4031	4031	4031	0
175	80420102	Indus Fan	Gas	0	0	0	3919	3919	3919	0
176	11090301	Offshore Prikumsk Zone	Oil	0	0	0	3910	3910	3910	0
177	60590102	Andean Fold Belt Structures	Oil	11	57	69	3655	3712	3723	2
178	11500201	Jurassic-Tertiary	Oil	0	1450	1450	2180	3630	3630	40
179	20480201	Jurassic-Cretaceous Structural/Stratigraphic	Oil	6	1295	1301	2280	3575	3581	36
180	20160201	Natih-Fiqa Structural/Stratigraphic	Oil	616	2485	3100	363	2848	3463	90
181	31440201	Structural Traps	Gas	0	145	145	3212	3357	3357	4
182	60990102	Southwest Maracaibo Basin Fold Belt	Oil	918	865	1783	1491	2356	3275	54
183	60450103	Foreland Central Chaco High	Oil	0	0	0	3211	3211	3211	0
184	80480101	Central Burma Basin	Oil	389	591	981	2008	2600	2989	33
185	11120105	Iran Onshore/Nearshore	Oil	0	0	0	2960	2960	2960	0
186	60900101	Northern	Oil	2186	605	2791	119	724	2910	96
187	40610101	Moesian Platform	Oil	1546	671	2217	561	1232	2778	80
188	31440101	Subtle Traps	Oil	3	909	913	1789	2698	2701	34
189	39100202	Vulcan Graben	Gas	0	0	0	2668	2668	2668	0
190	31150101	Upper Paleozoic/Lower Mesozoic Nonmarine Coarse Clastics	Oil	491	1757	2248	346	2104	2594	87
191	53050105	Tamaulipas-Like Basinal Limestone and Tertiary Strata Without Underlying Evaporites	Oil	293	1443	1736	804	2247	2540	68

Rank	Assessment Unit Code	Assessment Unit Name	Major Commodity	Cumulative Gas (BCFG)	Remaining Gas (BCFG)	Known Gas (BCFG)	Mean Undiscovered Gas (BCFG)	Future Gas (BCFG)	Gas Endowment (BCFG)	Discovery Maturity, Gas (%)
192	80260101	Kohat-Potwar Intrathrust Basin	Oil	807	1174	1981	551	1726	2533	78
193	11080103	Subsalt Jurassic	Gas	84	81	165	2219	2300	2384	7
194	60410101	Hollin-Napo	Oil	661	955	1616	755	1711	2372	68
195	52430601	Second White Specks-Cardium Gas	Gas	974	1097	2071	285	1382	2356	88
196	40610202	Romania Ploiesti Zone	Oil	1050	776	1826	441	1217	2267	81
197	60350102	Cretaceous Carbonates	Oil	79	403	482	1743	2146	2225	22
198	52430102	Keg River Oil and Gas	Oil	570	691	1261	907	1599	2169	58
199	20580301	Tanezzuft-Sbaa Structural/Stratigraphic	Oil	0	1490	1490	645	2134	2134	70
200	40600201	Thermal Triassic	Oil	154	257	412	1698	1955	2109	20
201	31420102	Northwestern Depression/Foldbelt	Gas	302	38	340	1650	1687	1990	17
202	37030102	North Malay Lacustrine	Oil	0	175	175	1762	1937	1937	9
203	53050101	El Abra-Like Reef and Backreef Limestone	Oil	209	1206	1415	507	1713	1922	74
204	11090103	Foreland Slope and Foredeep	Oil	0	0	0	1804	1804	1804	0
205	53050107	Tamabra-Like Debris-Flow-Breccia Limestone North of Campeche	Oil	0	0	0	1735	1735	1735	0
206	72030401	Cuanza-Namibe	Oil	46	13	59	1558	1571	1617	4
207	38240101	Sunda/Asri	Oil	233	381	614	874	1255	1488	41
208	31440102	Anticlinal	Oil	0	685	685	705	1390	1390	49
209	60290101	Western Pre-Aptian Reservoirs	Oil	282	890	1171	168	1057	1339	87
210	60580102	San Bernardo Fold Belt Structures	Oil	232	365	597	735	1100	1332	45
211	31150201	Jurassic/Tertiary Fluvial and Lacustrine Sandstone	Oil	10	332	342	950	1281	1292	26
212	31420402	Lower Paleozoic of Southeastern Fold Belt	Gas	0	0	0	1273	1273	1273	0
213	39100101	Barnett	Oil	0	48	48	1151	1199	1199	4
214	11500101	Mesozoic Sandstone Reservoirs	Oil	0	943	943	89	1033	1033	91
215	52440401	Lodgepole Oil	Oil	728	166	893	63	228	956	93
216	70130101	Coastal Plain and Offshore	Oil	4	45	49	856	901	905	5
217	60410201	Ene	Oil	0	0	0	841	841	841	0
218	60210103	Late Cretaceous-Tertiary Nearshore Sandstones	Oil	0	0	0	839	839	839	0
219	60830101	Neogene Pull-Apart Basin	Oil	0	340	340	471	811	811	42
220	61170101	North Cuba Fold and Thrust Belt	Oil	20	77	96	592	668	688	14
221	40470301	Paleozoic Reservoirs	Oil	0	233	233	383	616	616	38
222	20580601	Tanezzuft-Bechar/Abadla Structural/Stratigraphic	Gas	0	0	0	441	441	441	0
223	60900104	La Luna and Older	Oil	2	2	5	399	401	403	1
224	40480301	Danube Basin	Gas	7	240	247	146	386	393	63
225	60830201	Cretaceous-Paleogene Santa Elena Block	Oil	148	10	158	224	233	382	41
226	40480401	Transcarpathian Basin	Gas	0	261	261	112	373	373	70
227	40480601	Hungarian Paleogene Basin	Oil	15	1	16	339	340	355	4
228	20580401	Tanezzuft-Mouydir Structural/Stratigraphic	Gas	0	0	0	292	292	292	0
229	31420201	Jurassic Lacustrine	Oil	50	90	140	137	227	278	51
230	60900102	Southern	Oil	121	85	206	40	125	246	84
231	60340101	Espirito Santo Shelf	Oil	42	130	172	65	195	237	73
232	40360101	Southern Permian Basin-U.K. Onshore	Oil	18	81	99	31	112	130	76
233	20430104	Southeast Sirte Hypothetical	Oil	0	0	0	60	60	60	0

Rank	Assessment Unit Code	Assessment Unit Name	Major Commodity	Cumulative Gas (BCFG)	Remaining Gas (BCFG)	Known Gas (BCFG)	Mean Undiscovered Gas (BCFG)	Future Gas (BCFG)	Gas Endowment (BCFG)	Discovery Maturity, Gas (%)
234	20710102	Gulf of Suez Qaa Plain	Oil	0	0	0	46	46	46	0
235	31280101	Jurassic/Triassic Fluvial and Lacustrine Sandstone	Oil	13	7	20	20	27	40	50
236	52440201	Brightholme Oil	Oil	8	3	11	5	8	16	67
237	52440101	Yeoman Oil	Oil	0	0	0	4	4	4	3
238	52440301	Bakken Sandstone	Oil	0	0	0	3	3	4	9
NA	10080101	Northwest Izhma-Pechora Depression	NA	NA	NA	NA	120	NA	NA	NA
NA	10090103	Continuous Basin-Centered Gas Accumulation	NA	0	0	0	NA	NA	NA	NA
NA	10150104	Self-Sourced Domanik	NA	0	0	0	NA	NA	NA	NA
NA	10160101	North and West Margins Subsalt Pinnacle Reefs	NA	NA	NA	NA	30,544	NA	NA	NA
NA	10160105	Basin Center Subsalt	NA	0	0	0	NA	NA	NA	NA
NA	11500301	Upper Paleozoic Carbonates	NA	NA	NA	NA	8500	NA	NA	NA
NA	11740102	Self-Sourced Bazhenov Fractured Reservoirs	NA	0	0	0	NA	NA	NA	NA
NA	20190302	Paleozoic Reservoirs	NA	NA	NA	NA	323,892	NA	NA	NA
NA	31150301	Continuous-Type (Basin-Centered) Gas	NA	0	0	0	NA	NA	NA	NA
NA	31270201	Coal-Sourced Gas	NA	0	0	0	NA	NA	NA	NA
NA	31280201	Basin Centered Gas (Continuous) in Ordovician Carbonate and Permian Sandstone	NA	0	5590	5590	NA	NA	NA	NA
NA	31280202	Carboniferous and Permian Coalbed Methane (Continuous)	NA	0	0	0	NA	NA	NA	NA
NA	31420103	Continuous Gas in Central Uplift/Northwestern Depression	NA	38	13	51	NA	NA	NA	NA
NA	31420301	Continuous Gas in Northwestern Depression/Central Uplift	NA	NA	NA	NA	NA	NA	NA	NA
NA	38170103	Kutei Basin Fold and Thrust Belt	NA	NA	NA	NA	3313	NA	NA	NA
NA	38240301	Biogenic Gas	NA	0	119	119	NA	NA	NA	NA
NA	38240401	Billiton Basin Hypothetical	NA	0	0	0	NA	NA	NA	NA
NA	40480501	Central Carpathian Paleogene Basin	NA	NA	NA	NA	NA	NA	NA	NA
NA	40570102	Transylvanian Subsalt	NA	0	0	0	NA	NA	NA	NA
NA	52150201	Graben Systems Exclusive of Jeanne d'Arc	NA	0	4224	4224	NA	NA	NA	NA
NA	52430502	Mannville Continuous (Basin-Center) Gas	NA	2942	3335	6278	NA	NA	NA	NA
NA	52430603	Upper Cretaceous Eastern Shallow Gas	NA	11,738	7191	18,929	NA	NA	NA	NA
NA	60220101	Amazon Delta and Submarine Fan	NA	NA	NA	NA	29,840	NA	NA	NA
NA	60220102	Offshore Gas Hydrates	NA	0	0	0	NA	NA	NA	NA
NA	60340102	Late Cretaceous-Tertiary Slide Blocks and Turbidites	NA	NA	NA	NA	25,677	NA	NA	NA
NA	60370102	Offshore Gas Hydrates	NA	0	0	0	NA	NA	NA	NA
NA	60580201	Hypothetical Basin-Centered Gas	NA	0	0	0	NA	NA	NA	NA
NA	60980104	Orinoco Heavy Oil and Tar Belt	NA	3	257	260	NA	NA	NA	NA
NA	60990103	Basin-Centered Gas	NA	0	0	0	NA	NA	NA	NA
NA	61070102	Barbados Accreationary Prism	NA	16	7	22	NA	NA	NA	NA
NA	80470101	Coalbed Gas	NA	0	0	0	NA	NA	NA	NA
NA	80470201	Western Shelf and Slope	NA	NA	NA	NA	4613	NA	NA	NA

APPENDIX 1E. Assessment units ranked by future oil volume.

[MMBO = million barrels of oil. Major commodity is based on the ratio of gas endowment to oil endowment as ft^3 of gas per bbl of oil (CFG/BO). Major commodity is oil if this ratio is less than 20,000 CFG/BO and gas if this ratio is equal to or greater than 20,000 CFG/BO. Cumulative oil is the reported volume of oil that has been produced. Remaining oil is the calculated remaining reserves; the difference between known and cumulative oil. Known oil is the reported discovered volume (cumulative production plus remaining reserves). Mean undiscovered oil is the mean volume estimated in World Petroleum Assessment 2000 (USGS, 2000). Future oil is the remaining oil plus estimated mean undiscovered oil. Oil endowment is the known oil plus estimated mean undiscovered oil volumes. Discovery maturity is the percentage of oil discovered (known oil) with respect to oil endowment. Calculated values may not be equal to the sums of the component values because numbers have been independently rounded. Negative values result from inconsistent reporting of cumulative and known volumes. Values in parentheses are questionable data. NA means not applicable; it is shown for all values where there is only one field present, and used where the endowment could not be calculated.] Known volumes from Petroconsultants (1996) and NRG Associates (1995). Note that gas = natural gas.

Rank	Assessment Unit Code	Assessment Unit Name	Major Commodity	Cumulative Oil (MMBO)	Remaining Oil (MMBO)	Known Oil (MMBO)	Mean Undiscovered Oil (MMBO)	Future Oil (MMBO)	Oil Endowment (MMBO)	Discovery Maturity, Oil (%)
1	20300101	Cretaceous Reservoirs	Oil	53,182	209,735	262,917	74,911	284,646	337,828	78
2	20210201	Horst-Block Anticlinal Oil	Oil	53,906	101,239	155,145	26,715	127,954	181,860	85
3	11740101	Upper Jurassic-Cretaceous Sandstones	Oil	45,819	72,153	117,972	42,947	115,100	160,919	73
4	20300102	Tertiary Reservoirs	Oil	47,285	62,025	109,310	19,605	81,630	128,914	85
5	20210202	Salt-Involved Structural Oil	Oil	9948	33,902	43,850	14,989	48,891	58,839	75
6	52000101	Northeast Greenland Shelf Rift Systems	Oil	0	0	0	47,148	47,148	47,148	0
7	20190102	Cretaceous Reservoirs in South Gulf Suprasalt Structural	Oil	10,858	41,236	52,094	3764	45,000	55,858	93
8	71920101	Agbada Reservoirs	Oil	16,598	17,925	34,522	21,925	39,850	56,447	61
9	53050102	Tamabra-Like Debris-Flow-Breccia Limestone Overlying Evaporites	Oil	14,366	18,699	33,065	9857	28,556	42,922	77
10	20210101	Central Arch Horst-Block Anticlinal Oil and Gas	Oil	84	6142	6226	19,190	25,331	25,416	24
11	10160104	South Margin Subsalt	Oil	0	6245	6245	15,467	21,712	21,712	29
12	60990101	Main Maracaibo Basin	Oil	34,861	13,350	48,211	7842	21,192	56,053	86
13	20190101	Cretaceous Reservoirs in Northwest Desert Anticlines	Oil	2594	16,386	18,981	4702	21,088	23,683	80
14	40250101	Viking Graben	Oil	13,038	13,643	26,681	7412	21,055	34,093	78
15	60350101	Late Cretaceous-Tertiary Turbidites	Oil	1508	8122	9630	10,869	18,991	20,499	47
16	72030302	Central Congo Turbidites	Oil	0	205	205	18,522	18,727	18,727	1
17	11740201	Pre-Upper Jurassic	Oil	645	11,111	11,756	7474	18,584	19,230	61
18	71920102	Akata Reservoirs	Oil	0	0	0	18,562	18,562	18,562	0
19	20430101	Southeast Sirte Clastics	Oil	5274	15,622	20,895	2380	18,002	23,275	90
20	10150101	Supra-Domanik Carbonates/Clastics	Oil	19,973	14,931	34,904	1676	16,607	36,580	95
21	20190202	Jurassic Reservoirs in South Gulf Suprasalt/Qatar Arch Structural	Oil	5688	12,548	18,236	3845	16,394	22,081	83
22	10080102	Main Basin Platform	Oil	2543	10,442	12,985	4938	15,379	17,922	72
23	60980101	East Venezuela Fold and Thrust Belt	Oil	3253	9978	13,231	4928	14,905	18,159	73
24	60360101	Santos Shelf	Oil	6	279	285	13,723	14,002	14,008	2

Rank	Assessment Unit Code	Assessment Unit Name	Major Commodity	Cumulative Oil (MMBO)	Remaining Oil (MMBO)	Known Oil (MMBO)	Mean Undiscovered Oil (MMBO)	Future Oil (MMBO)	Oil Endowment (MMBO)	Discovery Maturity, Oil (%)
25	31270101	Tertiary Lacustrine	Oil	8944	10,993	19,937	2268	13,261	22,205	90
26	60210101	Late Cretaceous-Tertiary Turbidites	Oil	0	0	0	12,745	12,745	12,745	0
27	11740301	Northern West Siberian Onshore Gas	Gas	66	10,119	10,185	1304	11,424	11,489	89
28	20190201	Jurassic Reservoirs in Northwest Desert Anticlines	Oil	0	777	777	10,633	11,410	11,410	7
29	20430102	Central Sirte Carbonates	Oil	8712	7465	16,177	3840	11,305	20,017	81
30	72030301	Central Congo Delta and Carbonate Platform	Oil	3693	5847	9540	4677	10,523	14,216	67
31	40250103	Central Graben	Oil	7181	5588	12,769	4740	10,328	17,509	73
32	53050106	Tamaulipas-Like Basinal Limestone and Tertiary Strata Overlying Evaporites	Oil	1929	2339	4268	7823	10,163	12,092	35
33	60360102	Salt-Structured Deep-Water Sandstones	Oil	0	0	0	9486	9486	9486	0
34	40170101	Halten Terrace-Trondelag Platform	Oil	76	2585	2660	6204	8788	8864	30
35	11120101	Apsheron-Pribalkhan Zone	Oil	10,581	4512	15,093	3091	7602	18,183	83
36	20540301	Tanezzuft-Ghadames Structural/Stratigraphic	Oil	1629	2909	4538	4461	7370	8999	50
37	60410101	Hollin-Napo	Oil	2889	3712	6601	3023	6735	9625	69
38	11120104	Central Offshore	Oil	0	0	0	6410	6410	6410	0
39	10150103	Sub-Domanik Devonian Clastics	Oil	20,807	5632	26,439	682	6314	27,121	97
40	31150101	Upper Paleozoic/Lower Mesozoic Nonmarine Coarse Clastics	Oil	937	5687	6624	467	6154	7091	93
41	31540101	Tarim Basin Excluding Marginal Foldbelts	Oil	21	634	655	5364	5998	6019	11
42	72030201	Gabon Suprasalt	Oil	1278	836	2114	5031	5867	7145	30
43	20540101	Tanezzuft-Oued Mya Structural/Stratigraphic	Oil	5872	4971	10,843	830	5801	11,673	93
44	38080101	Pematang/Sihapas Siliciclastics	Oil	8349	4868	13,217	899	5768	14,117	94
45	20710101	Gulf of Suez Block-Fault Fairway	Oil	6310	3444	9754	2171	5615	11,925	82
46	80430101	Eocene-Miocene Bombay Shelf	Oil	2337	3920	6257	1612	5531	7868	80
47	31270102	Pre-Tertiary Buried Hills	Oil	1471	3145	4616	2144	5289	6760	68
48	20230201	Platform Horst/Graben-Related Oil	Oil	94	379	474	4682	5062	5156	9
49	11120103	Gograndag-Okarem Zone and Adjacent Shelf	Oil	64	242	306	4764	5005	5070	6
50	37010101	Brunei-Sabah Deltaics	Oil	3872	2780	6652	2137	4917	8789	76
51	11090101	Foldbelt-Foothills	Oil	752	2738	3490	2094	4832	5584	62
52	10160106	Suprasalt	Oil	690	1140	1830	3497	4637	5327	34
53	60350103	Salt Dome Province Tertiary Sandstones	Oil	0	0	0	4493	4493	4493	0
54	60960101	Central	Oil	152	2942	3094	1312	4254	4406	70
55	60980202	Orinoco Delta and Offshore	Oil	0	0	0	4228	4228	4228	0
56	20560101	Tanezzuft-Illizi Structural/Stratigraphic	Oil	2286	1384	3670	2814	4198	6484	57
57	20040101	Ma'Rib-Al Jawf/Shabwah/Masila	Oil	649	1284	1933	2740	4025	4673	41
58	13220101	Onshore and Offshore Northeastern Shelf	Oil	754	1428	2182	2582	4010	4764	46
59	60980103	Maturin Sub-Basin	Oil	7468	1978	9446	1919	3897	11,365	83
60	12100101	Baikal-Patom Foldbelt Riphean-Craton Margin Vendian	Gas	1	2005	2006	1611	3616	3617	55
61	60600101	North Falklands Basin	Oil	0	0	0	3603	3603	3603	0
62	60960102	Peripheral	Oil	1027	1282	2309	2319	3601	4627	50
63	20230101	Horst/Graben-Related Oil and Gas	Oil	269	944	1213	2632	3577	3845	32
64	38170102	Kutei Basin Turbidites	Oil	0	0	0	3569	3569	3569	0

Rank	Assessment Unit Code	Assessment Unit Name	Major Commodity	Cumulative Oil (MMBO)	Remaining Oil (MMBO)	Known Oil (MMBO)	Mean Undiscovered Oil (MMBO)	Future Oil (MMBO)	Oil Endowment (MMBO)	Discovery Maturity, Oil (%)
65	11740302	South Kara Sea Offshore	Gas	0	0	0	3510	3510	3510	0
66	11090201	South Mangyshlak (Entire)	Oil	2406	2835	5241	651	3486	5893	89
67	52150101	Jeanne d'Arc	Oil	0	1582	1582	1796	3378	3378	47
68	10160103	East and Southeast Margins Subsalt	Oil	131	1086	1217	2196	3283	3414	36
69	40250102	Moray Firth	Oil	2248	2196	4444	947	3143	5391	82
70	60370101	Pelotas Platform and Basin	Oil	0	0	0	2938	2938	2938	0
71	39480101	Dingo-Mungaroo/Barrow	Gas	489	660	1149	2249	2909	3398	34
72	20160101	Fahud-Huqf Combined Structural	Oil	1259	1426	2685	1304	2731	3990	67
73	11500101	Mesozoic Sandstone Reservoirs	Oil	459	1827	2286	814	2641	3100	74
74	37030101	South Malay Lacustrine	Oil	1459	1557	3017	1003	2560	4019	75
75	31440102	Anticlinal	Oil	12,228	2192	14,420	291	2483	14,711	98
76	20480101	Bou Dabbous-Tertiary Structural/Stratigraphic	Oil	481	1633	2114	667	2300	2782	76
77	11120102	Lower Kura Depression and Adjacent Shelf	Oil	1035	1004	2040	1275	2279	3315	62
78	20710201	Red Sea Coastal Block Faults	Gas	22	133	155	2136	2269	2291	7
79	60600201	South Falklands Basin	Oil	0	0	0	2229	2229	2229	0
80	60210102	Cretaceous Carbonates	Oil	0	0	0	2200	2200	2200	0
81	20140101	Ghaba-Makarem Combined Structural	Oil	715	1313	2028	743	2056	2771	73
82	60810101	Cretaceous-Paleogene Basin	Oil	1346	326	1672	1711	2038	3383	49
83	80340101	Sylhet-Kopili/Barail-Tipam Composite	Oil	911	1625	2536	399	2024	2935	86
84	53050104	Tamabra-Like Debris-Flow-Breccia Limestone of the Golden Lane	Oil	1917	1392	3308	557	1948	3865	86
85	38170101	Kutei Basin Deltaics	Oil	2195	684	2879	1258	1942	4137	70
86	20230202	Basinal Oil and Gas	Oil	0	0	0	1886	1886	1886	0
87	20540201	Tanezzuft-Melrhir Structural/Stratigraphic	Oil	0	6	6	1875	1881	1881	0
88	52430202	Leduc Oil and Gas	Oil	8113	1304	9417	501	1805	9918	95
89	72030101	Gabon Subsalt	Oil	708	1044	1752	727	1772	2480	71
90	10090101	Carboniferous-Lower Permian Clastics	Gas	110	1501	1611	251	1752	1862	87
91	31440101	Subtle Traps	Oil	171	979	1150	735	1714	1885	61
92	80430102	Eocene-Miocene Deltaics	Oil	1038	1146	2183	479	1625	2662	82
93	61070101	Inner Forearc Deformation Belt	Oil	0	0	0	1536	1536	1536	0
94	39100301	Malita	Oil	123	378	502	1146	1524	1648	30
95	60290102	Late Cretaceous-Tertiary Deep-Water Sandstones	Oil	55	67	122	1457	1523	1579	8
96	40610201	Romania Flysch Zone	Oil	1039	909	1948	564	1473	2512	78
97	60550101	Neuquen Extensional Structures	Oil	1111	605	1716	862	1467	2578	67
98	37010102	Brunei-Sabah Turbidites	Oil	81	165	246	1252	1417	1498	16
99	38240201	Ardjuna	Oil	727	1181	1908	195	1377	2104	91
100	60450102	Foreland Basins	Oil	0	2	2	1254	1256	1256	0
101	12070101	Yenisey Foldbelt Riphean-Craton Margin Riphean	Oil	0	34	34	1213	1246	1246	3
102	71830101	Coastal Plain and Offshore	Oil	74	162	236	1004	1166	1240	19
103	60580101	San Jorge Extensional Structures	Oil	2330	776	3106	347	1123	3453	90
104	60590101	Magallanes Extensional Structures	Gas	595	620	1216	494	1114	1709	71
105	60350102	Cretaceous Carbonates	Oil	261	166	426	932	1097	1358	31
106	39300101	Gippsland	Oil	3077	784	3860	310	1093	4170	93

Rank	Assessment Unit Code	Assessment Unit Name	Major Commodity	Cumulative Oil (MMBO)	Remaining Oil (MMBO)	Known Oil (MMBO)	Mean Undiscovered Oil (MMBO)	Future Oil (MMBO)	Oil Endowment (MMBO)	Discovery Maturity, Oil (%)
107	39130101	Late Jurassic/Early Cretaceous-Mesozoic	Gas	19	27	46	1055	1083	1101	4
108	60980201	Trinidad Basins	Oil	2780	667	3447	416	1083	3863	89
109	38280101	South Sumatra	Oil	1842	587	2429	469	1055	2898	84
110	60630101	Malvinas Extensional Structures	Oil	0	10	10	1031	1041	1041	1
111	37020102	Balingian	Oil	168	460	628	529	989	1157	54
112	20160201	Natih-Fiqa Structural/Stratigraphic	Oil	1299	553	1852	371	925	2224	83
113	38240101	Sunda/Asri	Oil	828	431	1259	462	894	1722	73
114	10150102	Lower Volga	Oil	1803	726	2529	148	874	2677	94
115	53050107	Tamabra-Like Debris-Flow-Breccia Limestone North of Campeche	Oil	0	0	0	871	871	871	0
116	53050103	Tamabra-Like Debris-Flow-Breccia Limestone and Overlying Strata	Oil	26	54	80	807	860	886	9
117	10090102	Devonian Synrift	Oil	0	0	0	848	848	848	0
118	72030401	Cuanza-Namibe	Oil	86	20	107	790	810	897	12
119	11090301	Offshore Prikumsk Zone	Oil	0	0	0	789	789	789	0
120	10160102	North and West Margins Subsalt Barrier Reefs	Oil	0	134	134	637	771	771	17
121	31280101	Jurassic/Triassic Fluvial and Lacustrine Sandstone	Oil	81	602	683	138	740	821	83
122	61170101	North Cuba Fold and Thrust Belt	Oil	65	232	297	494	726	790	38
123	40610101	Moesian Platform	Oil	1251	542	1793	171	713	1963	91
124	60450103	Foreland Central Chaco High	Oil	0	0	0	711	711	711	0
125	80480101	Central Burma Basin	Oil	494	142	637	516	658	1152	55
126	40610202	Romania Ploiesti Zone	Oil	1793	374	2167	278	652	2445	89
127	52440401	Lodgepole Oil	Oil	2239	539	2778	111	650	2889	96
128	53050105	Tamaulipas-Like Basinal Limestone and Tertiary Strata Without Underlying Evaporites	Oil	1354	273	1627	366	638	1993	82
129	37030201	South Malay Coaly	Gas	58	533	592	101	634	693	85
130	40470201	Deformed Belt	Oil	640	283	923	344	627	1267	73
131	10500101	Kolguyev Terrace	Gas	0	51	51	563	614	614	8
132	40600201	Thermal Triassic	Oil	78	254	332	360	614	692	48
133	11540104	Murgab Depression Subsalt	Gas	31	485	516	113	597	628	82
134	20190103	Mesozoic/Tertiary Foredeep Fold and Thrust	Gas	92	64	156	513	577	669	23
135	40480101	Greater Hungarian Plain Basins	Oil	731	358	1089	196	555	1285	85
136	52430302	Exshaw-Rundle Oil and Gas	Oil	1122	415	1537	126	541	1663	92
137	52430602	Second White Specks-Cardium Oil and Gas	Oil	1929	494	2423	36	529	2459	99
138	60900102	Southern	Oil	348	333	681	188	521	869	78
139	31540103	Southwest Foldbelt	Oil	17	24	41	495	519	536	8
140	20190301	Khuff Carbonates in Salt Structures	Gas	0	500	500	0	500	500	100
141	60990102	Southwest Maracaibo Basin Fold Belt	Oil	720	141	861	341	482	1202	72
142	60550102	Neuquen Foothills Structure	Oil	159	138	297	334	472	631	47
143	60340103	Abrolhos Sub-Volcanic Structures	Oil	0	0	0	468	468	468	0
144	20480201	Jurassic-Cretaceous Structural/Stratigraphic	Oil	39	64	103	403	467	506	20
145	11090103	Foreland Slope and Foredeep	Oil	0	0	0	452	452	452	0
146	20580301	Tanezzuft-Sbaa Structural/Stratigraphic	Oil	0	284	284	162	446	446	64

App-1e_Page 4

Rank	Assessment Unit Code	Assessment Unit Name	Major Commodity	Cumulative Oil (MMBO)	Remaining Oil (MMBO)	Known Oil (MMBO)	Mean Undiscovered Oil (MMBO)	Future Oil (MMBO)	Oil Endowment (MMBO)	Discovery Maturity, Oil (%)
147	40360102	Southern Permian Basin-Europe Onshore	Gas	2029	372	2400	74	446	2474	97
148	20710202	Red Sea Salt Basin	Oil	0	0	0	443	443	443	0
149	60980102	Guarico Sub-Basin	Gas	268	54	322	385	439	707	46
150	40680101	Subsalt	Gas	0	0	0	388	388	388	0
151	20430103	Offshore Sirte Hypothetical	Oil	0	0	0	382	382	382	0
152	40480201	Zala-Drava-Sava Basins	Oil	836	265	1101	114	379	1215	91
153	52430402	Combined Triassic/Jurassic Oil and Gas	Oil	965	277	1242	96	373	1338	93
154	60900101	Northern	Oil	1556	175	1730	191	366	1921	90
155	20710103	Southern Gulf of Suez	Oil	41	15	57	344	359	400	14
156	60210103	Late Cretaceous-Tertiary Nearshore Sandstones	Oil	17	46	63	301	347	364	17
157	40170102	Mid-Norway Continental Margin	Gas	0	0	0	334	334	334	0
158	11090303	Central Caspian Offshore	Oil	0	0	0	331	331	331	0
159	31150201	Jurassic/Tertiary Fluvial and Lacustrine Sandstone	Oil	16	193	210	96	289	306	69
160	38220101	North Sumatra	Gas	496	178	674	112	289	786	86
161	11090102	Terek-Sunzha Subsalt Jurassic	Gas	0	0	0	283	283	283	0
162	60580102	San Bernardo Fold Belt Structures	Oil	115	124	240	151	275	390	61
163	80260101	Kohat-Potwar Intrathrust Basin	Oil	204	121	324	145	266	469	69
164	52430102	Keg River Oil and Gas	Oil	835	176	1011	84	260	1095	92
165	10500102	South Barents and Ludlov Saddle	Gas	0	0	0	257	257	257	0
166	80480102	Irrawaddy-Andaman	Gas	34	46	79	209	254	288	28
167	20430104	Southeast Sirte Hypothetical	Oil	0	0	0	253	253	253	0
168	60830201	Cretaceous-Paleogene Santa Elena Block	Oil	118	53	171	200	252	370	46
169	60450101	Sub-Andean Fold and Thrust Belt	Gas	257	62	318	181	242	499	64
170	40360103	Southern Permian Basin-Offshore	Gas	232	176	408	62	238	470	87
171	60900104	La Luna and Older	Oil	11	4	15	231	235	246	6
172	11500201	Jurassic-Tertiary	Oil	0	80	80	139	220	220	37
173	60590102	Andean Fold Belt Structures	Oil	0	0	0	210	211	211	0
174	11540102	Karabil-Badkhyz (Southern Area)	Gas	0	110	111	88	198	198	56
175	60550103	Dorsal de Neuquen Structure	Oil	262	95	357	93	189	450	79
176	80420101	Greater Indus Foreland and Foldbelt	Gas	111	70	180	119	188	299	60
177	60290101	Western Pre-Aptian Reservoirs	Oil	413	157	570	31	187	601	95
178	11120105	Iran Onshore/Nearshore	Oil	0	0	0	185	185	185	0
179	37020101	Central Luconia	Gas	0	170	170	0	170	170	100
180	70130101	Coastal Plain and Offshore	Oil	0	10	10	157	167	168	6
181	11540101	Northern and Western Areas	Gas	48	92	140	70	162	210	67
182	39100101	Barnett	Oil	0	15	15	140	155	155	10
183	20210102	North Gulf Salt Basin Structural Gas	Gas	0	150	150	0	150	150	100
184	20580501	Tanezzuft-Benoud Structural/Stratigraphic	Gas	16	72	88	72	144	160	55
185	39480201	Locker-Mungaroo/Barrow	Gas	0	6	6	132	137	137	4
186	10500103	North Barents	Gas	0	0	0	135	135	135	0
187	37030102	North Malay Lacustrine	Oil	0	0	0	124	124	124	0
188	37020201	East Natuna	Gas	0	20	20	102	122	122	16
189	73030101	Offshore	Gas	0	0	0	116	116	116	0

Rank	Assessment Unit Code	Assessment Unit Name	Major Commodity	Cumulative Oil (MMBO)	Remaining Oil (MMBO)	Known Oil (MMBO)	Mean Undiscovered Oil (MMBO)	Future Oil (MMBO)	Oil Endowment (MMBO)	Discovery Maturity, Oil (%)
190	38220102	Mergui	Gas	0	0	0	101	101	101	0
191	31420201	Jurassic Lacustrine	Oil	38	71	109	27	99	136	80
192	60900103	Eastern	Gas	0	0	0	99	99	99	0
193	20710102	Gulf of Suez Qaa Plain	Oil	0	0	0	93	93	93	0
194	11080103	Subsalt Jurassic	Gas	4	1	5	90	91	95	5
195	60340101	Espirito Santo Shelf	Oil	93	37	130	54	91	184	71
196	31540102	Kuche (Northern) Foldbelt	Gas	7	0	7	89	89	96	8
197	60410201	Ene	Oil	0	0	0	74	74	74	0
198	11080102	Foreland Slope	Gas	1	19	20	48	67	68	30
199	40480601	Hungarian Paleogene Basin	Oil	10	1	11	49	50	60	18
200	40470301	Paleozoic Reservoirs	Oil	0	1	1	47	47	47	1
201	60830101	Neogene Pull-Apart Basin	Oil	4	0	4	37	37	41	11
202	20580201	Tanezzuft-Ahnet Structural/Stratigraphic	Gas	0	1	1	34	35	35	3
203	40360101	Southern Permian Basin-U.K. Onshore	Oil	39	25	64	7	32	71	90
204	20580101	Tanezzuft-Timimoun Structural/Stratigraphic	Gas	0	0	0	31	31	31	0
205	80470302	Eastern Fold Belt	Gas	3	2	4	24	26	28	16
206	52430301	Exshaw-Rundle Gas	Gas	170	20	191	0	20	191	100
207	61030101	Carupano Basin Gas	Gas	0	20	20	0	20	20	100
208	20580601	Tanezzuft-Bechar/Abadla Structural/Stratigraphic	Gas	0	0	0	16	16	16	0
209	80470301	Central Basin	Gas	0	0	0	15	15	15	0
210	52440301	Bakken Sandstone	Oil	5	2	7	11	13	18	40
211	20580401	Tanezzuft-Mouydir Structural/Stratigraphic	Gas	0	0	0	12	12	12	0
212	52440201	Brightholme Oil	Oil	11	4	14	8	11	22	66
213	52430401	Combined Triassic/Jurassic Gas	Gas	47	7	54	0	7	54	100
214	52440101	Yeoman Oil	Oil	1	0	1	5	6	6	15
215	31440201	Structural Traps	Gas	0	5	5	0	5	5	100
216	52430201	Leduc Gas	Gas	39	2	42	0	2	42	100
217	10150201	Permian Reefs/Thrust Folds	Gas	106	(-41)	65	42	(1)	106	61
218	40470101	Foreland Basin	Gas	1	1	1	0	1	1	100
219	20300201	Northern Qatar Arch Extension	Gas	0	0	0	0	0	0	NA
220	31420101	Southeastern Fold Belt	Gas	0	0	0	0	0	0	100
221	31420102	Northwestern Depression/Foldbelt	Gas	0	0	0	0	0	0	NA
222	31420402	Lower Paleozoic of Southeastern Fold Belt	Gas	0	0	0	0	0	0	NA
223	39100201	Petrel	Gas	0	0	0	0	0	0	NA
224	39100202	Vulcan Graben	Gas	0	0	0	0	0	0	NA
225	40480301	Danube Basin	Gas	0	0	0	0	0	0	NA
226	40480401	Transcarpathian Basin	Gas	0	0	0	0	0	0	NA
227	40570101	Transylvanian Neogene Suprasalt Gas	Gas	0	0	0	0	0	0	NA
228	40600101	Neogene Flysch Gas	Gas	13	0	14	0	0	14	100
229	52430101	Keg River Gas	Gas	0	0	0	0	0	0	NA
230	52430501	Mannville Gas	Gas	0	0	0	0	0	0	NA
231	52430601	Second White Specks-Cardium Gas	Gas	0	0	0	0	0	0	NA
232	80420102	Indus Fan	Gas	0	0	0	0	0	0	NA

Rank	Assessment Unit Code	Assessment Unit Name	Major Commodity	Cumulative Oil (MMBO)	Remaining Oil (MMBO)	Known Oil (MMBO)	Mean Undiscovered Oil (MMBO)	Future Oil (MMBO)	Oil Endowment (MMBO)	Discovery Maturity, Oil (%)
233	11540103	Murgab Depression Suprasalt	Gas	1	(-1)	0	0	(-1)	0	NA
234	31420401	Leshan-Longnusi Paleohigh	Gas	9	(-8)	1	0	(-8)	1	100
235	11090302	Onshore Stavropol-Prikumsk	Oil	1200	(-380)	820	287	(-93)	1107	74
236	53050101	El Abra-Like Reef and Backreef Limestone	Oil	2546	(-483)	2063	289	(-195)	2352	88
237	11080101	Tertiary Foredeep	Oil	1268	(-770)	498	280	(-490)	778	64
238	10080103	Foredeep Basins	Gas	803	(-719)	84	214	(-504)	299	28
NA	10080101	Northwest Izhma-Pechora Depression	NA	NA	NA	NA	18	NA	NA	NA
NA	10090103	Continuous Basin-Centered Gas Accumulation	NA	0	0	0	NA	NA	NA	NA
NA	10150104	Self-Sourced Domanik	NA	0	0	0	NA	NA	NA	NA
NA	10160101	North and West Margins Subsalt Pinnacle Reefs	NA	NA	NA	NA	1522	NA	NA	NA
NA	10160105	Basin Center Subsalt	NA	0	0	0	NA	NA	NA	NA
NA	11500301	Upper Paleozoic Carbonates	NA	NA	NA	NA	0	NA	NA	NA
NA	11740102	Self-Sourced Bazhenov Fractured Reservoirs	NA	0	0	0	NA	NA	NA	NA
NA	20190302	Paleozoic Reservoirs	NA	NA	NA	NA	13,235	NA	NA	NA
NA	31150301	Continuous-Type (Basin-Centered) Gas	NA	0	0	0	NA	NA	NA	NA
NA	31270201	Coal-Sourced Gas	NA	0	0	0	NA	NA	NA	NA
NA	31280201	Basin Centered Gas (Continuous) in Ordovician Carbonate and Permian Sandstone	NA	0	0	0	NA	NA	NA	NA
NA	31280202	Carboniferous and Permian Coalbed Methane (Continuous)	NA	0	0	0	NA	NA	NA	NA
NA	31420103	Continuous Gas in Central Uplift/Northwestern Depression	NA	0	0	0	NA	NA	NA	NA
NA	31420301	Continuous Gas in Northwestern Depression/Central Uplift	NA	NA	NA	NA	NA	NA	NA	NA
NA	38170103	Kutei Basin Fold and Thrust Belt	NA	NA	NA	NA	318	NA	NA	NA
NA	38240301	Biogenic Gas	NA	0	0	0	NA	NA	NA	NA
NA	38240401	Billiton Basin Hypothetical	NA	0	0	0	NA	NA	NA	NA
NA	40480501	Central Carpathian Paleogene Basin	NA	NA	NA	NA	NA	NA	NA	NA
NA	40570102	Transylvanian Subsalt	NA	0	0	0	NA	NA	NA	NA
NA	52150201	Graben Systems Exclusive of Jeanne d'Arc	NA	0	0	0	NA	NA	NA	NA
NA	52430502	Mannville Continuous (Basin-Center) Gas	NA	0	0	0	NA	NA	NA	NA
NA	52430603	Upper Cretaceous Eastern Shallow Gas	NA	216	49	265	NA	NA	NA	NA
NA	60220101	Amazon Delta and Submarine Fan	NA	NA	NA	NA	0	NA	NA	NA
NA	60220102	Offshore Gas Hydrates	NA	0	0	0	NA	NA	NA	NA
NA	60340102	Late Cretaceous-Tertiary Slide Blocks and Turbidites	NA	NA	NA	NA	2437	NA	NA	NA
NA	60370102	Offshore Gas Hydrates	NA	0	0	0	NA	NA	NA	NA
NA	60580201	Hypothetical Basin-Centered Gas	NA	0	0	0	NA	NA	NA	NA
NA	60980104	Orinoco Heavy Oil and Tar Belt	NA	38	3719	3757	NA	NA	NA	NA
NA	60990103	Basin-Centered Gas	NA	0	0	0	NA	NA	NA	NA
NA	61070102	Barbados Accreationary Prism	NA	8	2	11	NA	NA	NA	NA
NA	80470101	Coalbed Gas	NA	0	0	0	NA	NA	NA	NA
NA	80470201	Western Shelf and Slope	NA	NA	NA	NA	64	NA	NA	NA

APPENDIX 1F. Assessment units ranked by future natural gas volume.

[BCFG = billion cubic feet of gas. Major commodity is based on the ratio of gas endowment to oil endowment as ft^3 of gas per bbl of oil (CFG/BO). Major commodity is oil if this ratio is less than 20,000 CFG/BO and gas if this ratio is equal to or greater than 20,000 CFG/BO. Cumulative gas is the reported volume of gas that has been produced. Remaining gas is the calculated remaining reserves; the difference between known and cumulative gas. Known gas is the reported discovered volume (cumulative production plus remaining reserves). Mean undiscovered gas is the mean volume estimated in World Petroleum Assessment 2000 (USGS, 2000). Future gas is the remaining gas plus estimated mean undiscovered gas. Gas endowment is the known gas plus estimated mean undiscovered gas volumes. Discovery maturity is the percentage of gas discovered (known gas) with respect to gas endowment. Calculated values may not be equal to the sums of the component values because numbers have been independently rounded. NA means not applicable; it is shown for all values where there is only one field present, and used where the endowment could not be calculated.] Known volumes from Petroconsultants (1996) and NRG Associates (1995). Note that gas = natural gas.

Rank	Assessment Unit Code	Assessment Unit Name	Major Commodity	Cumulative Gas (BCFG)	Remaining Gas (BCFG)	Known Gas (BCFG)	Mean Undiscovered Gas (BCFG)	Future Gas (BCFG)	Gas Endowment (BCFG)	Discovery Maturity, Gas (%)
1	11740301	Northern West Siberian Onshore Gas	Gas	188,736	896,852	1,085,588	136,608	1,033,460	1,222,196	89
2	11740302	South Kara Sea Offshore	Gas	0	81,700	81,700	403,875	485,575	485,575	17
3	20190301	Khuff Carbonates in Salt Structures	Gas	1018	450,982	452,000	22,999	473,981	474,999	95
4	20300102	Tertiary Reservoirs	Oil	10,075	249,365	259,440	93,499	342,864	352,939	74
5	20300101	Cretaceous Reservoirs	Oil	15,185	218,613	233,798	104,098	322,711	337,896	69
6	10500102	South Barents and Ludlov Saddle	Gas	0	78,140	78,140	189,244	267,384	267,384	29
7	20210101	Central Arch Horst-Block Anticlinal Oil and Gas	Oil	0	12,115	12,115	230,387	242,502	242,502	5
8	20210201	Horst-Block Anticlinal Oil	Oil	19,990	209,625	229,615	16,055	225,680	245,670	93
9	20210102	North Gulf Salt Basin Structural Gas	Gas	0	67,000	67,000	152,952	219,952	219,952	30
10	20300201	Northern Qatar Arch Extension	Gas	0	131,220	131,220	69,031	200,251	200,251	66
11	11540104	Murgab Depression Subsalt	Gas	16,456	54,120	70,576	141,762	195,883	212,338	33
12	71920101	Agbada Reservoirs	Oil	6232	87,579	93,811	85,176	172,755	178,986	52
13	11740101	Upper Jurassic-Cretaceous Sandstones	Oil	29,287	68,295	97,582	96,934	165,229	194,516	50
14	10160104	South Margin Subsalt	Oil	428	97,672	98,100	67,521	165,193	165,621	59
15	40250101	Viking Graben	Oil	18,474	94,268	112,742	22,192	116,460	134,933	84
16	40170102	Mid-Norway Continental Margin	Gas	0	0	0	107,832	107,832	107,832	0
17	39480101	Dingo-Mungaroo/Barrow	Gas	1845	46,400	48,245	56,380	102,780	104,626	46
18	20580501	Tanezzuft-Benoud Structural/Stratigraphic	Gas	4980	100,070	105,050	2541	102,611	107,591	98
19	11120104	Central Offshore	Oil	0	0	0	82,836	82,836	82,836	0
20	20190202	Jurassic Reservoirs in South Gulf Suprasalt/Qatar Arch Structural	Oil	1951	69,061	71,012	12,457	81,518	83,469	85
21	52000101	Northeast Greenland Shelf Rift Systems	Oil	0	0	0	80,709	80,709	80,709	0
22	13220101	Onshore and Offshore Northeastern Shelf	Oil	162	22,221	22,383	57,451	79,673	79,835	28
23	40360102	Southern Permian Basin-Europe Onshore	Gas	81,742	66,484	148,227	12,730	79,214	160,957	92
24	12100101	Baikal-Patom Foldbelt Riphean-Craton Margin Vendian	Gas	0	30,210	30,210	46,039	76,249	76,249	40
25	10150101	Supra-Domanik Carbonates/Clastics	Oil	4508	72,570	77,078	1141	73,710	78,218	99
26	40170101	Halten Terrace-Trondelag Platform	Oil	0	15,662	15,662	57,369	73,031	73,031	21
27	20560101	Tanezzuft-Illizi Structural/Stratigraphic	Oil	750	44,311	45,061	27,785	72,096	72,846	62

Rank	Assessment Unit Code	Assessment Unit Name	Major Commodity	Cumulative Gas (BCFG)	Remaining Gas (BCFG)	Known Gas (BCFG)	Mean Undiscovered Gas (BCFG)	Future Gas (BCFG)	Gas Endowment (BCFG)	Discovery Maturity, Gas (%)
28	38170101	Kutei Basin Deltaics	Oil	8711	35,762	44,473	35,757	71,519	80,230	55
29	20190102	Cretaceous Reservoirs in South Gulf Suprasalt Structural	Oil	2705	65,090	67,795	4073	69,163	71,868	94
30	20230101	Horst/Graben-Related Oil and Gas	Oil	47	1614	1661	64,510	66,124	66,172	3
31	60360101	Santos Shelf	Oil	12	1113	1125	62,874	63,987	63,999	2
32	60980101	East Venezuela Fold and Thrust Belt	Oil	10,673	41,706	52,379	21,298	63,004	73,677	71
33	60980102	Guarico Sub-Basin	Gas	2933	44,646	47,579	16,995	61,641	64,574	74
34	10500103	North Barents	Gas	0	0	0	59,820	59,820	59,820	0
35	20210202	Salt-Involved Structural Oil	Oil	4831	40,849	45,680	17,991	58,840	63,671	72
36	11120103	Gograndag-Okarem Zone and Adjacent Shelf	Oil	173	8868	9042	47,515	56,383	56,557	16
37	72030302	Central Congo Turbidites	Oil	0	2	2	55,268	55,270	55,270	0
38	11540102	Karabil-Badkhyz (Southern Area)	Gas	11,053	47,057	58,110	7657	54,714	65,767	88
39	20190201	Jurassic Reservoirs in Northwest Desert Anticlines	Oil	0	6500	6500	46,058	52,558	52,558	12
40	40680101	Subsalt	Gas	0	0	0	51,210	51,210	51,210	0
41	71920102	Akata Reservoirs	Oil	0	0	0	47,540	47,540	47,540	0
42	20710201	Red Sea Coastal Block Faults	Gas	10	3517	3527	43,946	47,463	47,473	7
43	37020201	East Natuna	Gas	0	45,045	45,045	1963	47,008	47,008	96
44	31540101	Tarim Basin Excluding Marginal Foldbelts	Oil	14	5140	5154	41,454	46,594	46,608	11
45	37020101	Central Luconia	Gas	3731	31,304	35,035	15,067	46,370	50,102	70
46	80420101	Greater Indus Foreland and Foldbelt	Gas	15,101	20,272	35,373	25,405	45,677	60,778	58
47	37030101	South Malay Lacustrine	Oil	454	23,619	24,073	21,935	45,554	46,008	52
48	40360103	Southern Permian Basin-Offshore	Gas	42,288	31,546	73,834	13,029	44,575	86,862	85
49	53050102	Tamabra-Like Debris-Flow-Breccia Limestone Overlying Evaporites	Oil	13,276	22,402	35,678	20,450	42,851	56,127	64
50	61030101	Carupano Basin Gas	Gas	0	22,600	22,600	20,103	42,703	42,703	53
51	40250103	Central Graben	Oil	11,950	28,370	40,320	13,801	42,172	54,122	74
52	60990101	Main Maracaibo Basin	Oil	61	24,857	24,918	16,085	40,942	41,002	61
53	37010101	Brunei-Sabah Deltaics	Oil	9168	24,587	33,755	15,977	40,564	49,732	68
54	11120102	Lower Kura Depression and Adjacent Shelf	Oil	1285	6939	8224	32,091	39,030	40,315	20
55	10090101	Carboniferous-Lower Permian Clastics	Gas	35,453	23,645	59,098	14,735	38,380	73,833	80
56	39130101	Late Jurassic/Early Cretaceous-Mesozoic	Gas	0	17,960	17,960	20,093	38,053	38,053	47
57	11540101	Northern and Western Areas	Gas	24,483	27,929	52,412	9547	37,476	61,959	85
58	60980201	Trinidad Basins	Oil	3156	12,667	15,822	23,960	36,627	39,782	40
59	80470302	Eastern Fold Belt	Gas	2669	9620	12,289	26,494	36,114	38,783	32
60	60210101	Late Cretaceous-Tertiary Turbidites	Oil	0	0	0	34,625	34,625	34,625	0
61	10080102	Main Basin Platform	Oil	892	19,623	20,515	14,515	34,138	35,030	59
62	20040101	Ma'Rib-Al Jawf/Shabwah/Masila	Oil	0	17,120	17,120	16,982	34,102	34,102	50
63	20430102	Central Sirte Carbonates	Oil	4079	24,101	28,180	9918	34,019	38,098	74
64	20190103	Mesozoic/Tertiary Foredeep Fold and Thrust	Gas	2522	15,470	17,993	15,419	30,889	33,411	54
65	60450101	Sub-Andean Fold and Thrust Belt	Gas	6813	10,004	16,818	20,521	30,525	37,339	45
66	20140101	Ghaba-Makarem Combined Structural	Oil	149	20,189	20,339	10,112	30,301	30,450	67
67	60980202	Orinoco Delta and Offshore	Oil	0	0	0	28,049	28,049	28,049	0
68	20540301	Tanezzuft-Ghadames Structural/Stratigraphic	Oil	579	15,905	16,484	12,035	27,940	28,519	58

App-1f_Page 2

Rank	Assessment Unit Code	Assessment Unit Name	Major Commodity	Cumulative Gas (BCFG)	Remaining Gas (BCFG)	Known Gas (BCFG)	Mean Undiscovered Gas (BCFG)	Future Gas (BCFG)	Gas Endowment (BCFG)	Discovery Maturity, Gas (%)
69	11540103	Murgab Depression Suprasalt	Gas	26,549	22,967	49,516	4684	27,651	54,200	91
70	37030201	South Malay Coaly	Gas	38	23,863	23,901	3768	27,631	27,669	86
71	80480102	Irrawaddy-Andaman	Gas	333	8706	9040	18,512	27,219	27,552	33
72	38170102	Kutei Basin Turbidites	Oil	0	0	0	25,687	25,687	25,687	0
73	40600101	Neogene Flysch Gas	Gas	9373	9102	18,475	16,267	25,369	34,741	53
74	38280101	South Sumatra	Oil	3125	7080	10,204	18,250	25,330	28,455	36
75	60590101	Magallanes Extensional Structures	Gas	10,042	14,697	24,739	10,386	25,082	35,124	70
76	10080103	Foredeep Basins	Gas	11,482	4635	16,117	19,891	24,526	35,008	45
77	53050106	Tamaulipas-Like Basinal Limestone and Tertiary Strata Overlying Evaporites	Oil	1633	6007	7640	18,516	24,522	26,155	29
78	80470301	Central Basin	Gas	0	0	0	23,985	23,985	23,985	0
79	72030301	Central Congo Delta and Carbonate Platform	Oil	838	8603	9441	14,314	22,916	23,754	40
80	60370101	Pelotas Platform and Basin	Oil	0	0	0	22,806	22,806	22,806	0
81	11090101	Foldbelt-Foothills	Oil	192	7648	7840	14,512	22,159	22,351	35
82	12070101	Yenisey Foldbelt Riphean-Craton Margin Riphean	Oil	0	5052	5052	16,983	22,035	22,035	23
83	80430101	Eocene-Miocene Bombay Shelf	Oil	4887	12,641	17,528	9299	21,940	26,827	65
84	38220101	North Sumatra	Gas	13,977	11,582	25,559	9117	20,699	34,677	74
85	11120101	Apsheron-Pribalkhan Zone	Oil	7133	11,596	18,729	7908	19,503	26,636	70
86	60350101	Late Cretaceous-Tertiary Turbidites	Oil	224	5538	5761	13,917	19,455	19,679	29
87	60550101	Neuquen Extensional Structures	Oil	5293	11,407	16,699	7724	19,131	24,424	68
88	39100301	Malita	Oil	0	7526	7526	11,346	18,872	18,872	40
89	20480101	Bou Dabbous-Tertiary Structural/Stratigraphic	Oil	37	15,472	15,509	2746	18,218	18,256	85
90	20190101	Cretaceous Reservoirs in Northwest Desert Anticlines	Oil	1114	15,198	16,312	2819	18,017	19,131	85
91	60360102	Salt-Structured Deep-Water Sandstones	Oil	0	0	0	17,673	17,673	17,673	0
92	10500101	Kolguyev Terrace	Gas	0	3	3	17,554	17,556	17,556	0
93	60960101	Central	Oil	111	14,961	15,071	2180	17,141	17,251	87
94	39480201	Locker-Mungaroo/Barrow	Gas	27	8075	8102	8331	16,406	16,432	49
95	20160101	Fahud-Huqf Combined Structural	Oil	536	7097	7633	8206	15,303	15,839	48
96	11080102	Foreland Slope	Gas	6556	8039	14,595	6894	14,933	21,489	68
97	72030201	Gabon Suprasalt	Oil	203	1632	1835	13,230	14,862	15,065	12
98	52430501	Mannville Gas	Gas	14,229	10,225	24,453	4597	14,821	29,050	84
99	20230202	Basinal Oil and Gas	Oil	0	0	0	14,201	14,201	14,201	0
100	39100201	Petrel	Gas	0	5730	5730	8328	14,058	14,058	41
101	31270101	Tertiary Lacustrine	Oil	2955	8898	11,853	4639	13,537	16,492	72
102	38240201	Ardjuna	Oil	794	6529	7322	6579	13,108	13,901	53
103	10160103	East and Southeast Margins Subsalt	Oil	0	5364	5364	6781	12,145	12,145	44
104	71830101	Coastal Plain and Offshore	Oil	46	1903	1949	10,071	11,974	12,020	16
105	61070101	Inner Forearc Deformation Belt	Oil	0	0	0	11,957	11,957	11,957	0
106	39300101	Gippsland	Oil	3501	6274	9775	5658	11,932	15,433	63
107	40570101	Transylvanian Neogene Suprasalt Gas	Gas	21,194	9537	30,731	2083	11,620	32,814	94
108	11090102	Terek-Sunzha Subsalt Jurassic	Gas	0	0	0	11,464	11,464	11,464	0
109	10160106	Suprasalt	Oil	131	1496	1628	9934	11,430	11,562	14

App-1f_Page 3

Rank	Assessment Unit Code	Assessment Unit Name	Major Commodity	Cumulative Gas (BCFG)	Remaining Gas (BCFG)	Known Gas (BCFG)	Mean Undiscovered Gas (BCFG)	Future Gas (BCFG)	Gas Endowment (BCFG)	Discovery Maturity, Gas (%)
110	60630101	Malvinas Extensional Structures	Oil	0	75	75	11,327	11,402	11,402	1
111	52150101	Jeanne d'Arc	Oil	0	2406	2406	8893	11,298	11,298	21
112	20540101	Tanezzuft-Oued Mya Structural/Stratigraphic	Oil	68	8905	8973	2341	11,246	11,314	79
113	31540102	Kuche (Northern) Foldbelt	Gas	2	65	66	10,712	10,777	10,778	1
114	11740201	Pre-Upper Jurassic	Oil	443	4738	5180	5516	10,254	10,697	48
115	60980103	Maturin Sub-Basin	Oil	5288	6791	12,078	3258	10,049	15,337	79
116	31420101	Southeastern Fold Belt	Gas	5368	5313	10,681	4488	9801	15,169	70
117	37010102	Brunei-Sabah Turbidites	Oil	20	2425	2445	7249	9674	9694	25
118	73030101	Offshore	Gas	0	6015	6015	3603	9618	9618	63
119	10090102	Devonian Synrift	Oil	0	0	0	9317	9317	9317	0
120	20710101	Gulf of Suez Block-Fault Fairway	Oil	835	5125	5960	4172	9296	10,132	59
121	20430101	Southeast Sirte Clastics	Oil	1400	8187	9587	808	8995	10,395	92
122	80430102	Eocene-Miocene Deltaics	Oil	1342	5324	6665	3645	8969	10,310	65
123	60600101	North Falklands Basin	Oil	0	0	0	8871	8871	8871	0
124	20230201	Platform Horst/Graben-Related Oil	Oil	0	100	100	8741	8841	8841	1
125	60340103	Abrolhos Sub-Volcanic Structures	Oil	0	0	0	8545	8545	8545	0
126	60290102	Late Cretaceous-Tertiary Deep-Water Sandstones	Oil	66	169	236	8166	8336	8402	3
127	31540103	Southwest Foldbelt	Oil	192	368	560	7773	8141	8333	7
128	11090201	South Mangyshlak (Entire)	Oil	28	5679	5707	2262	7941	7969	72
129	60600201	South Falklands Basin	Oil	0	0	0	7710	7710	7710	0
130	31270102	Pre-Tertiary Buried Hills	Oil	32	3787	3819	3875	7662	7693	50
131	40470101	Foreland Basin	Gas	5455	5205	10,660	2321	7525	12,980	82
132	60450102	Foreland Basins	Oil	115	53	168	7375	7429	7543	2
133	52430301	Exshaw-Rundle Gas	Gas	10,268	5145	15,413	2197	7342	17,610	88
134	38080101	Pematang/Sihapas Siliciclastics	Oil	910	2956	3866	4078	7034	7944	49
135	40250102	Moray Firth	Oil	629	5223	5852	1752	6975	7604	77
136	52430201	Leduc Gas	Gas	4964	4257	9221	2442	6699	11,663	79
137	40480101	Greater Hungarian Plain Basins	Oil	5436	4221	9657	2395	6616	12,052	80
138	60210102	Cretaceous Carbonates	Oil	0	0	0	6595	6595	6595	0
139	53050103	Tamabra-Like Debris-Flow-Breccia Limestone and Overlying Strata	Oil	335	195	530	6147	6342	6677	8
140	11080101	Tertiary Foredeep	Oil	945	2999	3944	3139	6137	7083	56
141	20580201	Tanezzuft-Ahnet Structural/Stratigraphic	Gas	0	3117	3117	2973	6090	6090	51
142	38220102	Mergui	Gas	0	0	0	6005	6005	6005	0
143	20710103	Southern Gulf of Suez	Oil	4	30	35	5974	6004	6008	1
144	20710202	Red Sea Salt Basin	Oil	0	0	0	5875	5875	5875	0
145	60810101	Cretaceous-Paleogene Basin	Oil	1951	941	2892	4795	5736	7687	38
146	80340101	Sylhet-Kopili/Barail-Tipam Composite	Oil	2039	4495	6534	1178	5673	7712	85
147	10160102	North and West Margins Subsalt Barrier Reefs	Oil	8	1262	1270	4271	5533	5540	23
148	10150103	Sub-Domanik Devonian Clastics	Oil	4720	4728	9448	695	5423	10,144	93
149	20580101	Tanezzuft-Timimoun Structural/Stratigraphic	Gas	0	4200	4200	1128	5328	5328	79
150	20540201	Tanezzuft-Melrhir Structural/Stratigraphic	Oil	0	125	125	4887	5012	5012	2
151	11090302	Onshore Stavropol-Prikumsk	Oil	9670	3829	13,499	1163	4992	14,662	92

Rank	Assessment Unit Code	Assessment Unit Name	Major Commodity	Cumulative Gas (BCFG)	Remaining Gas (BCFG)	Known Gas (BCFG)	Mean Undiscovered Gas (BCFG)	Future Gas (BCFG)	Gas Endowment (BCFG)	Discovery Maturity, Gas (%)
152	40610201	Romania Flysch Zone	Oil	530	2692	3222	2177	4869	5399	60
153	60900103	Eastern	Gas	0	2000	2000	2735	4735	4735	42
154	11090303	Central Caspian Offshore	Oil	0	0	0	4572	4572	4572	0
155	60580101	San Jorge Extensional Structures	Oil	1435	1534	2968	3038	4572	6006	49
156	72030101	Gabon Subsalt	Oil	62	780	843	3674	4455	4517	19
157	40470201	Deformed Belt	Oil	3360	2033	5393	2345	4378	7738	70
158	52430402	Combined Triassic/Jurassic Oil and Gas	Oil	3119	3497	6615	877	4374	7493	88
159	60550102	Neuquen Foothills Structure	Oil	159	539	697	3764	4303	4462	16
160	37020102	Balingian	Oil	0	2084	2084	2217	4301	4301	48
161	60960102	Peripheral	Oil	28	214	242	4036	4251	4279	6
162	20430103	Offshore Sirte Hypothetical	Oil	0	0	0	4199	4199	4199	0
163	60350103	Salt Dome Province Tertiary Sandstones	Oil	0	0	0	4031	4031	4031	0
164	40480201	Zala-Drava-Sava Basins	Oil	1082	2847	3929	1151	3998	5080	77
165	80420102	Indus Fan	Gas	0	0	0	3919	3919	3919	0
166	11090301	Offshore Prikumsk Zone	Oil	0	0	0	3910	3910	3910	0
167	10150102	Lower Volga	Oil	6618	3314	9932	558	3873	10,490	95
168	52430401	Combined Triassic/Jurassic Gas	Gas	1465	2500	3965	1289	3789	5254	75
169	60590102	Andean Fold Belt Structures	Oil	11	57	69	3655	3712	3723	2
170	11500201	Jurassic-Tertiary	Oil	0	1450	1450	2180	3630	3630	40
171	31420401	Leshan-Longnusi Paleohigh	Gas	491	859	1350	2763	3622	4113	33
172	10150201	Permian Reefs/Thrust Folds	Gas	571	2067	2638	1547	3615	4185	63
173	20480201	Jurassic-Cretaceous Structural/Stratigraphic	Oil	6	1295	1301	2280	3575	3581	36
174	31440201	Structural Traps	Gas	0	145	145	3212	3357	3357	4
175	60450103	Foreland Central Chaco High	Oil	0	0	0	3211	3211	3211	0
176	53050104	Tamabra-Like Debris-Flow-Breccia Limestone of the Golden Lane	Oil	1862	1961	3824	1113	3074	4937	77
177	52430602	Second White Specks-Cardium Oil and Gas	Oil	3298	2151	5449	840	2991	6289	87
178	11120105	Iran Onshore/Nearshore	Oil	0	0	0	2960	2960	2960	0
179	20160201	Natih-Fiqa Structural/Stratigraphic	Oil	616	2485	3100	363	2848	3463	90
180	31440101	Subtle Traps	Oil	3	909	913	1789	2698	2701	34
181	39100202	Vulcan Graben	Gas	0	0	0	2668	2668	2668	0
182	52430101	Keg River Gas	Gas	4019	1793	5812	868	2661	6680	87
183	80480101	Central Burma Basin	Oil	389	591	981	2008	2600	2989	33
184	60990102	Southwest Maracaibo Basin Fold Belt	Oil	918	865	1783	1491	2356	3275	54
185	11080103	Subsalt Jurassic	Gas	84	81	165	2219	2300	2384	7
186	53050105	Tamaulipas-Like Basinal Limestone and Tertiary Strata Without Underlying Evaporites	Oil	293	1443	1736	804	2247	2540	68
187	60350102	Cretaceous Carbonates	Oil	79	403	482	1743	2146	2225	22
188	20580301	Tanezzuft-Sbaa Structural/Stratigraphic	Oil	0	1490	1490	645	2134	2134	70
189	31150101	Upper Paleozoic/Lower Mesozoic Nonmarine Coarse Clastics	Oil	491	1757	2248	346	2104	2594	87
190	60550103	Dorsal de Neuquen Structure	Oil	2184	1122	3307	927	2049	4234	78
191	52430302	Exshaw-Rundle Oil and Gas	Oil	4583	1572	6155	466	2038	6621	93

Rank	Assessment Unit Code	Assessment Unit Name	Major Commodity	Cumulative Gas (BCFG)	Remaining Gas (BCFG)	Known Gas (BCFG)	Mean Undiscovered Gas (BCFG)	Future Gas (BCFG)	Gas Endowment (BCFG)	Discovery Maturity, Gas (%)
192	40600201	Thermal Triassic	Oil	154	257	412	1698	1955	2109	20
193	37030102	North Malay Lacustrine	Oil	0	175	175	1762	1937	1937	9
194	52430202	Leduc Oil and Gas	Oil	4086	1050	5136	782	1832	5918	87
195	11090103	Foreland Slope and Foredeep	Oil	0	0	0	1804	1804	1804	0
196	53050107	Tamabra-Like Debris-Flow-Breccia Limestone North of Campeche	Oil	0	0	0	1735	1735	1735	0
197	80260101	Kohat-Potwar Intrathrust Basin	Oil	807	1174	1981	551	1726	2533	78
198	53050101	El Abra-Like Reef and Backreef Limestone	Oil	209	1206	1415	507	1713	1922	74
199	60410101	Hollin-Napo	Oil	661	955	1616	755	1711	2372	68
200	31420102	Northwestern Depression/Foldbelt	Gas	302	38	340	1650	1687	1990	17
201	52430102	Keg River Oil and Gas	Oil	570	691	1261	907	1599	2169	58
202	72030401	Cuanza-Namibe	Oil	46	13	59	1558	1571	1617	4
203	31440102	Anticlinal	Oil	0	685	685	705	1390	1390	49
204	52430601	Second White Specks-Cardium Gas	Gas	974	1097	2071	285	1382	2356	88
205	31150201	Jurassic/Tertiary Fluvial and Lacustrine Sandstone	Oil	10	332	342	950	1281	1292	26
206	31420402	Lower Paleozoic of Southeastern Fold Belt	Gas	0	0	0	1273	1273	1273	0
207	38240101	Sunda/Asri	Oil	233	381	614	874	1255	1488	41
208	40610101	Moesian Platform	Oil	1546	671	2217	561	1232	2778	80
209	40610202	Romania Ploiesti Zone	Oil	1050	776	1826	441	1217	2267	81
210	39100101	Barnett	Oil	0	48	48	1151	1199	1199	4
211	60580102	San Bernardo Fold Belt Structures	Oil	232	365	597	735	1100	1332	45
212	60290101	Western Pre-Aptian Reservoirs	Oil	282	890	1171	168	1057	1339	87
213	11500101	Mesozoic Sandstone Reservoirs	Oil	0	943	943	89	1033	1033	91
214	70130101	Coastal Plain and Offshore	Oil	4	45	49	856	901	905	5
215	60410201	Ene	Oil	0	0	0	841	841	841	0
216	60210103	Late Cretaceous-Tertiary Nearshore Sandstones	Oil	0	0	0	839	839	839	0
217	60830101	Neogene Pull-Apart Basin	Oil	0	340	340	471	811	811	42
218	60900101	Northern	Oil	2186	605	2791	119	724	2910	96
219	61170101	North Cuba Fold and Thrust Belt	Oil	20	77	96	592	668	688	14
220	40470301	Paleozoic Reservoirs	Oil	0	233	233	383	616	616	38
221	20580601	Tanezzuft-Bechar/Abadla Structural/Stratigraphic	Gas	0	0	0	441	441	441	0
222	60900104	La Luna and Older	Oil	2	2	5	399	401	403	1
223	40480301	Danube Basin	Gas	7	240	247	146	386	393	63
224	40480401	Transcarpathian Basin	Gas	0	261	261	112	373	373	70
225	40480601	Hungarian Paleogene Basin	Oil	15	1	16	339	340	355	4
226	20580401	Tanezzuft-Mouydir Structural/Stratigraphic	Gas	0	0	0	292	292	292	0
227	60830201	Cretaceous-Paleogene Santa Elena Block	Oil	148	10	158	224	233	382	41
228	52440401	Lodgepole Oil	Oil	728	166	893	63	228	956	93
229	31420201	Jurassic Lacustrine	Oil	50	90	140	137	227	278	51
230	60340101	Espirito Santo Shelf	Oil	42	130	172	65	195	237	73
231	60900102	Southern	Oil	121	85	206	40	125	246	84
232	40360101	Southern Permian Basin-U.K. Onshore	Oil	18	81	99	31	112	130	76
233	20430104	Southeast Sirte Hypothetical	Oil	0	0	0	60	60	60	0

Rank	Assessment Unit Code	Assessment Unit Name	Major Commodity	Cumulative Gas (BCFG)	Remaining Gas (BCFG)	Known Gas (BCFG)	Mean Undiscovered Gas (BCFG)	Future Gas (BCFG)	Gas Endowment (BCFG)	Discovery Maturity, Gas (%)
234	20710102	Gulf of Suez Qaa Plain	Oil	0	0	0	46	46	46	0
235	31280101	Jurassic/Triassic Fluvial and Lacustrine Sandstone	Oil	13	7	20	20	27	40	50
236	52440201	Brightholme Oil	Oil	8	3	11	5	8	16	67
237	52440101	Yeoman Oil	Oil	0	0	0	4	4	4	3
238	52440301	Bakken Sandstone	Oil	0	0	0	3	3	4	9
NA	10080101	Northwest Izhma-Pechora Depression	NA	NA	NA	NA	120	NA	NA	NA
NA	10090103	Continuous Basin-Centered Gas Accumulation	NA	0	0	0	NA	NA	NA	NA
NA	10150104	Self-Sourced Domanik	NA	0	0	0	NA	NA	NA	NA
NA	10160101	North and West Margins Subsalt Pinnacle Reefs	NA	NA	NA	NA	30,544	NA	NA	NA
NA	10160105	Basin Center Subsalt	NA	0	0	0	NA	NA	NA	NA
NA	11500301	Upper Paleozoic Carbonates	NA	NA	NA	NA	8500	NA	NA	NA
NA	11740102	Self-Sourced Bazhenov Fractured Reservoirs	NA	0	0	0	NA	NA	NA	NA
NA	20190302	Paleozoic Reservoirs	NA	NA	NA	NA	323,892	NA	NA	NA
NA	31150301	Continuous-Type (Basin-Centered) Gas	NA	0	0	0	NA	NA	NA	NA
NA	31270201	Coal-Sourced Gas	NA	0	0	0	NA	NA	NA	NA
NA	31280201	Basin Centered Gas (Continuous) in Ordovician Carbonate and Permian Sandstone	NA	0	5590	5590	NA	NA	NA	NA
NA	31280202	Carboniferous and Permian Coalbed Methane (Continuous)	NA	0	0	0	NA	NA	NA	NA
NA	31420103	Continuous Gas in Central Uplift/Northwestern Depression	NA	38	13	51	NA	NA	NA	NA
NA	31420301	Continuous Gas in Northwestern Depression/Central Uplift	NA	NA	NA	NA	NA	NA	NA	NA
NA	38170103	Kutei Basin Fold and Thrust Belt	NA	NA	NA	NA	3313	NA	NA	NA
NA	38240301	Biogenic Gas	NA	0	119	119	NA	NA	NA	NA
NA	38240401	Billiton Basin Hypothetical	NA	0	0	0	NA	NA	NA	NA
NA	40480501	Central Carpathian Paleogene Basin	NA	NA	NA	NA	NA	NA	NA	NA
NA	40570102	Transylvanian Subsalt	NA	0	0	0	NA	NA	NA	NA
NA	52150201	Graben Systems Exclusive of Jeanne d'Arc	NA	0	4224	4224	NA	NA	NA	NA
NA	52430502	Mannville Continuous (Basin-Center) Gas	NA	2942	3335	6278	NA	NA	NA	NA
NA	52430603	Upper Cretaceous Eastern Shallow Gas	NA	11,738	7191	18,929	NA	NA	NA	NA
NA	60220101	Amazon Delta and Submarine Fan	NA	NA	NA	NA	29,840	NA	NA	NA
NA	60220102	Offshore Gas Hydrates	NA	0	0	0	NA	NA	NA	NA
NA	60340102	Late Cretaceous-Tertiary Slide Blocks and Turbidites	NA	NA	NA	NA	25,677	NA	NA	NA
NA	60370102	Offshore Gas Hydrates	NA	0	0	0	NA	NA	NA	NA
NA	60580201	Hypothetical Basin-Centered Gas	NA	0	0	0	NA	NA	NA	NA
NA	60980104	Orinoco Heavy Oil and Tar Belt	NA	3	257	260	NA	NA	NA	NA
NA	60990103	Basin-Centered Gas	NA	0	0	0	NA	NA	NA	NA
NA	61070102	Barbados Accreationary Prism	NA	16	7	22	NA	NA	NA	NA
NA	80470101	Coalbed Gas	NA	0	0	0	NA	NA	NA	NA
NA	80470201	Western Shelf and Slope	NA	NA	NA	NA	4613	NA	NA	NA

APPENDIX 2A. Total petroleum systems ranked by mean (expected) undiscovered oil.

[MMBO = million barrels of oil. Major commodity is based on the ratio of gas endowment to oil endowment as ft^3 of gas per bbl of oil (CFG/BO). Major commodity is oil if this ratio is less than 20,000 CFG/BO and gas if this ratio is equal to or greater than 20,000 CFG/BO. Cumulative oil is the reported volume of oil that has been produced. Remaining oil is the calculated remaining reserves; the difference between known and cumulative oil. Known oil is the reported discovered volume (cumulative production plus remaining reserves). Mean undiscovered oil is the mean volume estimated in World Petroleum Assessment 2000 (USGS, 2000). Future oil is the remaining oil plus estimated mean undiscovered oil. Oil endowment is the known oil plus estimated mean undiscovered oil volumes. Discovery maturity is the percentage of oil discovered (known oil) with respect to oil endowment. Calculated values may not be equal to the sums of the component values because numbers have been independently rounded. Negative values result from inconsistent reporting of cumulative and known volumes. Values in parentheses are questionable data. NA means not applicable; it is shown for all values where there is only one field present, and used where the endowment could not be calculated.] Known volumes from Petroconsultants (1996) and NRG Associates (1995). Note that gas = natural gas.

Rank	Total Petroleum System Code	Total Petroleum System Name	Major Commodity	Cumulative Oil (MMBO)	Remaining Oil (MMBO)	Known Oil (MMBO)	Mean Undiscovered Oil (MMBO)	Future Oil (MMBO)	Oil Endowment (MMBO)	Discovery Maturity, Oil (%)
1	203001	Zagros-Mesopotamian Cretaceous-Tertiary	Oil	100,467	271,760	372,226	94,516	366,275	466,742	80
2	520001	Permian/Upper Jurassic Composite	Oil	0	0	0	47,148	47,148	47,148	0
3	117401	Bazhenov-Neocomian	Oil	45,819	72,153	117,972	42,947	115,100	160,919	73
4	202102	Arabian Sub-Basin Tuwaiq/Hanifa-Arab	Oil	63,853	135,142	198,995	41,703	176,845	240,698	83
5	719201	Tertiary Niger Delta (Agbada/Akata)	Oil	16,598	17,925	34,522	40,487	58,412	75,010	46
6	101601	Paleozoic North Caspian	Oil	822	9987	10,808	23,320	33,307	34,128	32
7	603601	Guaratiba-Guaruja (Cretaceous) Composite	Oil	6	279	285	23,209	23,488	23,494	1
8	720303	Congo Delta Composite	Oil	3693	6052	9745	23,198	29,250	32,943	30
9	530501	Pimienta-Tamabra	Oil	22,139	22,273	44,412	20,569	42,841	64,980	68
10	202101	Central Arabia Qusaiba Paleozoic	Oil	84	6292	6376	19,190	25,481	25,566	25
11	603501	Lagoa Feia-Carapebus	Oil	1769	8287	10,056	16,293	24,580	26,349	38
12	111201	Oligocene-Miocene Maykop/Diatom	Oil	11,681	5758	17,438	15,725	21,483	33,163	53
13	602101	Cenomanian-Turonian	Oil	17	46	63	15,247	15,293	15,310	0
14	201902	Jurassic Hanifa/Diyab-Arab	Oil	5688	13,325	19,013	14,478	27,803	33,491	57
15	201903	Silurian Qusaiba	Gas	41	514	555	13,235	13,749	13,790	4
16	402501	Kimmeridgian Shales	Oil	22,467	21,427	43,895	13,098	34,526	56,993	77
17	201901	Cretaceous Thamama/Wasia	Oil	13,544	57,686	71,231	8979	66,665	80,209	89
18	609901	La Luna/Maracaibo	Oil	35,581	13,491	49,072	8183	21,675	57,256	86
19	117402	Togur-Tyumen	Oil	645	11,111	11,756	7474	18,584	19,230	61
20	609801	Querecual	Oil	11,027	15,729	26,756	7231	22,960	33,988	79
21	204301	Sirte-Zelten	Oil	13,986	23,087	37,072	6854	29,941	43,927	84
22	202302	Jurassic Gotnia/Barsarin/Sargelu/Najmah	Oil	94	379	474	6569	6948	7042	7
23	401701	Upper Jurassic Spekk	Oil	76	2585	2660	6538	9123	9199	29
24	315401	Ordovician/Jurassic-Phanerozoic	Oil	45	658	704	5948	6606	6652	11
25	100801	Domanik-Paleozoic	Oil	3346	9723	13,069	5170	14,892	18,239	72
26	381701	Kutei Basin	Oil	2195	684	2879	5145	5829	8024	36
27	720302	Azile-Senonian	Oil	1278	836	2114	5031	5867	7145	30
28	117403	Northern West Siberian Mesozoic Composite	Gas	66	10,119	10,185	4814	14,933	14,999	68

App-2a_Page 1

Rank	Total Petroleum System Code	Total Petroleum System Name	Major Commodity	Cumulative Oil (MMBO)	Remaining Oil (MMBO)	Known Oil (MMBO)	Mean Undiscovered Oil (MMBO)	Future Oil (MMBO)	Oil Endowment (MMBO)	Discovery Maturity, Oil (%)
29	609802	Upper Cretaceous/Tertiary	Oil	2780	667	3447	4644	5311	8091	43
30	205403	Tanezzuft-Ghadames	Oil	1629	2909	4538	4461	7370	8999	50
31	312701	Shahejie-Shahejie/Guantao/Wumishan	Oil	10,415	14,139	24,553	4412	18,550	28,965	85
32	609601	Gacheta-Mirador	Oil	1179	4224	5402	3631	7855	9033	60
33	606001	Neocomian Lacustrine	Oil	0	0	0	3603	3603	3603	0
34	370101	Brunei-Sabah	Oil	3953	2945	6898	3390	6335	10,288	67
35	604101	Mesozoic-Cenozoic	Oil	2889	3712	6601	3023	6735	9625	69
36	603401	Cretaceous Composite	Oil	93	37	130	2959	2995	3089	4
37	603701	Cenomanian-Turonian-Tertiary Composite	Oil	0	0	0	2938	2938	2938	0
38	110901	Terek-Caspian	Oil	752	2738	3490	2829	5567	6319	55
39	205601	Tanezzuft-Illizi	Oil	2286	1384	3670	2814	4198	6484	57
40	200401	Madbi Amran/Qishn	Oil	649	1284	1933	2740	4025	4673	41
41	202301	Paleozoic Qusaiba/Akkas/Abba/Mudawwara	Oil	269	944	1213	2632	3577	3845	32
42	207101	Sudr-Nubia	Oil	6351	3459	9810	2608	6067	12,418	79
43	132201	North Sakhalin Neogene	Oil	754	1428	2182	2582	4010	4764	46
44	207102	Maqna	Oil	22	133	155	2579	2712	2734	6
45	101501	Volga-Ural Domanik-Paleozoic	Oil	42,584	21,288	63,872	2507	23,795	66,379	96
46	394801	Dingo-Mungaroo/Barrow	Gas	489	660	1149	2249	2909	3398	34
47	606002	Lower Cretaceous	Oil	0	0	0	2229	2229	2229	0
48	604501	Los Monos-Machareti	Oil	257	63	320	2145	2209	2466	13
49	804301	Eocene-Miocene Composite	Oil	3375	5065	8440	2091	7156	10,531	80
50	205402	Tanezzuft-Melrhir	Oil	0	6	6	1875	1881	1881	0
51	521501	Egret-Hibernia	Oil	0	1582	1582	1796	3378	3378	47
52	608101	Cretaceous-Tertiary	Oil	1346	326	1672	1711	2038	3383	49
53	121001	Baikal-Patom Foldbelt Riphean-Craton Margin Vendian	Gas	1	2005	2006	1611	3616	3617	55
54	610701	Tobago Trough Paleogene	Oil	8	2	11	1536	1538	1546	1
55	602901	Neocomian to Turonian Composite	Oil	469	223	692	1487	1711	2179	32
56	110903	Stavropol-Prikumsk	Oil	1200	(-380)	820	1407	(1027)	2227	37
57	201601	North Oman Huqf-Shu'aiba	Oil	1259	1426	2685	1304	2731	3990	67
58	605501	Neuquen Hybrid	Oil	1532	838	2370	1290	2128	3659	65
59	120701	Yenisey Foldbelt Riphean-Craton Margin Riphean	Oil	0	34	34	1213	1246	1246	3
60	391003	Jurassic/Early Cretaceous-Mesozoic	Oil	123	378	502	1146	1524	1648	30
61	370301	Oligocene-Miocene Lacustrine	Oil	1459	1557	3017	1127	2684	4143	73
62	100901	Dnieper-Donets Paleozoic	Gas	110	1501	1611	1098	2599	2709	59
63	391301	Late Jurassic/Early Cretaceous-Mesozoic	Gas	19	27	46	1055	1083	1101	4
64	606301	Lower Cretaceous Marine	Oil	0	10	10	1031	1041	1041	1
65	314401	Qingshankou-Putaohua/Shaertu	Oil	12,399	3171	15,570	1026	4198	16,596	94
66	718301	Cretaceous Composite	Oil	74	162	236	1004	1166	1240	19
67	105001	South and North Barents Triassic-Jurassic	Gas	0	51	51	955	1006	1007	5
68	380801	Brown Shale-Sihapas	Oil	8349	4868	13,217	899	5768	14,117	94
69	406102	Dysodile Schist-Tertiary	Oil	2832	1283	4115	842	2125	4957	83
70	205401	Tanezzuft-Oued Mya	Oil	5872	4971	10,843	830	5801	11,673	93

Rank	Total Petroleum System Code	Total Petroleum System Name	Major Commodity	Cumulative Oil (MMBO)	Remaining Oil (MMBO)	Known Oil (MMBO)	Mean Undiscovered Oil (MMBO)	Future Oil (MMBO)	Oil Endowment (MMBO)	Discovery Maturity, Oil (%)
71	115001	Buzachi Arch and Surrounding Areas Composite	Oil	459	1827	2286	814	2641	3100	74
72	720304	Cuanza Composite	Oil	86	20	107	790	810	897	12
73	201401	North Oman Huqf/'Q'-Haushi	Oil	715	1313	2028	743	2056	2771	73
74	720301	Melania-Gamba	Oil	708	1044	1752	727	1772	2480	71
75	804801	Eocene to Miocene Composite	Gas	528	188	716	724	912	1441	50
76	609001	La Luna-La Paz	Oil	1915	512	2426	709	1220	3135	77
77	605901	Lower Inoceramus	Gas	595	620	1216	704	1325	1920	63
78	204801	Bou Dabbous-Tertiary	Oil	481	1633	2114	667	2300	2782	76
79	110902	South Mangyshlak	Oil	2406	2835	5241	651	3486	5893	89
80	370201	Sarawak Basin	Gas	168	629	797	529	1159	1327	60
81	524302	Duvernay-Leduc	Oil	8153	1306	9459	501	1807	9960	95
82	605801	D-129	Oil	2445	900	3346	498	1398	3844	87
83	611701	Upper Jurassic-Neocomian	Oil	65	232	297	494	726	790	38
84	382801	Lahat/Talang Akar-Cenozoic	Oil	1842	587	2429	469	1055	2898	84
85	311501	Lucaogou-Karamay/Ulho/Pindequan	Oil	937	5687	6624	467	6154	7091	93
86	382401	Banuwati-Oligocene/Miocene	Oil	828	431	1259	462	894	1722	73
87	110801	Azov-Kuban Mesozoic-Cenozoic	Gas	1273	(-750)	524	417	(-332)	941	56
88	204802	Jurassic-Cretaceous Composite	Oil	39	64	103	403	467	506	20
89	803401	Sylhet-Kopili/Barail-Tipam Composite	Oil	911	1625	2536	399	2024	2935	86
90	406801	Pre-Messinian	Gas	0	0	0	388	388	388	0
91	201602	Middle Cretaceous Natih	Oil	1299	553	1852	371	925	2224	83
92	406002	Meride/Riva di Solto	Oil	78	254	332	360	614	692	48
93	404702	Mesozoic/Paleogene Composite	Oil	640	283	923	344	627	1267	73
94	393001	Latrobe	Oil	3077	784	3860	310	1093	4170	93
95	115401	Amu-Darya Jurassic-Cretaceous	Gas	80	686	766	270	957	1037	74
96	382201	Bampo-Cenozoic	Gas	496	178	674	213	391	887	76
97	608302	Cretaceous-Paleogene	Oil	118	53	171	200	252	370	46
98	404801	Greater Hungarian Plain Neogene	Oil	731	358	1089	196	555	1285	85
99	382402	Jatibarang/Talang Akar-Oligocene/Miocene	Oil	727	1181	1908	195	1377	2104	91
100	406101	Moesian Platform Composite	Oil	1251	542	1793	171	713	1963	91
101	205803	Tanezzuft-Sbaa	Oil	0	284	284	162	446	446	64
102	701301	Cretaceous-Tertiary Composite	Oil	0	10	10	157	167	168	6
103	802601	Patala-Namal	Oil	204	121	324	145	266	469	69
104	403601	Carboniferous-Rotliegend	Gas	2300	572	2872	143	715	3015	95
105	391001	Milligans-Carboniferous/Permian	Oil	0	15	15	140	155	155	10
106	115002	North Ustyurt Jurassic	Oil	0	80	80	139	220	220	37
107	312801	Yanchang-Yanan	Oil	81	602	683	138	740	821	83
108	394802	Locker-Mungaroo/Barrow	Gas	0	6	6	132	137	137	4
109	524303	Exshaw-Rundle	Oil	1292	435	1728	126	561	1854	93
110	804201	Sembar-Goru/Ghazij	Gas	111	70	180	119	188	299	60
111	730301	Cretaceous Composite	Gas	0	0	0	116	116	116	0
112	404802	Zala-Drava-Sava Mesozoic/Neogene	Oil	836	265	1101	114	379	1215	91
113	524404	Lodgepole	Oil	2239	539	2778	111	650	2889	96

Rank	Total Petroleum System Code	Total Petroleum System Name	Major Commodity	Cumulative Oil (MMBO)	Remaining Oil (MMBO)	Known Oil (MMBO)	Mean Undiscovered Oil (MMBO)	Future Oil (MMBO)	Oil Endowment (MMBO)	Discovery Maturity, Oil (%)
114	370202	East Natuna	Gas	0	20	20	102	122	122	16
115	370302	Miocene Coaly Strata	Gas	58	533	592	101	634	693	85
116	311502	Jurassic Coal-Jurassic/Tertiary	Oil	16	193	210	96	289	306	69
117	524304	Combined Triassic/Jurassic	Oil	1012	284	1296	96	380	1393	93
118	524301	Keg River-Keg River	Oil	835	176	1011	84	260	1095	92
119	604102	Paleozoic	Oil	0	0	0	74	74	74	0
120	205805	Tanezzuft-Benoud	Gas	16	72	88	72	144	160	55
121	804702	Jalangi-Sylhet/Burdwan Composite	NA	NA	NA	NA	64	NA	NA	NA
122	404806	Hungarian Paleogene	Oil	10	1	11	49	50	60	18
123	404703	Paleozoic Composite	Oil	0	1	1	47	47	47	1
124	101502	Belsk Basin	Gas	106	(-41)	65	42	(1)	106	61
125	804703	Jenam/Bhuban-Bokabil	Gas	3	2	4	39	41	43	10
126	608301	Neogene	Oil	4	0	4	37	37	41	11
127	524306	Second White Specks-Cardium	Oil	2146	543	2688	36	578	2724	99
128	205802	Tanezzuft-Ahnet	Gas	0	1	1	34	35	35	3
129	205801	Tanezzuft-Timimoun	Gas	0	0	0	31	31	31	0
130	314202	Daanzhai-Daanzhai/Lianggaoshan	Oil	38	71	109	27	99	136	80
131	205806	Tanezzuft-Bechar/Abadla	Gas	0	0	0	16	16	16	0
132	205804	Tanezzuft-Mouydir	Gas	0	0	0	12	12	12	0
133	524403	Bakken	Oil	5	2	7	11	13	18	40
134	524402	Brightholme	Oil	11	4	14	8	11	22	66
135	524401	Yeoman	Oil	1	0	1	5	6	6	15
136	115003	North Ustyurt Paleozoic	NA	NA	NA	NA	0	NA	NA	NA
137	203002	Paleozoic-Permian/Triassic	Gas	0	0	0	0	0	0	NA
138	314201	Maokou/Longtang-Jialingjiang/Maokou/Huanglong	Gas	0	0	0	0	0	0	100
139	314204	Cambrian/Silurian Marine Shale-Dengying/Lower Paleozoic	Gas	9	(-8)	1	0	(-8)	1	100
140	314402	Jurassic Coal-Denglouku/Nongan	Gas	0	5	5	0	5	5	100
141	391002	Keyling/Hyland Bay-Permian	Gas	0	0	0	0	0	0	NA
142	404701	Isotopically Light Gas	Gas	1	1	1	0	1	1	100
143	404803	Danube Neogene	Gas	0	0	0	0	0	0	NA
144	404804	Transcarpathian Neogene	Gas	0	0	0	0	0	0	NA
145	405701	Transylvanian Composite	Gas	0	0	0	0	0	0	NA
146	406001	Porto Garibaldi	Gas	13	0	14	0	0	14	100
147	524305	Mannville-Upper Mannville	Gas	0	0	0	0	0	0	NA
148	602201	Neogene	NA	NA	NA	NA	0	NA	NA	NA
149	610301	Lower Cruse	Gas	0	20	20	0	20	20	100
NA	311503	Lucaogou/Jurassic Coal-Paleozoic/Mesozoic	NA	0	0	0	NA	NA	NA	NA
NA	312702	Carboniferous/Permian Coal-Paleozoic	NA	0	0	0	NA	NA	NA	NA
NA	312802	Taiyuan/Shanxi-Majiagou/Shihezi	NA	0	0	0	NA	NA	NA	NA
NA	314203	Xujiahe-Xujiahe/Shaximiao	NA	NA	NA	NA	NA	NA	NA	NA
NA	382403	Tertiary-Parigi	NA	0	0	0	NA	NA	NA	NA
NA	382404	Tertiary-Cenozoic	NA	0	0	0	NA	NA	NA	NA

Rank	Total Petroleum System Code	Total Petroleum System Name	Major Commodity	Cumulative Oil (MMBO)	Remaining Oil (MMBO)	Known Oil (MMBO)	Mean Undiscovered Oil (MMBO)	Future Oil (MMBO)	Oil Endowment (MMBO)	Discovery Maturity, Oil (%)
NA	404805	Central Carpathian Paleogene	NA	NA	NA	NA	NA	NA	NA	NA
NA	521502	Mesozoic Composite	NA	0	0	0	NA	NA	NA	NA
NA	605802	Aguada Bandera	NA	0	0	0	NA	NA	NA	NA
NA	804701	Permian Coal	NA	0	0	0	NA	NA	NA	NA

APPENDIX 2B. Total petroleum systems ranked by mean (expected) undiscovered natural gas.

[BCFG = billion cubic feet of gas. Major commodity is based on the ratio of gas endowment to oil endowment as ft^3 of gas per bbl of oil (CFG/BO). Major commodity is oil if this ratio is less than 20,000 CFG/BO and gas if this ratio is equal to or greater than 20,000 CFG/BO. Cumulative gas is the reported volume of gas that has been produced. Remaining gas is the calculated remaining reserves; the difference between known and cumulative gas. Known gas is the reported discovered volume (cumulative production plus remaining reserves). Mean undiscovered gas is the mean volume estimated in World Petroleum Assessment 2000 (USGS, 2000). Future gas is the remaining gas plus estimated mean undiscovered gas. Gas endowment is the known gas plus estimated mean undiscovered gas volumes. Discovery maturity is the percentage of gas discovered (known gas) with respect to gas endowment. Calculated values may not be equal to the sums of the component values because numbers have been independently rounded. NA means not applicable; it is shown for all values where there is only one field present, and used where the endowment could not be calculated.] Known volumes from Petroconsultants (1996) and NRG Associates (1995). Note that gas = natural gas.

Rank	Total Petroleum System Code	Total Petroleum System Name	Major Commodity	Cumulative Gas (BCFG)	Remaining Gas (BCFG)	Known Gas (BCFG)	Mean Undiscovered Gas (BCFG)	Future Gas (BCFG)	Gas Endowment (BCFG)	Discovery Maturity, Gas (%)
1	117403	Northern West Siberian Mesozoic Composite	Gas	188,736	978,552	1,167,288	540,483	1,519,035	1,707.771	68
2	202101	Central Arabia Qusaiba Paleozoic	Oil	0	79,115	79,115	383,339	462,454	462,454	17
3	201903	Silurian Qusaiba	Gas	1018	451,012	452,030	346,892	797,903	798,922	57
4	105001	South and North Barents Triassic-Jurassic	Gas	0	78,143	78,143	266,618	344,761	344,761	23
5	203001	Zagros-Mesopotamian Cretaceous-Tertiary	Oil	25,260	467,978	493,238	197,597	665,575	690,835	71
6	111201	Oligocene-Miocene Maykop/Diatom	Oil	8591	27,403	35,994	173,310	200,713	209,304	17
7	401701	Upper Jurassic Spekk	Oil	0	15,662	15,662	165,201	180,863	180,863	9
8	115401	Amu-Darya Jurassic-Cretaceous	Gas	78,540	152,073	230,614	163,650	315,724	394 264	58
9	719201	Tertiary Niger Delta (Agbada/Akata)	Oil	6232	87,579	93,811	132,716	220,295	226 526	41
10	101601	Paleozoic North Caspian	Oil	1745	155,231	156,976	119,051	274,282	276 027	57
11	117401	Bazhenov-Neocomian	Oil	29,287	68,295	97,582	96,934	165,229	194 516	50
12	520001	Permian/Upper Jurassic Composite	Oil	0	0	0	80,709	80,709	80 709	0
13	603601	Guaratiba-Guaruja (Cretaceous) Composite	Oil	12	1113	1125	80,547	81,660	81 672	1
14	720303	Congo Delta Composite	Oil	838	8605	9443	69,582	78,186	79 025	12
15	203002	Paleozoic-Permian/Triassic	Gas	0	131,220	131,220	69,031	200,251	200 251	66
16	381701	Kutei Basin	Oil	8711	36,762	45,473	64,757	101,519	110 230	41
17	202301	Paleozoic Qusaiba/Akkas/Abba/Mudawwara	Oil	47	1614	1661	64,510	66,124	66 172	3
18	315401	Ordovician/Jurassic-Phanerozoic	Oil	208	5572	5780	59,939	65,511	65 719	9
19	201902	Jurassic Hanifa/Diyab-Arab	Oil	1951	75,561	77,512	58,515	134,076	136 027	57
20	132201	North Sakhalin Neogene	Oil	162	22,221	22,383	57,451	79,673	79 835	28
21	394801	Dingo-Mungaroo/Barrow	Gas	1845	46,400	48,245	56,380	102,780	104 626	46
22	609802	Upper Cretaceous/Tertiary	Oil	3156	12,667	15,822	52,009	64,676	67 832	23
23	406801	Pre-Messinian	Gas	0	0	0	51,210	51,210	51 210	0
24	804703	Jenam/Bhuban-Bokabil	Gas	2669	9620	12,289	50,479	60,098	62 767	20
25	207102	Maqna	Oil	10	3517	3527	49,820	53,338	53 347	7
26	530501	Pimienta-Tamabra	Oil	17,609	33,213	50,822	49,272	82,485	100 094	51
27	121001	Baikal-Patom Foldbelt Riphean-Craton Margin Vendian	Gas	0	30,210	30,210	46,039	76,249	76 249	40

Rank	Total Petroleum System Code	Total Petroleum System Name	Major Commodity	Cumulative Gas (BCFG)	Remaining Gas (BCFG)	Known Gas (BCFG)	Mean Undiscovered Gas (BCFG)	Future Gas (BCFG)	Gas Endowment (BCFG)	Discovery Maturity, Gas (%)
28	602101	Cenomanian-Turonian	Oil	0	0	0	42,058	42,058	42,058	0
29	609801	Querecual	Oil	18,897	93,399	112,296	41,552	134,951	153,848	73
30	402501	Kimmeridgian Shales	Oil	31,053	127,861	158,914	37,745	165,606	196,659	81
31	100801	Domanik-Paleozoic	Oil	12,374	24,259	36,632	34,525	58,784	71,158	51
32	603401	Cretaceous Composite	Oil	42	180	222	34,287	34,467	34,509	1
33	202102	Arabian Sub-Basin Tuwaiq/Hanifa-Arab	Oil	24,821	250,474	275,295	34,046	284,520	309,341	89
34	604501	Los Monos-Machareti	Oil	6928	10,058	16,985	31,107	41,165	48,092	35
35	602201	Neogene	NA	NA	NA	NA	29,840	NA	NA	NA
36	804201	Sembar-Goru/Ghazij	Gas	15,101	20,272	35,373	29,325	49,596	64,697	55
37	205601	Tanezzuft-Illizi	Oil	750	44,311	45,061	27,785	72,096	72,846	62
38	110901	Terek-Caspian	Oil	192	7648	7840	27,779	35,427	35,619	22
39	403601	Carboniferous-Rotliegend	Gas	124,048	98,111	222,159	25,790	123,901	247,948	90
40	100901	Dnieper-Donets Paleozoic	Gas	35,453	23,645	59,098	24,051	47,697	83,149	71
41	370301	Oligocene-Miocene Lacustrine	Oil	454	23,794	24,248	23,696	47,491	47,944	51
42	370101	Brunei-Sabah	Oil	9188	27,012	36,200	23,226	50,238	59,426	61
43	202302	Jurassic Gotnia/Barsarin/Sargelu/Najmah	Oil	0	100	100	22,942	23,042	23,042	0
44	603701	Cenomanian-Turonian-Tertiary Composite	Oil	0	0	0	22,806	22,806	22,806	0
45	201901	Cretaceous Thamama/Wasia	Oil	6342	95,758	102,100	22,311	118,069	124,410	82
46	804801	Eocene to Miocene Composite	Gas	722	9298	10,020	20,521	29,819	30,541	33
47	610301	Lower Cruse	Gas	0	22,600	22,600	20,103	42,703	42,703	53
48	391301	Late Jurassic/Early Cretaceous-Mesozoic	Gas	0	17,960	17,960	20,093	38,053	38,053	47
49	603501	Lagoa Feia-Carapebus	Oil	303	5941	6244	19,691	25,632	25,935	24
50	382801	Lahat/Talang Akar-Cenozoic	Oil	3125	7080	10,204	18,250	25,330	28,455	36
51	609901	La Luna/Maracaibo	Oil	979	25,722	26,701	17,576	43,298	44,277	60
52	370201	Sarawak Basin	Gas	3731	33,388	37,119	17,284	50,672	54,403	68
53	120701	Yenisey Foldbelt Riphean-Craton Margin Riphean	Oil	0	5052	5052	16,983	22,035	22,035	23
54	200401	Madbi Amran/Qishn	Oil	0	17,120	17,120	16,982	34,102	34,102	50
55	406001	Porto Garibaldi	Gas	9373	9102	18,475	16,267	25,369	34,741	53
56	382201	Bampo-Cenozoic	Gas	13,977	11,582	25,559	15,123	26,705	40,682	63
57	204301	Sirte-Zelten	Oil	5479	32,288	37,767	14,984	47,272	52,752	72
58	605901	Lower Inoceramus	Gas	10,053	14,754	24,807	14,040	28,794	38,848	64
59	720302	Azile-Senonian	Oil	203	1632	1835	13,230	14,862	15,065	12
60	804301	Eocene-Miocene Composite	Oil	6228	17,965	24,193	12,944	30,909	37,137	65
61	605501	Neuquen Hybrid	Oil	7636	13,068	20,704	12,416	25,483	33,119	63
62	110801	Azov-Kuban Mesozoic-Cenozoic	Gas	7586	11,118	18,704	12,252	23,369	30,955	60
63	205403	Tanezzuft-Ghadames	Oil	579	15,905	16,484	12,035	27,940	28,519	58
64	610701	Tobago Trough Paleogene	Oil	16	7	22	11,957	11,963	11,979	0
65	391003	Jurassic/Early Cretaceous-Mesozoic	Oil	0	7526	7526	11,346	18,872	18,872	40
66	606301	Lower Cretaceous Marine	Oil	0	75	75	11,327	11,402	11,402	1
67	391002	Keyling/Hyland Bay-Permian	Gas	0	5730	5730	10,995	16,725	16,725	34
68	207101	Sudr-Nubia	Oil	840	5155	5995	10,192	15,347	16,186	37
69	201401	North Oman Huqf/'Q'-Haushi	Oil	149	20,189	20,339	10,112	30,301	30,450	67
70	718301	Cretaceous Composite	Oil	46	1903	1949	10,071	11,974	12,020	16

Rank	Total Petroleum System Code	Total Petroleum System Name	Major Commodity	Cumulative Gas (BCFG)	Remaining Gas (BCFG)	Known Gas (BCFG)	Mean Undiscovered Gas (BCFG)	Future Gas (BCFG)	Gas Endowment (BCFG)	Discovery Maturity, Gas (%)
71	110903	Stavropol-Prikumsk	Oil	9670	3829	13,499	9644	13,473	23,143	58
72	521501	Egret-Hibernia	Oil	0	2406	2406	8893	11,298	11,298	21
73	606001	Neocomian Lacustrine	Oil	0	0	0	8871	8871	8871	0
74	312701	Shahejie-Shahejie/Guantao/Wumishan	Oil	2987	12,685	15,672	8514	21,199	24,186	65
75	115003	North Ustyurt Paleozoic	NA	NA	NA	NA	8500	NA	NA	NA
76	602901	Neocomian to Turonian Composite	Oil	348	1059	1407	8334	9393	9741	14
77	394802	Locker-Mungaroo/Barrow	Gas	27	8075	8102	8331	16,406	16,432	49
78	201601	North Oman Huqf-Shu'aiba	Oil	536	7097	7633	8206	15,303	15,839	48
79	606002	Lower Cretaceous	Oil	0	0	0	7710	7710	7710	0
80	382402	Jatibarang/Talang Akar-Oligocene/Miocene	Oil	794	6529	7322	6579	13,108	13,901	53
81	609601	Gacheta-Mirador	Oil	139	15,175	15,314	6217	21,391	21,530	71
82	314201	Maokou/Longtang-Jialingjiang/Maokou/Huanglong	Gas	5709	5364	11,072	6137	11,501	17,210	64
83	393001	Latrobe	Oil	3501	6274	9775	5658	11,932	15,433	63
84	117402	Togur-Tyumen	Oil	443	4738	5180	5516	10,254	10,697	48
85	205402	Tanezzuft-Melrhir	Oil	0	125	125	4887	5012	5012	2
86	608101	Cretaceous-Tertiary	Oil	1951	941	2892	4795	5736	7687	38
87	804702	Jalangi-Sylhet/Burdwan Composite	NA	NA	NA	NA	4613	NA	NA	NA
88	524305	Mannville-Upper Mannville	Gas	17,171	13,560	30,731	4597	18,157	35,328	87
89	380801	Brown Shale-Sihapas	Oil	910	2956	3866	4078	7034	7944	49
90	314204	Cambrian/Silurian Marine Shale-Dengying/Lower Paleozoic	Gas	491	859	1350	4036	4895	5386	25
91	605801	D-129	Oil	1667	1898	3565	3774	5672	7339	49
92	370302	Miocene Coaly Strata	Gas	38	23,863	23,901	3768	27,631	27,669	86
93	720301	Melania-Gamba	Oil	62	780	843	3674	4455	4517	19
94	730301	Cretaceous Composite	Gas	0	6015	6015	3603	9618	9618	63
95	609001	La Luna-La Paz	Oil	2310	2692	5002	3292	5985	8294	60
96	524302	Duvernay-Leduc	Oil	9050	5307	14,357	3224	8531	17,581	82
97	314402	Jurassic Coal-Denglouku/Nongan	Gas	0	145	145	3212	3357	3357	4
98	205802	Tanezzuft-Ahnet	Gas	0	3117	3117	2973	6090	6090	51
99	204801	Bou Dabbous-Tertiary	Oil	37	15,472	15,509	2746	18,218	18,256	85
100	524303	Exshaw-Rundle	Oil	14,851	6717	21,568	2663	9380	24,232	89
101	406102	Dysodile Schist-Tertiary	Oil	1580	3468	5048	2618	6086	7666	66
102	205805	Tanezzuft-Benoud	Gas	4980	100,070	105,050	2541	102,611	107,591	98
103	314401	Qingshankou-Putaohua/Shaertu	Oil	3	1594	1598	2494	4088	4091	39
104	404801	Greater Hungarian Plain Neogene	Oil	5436	4221	9657	2395	6616	12,052	80
105	101501	Volga-Ural Domanik-Paleozoic	Oil	15,846	80,612	96,458	2394	83,006	98,852	98
106	404702	Mesozoic/Paleogene Composite	Oil	3360	2033	5393	2345	4378	7738	70
107	205401	Tanezzuft-Oued Mya	Oil	68	8905	8973	2341	11,246	11,314	79
108	404701	Isotopically Light Gas	Gas	5455	5205	10,660	2321	7525	12,980	82
109	204802	Jurassic-Cretaceous Composite	Oil	6	1295	1301	2280	3575	3581	36
110	110902	South Mangyshlak	Oil	28	5679	5707	2262	7941	7969	72
111	115002	North Ustyurt Jurassic	Oil	0	1450	1450	2180	3630	3630	40
112	524304	Combined Triassic/Jurassic	Oil	4584	5996	10,580	2166	8163	12,747	83

Rank	Total Petroleum System Code	Total Petroleum System Name	Major Commodity	Cumulative Gas (BCFG)	Remaining Gas (BCFG)	Known Gas (BCFG)	Mean Undiscovered Gas (BCFG)	Future Gas (BCFG)	Gas Endowment (BCFG)	Discovery Maturity, Gas (%)
113	405701	Transylvanian Composite	Gas	21,194	9537	30,731	2083	11,620	32,814	94
114	370202	East Natuna	Gas	0	45,045	45,045	1963	47,008	47,008	96
115	524301	Keg River-Keg River	Oil	4589	2484	7073	1776	4260	8849	80
116	406002	Meride/Riva di Solto	Oil	154	257	412	1698	1955	2109	20
117	720304	Cuanza Composite	Oil	46	13	59	1558	1571	1617	4
118	101502	Belsk Basin	Gas	571	2067	2638	1547	3615	4185	63
119	803401	Sylhet-Kopili/Barail-Tipam Composite	Oil	2039	4495	6534	1178	5673	7712	85
120	391001	Milligans-Carboniferous/Permian	Oil	0	48	48	1151	1199	1199	4
121	404802	Zala-Drava-Sava Mesozoic/Neogene	Oil	1082	2847	3929	1151	3998	5080	77
122	205801	Tanezzuft-Timimoun	Gas	0	4200	4200	1128	5328	5328	79
123	524306	Second White Specks-Cardium	Oil	16,010	10,439	26,449	1126	11,565	27,574	96
124	311502	Jurassic Coal-Jurassic/Tertiary	Oil	10	332	342	950	1281	1292	26
125	382401	Banuwati-Oligocene/Miocene	Oil	233	381	614	874	1255	1488	41
126	701301	Cretaceous-Tertiary Composite	Oil	4	45	49	856	901	905	5
127	604102	Paleozoic	Oil	0	0	0	841	841	841	0
128	604101	Mesozoic-Cenozoic	Oil	661	955	1616	755	1711	2372	68
129	205803	Tanezzuft-Sbaa	Oil	0	1490	1490	645	2134	2134	70
130	611701	Upper Jurassic-Neocomian	Oil	20	77	96	592	668	688	14
131	406101	Moesian Platform Composite	Oil	1546	671	2217	561	1232	2778	80
132	802601	Patala-Namal	Oil	807	1174	1981	551	1726	2533	78
133	608301	Neogene	Oil	0	340	340	471	811	811	42
134	205806	Tanezzuft-Bechar/Abadla	Gas	0	0	0	441	441	441	0
135	404703	Paleozoic Composite	Oil	0	233	233	383	616	616	38
136	201602	Middle Cretaceous Natih	Oil	616	2485	3100	363	2848	3463	90
137	311501	Lucaogou-Karamay/Ulho/Pindequan	Oil	491	1757	2248	346	2104	2594	87
138	404806	Hungarian Paleogene	Oil	15	1	16	339	340	355	4
139	205804	Tanezzuft-Mouydir	Gas	0	0	0	292	292	292	0
140	608302	Cretaceous-Paleogene	Oil	148	10	158	224	233	382	41
141	404803	Danube Neogene	Gas	7	240	247	146	386	393	63
142	314202	Daanzhai-Daanzhai/Lianggaoshan	Oil	50	90	140	137	227	278	51
143	404804	Transcarpathian Neogene	Gas	0	261	261	112	373	373	70
144	115001	Buzachi Arch and Surrounding Areas Composite	Oil	0	943	943	89	1033	1033	91
145	524404	Lodgepole	Oil	728	166	893	63	228	956	93
146	312801	Yanchang-Yanan	Oil	13	7	20	20	27	40	50
147	524402	Brightholme	Oil	8	3	11	5	8	16	67
148	524401	Yeoman	Oil	0	0	0	4	4	4	3
149	524403	Bakken	Oil	0	0	0	3	3	4	9
NA	311503	Lucaogou/Jurassic Coal-Paleozoic/Mesozoic	NA	0	0	0	NA	NA	NA	NA
NA	312702	Carboniferous/Permian Coal-Paleozoic	NA	0	0	0	NA	NA	NA	NA
NA	312802	Taiyuan/Shanxi-Majiagou/Shihezi	NA	0	5590	5590	NA	NA	NA	NA
NA	314203	Xujiahe-Xujiahe/Shaximiao	NA	NA	NA	NA	NA	NA	NA	NA
NA	382403	Tertiary-Parigi	NA	0	119	119	NA	NA	NA	NA
NA	382404	Tertiary-Cenozoic	NA	0	0	0	NA	NA	NA	NA

Rank	Total Petroleum System Code	Total Petroleum System Name	Major Commodity	Cumulative Gas (BCFG)	Remaining Gas (BCFG)	Known Gas (BCFG)	Mean Undiscovered Gas (BCFG)	Future Gas (BCFG)	Gas Endowment (BCFG)	Discovery Maturity, Gas (%)
NA	404805	Central Carpathian Paleogene	NA	NA	NA	NA	NA	NA	NA	NA
NA	521502	Mesozoic Composite	NA	0	4224	4224	NA	NA	NA	NA
NA	605802	Aguada Bandera	NA	0	0	0	NA	NA	NA	NA
NA	804701	Permian Coal	NA	0	0	0	NA	NA	NA	NA

APPENDIX 2C. Total petroleum systems ranked by conventional oil endowment.

[MMBO = million barrels of oil. Major commodity is based on the ratio of gas endowment to oil endowment as ft^3 of gas per bbl of oil (CFG/BO). Major commodity is oil if this ratio is less than 20,000 CFG/BO and gas if this ratio is equal to or greater than 20,000 CFG/BO. Cumulative oil is the reported volume of oil that has been produced. Remaining oil is the calculated remaining reserves; the difference between known and cumulative oil. Known oil is the reported discovered volume (cumulative production plus remaining reserves). Mean undiscovered oil is the mean volume estimated in World Petroleum Assessment 2000 (USGS, 2000). Future oil is the remaining oil plus estimated mean undiscovered oil. Oil endowment is the known oil plus estimated mean undiscovered oil volumes. Discovery maturity is the percentage of oil discovered (known oil) with respect to oil endowment. Calculated values may not be equal to the sums of the component values because numbers have been independently rounded. Negative values result from inconsistent reporting of cumulative and known volumes. Values in parentheses are questionable data. NA means not applicable; it is shown for all values where there is only one field present, and used where the endowment could not be calculated.] Known volumes from Petroconsultants (1996) and NRG Associates (1995). Note that gas = natural gas.

Rank	Total Petroleum System Code	Total Petroleum System Name	Major Commodity	Cumulative Oil (MMBO)	Remaining Oil (MMBO)	Known Oil (MMBO)	Mean Undiscovered Oil (MMBO)	Future Oil (MMBO)	Oil Endowment (MMBO)	Discovery Maturity, Oil (%)
1	203001	Zagros-Mesopotamian Cretaceous-Tertiary	Oil	100,467	271,760	372,226	94,516	366,275	466,742	80
2	202102	Arabian Sub-Basin Tuwaiq/Hanifa-Arab	Oil	63,853	135,142	198,995	41,703	176,845	240,698	83
3	117401	Bazhenov-Neocomian	Oil	45,819	72,153	117,972	42,947	115,100	160,919	73
4	201901	Cretaceous Thamama/Wasia	Oil	13,544	57,686	71,231	8979	66,665	80,209	89
5	719201	Tertiary Niger Delta (Agbada/Akata)	Oil	16,598	17,925	34,522	40,487	58,412	75,010	46
6	101501	Volga-Ural Domanik-Paleozoic	Oil	42,584	21,288	63,872	2507	23,795	66,379	96
7	530501	Pimienta-Tamabra	Oil	22,139	22,273	44,412	20,569	42,841	64,980	68
8	609901	La Luna/Maracaibo	Oil	35,581	13,491	49,072	8183	21,675	57,256	86
9	402501	Kimmeridgian Shales	Oil	22,467	21,427	43,895	13,098	34,526	56,993	77
10	520001	Permian/Upper Jurassic Composite	Oil	0	0	0	47,148	47,148	47,148	0
11	204301	Sirte-Zelten	Oil	13,986	23,087	37,072	6854	29,941	43,927	84
12	101601	Paleozoic North Caspian	Oil	822	9987	10,808	23,320	33,307	34,128	32
13	609801	Querecual	Oil	11,027	15,729	26,756	7231	22,960	33,988	79
14	201902	Jurassic Hanifa/Diyab-Arab	Oil	5688	13,325	19,013	14,478	27,803	33,491	57
15	111201	Oligocene-Miocene Maykop/Diatom	Oil	11,681	5758	17,438	15,725	21,483	33,163	53
16	720303	Congo Delta Composite	Oil	3693	6052	9745	23,198	29,250	32,943	30
17	312701	Shahejie-Shahejie/Guantao/Wumishan	Oil	10,415	14,139	24,553	4412	18,550	28,965	85
18	603501	Lagoa Feia-Carapebus	Oil	1769	8287	10,056	16,293	24,580	26,349	38
19	202101	Central Arabia Qusaiba Paleozoic	Oil	84	6292	6376	19,190	25,481	25,566	25
20	603601	Guaratiba-Guaruja (Cretaceous) Composite	Oil	6	279	285	23,209	23,488	23,494	1
21	117402	Togur-Tyumen	Oil	645	11,111	11,756	7474	18,584	19,230	61
22	100801	Domanik-Paleozoic	Oil	3346	9723	13,069	5170	14,892	18,239	72
23	314401	Qingshankou-Putaohua/Shaertu	Oil	12,399	3171	15,570	1026	4198	16,596	94
24	602101	Cenomanian-Turonian	Oil	17	46	63	15,247	15,293	15,310	0
25	117403	Northern West Siberian Mesozoic Composite	Gas	66	10,119	10,185	4814	14,933	14,999	68
26	380801	Brown Shale-Sihapas	Oil	8349	4868	13,217	899	5768	14,117	94
27	201903	Silurian Qusaiba	Gas	41	514	555	13,235	13,749	13,790	4
28	207101	Sudr-Nubia	Oil	6351	3459	9810	2608	6067	12,418	79

Rank	Total Petroleum System Code	Total Petroleum System Name	Major Commodity	Cumulative Oil (MMBO)	Remaining Oil (MMBO)	Known Oil (MMBO)	Mean Undiscovered Oil (MMBO)	Future Oil (MMBO)	Oil Endowment (MMBO)	Discovery Maturity, Oil (%)
29	205401	Tanezzuft-Oued Mya	Oil	5872	4971	10,843	830	5801	11,673	93
30	804301	Eocene-Miocene Composite	Oil	3375	5065	8440	2091	7156	10,531	80
31	370101	Brunei-Sabah	Oil	3953	2945	6898	3390	6335	10,288	67
32	524302	Duvernay-Leduc	Oil	8153	1306	9459	501	1807	9960	95
33	604101	Mesozoic-Cenozoic	Oil	2889	3712	6601	3023	6735	9625	69
34	401701	Upper Jurassic Spekk	Oil	76	2585	2660	6538	9123	9199	29
35	609601	Gacheta-Mirador	Oil	1179	4224	5402	3631	7855	9033	60
36	205403	Tanezzuft-Ghadames	Oil	1629	2909	4538	4461	7370	8999	50
37	609802	Upper Cretaceous/Tertiary	Oil	2780	667	3447	4644	5311	8091	43
38	381701	Kutei Basin	Oil	2195	684	2879	5145	5829	8024	36
39	720302	Azile-Senonian	Oil	1278	836	2114	5031	5867	7145	30
40	311501	Lucaogou-Karamay/Ulho/Pindequan	Oil	937	5687	6624	467	6154	7091	93
41	202302	Jurassic Gotnia/Barsarin/Sargelu/Najmah	Oil	94	379	474	6569	6948	7042	7
42	315401	Ordovician/Jurassic-Phanerozoic	Oil	45	658	704	5948	6606	6652	11
43	205601	Tanezzuft-Illizi	Oil	2286	1384	3670	2814	4198	6484	57
44	110901	Terek-Caspian	Oil	752	2738	3490	2829	5567	6319	55
45	110902	South Mangyshlak	Oil	2406	2835	5241	651	3486	5893	89
46	406102	Dysodile Schist-Tertiary	Oil	2832	1283	4115	842	2125	4957	83
47	132201	North Sakhalin Neogene	Oil	754	1428	2182	2582	4010	4764	46
48	200401	Madbi Amran/Qishn	Oil	649	1284	1933	2740	4025	4673	41
49	393001	Latrobe	Oil	3077	784	3860	310	1093	4170	93
50	370301	Oligocene-Miocene Lacustrine	Oil	1459	1557	3017	1127	2684	4143	73
51	201601	North Oman Huqf-Shu'aiba	Oil	1259	1426	2685	1304	2731	3990	67
52	202301	Paleozoic Qusaiba/Akkas/Abba/Mudawwara	Oil	269	944	1213	2632	3577	3845	32
53	605801	D-129	Oil	2445	900	3346	498	1398	3844	87
54	605501	Neuquen Hybrid	Oil	1532	838	2370	1290	2128	3659	65
55	121001	Baikal-Patom Foldbelt Riphean-Craton Margin Vendian	Gas	1	2005	2006	1611	3616	3617	55
56	606001	Neocomian Lacustrine	Oil	0	0	0	3603	3603	3603	0
57	394801	Dingo-Mungaroo/Barrow	Gas	489	660	1149	2249	2909	3398	34
58	608101	Cretaceous-Tertiary	Oil	1346	326	1672	1711	2038	3383	49
59	521501	Egret-Hibernia	Oil	0	1582	1582	1796	3378	3378	47
60	609001	La Luna-La Paz	Oil	1915	512	2426	709	1220	3135	77
61	115001	Buzachi Arch and Surrounding Areas Composite	Oil	459	1827	2286	814	2641	3100	74
62	603401	Cretaceous Composite	Oil	93	37	130	2959	2995	3089	4
63	403601	Carboniferous-Rotliegend	Gas	2300	572	2872	143	715	3015	95
64	603701	Cenomanian-Turonian-Tertiary Composite	Oil	0	0	0	2938	2938	2938	0
65	803401	Sylhet-Kopili/Barail-Tipam Composite	Oil	911	1625	2536	399	2024	2935	86
66	382801	Lahat/Talang Akar-Cenozoic	Oil	1842	587	2429	469	1055	2898	84
67	524404	Lodgepole	Oil	2239	539	2778	111	650	2889	96
68	204801	Bou Dabbous-Tertiary	Oil	481	1633	2114	667	2300	2782	76
69	201401	North Oman Huqf/'Q'-Haushi	Oil	715	1313	2028	743	2056	2771	73
70	207102	Maqna	Oil	22	133	155	2579	2712	2734	6

App-2c_Page 2

Rank	Total Petroleum System Code	Total Petroleum System Name	Major Commodity	Cumulative Oil (MMBO)	Remaining Oil (MMBO)	Known Oil (MMBO)	Mean Undiscovered Oil (MMBO)	Future Oil (MMBO)	Oil Endowment (MMBO)	Discovery Maturity, Oil (%)
71	524306	Second White Specks-Cardium	Oil	2146	543	2688	36	578	2724	99
72	100901	Dnieper-Donets Paleozoic	Gas	110	1501	1611	1098	2599	2709	59
73	720301	Melania-Gamba	Oil	708	1044	1752	727	1772	2480	71
74	604501	Los Monos-Machareti	Oil	257	63	320	2145	2209	2466	13
75	606002	Lower Cretaceous	Oil	0	0	0	2229	2229	2229	0
76	110903	Stavropol-Prikumsk	Oil	1200	(-380)	820	1407	(1027)	2227	37
77	201602	Middle Cretaceous Natih	Oil	1299	553	1852	371	925	2224	83
78	602901	Neocomian to Turonian Composite	Oil	469	223	692	1487	1711	2179	32
79	382402	Jatibarang/Talang Akar-Oligocene/Miocene	Oil	727	1181	1908	195	1377	2104	91
80	406101	Moesian Platform Composite	Oil	1251	542	1793	171	713	1963	91
81	605901	Lower Inoceramus	Gas	595	620	1216	704	1325	1920	63
82	205402	Tanezzuft-Melrhir	Oil	0	6	6	1875	1881	1881	0
83	524303	Exshaw-Rundle	Oil	1292	435	1728	126	561	1854	93
84	382401	Banuwati-Oligocene/Miocene	Oil	828	431	1259	462	894	1722	73
85	391003	Jurassic/Early Cretaceous-Mesozoic	Oil	123	378	502	1146	1524	1648	30
86	610701	Tobago Trough Paleogene	Oil	8	2	11	1536	1538	1546	1
87	804801	Eocene to Miocene Composite	Gas	528	188	716	724	912	1441	50
88	524304	Combined Triassic/Jurassic	Oil	1012	284	1296	96	380	1393	93
89	370201	Sarawak Basin	Gas	168	629	797	529	1159	1327	60
90	404801	Greater Hungarian Plain Neogene	Oil	731	358	1089	196	555	1285	85
91	404702	Mesozoic/Paleogene Composite	Oil	640	283	923	344	627	1267	73
92	120701	Yenisey Foldbelt Riphean-Craton Margin Riphean	Oil	0	34	34	1213	1246	1246	3
93	718301	Cretaceous Composite	Oil	74	162	236	1004	1166	1240	19
94	404802	Zala-Drava-Sava Mesozoic/Neogene	Oil	836	265	1101	114	379	1215	91
95	391301	Late Jurassic/Early Cretaceous-Mesozoic	Gas	19	27	46	1055	1083	1101	4
96	524301	Keg River-Keg River	Oil	835	176	1011	84	260	1095	92
97	606301	Lower Cretaceous Marine	Oil	0	10	10	1031	1041	1041	1
98	115401	Amu-Darya Jurassic-Cretaceous	Gas	80	686	766	270	957	1037	74
99	105001	South and North Barents Triassic-Jurassic	Gas	0	51	51	955	1006	1007	5
100	110801	Azov-Kuban Mesozoic-Cenozoic	Gas	1273	(-750)	524	417	(-332)	941	56
101	720304	Cuanza Composite	Oil	86	20	107	790	810	897	12
102	382201	Bampo-Cenozoic	Gas	496	178	674	213	391	887	76
103	312801	Yanchang-Yanan	Oil	81	602	683	138	740	821	83
104	611701	Upper Jurassic-Neocomian	Oil	65	232	297	494	726	790	38
105	370302	Miocene Coaly Strata	Gas	58	533	592	101	634	693	85
106	406002	Meride/Riva di Solto	Oil	78	254	332	360	614	692	48
107	204802	Jurassic-Cretaceous Composite	Oil	39	64	103	403	467	506	20
108	802601	Patala-Namal	Oil	204	121	324	145	266	469	69
109	205803	Tanezzuft-Sbaa	Oil	0	284	284	162	446	446	64
110	406801	Pre-Messinian	Gas	0	0	0	388	388	388	0
111	608302	Cretaceous-Paleogene	Oil	118	53	171	200	252	370	46
112	311502	Jurassic Coal-Jurassic/Tertiary	Oil	16	193	210	96	289	306	69
113	804201	Sembar-Goru/Ghazij	Gas	111	70	180	119	188	299	60

App-2c_Page 3

Rank	Total Petroleum System Code	Total Petroleum System Name	Major Commodity	Cumulative Oil (MMBO)	Remaining Oil (MMBO)	Known Oil (MMBO)	Mean Undiscovered Oil (MMBO)	Future Oil (MMBO)	Oil Endowment (MMBO)	Discovery Maturity, Oil (%)
114	115002	North Ustyurt Jurassic	Oil	0	80	80	139	220	220	37
115	701301	Cretaceous-Tertiary Composite	Oil	0	10	10	157	167	168	6
116	205805	Tanezzuft-Benoud	Gas	16	72	88	72	144	160	55
117	391001	Milligans-Carboniferous/Permian	Oil	0	15	15	140	155	155	10
118	394802	Locker-Mungaroo/Barrow	Gas	0	6	6	132	137	137	4
119	314202	Daanzhai-Daanzhai/Lianggaoshan	Oil	38	71	109	27	99	136	80
120	370202	East Natuna	Gas	0	20	20	102	122	122	16
121	730301	Cretaceous Composite	Gas	0	0	0	116	116	116	0
122	101502	Belsk Basin	Gas	106	(-41)	65	42	(1)	106	61
123	604102	Paleozoic	Oil	0	0	0	74	74	74	0
124	404806	Hungarian Paleogene	Oil	10	1	11	49	50	60	18
125	404703	Paleozoic Composite	Oil	0	1	1	47	47	47	1
126	804703	Jenam/Bhuban-Bokabil	Gas	3	2	4	39	41	43	10
127	608301	Neogene	Oil	4	0	4	37	37	41	11
128	205802	Tanezzuft-Ahnet	Gas	0	1	1	34	35	35	3
129	205801	Tanezzuft-Timimoun	Gas	0	0	0	31	31	31	0
130	524402	Brightholme	Oil	11	4	14	8	11	22	66
131	610301	Lower Cruse	Gas	0	20	20	0	20	20	100
132	524403	Bakken	Oil	5	2	7	11	13	18	40
133	205806	Tanezzuft-Bechar/Abadla	Gas	0	0	0	16	16	16	0
134	406001	Porto Garibaldi	Gas	13	0	14	0	0	14	100
135	205804	Tanezzuft-Mouydir	Gas	0	0	0	12	12	12	0
136	524401	Yeoman	Oil	1	0	1	5	6	6	15
137	314402	Jurassic Coal-Denglouku/Nongan	Gas	0	5	5	0	5	5	100
138	314204	Cambrian/Silurian Marine Shale-Dengying/Lower Paleozoic	Gas	9	(-8)	1	0	(-8)	1	100
139	404701	Isotopically Light Gas	Gas	1	1	1	0	1	1	100
140	203002	Paleozoic-Permian/Triassic	Gas	0	0	0	0	0	0	NA
141	314201	Maokou/Longtang-Jialingjiang/Maokou/Huanglong	Gas	0	0	0	0	0	0	100
142	391002	Keyling/Hyland Bay-Permian	Gas	0	0	0	0	0	0	NA
143	404803	Danube Neogene	Gas	0	0	0	0	0	0	NA
144	404804	Transcarpathian Neogene	Gas	0	0	0	0	0	0	NA
145	405701	Transylvanian Composite	Gas	0	0	0	0	0	0	NA
146	524305	Mannville-Upper Mannville	Gas	0	0	0	0	0	0	NA
NA	115003	North Ustyurt Paleozoic	NA	NA	NA	NA	0	NA	NA	NA
NA	311503	Lucaogou/Jurassic Coal-Paleozoic/Mesozoic	NA	0	0	0	NA	NA	NA	NA
NA	312702	Carboniferous/Permian Coal-Paleozoic	NA	0	0	0	NA	NA	NA	NA
NA	312802	Taiyuan/Shanxi-Majiagou/Shihezi	NA	0	0	0	NA	NA	NA	NA
NA	314203	Xujiahe-Xujiahe/Shaximiao	NA	NA	NA	NA	NA	NA	NA	NA
NA	382403	Tertiary-Parigi	NA	0	0	0	NA	NA	NA	NA
NA	382404	Tertiary-Cenozoic	NA	0	0	0	NA	NA	NA	NA
NA	404805	Central Carpathian Paleogene	NA	NA	NA	NA	NA	NA	NA	NA
NA	521502	Mesozoic Composite	NA	0	0	0	NA	NA	NA	NA

Rank	Total Petroleum System Code	Total Petroleum System Name	Major Commodity	Cumulative Oil (MMBO)	Remaining Oil (MMBO)	Known Oil (MMBO)	Mean Undiscovered Oil (MMBO)	Future Oil (MMBO)	Oil Endowment (MMBO)	Discovery Maturity, Oil (%)
NA	602201	Neogene	NA	NA	NA	NA	0	NA	NA	NA
NA	605802	Aguada Bandera	NA	0	0	0	NA	NA	NA	NA
NA	804701	Permian Coal	NA	0	0	0	NA	NA	NA	NA
NA	804702	Jalangi-Sylhet/Burdwan Composite	NA	NA	NA	NA	64	NA	NA	NA

APPENDIX 2D. Total petroleum systems ranked by conventional natural gas endowment.

[BCFG = billion cubic feet of gas. Major commodity is based on the ratio of gas endowment to oil endowment as ft^3 of gas per bbl of oil (CFG/BO). Major commodity is oil if this ratio is less than 20,000 CFG/BO and gas if this ratio is equal to or greater than 20,000 CFG/BO. Cumulative gas is the reported volume of gas that has been produced. Remaining gas is the calculated remaining reserves; the difference between known and cumulative gas. Known gas is the reported discovered volume (cumulative production plus remaining reserves). Mean undiscovered gas is the mean volume estimated in World Petroleum Assessment 2000 (USGS, 2000). Future gas is the remaining gas plus estimated mean undiscovered gas. Gas endowment is the known gas plus estimated mean undiscovered gas volumes. Discovery maturity is the percentage of gas discovered (known gas) with respect to gas endowment. Calculated values may not be equal to the sums of the component values because numbers have been independently rounded. NA means not applicable; it is shown for all values where there is only one field present, and used where the endowment could not be calculated.] Known volumes from Petroconsultants (1996) and NRG Associates (1995). Note that gas = natural gas.

Rank	Total Petroleum System Code	Total Petroleum System Name	Major Commodity	Cumulative Gas (BCFG)	Remaining Gas (BCFG)	Known Gas (BCFG)	Mean Undiscovered Gas (BCFG)	Future Gas (BCFG)	Gas Endowment (BCFG)	Discovery Maturity, Gas (%)
1	117403	Northern West Siberian Mesozoic Composite	Gas	188,736	978,552	1,167,288	540,483	1,519,035	1,707,771	68
2	201903	Silurian Qusaiba	Gas	1018	451,012	452,030	346,892	797,903	798,922	57
3	203001	Zagros-Mesopotamian Cretaceous-Tertiary	Oil	25,260	467,978	493,238	197,597	665,575	690,835	71
4	202101	Central Arabia Qusaiba Paleozoic	Oil	0	79,115	79,115	383,339	462,454	462,454	17
5	115401	Amu-Darya Jurassic-Cretaceous	Gas	78,540	152,073	230,614	163,650	315,724	394,264	58
6	105001	South and North Barents Triassic-Jurassic	Gas	0	78,143	78,143	266,618	344,761	344,761	23
7	202102	Arabian Sub-Basin Tuwaiq/Hanifa-Arab	Oil	24,821	250,474	275,295	34,046	284,520	309,341	89
8	101601	Paleozoic North Caspian	Oil	1745	155,231	156,976	119,051	274,282	276,027	57
9	403601	Carboniferous-Rotliegend	Gas	124,048	98,111	222,159	25,790	123,901	247,948	90
10	719201	Tertiary Niger Delta (Agbada/Akata)	Oil	6232	87,579	93,811	132,716	220,295	226,526	41
11	111201	Oligocene-Miocene Maykop/Diatom	Oil	8591	27,403	35,994	173,310	200,713	209,304	17
12	203002	Paleozoic-Permian/Triassic	Gas	0	131,220	131,220	69,031	200,251	200,251	66
13	402501	Kimmeridgian Shales	Oil	31,053	127,861	158,914	37,745	165,606	196,659	81
14	117401	Bazhenov-Neocomian	Oil	29,287	68,295	97,582	96,934	165,229	194,516	50
15	401701	Upper Jurassic Spekk	Oil	0	15,662	15,662	165,201	180,863	180,863	9
16	609801	Querecual	Oil	18,897	93,399	112,296	41,552	134,951	153,848	73
17	201902	Jurassic Hanifa/Diyab-Arab	Oil	1951	75,561	77,512	58,515	134,076	136,027	57
18	201901	Cretaceous Thamama/Wasia	Oil	6342	95,758	102,100	22,311	118,069	124,410	82
19	381701	Kutei Basin	Oil	8711	36,762	45,473	64,757	101,519	110,230	41
20	205805	Tanezzuft-Benoud	Gas	4980	100,070	105,050	2541	102,611	107,591	98
21	394801	Dingo-Mungaroo/Barrow	Gas	1845	46,400	48,245	56,380	102,780	104,626	46
22	530501	Pimienta-Tamabra	Oil	17,609	33,213	50,822	49,272	82,485	100,094	51
23	101501	Volga-Ural Domanik-Paleozoic	Oil	15,846	80,612	96,458	2394	83,006	98,852	98
24	100901	Dnieper-Donets Paleozoic	Gas	35,453	23,645	59,098	24,051	47,697	83,149	71
25	603601	Guaratiba-Guaruja (Cretaceous) Composite	Oil	12	1113	1125	80,547	81,660	81,672	1
26	520001	Permian/Upper Jurassic Composite	Oil	0	0	0	80,709	80,709	80,709	0
27	132201	North Sakhalin Neogene	Oil	162	22,221	22,383	57,451	79,673	79,835	28
28	720303	Congo Delta Composite	Oil	838	8605	9443	69,582	78,186	79,025	12

Rank	Total Petroleum System Code	Total Petroleum System Name	Major Commodity	Cumulative Gas (BCFG)	Remaining Gas (BCFG)	Known Gas (BCFG)	Mean Undiscovered Gas (BCFG)	Future Gas (BCFG)	Gas Endowment (BCFG)	Discovery Maturity, Gas (%)
29	121001	Baikal-Patom Foldbelt Riphean-Craton Margin Vendian	Gas	0	30,210	30,210	46,039	76,249	76,249	40
30	205601	Tanezzuft-Illizi	Oil	750	44,311	45,061	27,785	72,096	72,846	62
31	100801	Domanik-Paleozoic	Oil	12,374	24,259	36,632	34,525	58,784	71,158	51
32	609802	Upper Cretaceous/Tertiary	Oil	3156	12,667	15,822	52,009	64,676	67,832	23
33	202301	Paleozoic Qusaiba/Akkas/Abba/Mudawwara	Oil	47	1614	1661	64,510	66,124	66,172	3
34	315401	Ordovician/Jurassic-Phanerozoic	Oil	208	5572	5780	59,939	65,511	65,719	9
35	804201	Sembar-Goru/Ghazij	Gas	15,101	20,272	35,373	29,325	49,596	64,697	55
36	804703	Jenam/Bhuban-Bokabil	Gas	2669	9620	12,289	50,479	60,098	62,767	20
37	370101	Brunei-Sabah	Oil	9188	27,012	36,200	23,226	50,238	59,426	61
38	370201	Sarawak Basin	Gas	3731	33,388	37,119	17,284	50,672	54,403	68
39	207102	Maqna	Oil	10	3517	3527	49,820	53,338	53,347	7
40	204301	Sirte-Zelten	Oil	5479	32,288	37,767	14,984	47,272	52,752	72
41	406801	Pre-Messinian	Gas	0	0	0	51,210	51,210	51,210	0
42	604501	Los Monos-Machareti	Oil	6928	10,058	16,985	31,107	41,165	48,092	35
43	370301	Oligocene-Miocene Lacustrine	Oil	454	23,794	24,248	23,696	47,491	47,944	51
44	370202	East Natuna	Gas	0	45,045	45,045	1963	47,008	47,008	96
45	609901	La Luna/Maracaibo	Oil	979	25,722	26,701	17,576	43,298	44,277	60
46	610301	Lower Cruse	Gas	0	22,600	22,600	20,103	42,703	42,703	53
47	602101	Cenomanian-Turonian	Oil	0	0	0	42,058	42,058	42,058	0
48	382201	Bampo-Cenozoic	Gas	13,977	11,582	25,559	15,123	26,705	40,682	63
49	605901	Lower Inoceramus	Gas	10,053	14,754	24,807	14,040	28,794	38,848	64
50	391301	Late Jurassic/Early Cretaceous-Mesozoic	Gas	0	17,960	17,960	20,093	38,053	38,053	47
51	804301	Eocene-Miocene Composite	Oil	6228	17,965	24,193	12,944	30,909	37,137	65
52	110901	Terek-Caspian	Oil	192	7648	7840	27,779	35,427	35,619	22
53	524305	Mannville-Upper Mannville	Gas	17,171	13,560	30,731	4597	18,157	35,328	87
54	406001	Porto Garibaldi	Gas	9373	9102	18,475	16,267	25,369	34,741	53
55	603401	Cretaceous Composite	Oil	42	180	222	34,287	34,467	34,509	1
56	200401	Madbi Amran/Qishn	Oil	0	17,120	17,120	16,982	34,102	34,102	50
57	605501	Neuquen Hybrid	Oil	7636	13,068	20,704	12,416	25,483	33,119	63
58	405701	Transylvanian Composite	Gas	21,194	9537	30,731	2083	11,620	32,814	94
59	110801	Azov-Kuban Mesozoic-Cenozoic	Gas	7586	11,118	18,704	12,252	23,369	30,955	60
60	804801	Eocene to Miocene Composite	Gas	722	9298	10,020	20,521	29,819	30,541	33
61	201401	North Oman Huqf/'Q'-Haushi	Oil	149	20,189	20,339	10,112	30,301	30,450	67
62	205403	Tanezzuft-Ghadames	Oil	579	15,905	16,484	12,035	27,940	28,519	58
63	382801	Lahat/Talang Akar-Cenozoic	Oil	3125	7080	10,204	18,250	25,330	28,455	36
64	370302	Miocene Coaly Strata	Gas	38	23,863	23,901	3768	27,631	27,669	86
65	524306	Second White Specks-Cardium	Oil	16,010	10,439	26,449	1126	11,565	27,574	96
66	603501	Lagoa Feia-Carapebus	Oil	303	5941	6244	19,691	25,632	25,935	24
67	524303	Exshaw-Rundle	Oil	14,851	6717	21,568	2663	9380	24,232	89
68	312701	Shahejie-Shahejie/Guantao/Wumishan	Oil	2987	12,685	15,672	8514	21,199	24,186	65
69	110903	Stavropol-Prikumsk	Oil	9670	3829	13,499	9644	13,473	23,143	58
70	202302	Jurassic Gotnia/Barsarin/Sargelu/Najmah	Oil	0	100	100	22,942	23,042	23,042	0

App-2d_Page 2

Rank	Total Petroleum System Code	Total Petroleum System Name	Major Commodity	Cumulative Gas (BCFG)	Remaining Gas (BCFG)	Known Gas (BCFG)	Mean Undiscovered Gas (BCFG)	Future Gas (BCFG)	Gas Endowment (BCFG)	Discovery Maturity, Gas (%)
71	603701	Cenomanian-Turonian-Tertiary Composite	Oil	0	0	0	22,806	22,806	22,806	0
72	120701	Yenisey Foldbelt Riphean-Craton Margin Riphean	Oil	0	5052	5052	16,983	22,035	22,035	23
73	609601	Gacheta-Mirador	Oil	139	15,175	15,314	6217	21,391	21,530	71
74	391003	Jurassic/Early Cretaceous-Mesozoic	Oil	0	7526	7526	11,346	18,872	18,872	40
75	204801	Bou Dabbous-Tertiary	Oil	37	15,472	15,509	2746	18,218	18,256	85
76	524302	Duvernay-Leduc	Oil	9050	5307	14,357	3224	8531	17,581	82
77	314201	Maokou/Longtang-Jialingjiang/Maokou/Huanglong	Gas	5709	5364	11,072	6137	11,501	17,210	64
78	391002	Keyling/Hyland Bay-Permian	Gas	0	5730	5730	10,995	16,725	16,725	34
79	394802	Locker-Mungaroo/Barrow	Gas	27	8075	8102	8331	16,406	16,432	49
80	207101	Sudr-Nubia	Oil	840	5155	5995	10,192	15,347	16,186	37
81	201601	North Oman Huqf-Shu'aiba	Oil	536	7097	7633	8206	15,303	15,839	48
82	393001	Latrobe	Oil	3501	6274	9775	5658	11,932	15,433	63
83	720302	Azile-Senonian	Oil	203	1632	1835	13,230	14,862	15,065	12
84	382402	Jatibarang/Talang Akar-Oligocene/Miocene	Oil	794	6529	7322	6579	13,108	13,901	53
85	404701	Isotopically Light Gas	Gas	5455	5205	10,660	2321	7525	12,980	82
86	524304	Combined Triassic/Jurassic	Oil	4584	5996	10,580	2166	8163	12,747	83
87	404801	Greater Hungarian Plain Neogene	Oil	5436	4221	9657	2395	6616	12,052	80
88	718301	Cretaceous Composite	Oil	46	1903	1949	10,071	11,974	12,020	16
89	610701	Tobago Trough Paleogene	Oil	16	7	22	11,957	11,963	11,979	0
90	606301	Lower Cretaceous Marine	Oil	0	75	75	11,327	11,402	11,402	1
91	205401	Tanezzuft-Oued Mya	Oil	68	8905	8973	2341	11,246	11,314	79
92	521501	Egret-Hibernia	Oil	0	2406	2406	8893	11,298	11,298	21
93	117402	Togur-Tyumen	Oil	443	4738	5180	5516	10,254	10,697	48
94	602901	Neocomian to Turonian Composite	Oil	348	1059	1407	8334	9393	9741	14
95	730301	Cretaceous Composite	Gas	0	6015	6015	3603	9618	9618	63
96	606001	Neocomian Lacustrine	Oil	0	0	0	8871	8871	8871	0
97	524301	Keg River-Keg River	Oil	4589	2484	7073	1776	4260	8849	80
98	609001	La Luna-La Paz	Oil	2310	2692	5002	3292	5985	8294	60
99	110902	South Mangyshlak	Oil	28	5679	5707	2262	7941	7969	72
100	380801	Brown Shale-Sihapas	Oil	910	2956	3866	4078	7034	7944	49
101	404702	Mesozoic/Paleogene Composite	Oil	3360	2033	5393	2345	4378	7738	70
102	803401	Sylhet-Kopili/Barail-Tipam Composite	Oil	2039	4495	6534	1178	5673	7712	85
103	606002	Lower Cretaceous	Oil	0	0	0	7710	7710	7710	0
104	608101	Cretaceous-Tertiary	Oil	1951	941	2892	4795	5736	7687	38
105	406102	Dysodile Schist-Tertiary	Oil	1580	3468	5048	2618	6086	7666	66
106	605801	D-129	Oil	1667	1898	3565	3774	5672	7339	49
107	205802	Tanezzuft-Ahnet	Gas	0	3117	3117	2973	6090	6090	51
108	314204	Cambrian/Silurian Marine Shale-Dengying/Lower Paleozoic	Gas	491	859	1350	4036	4895	5386	25
109	205801	Tanezzuft-Timimoun	Gas	0	4200	4200	1128	5328	5328	79
110	404802	Zala-Drava-Sava Mesozoic/Neogene	Oil	1082	2847	3929	1151	3998	5080	77
111	205402	Tanezzuft-Melrhir	Oil	0	125	125	4887	5012	5012	2
112	720301	Melania-Gamba	Oil	62	780	843	3674	4455	4517	19

Rank	Total Petroleum System Code	Total Petroleum System Name	Major Commodity	Cumulative Gas (BCFG)	Remaining Gas (BCFG)	Known Gas (BCFG)	Mean Undiscovered Gas (BCFG)	Future Gas (BCFG)	Gas Endowment (BCFG)	Discovery Maturity, Gas (%)
113	101502	Belsk Basin	Gas	571	2067	2638	1547	3615	4185	63
114	314401	Qingshankou-Putaohua/Shaertu	Oil	3	1594	1598	2494	4088	4091	39
115	115002	North Ustyurt Jurassic	Oil	0	1450	1450	2180	3630	3630	40
116	204802	Jurassic-Cretaceous Composite	Oil	6	1295	1301	2280	3575	3581	36
117	201602	Middle Cretaceous Natih	Oil	616	2485	3100	363	2848	3463	90
118	314402	Jurassic Coal-Denglouku/Nongan	Gas	0	145	145	3212	3357	3357	4
119	406101	Moesian Platform Composite	Oil	1546	671	2217	561	1232	2778	80
120	311501	Lucaogou-Karamay/Ulho/Pindequan	Oil	491	1757	2248	346	2104	2594	87
121	802601	Patala-Namal	Oil	807	1174	1981	551	1726	2533	78
122	604101	Mesozoic-Cenozoic	Oil	661	955	1616	755	1711	2372	68
123	205803	Tanezzuft-Sbaa	Oil	0	1490	1490	645	2134	2134	70
124	406002	Meride/Riva di Solto	Oil	154	257	412	1698	1955	2109	20
125	720304	Cuanza Composite	Oil	46	13	59	1558	1571	1617	4
126	382401	Banuwati-Oligocene/Miocene	Oil	233	381	614	874	1255	1488	41
127	311502	Jurassic Coal-Jurassic/Tertiary	Oil	10	332	342	950	1281	1292	26
128	391001	Milligans-Carboniferous/Permian	Oil	0	48	48	1151	1199	1199	4
129	115001	Buzachi Arch and Surrounding Areas Composite	Oil	0	943	943	89	1033	1033	91
130	524404	Lodgepole	Oil	728	166	893	63	228	956	93
131	701301	Cretaceous-Tertiary Composite	Oil	4	45	49	856	901	905	5
132	604102	Paleozoic	Oil	0	0	0	841	841	841	0
133	608301	Neogene	Oil	0	340	340	471	811	811	42
134	611701	Upper Jurassic-Neocomian	Oil	20	77	96	592	668	688	14
135	404703	Paleozoic Composite	Oil	0	233	233	383	616	616	38
136	205806	Tanezzuft-Bechar/Abadla	Gas	0	0	0	441	441	441	0
137	404803	Danube Neogene	Gas	7	240	247	146	386	393	63
138	608302	Cretaceous-Paleogene	Oil	148	10	158	224	233	382	41
139	404804	Transcarpathian Neogene	Gas	0	261	261	112	373	373	70
140	404806	Hungarian Paleogene	Oil	15	1	16	339	340	355	4
141	205804	Tanezzuft-Mouydir	Gas	0	0	0	292	292	292	0
142	314202	Daanzhai-Daanzhai/Lianggaoshan	Oil	50	90	140	137	227	278	51
143	312801	Yanchang-Yanan	Oil	13	7	20	20	27	40	50
144	524402	Brightholme	Oil	8	3	11	5	8	16	67
145	524401	Yeoman	Oil	0	0	0	4	4	4	3
146	524403	Bakken	Oil	0	0	0	3	3	4	9
NA	115003	North Ustyurt Paleozoic	NA	NA	NA	NA	8500	NA	NA	NA
NA	311503	Lucaogou/Jurassic Coal-Paleozoic/Mesozoic	NA	0	0	0	NA	NA	NA	NA
NA	312702	Carboniferous/Permian Coal-Paleozoic	NA	0	0	0	NA	NA	NA	NA
NA	312802	Taiyuan/Shanxi-Majiagou/Shihezi	NA	0	5590	5590	NA	NA	NA	NA
NA	314203	Xujiahe-Xujiahe/Shaximiao	NA	NA	NA	NA	NA	NA	NA	NA
NA	382403	Tertiary-Parigi	NA	0	119	119	NA	NA	NA	NA
NA	382404	Tertiary-Cenozoic	NA	0	0	0	NA	NA	NA	NA
NA	404805	Central Carpathian Paleogene	NA	NA	NA	NA	NA	NA	NA	NA
NA	521502	Mesozoic Composite	NA	0	4224	4224	NA	NA	NA	NA

Rank	Total Petroleum System Code	Total Petroleum System Name	Major Commodity	Cumulative Gas (BCFG)	Remaining Gas (BCFG)	Known Gas (BCFG)	Mean Undiscovered Gas (BCFG)	Future Gas (BCFG)	Gas Endowment (BCFG)	Discovery Maturity, Gas (%)
NA	602201	Neogene	NA	NA	NA	NA	29,840	NA	NA	NA
NA	605802	Aguada Bandera	NA	0	0	0	NA	NA	NA	NA
NA	804701	Permian Coal	NA	0	0	0	NA	NA	NA	NA
NA	804702	Jalangi-Sylhet/Burdwan Composite	NA	NA	NA	NA	4613	NA	NA	NA

APPENDIX 2E. Total petroleum systems ranked by future oil volume.

[MMBO = million barrels of oil. Major commodity is based on the ratio of gas endowment to oil endowment as ft^3 of gas per bbl of oil (CFG/BO). Major commodity is oil if this ratio is less than 20,000 CFG/BO and gas if this ratio is equal to or greater than 20,000 CFG/BO. Cumulative oil is the reported volume of oil that has been produced. Remaining oil is the calculated remaining reserves; the difference between known and cumulative oil. Known oil is the reported discovered volume (cumulative production plus remaining reserves). Mean undiscovered oil is the mean volume estimated in World Petroleum Assessment 2000 (USGS, 2000). Future oil is the remaining oil plus estimated mean undiscovered oil. Oil endowment is the known oil plus estimated mean undiscovered oil volumes. Discovery maturity is the percentage of oil discovered (known oil) with respect to oil endowment. Calculated values may not be equal to the sums of the component values because numbers have been independently rounded. Negative values result from inconsistent reporting of cumulative and known volumes. Values in parentheses are questionable data. NA means not applicable; it is shown for all values where there is only one field present, and used where the endowment could not be calculated.] Known volumes from Petroconsultants (1996) and NRG Associates (1995). Note that gas = natural gas.

Rank	Total Petroleum System Code	Total Petroleum System Name	Major Commodity	Cumulative Oil (MMBO)	Remaining Oil (MMBO)	Known Oil (MMBO)	Mean Undiscovered Oil (MMBO)	Future Oil (MMBO)	Oil Endowment (MMBO)	Discovery Maturity, Oil (%)
1	203001	Zagros-Mesopotamian Cretaceous-Tertiary	Oil	100,467	271,760	372,226	94,516	366,275	466,742	80
2	202102	Arabian Sub-Basin Tuwaiq/Hanifa-Arab	Oil	63,853	135,142	198,995	41,703	176,845	240,698	83
3	117401	Bazhenov-Neocomian	Oil	45,819	72,153	117,972	42,947	115,100	160,919	73
4	201901	Cretaceous Thamama/Wasia	Oil	13,544	57,686	71,231	8979	66,665	80,209	89
5	719201	Tertiary Niger Delta (Agbada/Akata)	Oil	16,598	17,925	34,522	40,487	58,412	75,010	46
6	520001	Permian/Upper Jurassic Composite	Oil	0	0	0	47,148	47,148	47,148	0
7	530501	Pimienta-Tamabra	Oil	22,139	22,273	44,412	20,569	42,841	64,980	68
8	402501	Kimmeridgian Shales	Oil	22,467	21,427	43,895	13,098	34,526	56,993	77
9	101601	Paleozoic North Caspian	Oil	822	9987	10,808	23,320	33,307	34,128	32
10	204301	Sirte-Zelten	Oil	13,986	23,087	37,072	6854	29,941	43,927	84
11	720303	Congo Delta Composite	Oil	3693	6052	9745	23,198	29,250	32,943	30
12	201902	Jurassic Hanifa/Diyab-Arab	Oil	5688	13,325	19,013	14,478	27,803	33,491	57
13	202101	Central Arabia Qusaiba Paleozoic	Oil	84	6292	6376	19,190	25,481	25,566	25
14	603501	Lagoa Feia-Carapebus	Oil	1769	8287	10,056	16,293	24,580	26,349	38
15	101501	Volga-Ural Domanik-Paleozoic	Oil	42,584	21,288	63,872	2507	23,795	66,379	96
16	603601	Guaratiba-Guaruja (Cretaceous) Composite	Oil	6	279	285	23,209	23,488	23,494	1
17	609801	Querecual	Oil	11,027	15,729	26,756	7231	22,960	33,988	79
18	609901	La Luna/Maracaibo	Oil	35,581	13,491	49,072	8183	21,675	57,256	86
19	111201	Oligocene-Miocene Maykop/Diatom	Oil	11,681	5758	17,438	15,725	21,483	33,163	53
20	117402	Togur-Tyumen	Oil	645	11,111	11,756	7474	18,584	19,230	61
21	312701	Shahejie-Shahejie/Guantao/Wumishan	Oil	10,415	14,139	24,553	4412	18,550	28,965	85
22	602101	Cenomanian-Turonian	Oil	17	46	63	15,247	15,293	15,310	0
23	117403	Northern West Siberian Mesozoic Composite	Gas	66	10,119	10,185	4814	14,933	14,999	68
24	100801	Domanik-Paleozoic	Oil	3346	9723	13,069	5170	14,892	18,239	72
25	201903	Silurian Qusaiba	Gas	41	514	555	13,235	13,749	13,790	4
26	401701	Upper Jurassic Spekk	Oil	76	2585	2660	6538	9123	9199	29
27	609601	Gacheta-Mirador	Oil	1179	4224	5402	3631	7855	9033	60
28	205403	Tanezzuft-Ghadames	Oil	1629	2909	4538	4461	7370	8999	50

Rank	Total Petroleum System Code	Total Petroleum System Name	Major Commodity	Cumulative Oil (MMBO)	Remaining Oil (MMBO)	Known Oil (MMBO)	Mean Undiscovered Oil (MMBO)	Future Oil (MMBO)	Oil Endowment (MMBO)	Discovery Maturity, Oil (%)
29	804301	Eocene-Miocene Composite	Oil	3375	5065	8440	2091	7156	10,531	80
30	202302	Jurassic Gotnia/Barsarin/Sargelu/Najmah	Oil	94	379	474	6569	6948	7042	7
31	604101	Mesozoic-Cenozoic	Oil	2889	3712	6601	3023	6735	9625	69
32	315401	Ordovician/Jurassic-Phanerozoic	Oil	45	658	704	5948	6606	6652	11
33	370101	Brunei-Sabah	Oil	3953	2945	6898	3390	6335	10,288	67
34	311501	Lucaogou-Karamay/Ulho/Pindequan	Oil	937	5687	6624	467	6154	7091	93
35	207101	Sudr-Nubia	Oil	6351	3459	9810	2608	6067	12,418	79
36	720302	Azile-Senonian	Oil	1278	836	2114	5031	5867	7145	30
37	381701	Kutei Basin	Oil	2195	684	2879	5145	5829	8024	36
38	205401	Tanezzuft-Oued Mya	Oil	5872	4971	10,843	830	5801	11,673	93
39	380801	Brown Shale-Sihapas	Oil	8349	4868	13,217	899	5768	14,117	94
40	110901	Terek-Caspian	Oil	752	2738	3490	2829	5567	6319	55
41	609802	Upper Cretaceous/Tertiary	Oil	2780	667	3447	4644	5311	8091	43
42	205601	Tanezzuft-Illizi	Oil	2286	1384	3670	2814	4198	6484	57
43	314401	Qingshankou-Putaohua/Shaertu	Oil	12,399	3171	15,570	1026	4198	15,596	94
44	200401	Madbi Amran/Qishn	Oil	649	1284	1933	2740	4025	4673	41
45	132201	North Sakhalin Neogene	Oil	754	1428	2182	2582	4010	4764	46
46	121001	Baikal-Patom Foldbelt Riphean-Craton Margin Vendian	Gas	1	2005	2006	1611	3616	3617	55
47	606001	Neocomian Lacustrine	Oil	0	0	0	3603	3603	3603	0
48	202301	Paleozoic Qusaiba/Akkas/Abba/Mudawwara	Oil	269	944	1213	2632	3577	3845	32
49	110902	South Mangyshlak	Oil	2406	2835	5241	651	3486	5893	89
50	521501	Egret-Hibernia	Oil	0	1582	1582	1796	3378	3378	47
51	603401	Cretaceous Composite	Oil	93	37	130	2959	2995	3089	4
52	603701	Cenomanian-Turonian-Tertiary Composite	Oil	0	0	0	2938	2938	2938	0
53	394801	Dingo-Mungaroo/Barrow	Gas	489	660	1149	2249	2909	3398	34
54	201601	North Oman Huqf-Shu'aiba	Oil	1259	1426	2685	1304	2731	3990	67
55	207102	Maqna	Oil	22	133	155	2579	2712	2734	6
56	370301	Oligocene-Miocene Lacustrine	Oil	1459	1557	3017	1127	2684	4143	73
57	115001	Buzachi Arch and Surrounding Areas Composite	Oil	459	1827	2286	814	2641	3100	74
58	100901	Dnieper-Donets Paleozoic	Gas	110	1501	1611	1098	2599	2709	59
59	204801	Bou Dabbous-Tertiary	Oil	481	1633	2114	667	2300	2782	76
60	606002	Lower Cretaceous	Oil	0	0	0	2229	2229	2229	0
61	604501	Los Monos-Machareti	Oil	257	63	320	2145	2209	2466	13
62	605501	Neuquen Hybrid	Oil	1532	838	2370	1290	2128	3659	65
63	406102	Dysodile Schist-Tertiary	Oil	2832	1283	4115	842	2125	4957	83
64	201401	North Oman Huqf/'Q'-Haushi	Oil	715	1313	2028	743	2056	2771	73
65	608101	Cretaceous-Tertiary	Oil	1346	326	1672	1711	2038	3383	49
66	803401	Sylhet-Kopili/Barail-Tipam Composite	Oil	911	1625	2536	399	2024	2935	86
67	205402	Tanezzuft-Melrhir	Oil	0	6	6	1875	1881	1881	0
68	524302	Duvernay-Leduc	Oil	8153	1306	9459	501	1807	9960	95
69	720301	Melania-Gamba	Oil	708	1044	1752	727	1772	2480	71
70	602901	Neocomian to Turonian Composite	Oil	469	223	692	1487	1711	2179	32

App-2e_Page 2

Rank	Total Petroleum System Code	Total Petroleum System Name	Major Commodity	Cumulative Oil (MMBO)	Remaining Oil (MMBO)	Known Oil (MMBO)	Mean Undiscovered Oil (MMBO)	Future Oil (MMBO)	Oil Endowment (MMBO)	Discovery Maturity, Oil (%)
71	610701	Tobago Trough Paleogene	Oil	8	2	11	1536	1538	1546	1
72	391003	Jurassic/Early Cretaceous-Mesozoic	Oil	123	378	502	1146	1524	1648	30
73	605801	D-129	Oil	2445	900	3346	498	1398	3844	87
74	382402	Jatibarang/Talang Akar-Oligocene/Miocene	Oil	727	1181	1908	195	1377	2104	91
75	605901	Lower Inoceramus	Gas	595	620	1216	704	1325	1920	63
76	120701	Yenisey Foldbelt Riphean-Craton Margin Riphean	Oil	0	34	34	1213	1246	1246	3
77	609001	La Luna-La Paz	Oil	1915	512	2426	709	1220	3135	77
78	718301	Cretaceous Composite	Oil	74	162	236	1004	1166	1240	19
79	370201	Sarawak Basin	Gas	168	629	797	529	1159	1327	60
80	393001	Latrobe	Oil	3077	784	3860	310	1093	4170	93
81	391301	Late Jurassic/Early Cretaceous-Mesozoic	Gas	19	27	46	1055	1083	1101	4
82	382801	Lahat/Talang Akar-Cenozoic	Oil	1842	587	2429	469	1055	2898	84
83	606301	Lower Cretaceous Marine	Oil	0	10	10	1031	1041	1041	1
84	110903	Stavropol-Prikumsk	Oil	1200	(-380)	820	1407	(1027)	2227	37
85	105001	South and North Barents Triassic-Jurassic	Gas	0	51	51	955	1006	1007	5
86	115401	Amu-Darya Jurassic-Cretaceous	Gas	80	686	766	270	957	1037	74
87	201602	Middle Cretaceous Natih	Oil	1299	553	1852	371	925	2224	83
88	804801	Eocene to Miocene Composite	Gas	528	188	716	724	912	1441	50
89	382401	Banuwati-Oligocene/Miocene	Oil	828	431	1259	462	894	1722	73
90	720304	Cuanza Composite	Oil	86	20	107	790	810	897	12
91	312801	Yanchang-Yanan	Oil	81	602	683	138	740	821	83
92	611701	Upper Jurassic-Neocomian	Oil	65	232	297	494	726	790	38
93	403601	Carboniferous-Rotliegend	Gas	2300	572	2872	143	715	3015	95
94	406101	Moesian Platform Composite	Oil	1251	542	1793	171	713	1963	91
95	524404	Lodgepole	Oil	2239	539	2778	111	650	2889	96
96	370302	Miocene Coaly Strata	Gas	58	533	592	101	634	693	85
97	404702	Mesozoic/Paleogene Composite	Oil	640	283	923	344	627	1267	73
98	406002	Meride/Riva di Solto	Oil	78	254	332	360	614	692	48
99	524306	Second White Specks-Cardium	Oil	2146	543	2688	36	578	2724	99
100	524303	Exshaw-Rundle	Oil	1292	435	1728	126	561	1854	93
101	404801	Greater Hungarian Plain Neogene	Oil	731	358	1089	196	555	1285	85
102	204802	Jurassic-Cretaceous Composite	Oil	39	64	103	403	467	506	20
103	205803	Tanezzuft-Sbaa	Oil	0	284	284	162	446	446	64
104	382201	Bampo-Cenozoic	Gas	496	178	674	213	391	887	76
105	406801	Pre-Messinian	Gas	0	0	0	388	388	388	0
106	524304	Combined Triassic/Jurassic	Oil	1012	284	1296	96	380	1393	93
107	404802	Zala-Drava-Sava Mesozoic/Neogene	Oil	836	265	1101	114	379	1215	91
108	311502	Jurassic Coal-Jurassic/Tertiary	Oil	16	193	210	96	289	306	69
109	802601	Patala-Namal	Oil	204	121	324	145	266	469	69
110	524301	Keg River-Keg River	Oil	835	176	1011	84	260	1095	92
111	608302	Cretaceous-Paleogene	Oil	118	53	171	200	252	370	46
112	115002	North Ustyurt Jurassic	Oil	0	80	80	139	220	220	37
113	804201	Sembar-Goru/Ghazij	Gas	111	70	180	119	188	299	60

Rank	Total Petroleum System Code	Total Petroleum System Name	Major Commodity	Cumulative Oil (MMBO)	Remaining Oil (MMBO)	Known Oil (MMBO)	Mean Undiscovered Oil (MMBO)	Future Oil (MMBO)	Oil Endowment (MMBO)	Discovery Maturity, Oil (%)
114	701301	Cretaceous-Tertiary Composite	Oil	0	10	10	157	167	168	6
115	391001	Milligans-Carboniferous/Permian	Oil	0	15	15	140	155	155	10
116	205805	Tanezzuft-Benoud	Gas	16	72	88	72	144	160	55
117	394802	Locker-Mungaroo/Barrow	Gas	0	6	6	132	137	137	4
118	370202	East Natuna	Gas	0	20	20	102	122	122	16
119	730301	Cretaceous Composite	Gas	0	0	0	116	116	116	0
120	314202	Daanzhai-Daanzhai/Lianggaoshan	Oil	38	71	109	27	99	136	80
121	604102	Paleozoic	Oil	0	0	0	74	74	74	0
122	404806	Hungarian Paleogene	Oil	10	1	11	49	50	60	18
123	404703	Paleozoic Composite	Oil	0	1	1	47	47	47	1
124	804703	Jenam/Bhuban-Bokabil	Gas	3	2	4	39	41	43	10
125	608301	Neogene	Oil	4	0	4	37	37	41	11
126	205802	Tanezzuft-Ahnet	Gas	0	1	1	34	35	35	3
127	205801	Tanezzuft-Timimoun	Gas	0	0	0	31	31	31	0
128	610301	Lower Cruse	Gas	0	20	20	0	20	20	100
129	205806	Tanezzuft-Bechar/Abadla	Gas	0	0	0	16	16	16	0
130	524403	Bakken	Oil	5	2	7	11	13	18	40
131	205804	Tanezzuft-Mouydir	Gas	0	0	0	12	12	12	0
132	524402	Brightholme	Oil	11	4	14	8	11	22	66
133	524401	Yeoman	Oil	1	0	1	5	6	6	15
134	314402	Jurassic Coal-Denglouku/Nongan	Gas	0	5	5	0	5	5	100
135	101502	Belsk Basin	Gas	106	(-41)	65	42	(1)	106	61
136	404701	Isotopically Light Gas	Gas	1	1	1	0	1	1	100
137	203002	Paleozoic-Permian/Triassic	Gas	0	0	0	0	0	0	NA
138	314201	Maokou/Longtang-Jialingjiang/Maokou/Huanglong	Gas	0	0	0	0	0	0	100
139	391002	Keyling/Hyland Bay-Permian	Gas	0	0	0	0	0	0	NA
140	404803	Danube Neogene	Gas	0	0	0	0	0	0	NA
141	404804	Transcarpathian Neogene	Gas	0	0	0	0	0	0	NA
142	405701	Transylvanian Composite	Gas	0	0	0	0	0	0	NA
143	406001	Porto Garibaldi	Gas	13	0	14	0	0	14	100
144	524305	Mannville-Upper Mannville	Gas	0	0	0	0	0	0	NA
145	314204	Cambrian/Silurian Marine Shale-Dengying/Lower Paleozoic	Gas	9	(-8)	1	0	(-8)	1	100
146	110801	Azov-Kuban Mesozoic-Cenozoic	Gas	1273	(-750)	524	417	(-332)	941	56
NA	115003	North Ustyurt Paleozoic	NA	NA	NA	NA	0	NA	NA	NA
NA	311503	Lucaogou/Jurassic Coal-Paleozoic/Mesozoic	NA	0	0	0	NA	NA	NA	NA
NA	312702	Carboniferous/Permian Coal-Paleozoic	NA	0	0	0	NA	NA	NA	NA
NA	312802	Taiyuan/Shanxi-Majiagou/Shihezi	NA	0	0	0	NA	NA	NA	NA
NA	314203	Xujiahe-Xujiahe/Shaximiao	NA	NA	NA	NA	NA	NA	NA	NA
NA	382403	Tertiary-Parigi	NA	0	0	0	NA	NA	NA	NA
NA	382404	Tertiary-Cenozoic	NA	0	0	0	NA	NA	NA	NA
NA	404805	Central Carpathian Paleogene	NA	NA	NA	NA	NA	NA	NA	NA
NA	521502	Mesozoic Composite	NA	0	0	0	NA	NA	NA	NA

Rank	Total Petroleum System Code	Total Petroleum System Name	Major Commodity	Cumulative Oil (MMBO)	Remaining Oil (MMBO)	Known Oil (MMBO)	Mean Undiscovered Oil (MMBO)	Future Oil (MMBO)	Oil Endowment (MMBO)	Discovery Maturity, Oil (%)
NA	602201	Neogene	NA	NA	NA	NA	0	NA	NA	NA
NA	605802	Aguada Bandera	NA	0	0	0	NA	NA	NA	NA
NA	804701	Permian Coal	NA	0	0	0	NA	NA	NA	NA
NA	804702	Jalangi-Sylhet/Burdwan Composite	NA	NA	NA	NA	64	NA	NA	NA

APPENDIX 2F. Total petroleum systems ranked by future natural gas volume.

[BCFG = billion cubic feet of gas. Major commodity is based on the ratio of gas endowment to oil endowment as ft^3 of gas per bbl of oil (CFG/BO). Major commodity is oil if this ratio is less than 20,000 CFG/BO and gas if this ratio is equal to or greater than 20,000 CFG/BO. Cumulative gas is the reported volume of gas that has been produced. Remaining gas is the calculated remaining reserves; the difference between known and cumulative gas. Known gas is the reported discovered volume (cumulative production plus remaining reserves). Mean undiscovered gas is the mean volume estimated in World Petroleum Assessment 2000 (USGS, 2000). Future gas is the remaining gas plus estimated mean undiscovered gas. Gas endowment is the known gas plus estimated mean undiscovered gas volumes. Discovery maturity is the percentage of gas discovered (known gas) with respect to gas endowment. Calculated values may not be equal to the sums of the component values because numbers have been independently rounded. NA means not applicable; it is shown for all values where there is only one field present, and used where the endowment could not be calculated.] Known volumes from Petroconsultants (1996) and NRG Associates (1995). Note that gas = natural gas.

Rank	Total Petroleum System Code	Total Petroleum System Name	Major Commodity	Cumulative Gas (BCFG)	Remaining Gas (BCFG)	Known Gas (BCFG)	Mean Undiscovered Gas (BCFG)	Future Gas (BCFG)	Gas Endowment (BCFG)	Discovery Maturity, Gas (%)
1	117403	Northern West Siberian Mesozoic Composite	Gas	188,736	978,552	1,167,288	540,483	1,519,035	1,707,771	68
2	201903	Silurian Qusaiba	Gas	1018	451,012	452,030	346,892	797,903	798,922	57
3	203001	Zagros-Mesopotamian Cretaceous-Tertiary	Oil	25,260	467,978	493,238	197,597	665,575	690,835	71
4	202101	Central Arabia Qusaiba Paleozoic	Oil	0	79,115	79,115	383,339	462,454	462,454	17
5	105001	South and North Barents Triassic-Jurassic	Gas	0	78,143	78,143	266,618	344,761	344,761	23
6	115401	Amu-Darya Jurassic-Cretaceous	Gas	78,540	152,073	230,614	163,650	315,724	394,264	58
7	202102	Arabian Sub-Basin Tuwaiq/Hanifa-Arab	Oil	24,821	250,474	275,295	34,046	284,520	309,341	89
8	101601	Paleozoic North Caspian	Oil	1745	155,231	156,976	119,051	274,282	276,027	57
9	719201	Tertiary Niger Delta (Agbada/Akata)	Oil	6232	87,579	93,811	132,716	220,295	226,526	41
10	111201	Oligocene-Miocene Maykop/Diatom	Oil	8591	27,403	35,994	173,310	200,713	209,304	17
11	203002	Paleozoic-Permian/Triassic	Gas	0	131,220	131,220	69,031	200,251	200,251	66
12	401701	Upper Jurassic Spekk	Oil	0	15,662	15,662	165,201	180,863	180,863	9
13	402501	Kimmeridgian Shales	Oil	31,053	127,861	158,914	37,745	165,606	196,659	81
14	117401	Bazhenov-Neocomian	Oil	29,287	68,295	97,582	96,934	165,229	194,516	50
15	609801	Querecual	Oil	18,897	93,399	112,296	41,552	134,951	153,848	73
16	201902	Jurassic Hanifa/Diyab-Arab	Oil	1951	75,561	77,512	58,515	134,076	136,027	57
17	403601	Carboniferous-Rotliegend	Gas	124,048	98,111	222,159	25,790	123,901	247,948	90
18	201901	Cretaceous Thamama/Wasia	Oil	6342	95,758	102,100	22,311	118,069	124,410	82
19	394801	Dingo-Mungaroo/Barrow	Gas	1845	46,400	48,245	56,380	102,780	104,626	46
20	205805	Tanezzuft-Benoud	Gas	4980	100,070	105,050	2541	102,611	107,591	98
21	381701	Kutei Basin	Oil	8711	36,762	45,473	64,757	101,519	110,230	41
22	101501	Volga-Ural Domanik-Paleozoic	Oil	15,846	80,612	96,458	2394	83,006	98,852	98
23	530501	Pimienta-Tamabra	Oil	17,609	33,213	50,822	49,272	82,485	100,094	51
24	603601	Guaratiba-Guaruja (Cretaceous) Composite	Oil	12	1113	1125	80,547	81,660	81,672	1
25	520001	Permian/Upper Jurassic Composite	Oil	0	0	0	80,709	80,709	80,709	0
26	132201	North Sakhalin Neogene	Oil	162	22,221	22,383	57,451	79,673	79,835	28
27	720303	Congo Delta Composite	Oil	838	8605	9443	69,582	78,186	79,025	12

Rank	Total Petroleum System Code	Total Petroleum System Name	Major Commodity	Cumulative Gas (BCFG)	Remaining Gas (BCFG)	Known Gas (BCFG)	Mean Undiscovered Gas (BCFG)	Future Gas (BCFG)	Gas Endowment (BCFG)	Discovery Maturity, Gas (%)
28	121001	Baikal-Patom Foldbelt Riphean-Craton Margin Vendian	Gas	0	30,210	30,210	46,039	76,249	76,249	40
29	205601	Tanezzuft-Illizi	Oil	750	44,311	45,061	27,785	72,096	72,846	62
30	202301	Paleozoic Qusaiba/Akkas/Abba/Mudawwara	Oil	47	1614	1661	64,510	66,124	66,172	3
31	315401	Ordovician/Jurassic-Phanerozoic	Oil	208	5572	5780	59,939	65,511	65,719	9
32	609802	Upper Cretaceous/Tertiary	Oil	3156	12,667	15,822	52,009	64,676	67,832	23
33	804703	Jenam/Bhuban-Bokabil	Gas	2669	9620	12,289	50,479	60,098	62,767	20
34	100801	Domanik-Paleozoic	Oil	12,374	24,259	36,632	34,525	58,784	71,158	51
35	207102	Maqna	Oil	10	3517	3527	49,820	53,338	53,347	7
36	406801	Pre-Messinian	Gas	0	0	0	51,210	51,210	51,210	0
37	370201	Sarawak Basin	Gas	3731	33,388	37,119	17,284	50,672	54,403	68
38	370101	Brunei-Sabah	Oil	9188	27,012	36,200	23,226	50,238	59,426	61
39	804201	Sembar-Goru/Ghazij	Gas	15,101	20,272	35,373	29,325	49,596	64,697	55
40	100901	Dnieper-Donets Paleozoic	Gas	35,453	23,645	59,098	24,051	47,697	83,149	71
41	370301	Oligocene-Miocene Lacustrine	Oil	454	23,794	24,248	23,696	47,491	47,944	51
42	204301	Sirte-Zelten	Oil	5479	32,288	37,767	14,984	47,272	52,752	72
43	370202	East Natuna	Gas	0	45,045	45,045	1963	47,008	47,008	96
44	609901	La Luna/Maracaibo	Oil	979	25,722	26,701	17,576	43,298	44,277	60
45	610301	Lower Cruse	Gas	0	22,600	22,600	20,103	42,703	42,703	53
46	602101	Cenomanian-Turonian	Oil	0	0	0	42,058	42,058	42,058	0
47	604501	Los Monos-Machareti	Oil	6928	10,058	16,985	31,107	41,165	48,092	35
48	391301	Late Jurassic/Early Cretaceous-Mesozoic	Gas	0	17,960	17,960	20,093	38,053	38,053	47
49	110901	Terek-Caspian	Oil	192	7648	7840	27,779	35,427	35,619	22
50	603401	Cretaceous Composite	Oil	42	180	222	34,287	34,467	34,509	1
51	200401	Madbi Amran/Qishn	Oil	0	17,120	17,120	16,982	34,102	34,102	50
52	804301	Eocene-Miocene Composite	Oil	6228	17,965	24,193	12,944	30,909	37,137	65
53	201401	North Oman Huqf/'Q'-Haushi	Oil	149	20,189	20,339	10,112	30,301	30,450	67
54	804801	Eocene to Miocene Composite	Gas	722	9298	10,020	20,521	29,819	30,541	33
55	605901	Lower Inoceramus	Gas	10,053	14,754	24,807	14,040	28,794	38,848	64
56	205403	Tanezzuft-Ghadames	Oil	579	15,905	16,484	12,035	27,940	28,519	58
57	370302	Miocene Coaly Strata	Gas	38	23,863	23,901	3768	27,631	27,669	86
58	382201	Bampo-Cenozoic	Gas	13,977	11,582	25,559	15,123	26,705	40,682	63
59	603501	Lagoa Feia-Carapebus	Oil	303	5941	6244	19,691	25,632	25,935	24
60	605501	Neuquen Hybrid	Oil	7636	13,068	20,704	12,416	25,483	33,119	63
61	406001	Porto Garibaldi	Gas	9373	9102	18,475	16,267	25,369	34,741	53
62	382801	Lahat/Talang Akar-Cenozoic	Oil	3125	7080	10,204	18,250	25,330	28,455	36
63	110801	Azov-Kuban Mesozoic-Cenozoic	Gas	7586	11,118	18,704	12,252	23,369	30,955	60
64	202302	Jurassic Gotnia/Barsarin/Sargelu/Najmah	Oil	0	100	100	22,942	23,042	23,042	0
65	603701	Cenomanian-Turonian-Tertiary Composite	Oil	0	0	0	22,806	22,806	22,806	0
66	120701	Yenisey Foldbelt Riphean-Craton Margin Riphean	Oil	0	5052	5052	16,983	22,035	22,035	23
67	609601	Gacheta-Mirador	Oil	139	15,175	15,314	6217	21,391	21,530	71
68	312701	Shahejie-Shahejie/Guantao/Wumishan	Oil	2987	12,685	15,672	8514	21,199	24,186	65
69	391003	Jurassic/Early Cretaceous-Mesozoic	Oil	0	7526	7526	11,346	18,872	18,872	40

Rank	Total Petroleum System Code	Total Petroleum System Name	Major Commodity	Cumulative Gas (BCFG)	Remaining Gas (BCFG)	Known Gas (BCFG)	Mean Undiscovered Gas (BCFG)	Future Gas (BCFG)	Gas Endowment (BCFG)	Discovery Maturity, Gas (%)
70	204801	Bou Dabbous-Tertiary	Oil	37	15,472	15,509	2746	18,218	18,256	85
71	524305	Mannville-Upper Mannville	Gas	17,171	13,560	30,731	4597	18,157	35,328	87
72	391002	Keyling/Hyland Bay-Permian	Gas	0	5730	5730	10,995	16,725	16,725	34
73	394802	Locker-Mungaroo/Barrow	Gas	27	8075	8102	8331	16,406	16,432	49
74	207101	Sudr-Nubia	Oil	840	5155	5995	10,192	15,347	16,186	37
75	201601	North Oman Huqf-Shu'aiba	Oil	536	7097	7633	8206	15,303	15,839	48
76	720302	Azile-Senonian	Oil	203	1632	1835	13,230	14,862	15,065	12
77	110903	Stavropol-Prikumsk	Oil	9670	3829	13,499	9644	13,473	23,143	58
78	382402	Jatibarang/Talang Akar-Oligocene/Miocene	Oil	794	6529	7322	6579	13,108	13,901	53
79	718301	Cretaceous Composite	Oil	46	1903	1949	10,071	11,974	12,020	16
80	610701	Tobago Trough Paleogene	Oil	16	7	22	11,957	11,963	11,979	0
81	393001	Latrobe	Oil	3501	6274	9775	5658	11,932	15,433	63
82	405701	Transylvanian Composite	Gas	21,194	9537	30,731	2083	11,620	32,814	94
83	524306	Second White Specks-Cardium	Oil	16,010	10,439	26,449	1126	11,565	27,574	96
84	314201	Maokou/Longtang-Jialingjiang/Maokou/Huanglong	Gas	5709	5364	11,072	6137	11,501	17,210	64
85	606301	Lower Cretaceous Marine	Oil	0	75	75	11,327	11,402	11,402	1
86	521501	Egret-Hibernia	Oil	0	2406	2406	8893	11,298	11,298	21
87	205401	Tanezzuft-Oued Mya	Oil	68	8905	8973	2341	11,246	11,314	79
88	117402	Togur-Tyumen	Oil	443	4738	5180	5516	10,254	10,697	48
89	730301	Cretaceous Composite	Gas	0	6015	6015	3603	9618	9618	63
90	602901	Neocomian to Turonian Composite	Oil	348	1059	1407	8334	9393	9741	14
91	524303	Exshaw-Rundle	Oil	14,851	6717	21,568	2663	9380	24,232	89
92	606001	Neocomian Lacustrine	Oil	0	0	0	8871	8871	8871	0
93	524302	Duvernay-Leduc	Oil	9050	5307	14,357	3224	8531	17,581	82
94	524304	Combined Triassic/Jurassic	Oil	4584	5996	10,580	2166	8163	12,747	83
95	110902	South Mangyshlak	Oil	28	5679	5707	2262	7941	7969	72
96	606002	Lower Cretaceous	Oil	0	0	0	7710	7710	7710	0
97	404701	Isotopically Light Gas	Gas	5455	5205	10,660	2321	7525	12,980	82
98	380801	Brown Shale-Sihapas	Oil	910	2956	3866	4078	7034	7944	49
99	404801	Greater Hungarian Plain Neogene	Oil	5436	4221	9657	2395	6616	12,052	80
100	205802	Tanezzuft-Ahnet	Gas	0	3117	3117	2973	6090	6090	51
101	406102	Dysodile Schist-Tertiary	Oil	1580	3468	5048	2618	6086	7666	66
102	609001	La Luna-La Paz	Oil	2310	2692	5002	3292	5985	8294	60
103	608101	Cretaceous-Tertiary	Oil	1951	941	2892	4795	5736	7687	38
104	803401	Sylhet-Kopili/Barail-Tipam Composite	Oil	2039	4495	6534	1178	5673	7712	85
105	605801	D-129	Oil	1667	1898	3565	3774	5672	7339	49
106	205801	Tanezzuft-Timimoun	Gas	0	4200	4200	1128	5328	5328	79
107	205402	Tanezzuft-Melrhir	Oil	0	125	125	4887	5012	5012	2
108	314204	Cambrian/Silurian Marine Shale-Dengying/Lower Paleozoic	Gas	491	859	1350	4036	4895	5386	25
109	720301	Melania-Gamba	Oil	62	780	843	3674	4455	4517	19
110	404702	Mesozoic/Paleogene Composite	Oil	3360	2033	5393	2345	4378	7738	70
111	524301	Keg River-Keg River	Oil	4589	2484	7073	1776	4260	8849	80

Rank	Total Petroleum System Code	Total Petroleum System Name	Major Commodity	Cumulative Gas (BCFG)	Remaining Gas (BCFG)	Known Gas (BCFG)	Mean Undiscovered Gas (BCFG)	Future Gas (BCFG)	Gas Endowment (BCFG)	Discovery Maturity, Gas (%)
112	314401	Qingshankou-Putaohua/Shaertu	Oil	3	1594	1598	2494	4088	4091	39
113	404802	Zala-Drava-Sava Mesozoic/Neogene	Oil	1082	2847	3929	1151	3998	5080	77
114	115002	North Ustyurt Jurassic	Oil	0	1450	1450	2180	3630	3630	40
115	101502	Belsk Basin	Gas	571	2067	2638	1547	3615	4185	63
116	204802	Jurassic-Cretaceous Composite	Oil	6	1295	1301	2280	3575	3581	36
117	314402	Jurassic Coal-Denglouku/Nongan	Gas	0	145	145	3212	3357	3357	4
118	201602	Middle Cretaceous Natih	Oil	616	2485	3100	363	2848	3463	90
119	205803	Tanezzuft-Sbaa	Oil	0	1490	1490	645	2134	2134	70
120	311501	Lucaogou-Karamay/Ulho/Pindequan	Oil	491	1757	2248	346	2104	2594	87
121	406002	Meride/Riva di Solto	Oil	154	257	412	1698	1955	2109	20
122	802601	Patala-Namal	Oil	807	1174	1981	551	1726	2533	78
123	604101	Mesozoic-Cenozoic	Oil	661	955	1616	755	1711	2372	68
124	720304	Cuanza Composite	Oil	46	13	59	1558	1571	1617	4
125	311502	Jurassic Coal-Jurassic/Tertiary	Oil	10	332	342	950	1281	1292	26
126	382401	Banuwati-Oligocene/Miocene	Oil	233	381	614	874	1255	1488	41
127	406101	Moesian Platform Composite	Oil	1546	671	2217	561	1232	2778	80
128	391001	Milligans-Carboniferous/Permian	Oil	0	48	48	1151	1199	1199	4
129	115001	Buzachi Arch and Surrounding Areas Composite	Oil	0	943	943	89	1033	1033	91
130	701301	Cretaceous-Tertiary Composite	Oil	4	45	49	856	901	905	5
131	604102	Paleozoic	Oil	0	0	0	841	841	841	0
132	608301	Neogene	Oil	0	340	340	471	811	811	42
133	611701	Upper Jurassic-Neocomian	Oil	20	77	96	592	668	688	14
134	404703	Paleozoic Composite	Oil	0	233	233	383	616	616	38
135	205806	Tanezzuft-Bechar/Abadla	Gas	0	0	0	441	441	441	0
136	404803	Danube Neogene	Gas	7	240	247	146	386	393	63
137	404804	Transcarpathian Neogene	Gas	0	261	261	112	373	373	70
138	404806	Hungarian Paleogene	Oil	15	1	16	339	340	355	4
139	205804	Tanezzuft-Mouydir	Gas	0	0	0	292	292	292	0
140	608302	Cretaceous-Paleogene	Oil	148	10	158	224	233	382	41
141	524404	Lodgepole	Oil	728	166	893	63	228	956	93
142	314202	Daanzhai-Daanzhai/Lianggaoshan	Oil	50	90	140	137	227	278	51
143	312801	Yanchang-Yanan	Oil	13	7	20	20	27	40	50
144	524402	Brightholme	Oil	8	3	11	5	8	16	67
145	524401	Yeoman	Oil	0	0	0	4	4	4	3
146	524403	Bakken	Oil	0	0	0	3	3	4	9
NA	115003	North Ustyurt Paleozoic	NA	NA	NA	NA	8500	NA	NA	NA
NA	311503	Lucaogou/Jurassic Coal-Paleozoic/Mesozoic	NA	0	0	0	NA	NA	NA	NA
NA	312702	Carboniferous/Permian Coal-Paleozoic	NA	0	0	0	NA	NA	NA	NA
NA	312802	Taiyuan/Shanxi-Majiagou/Shihezi	NA	0	5590	5590	NA	NA	NA	NA
NA	314203	Xujiahe-Xujiahe/Shaximiao	NA	NA	NA	NA	NA	NA	NA	NA
NA	382403	Tertiary-Parigi	NA	0	119	119	NA	NA	NA	NA
NA	382404	Tertiary-Cenozoic	NA	0	0	0	NA	NA	NA	NA
NA	404805	Central Carpathian Paleogene	NA	NA	NA	NA	NA	NA	NA	NA

Rank	Total Petroleum System Code	Total Petroleum System Name	Major Commodity	Cumulative Gas (BCFG)	Remaining Gas (BCFG)	Known Gas (BCFG)	Mean Undiscovered Gas (BCFG)	Future Gas (BCFG)	Gas Endowment (BCFG)	Discovery Maturity, Gas (%)
NA	521502	Mesozoic Composite	NA	0	4224	4224	NA	NA	NA	NA
NA	602201	Neogene	NA	NA	NA	NA	29,840	NA	NA	NA
NA	605802	Aguada Bandera	NA	0	0	0	NA	NA	NA	NA
NA	804701	Permian Coal	NA	0	0	0	NA	NA	NA	NA
NA	804702	Jalangi-Sylhet/Burdwan Composite	NA	NA	NA	NA	4613	NA	NA	NA

APPENDIX 3A. Provinces ranked by mean (expected) undiscovered oil.

[MMBO = million barrels of oil. Values given are not necessarily those for the entire province, but only for that part assessed in World Petroleum Assessment 2000 (USGS, 2000). Major commodity is based on the ratio of gas endowment to oil endowment as ft^3 of gas per bbl of oil (CFG/BO). Major commodity is oil if this ratio is less than 20,000 CFG/BO and gas if this ratio is equal to or greater than 20,000 CFG/BO. Cumulative oil is the reported volume of oil that has been produced. Remaining oil is the calculated remaining reserves; the difference between known and cumulative oil. Known oil is the reported discovered volume (cumulative production plus remaining reserves). Mean undiscovered oil is the mean volume estimated in World Petroleum Assessment 2000 (USGS, 2000). Future oil is the remaining oil plus estimated mean undiscovered oil. Oil endowment is the known oil plus estimated mean undiscovered oil volumes. Discovery maturity is the percentage of oil discovered (known oil) with respect to oil endowment. Calculated values may not be equal to the sums of the component values because numbers have been independently rounded. Negative values result from inconsistent reporting of cumulative and known volumes. Values in parentheses are questionable data. NA means not applicable; it is shown for all values where there is only one field present, and used where the endowment could not be calculated.] Known volumes from Petroconsultants (1996) and NRG Associates (1995). Note that gas = natural gas.

Rank	Province Code	Province Name	Major Commodity	Cumulative Oil (MMBO)	Remaining Oil (MMBO)	Known Oil (MMBO)	Mean Undiscovered Oil (MMBO)	Future Oil (MMBO)	Oil Endowment (MMBO)	Discovery Maturity, Oil (%)
1	2024	Mesopotamian Foredeep Basin	Oil	56,746	235,696	292,442	61,134	296,830	353,576	83
2	1174	West Siberian Basin	Oil	46,530	93,383	139,913	55,235	148,618	195,148	72
3	5200	East Greenland Rift Basins	Oil	0	0	0	47,148	47,148	47,148	0
4	2030	Zagros Fold Belt	Oil	49,367	72,234	121,601	45,459	117,693	167,060	73
5	7192	Niger Delta	Oil	16,598	17,925	34,522	40,487	58,412	75,010	46
6	2019	Rub Al Khali Basin	Oil	19,320	70,507	89,827	36,647	107,154	126,474	71
7	7203	West-Central Coastal	Oil	5765	7952	13,717	29,747	37,699	43,464	32
8	1016	North Caspian Basin	Oil	822	9987	10,808	23,320	33,307	34,128	32
9	6036	Santos Basin	Oil	6	279	285	23,209	23,488	23,494	1
10	2023	Widyan Basin-Interior Platform	Oil	219	17,216	17,435	21,215	38,431	38,650	45
11	2020	Interior Homocline-Central Arch	Oil	84	4616	4700	16,495	21,111	21,195	22
12	6035	Campos Basin	Oil	1769	8287	10,056	16,293	24,580	26,349	38
13	1112	South Caspian Basin	Oil	11,681	5758	17,438	15,725	21,483	33,163	53
14	6021	Guyana-Suriname Basin	Oil	17	46	63	15,247	15,293	15,310	0
15	2021	Greater Ghawar Uplift	Oil	57,744	83,956	141,700	13,830	97,786	155,530	91
16	4025	North Sea Graben	Oil	22,467	21,427	43,895	13,098	34,526	56,993	77
17	6098	East Venezuela Basin	Oil	13,808	16,395	30,203	11,875	28,271	42,078	72
18	6099	Maracaibo Basin	Oil	35,581	13,491	49,072	8183	21,675	57,256	86
19	5305	Villahermosa Uplift	Oil	14,640	20,382	35,022	7436	27,818	42,457	82
20	5304	Saline-Comalcalco Basin	Oil	1673	686	2359	7018	7703	9377	25
21	2043	Sirte Basin	Oil	13,986	23,087	37,072	6854	29,941	43,927	84
22	4017	Vestford-Helgeland	Oil	76	2585	2660	6538	9123	9199	29
23	2054	Trias/Ghadames Basin	Oil	7410	7793	15,203	6477	14,270	21,680	70
24	3154	Tarim Basin	Oil	45	658	704	5948	6606	6652	11
25	6060	Falklands Plateau	Oil	0	0	0	5833	5833	5833	0
26	1008	Timan-Pechora Basin	Oil	3346	9774	13,120	5732	15,506	18,852	70
27	2071	Red Sea Basin	Oil	6280	3580	9861	5187	8767	15,048	66
28	3817	Kutei Basin	Oil	2195	684	2879	5145	5829	8024	36

Rank	Province Code	Province Name	Major Commodity	Cumulative Oil (MMBO)	Remaining Oil (MMBO)	Known Oil (MMBO)	Mean Undiscovered Oil (MMBO)	Future Oil (MMBO)	Oil Endowment (MMBO)	Discovery Maturity, Oil (%)
29	2022	Qatar Arch	Gas	0	1150	1150	5026	6176	6176	19
30	1109	Middle Caspian Basin	Oil	4359	5193	9551	4887	10,080	14,439	66
31	3127	Bohaiwan Basin	Oil	10,415	14,139	24,553	4412	18,550	28,965	85
32	6096	Llanos Basin	Oil	1179	4224	5402	3631	7855	9033	60
33	3701	Baram Delta/Brunei-Sabah Basin	Oil	3953	2945	6898	3390	6335	10,288	67
34	6041	Putumayo-Oriente-Maranon Basin	Oil	2889	3712	6601	3098	6810	9699	68
35	6034	Espirito Santo Basin	Oil	93	37	130	2959	2995	3089	4
36	2056	Illizi Basin	Oil	2286	1384	3670	2947	4331	6617	55
37	6037	Pelotas Basin	Oil	0	0	0	2938	2938	2938	0
38	1322	North Sakhalin Basin	Oil	754	1428	2182	2582	4010	4764	46
39	1015	Volga-Ural Region	Oil	42,690	21,247	63,937	2549	23,796	66,485	96
40	5307	Campeche-Sigsbe Salt Basin	Oil	0	0	0	2478	2478	2478	0
41	3948	Northwest Shelf	Gas	473	647	1119	2381	3028	3500	32
42	6045	Santa Cruz-Tarija Basin	Oil	257	63	320	2145	2209	2466	13
43	8043	Bombay	Oil	3375	5065	8440	2091	7156	10,531	80
44	5215	Labrador-Newfoundland Shelf	Oil	0	1582	1582	1796	3378	3378	47
45	6081	Talara Basin	Oil	1346	326	1672	1711	2038	3383	49
46	6107	Lesser Antilles Deformed Belt	Oil	8	2	11	1536	1538	1546	1
47	6029	Sergipe-Alagoas Basin	Oil	469	223	692	1487	1711	2179	32
48	6055	Neuquen Basin	Oil	1532	838	2370	1290	2128	3659	65
49	3910	Bonaparte Gulf Basin	Gas	123	393	517	1286	1679	1803	29
50	2048	Pelagian Basin	Oil	521	1719	2239	1268	2986	3507	64
51	3703	Malay Basin	Oil	1518	2090	3608	1228	3318	4836	75
52	1207	Baykit Arch	Oil	0	34	34	1213	1246	1246	3
53	1210	Nepa-Botuoba Arch	Oil	1	2005	2006	1208	3213	3214	62
54	1009	Dnieper-Donets Basin	Gas	110	1501	1611	1098	2599	2709	59
55	2009	Masila-Jeza Basin	Oil	144	488	632	1096	1584	1728	37
56	3913	Browse Basin	Gas	19	27	46	1055	1083	1101	4
57	6063	Malvinas Basin	Oil	0	10	10	1031	1041	1041	1
58	3144	Songliao Basin	Oil	12,399	3176	15,575	1026	4203	16,601	94
59	2016	Fahud Salt Basin	Oil	2555	1918	4473	1020	2938	5493	81
60	4061	Carpathian-Balkanian Basin	Oil	4083	1825	5908	1013	2838	6921	85
61	7183	Gulf of Guinea	Oil	74	162	236	1004	1166	1240	19
62	2004	Ma'Rib-Al Jawf/Masila Basin	Oil	504	683	1188	959	1643	2147	55
63	5302	Veracruz Basin	Oil	51	79	130	955	1034	1085	12
64	1150	North Ustyurt Basin	Oil	459	1907	2366	954	2861	3320	71
65	2010	Ghudun-Khasfeh Flank Province	Oil	0	0	0	926	926	926	0
66	3808	Central Sumatra Basin	Oil	8349	4868	13,217	899	5768	14,117	94
67	5301	Tampico-Misantla Basin	Oil	5773	1122	6895	893	2015	7787	89
68	5310	Sierra Madre de Chiapas-Peten Foldbelt	NA	NA	NA	NA	890	NA	NA	NA
69	5243	Alberta Basin	Oil	12,821	2458	15,279	818	3276	16,097	95
70	5308	Yucatan Platform	Oil	0	0	0	754	754	754	0
71	2075	Euphrates/Mardin	Oil	282	312	594	749	1061	1343	44

Rank	Province Code	Province Name	Major Commodity	Cumulative Oil (MMBO)	Remaining Oil (MMBO)	Known Oil (MMBO)	Mean Undiscovered Oil (MMBO)	Future Oil (MMBO)	Oil Endowment (MMBO)	Discovery Maturity, Oil (%)
72	8048	Irrawaddy	Gas	527	188	715	724	912	1439	50
73	6090	Middle Magdalena	Oil	1915	512	2426	709	1220	3135	77
74	6059	Magallanes Basin	Gas	595	620	1216	704	1325	1920	63
75	2006	Shabwah Basin	Oil	0	113	113	685	798	798	14
76	3824	Northwest Java Basin	Oil	1555	1613	3168	658	2271	3825	83
77	3702	Greater Sarawak Basin	Gas	168	649	817	631	1281	1448	56
78	2058	Grand Erg/Ahnet Basin	Gas	106	388	494	599	986	1093	45
79	3115	Junggar Basin	Oil	952	5858	6810	563	6422	7373	92
80	6058	San Jorge Basin	Oil	2445	900	3346	498	1398	3844	87
81	6117	Greater Antilles Deformed Belt	Oil	65	232	297	494	726	790	38
82	3828	South Sumatra Basin	Oil	1842	587	2429	469	1055	2898	84
83	2015	Central Oman Platform	Oil	3	68	71	424	493	495	14
84	1108	Azov-Kuban Basin	Gas	1273	(-750)	524	417	(-332)	941	56
85	8034	Assam	Oil	911	1625	2536	399	2024	2935	86
86	4047	North Carpathian Basin	Oil	649	286	935	391	677	1326	71
87	4068	Provence Basin	Gas	0	0	0	388	388	388	0
88	2014	Ghaba Salt Basin	Oil	466	878	1344	371	1250	1715	78
89	4060	Po Basin	Gas	91	254	346	360	615	706	49
90	4048	Pannonian Basin	Oil	1567	623	2190	359	982	2550	86
91	3930	Gippsland Basin	Oil	3077	784	3860	310	1093	4170	93
92	1154	Amu-Darya Basin	Gas	80	686	766	270	957	1037	74
93	1209	Angara-Lena Terrace	Gas	0	0	0	242	242	242	0
94	6083	Progreso Basin	Oil	122	53	175	237	290	412	43
95	1050	South Barents Basin	Gas	0	0	0	219	219	219	0
96	3822	North Sumatra Basin	Gas	496	178	674	213	391	887	76
97	1211	Cis-Patom Foredeep	Gas	0	0	0	161	161	161	0
98	7013	Senegal	Oil	0	10	10	157	167	168	6
99	5244	Williston Basin, Canada	Oil	2696	809	3506	154	963	3660	96
100	8026	Kohat-Potwar	Oil	201	120	321	149	268	469	68
101	3128	Ordos Basin	Oil	81	602	683	138	740	821	83
102	1060	North Barents Basin	Gas	0	0	0	135	135	135	0
103	7303	Orange River Coastal	Gas	0	0	0	116	116	116	0
104	2074	Khleisha Uplift	Oil	250	407	657	105	512	762	86
105	8042	Indus	Gas	111	70	180	103	173	283	64
106	8047	Ganges-Brahmaputra Delta	Gas	3	3	5	103	105	108	5
107	5306	Macuspana Basin	Oil	2	4	6	101	105	108	6
108	2011	South Oman Salt Basin	Oil	203	297	500	97	394	596	84
109	2047	Hamra Basin	Oil	0	0	0	89	89	89	0
110	4036	Anglo-Dutch Basin	Gas	432	159	592	88	247	679	87
111	4035	Northwest German Basin	Gas	1867	413	2280	56	468	2336	98
112	2025	Hail-Ga'Ara Arch	Gas	0	0	0	53	53	53	0
113	2028	Rutbah Uplift	Gas	0	0	0	53	53	53	0
114	1059	Ludlov Saddle	NA	NA	NA	NA	39	NA	NA	NA

Rank	Province Code	Province Name	Major Commodity	Cumulative Oil (MMBO)	Remaining Oil (MMBO)	Known Oil (MMBO)	Mean Undiscovered Oil (MMBO)	Future Oil (MMBO)	Oil Endowment (MMBO)	Discovery Maturity, Oil (%)
115	2031	Zagros Thrust Zone	Gas	0	0	0	38	38	38	0
116	2013	Huqf-Haushi Uplift	Oil	0	0	0	30	30	30	0
117	5316	Trans-Mexican Neovolcanic Axis	NA	NA	NA	NA	30	NA	NA	NA
118	3142	Sichuan Basin	Gas	47	63	110	27	91	138	80
119	2089	Anah Graben	Oil	75	80	155	26	106	181	85
120	2017	Oman Mountains	Oil	0	0	0	22	22	22	0
121	5317	Chicontepec Basin	Oil	0	0	0	15	15	15	1
122	8025	Sulaiman-Kirthar	Gas	0	0	0	12	12	12	0
123	5245	Rocky Mountain Deformed Belt	Gas	176	22	198	6	28	204	97
124	1158	Tian Shan Foldbelt	NA	NA	NA	NA	0	NA	NA	NA
125	2029	Wadi-Surhan Basin	Gas	0	0	0	0	0	0	NA
126	2033	Sinai Basin	Oil	94	11	105	0	11	105	100
127	2061	Ougarta Uplift	Oil	0	41	41	0	41	41	100
128	3161	Yunnan Guizhou Hubei Fold Belt	NA	NA	NA	NA	0	NA	NA	NA
129	3916	Carnarvon Basin	Oil	16	19	35	0	19	35	100
130	4028	Mid-North Sea High	NA	NA	NA	NA	0	NA	NA	NA
131	4033	German-Polish Basin	Gas	0	0	0	0	0	0	100
132	4038	Munsterland Basin	Gas	0	0	0	0	0	0	NA
133	4039	Southwest German Basin	NA	NA	NA	NA	0	NA	NA	NA
134	4046	Bohemia	Gas	1	0	1	0	0	1	100
135	4057	Transylvanian Basin	Gas	0	0	0	0	0	0	NA
136	5246	Mackenzie Foldbelt	Gas	0	0	0	0	0	0	NA
137	6022	Foz do Amazonas Basin	NA	NA	NA	NA	0	NA	NA	NA
138	6103	Tobago Trough	Gas	0	20	20	0	20	20	100
139	8002	Himalayan	NA	NA	NA	NA	0	NA	NA	NA
140	8023	Central Afghanistan	Gas	0	0	0	0	0	0	NA
141	8035	North Burma	Oil	1	0	1	0	0	1	100

APPENDIX 3B. Provinces ranked by mean (expected) undiscovered natural gas.

[BCFG = billion cubic feet of gas. Values given are not necessarily those for the entire province, but only for that part assessed in World Petroleum Assessment 2000 (USGS, 2000). Major commodity is based on the ratio of gas endowment to oil endowment as ft^3 of gas per bbl of oil (CFG/BO). Major commodity is oil if this ratio is less than 20,000 CFG/BO and gas if this ratio is equal to or greater than 20,000 CFG/BO. Cumulative gas is the reported volume of gas that has been produced. Remaining gas is the calculated remaining reserves; the difference between known and cumulative gas. Known gas is the reported discovered volume (cumulative production plus remaining reserves). Mean undiscovered gas is the mean volume estimated in World Petroleum Assessment 2000 (USGS, 2000). Future gas is the remaining gas plus estimated mean undiscovered gas. Gas endowment is the known gas plus estimated mean undiscovered gas volumes. Discovery maturity is the percentage of gas discovered (known gas) with respect to gas endowment. Calculated values may not be equal to the sums of the component values because numbers have been independently rounded. Negative values result from inconsistent reporting of cumulative and known volumes. Values in parentheses are questionable data. NA means not applicable; it is shown for all values where there is only one field present, and used where the endowment could not be calculated.] Known volumes from Petroconsultants (1996) and NRG Associates (1995). Note that gas = natural gas.

Rank	Province Code	Province Name	Major Commodity	Cumulative Gas (BCFG)	Remaining Gas (BCFG)	Known Gas (BCFG)	Mean Undiscovered Gas (BCFG)	Future Gas (BCFG)	Gas Endowment (BCFG)	Discovery Maturity, Gas (%)
1	1174	West Siberian Basin	Oil	218,465	1,051,584	1,270,049	642,934	1,694,518	1,912,983	66
2	2019	Rub Al Khali Basin	Oil	8303	173,934	182,237	425,712	599,646	607,949	30
3	2021	Greater Ghawar Uplift	Oil	26,493	222,057	248,550	226,969	449,026	475,519	52
4	2030	Zagros Fold Belt	Oil	10,333	389,111	399,444	212,008	601,119	611,451	65
5	1112	South Caspian Basin	Oil	8591	27,403	35,994	173,310	200,713	209,304	17
6	4017	Vestford-Helgeland	Oil	0	15,662	15,662	165,201	180,863	180,863	9
7	1154	Amu-Darya Basin	Gas	78,513	151,830	230,343	163,650	315,480	393,993	58
8	1050	South Barents Basin	Gas	0	70,020	70,020	160,857	230,877	230,877	30
9	7192	Niger Delta	Oil	6232	87,579	93,811	132,716	220,295	226,526	41
10	2020	Interior Homocline-Central Arch	Oil	0	9825	9825	122,074	131,899	131,899	7
11	1016	North Caspian Basin	Oil	1745	155,161	156,906	119,051	274,212	275,957	57
12	2023	Widyan Basin-Interior Platform	Oil	60	7295	7355	94,746	102,041	102,101	7
13	6098	East Venezuela Basin	Oil	22,053	106,066	128,119	93,561	199,627	221,679	58
14	7203	West-Central Coastal	Oil	1150	11,030	12,180	88,044	99,074	100,223	12
15	2024	Mesopotamian Foredeep Basin	Oil	13,129	285,156	298,285	83,722	368,878	382,007	78
16	5200	East Greenland Rift Basins	Oil	0	0	0	80,709	80,709	80,709	0
17	6036	Santos Basin	Oil	12	1113	1125	80,547	81,660	81,672	1
18	3817	Kutei Basin	Oil	8711	36,762	45,473	64,757	101,519	110,230	41
19	3948	Northwest Shelf	Gas	1840	54,405	56,245	64,711	119,116	120,956	47
20	2071	Red Sea Basin	Oil	849	8668	9517	60,012	68,680	69,529	14
21	3154	Tarim Basin	Oil	208	5572	5780	59,939	65,511	65,719	9
22	1060	North Barents Basin	Gas	0	0	0	59,820	59,820	59,820	0
23	1322	North Sakhalin Basin	Oil	162	22,221	22,383	57,451	79,673	79,835	28
24	8047	Ganges-Brahmaputra Delta	Gas	2669	9620	12,289	55,091	64,711	67,380	18
25	1008	Timan-Pechora Basin	Oil	12,374	24,261	36,635	52,079	76,340	88,714	41
26	4068	Provence Basin	Gas	0	0	0	51,210	51,210	51,210	0
27	6021	Guyana-Suriname Basin	Oil	0	0	0	42,058	42,058	42,058	0
28	1109	Middle Caspian Basin	Oil	9984	17,156	27,140	39,685	56,841	66,825	41

Rank	Province Code	Province Name	Major Commodity	Cumulative Gas (BCFG)	Remaining Gas (BCFG)	Known Gas (BCFG)	Mean Undiscovered Gas (BCFG)	Future Gas (BCFG)	Gas Endowment (BCFG)	Discovery Maturity, Gas (%)
29	4025	North Sea Graben	Oil	31,235	128,603	159,839	37,745	166,348	197,584	81
30	6034	Espirito Santo Basin	Oil	42	180	222	34,287	34,467	34,509	1
31	6045	Santa Cruz-Tarija Basin	Oil	6928	10,058	16,985	31,107	41,165	48,092	35
32	1210	Nepa-Botuoba Arch	Oil	0	22,035	22,035	29,971	52,006	52,006	42
33	6022	Foz do Amazonas Basin	NA	NA	NA	NA	29,840	NA	NA	NA
34	1059	Ludlov Saddle	NA	NA	NA	NA	28,387	NA	NA	NA
35	2056	Illizi Basin	Oil	750	44,311	45,061	28,138	72,449	73,199	62
36	3703	Malay Basin	Oil	492	47,657	48,149	27,465	75,122	75,614	64
37	1009	Dnieper-Donets Basin	Gas	35,453	23,645	59,098	24,051	47,697	83,149	71
38	3910	Bonaparte Gulf Basin	Gas	0	13,304	13,304	23,492	36,796	36,796	36
39	3701	Baram Delta/Brunei-Sabah Basin	Oil	9188	27,012	36,200	23,226	50,238	59,426	61
40	6037	Pelotas Basin	Oil	0	0	0	22,806	22,806	22,806	0
41	8048	Irrawaddy	Gas	721	9288	10,009	20,521	29,809	30,530	33
42	6103	Tobago Trough	Gas	0	22,600	22,600	20,103	42,703	42,703	53
43	3913	Browse Basin	Gas	0	17,960	17,960	20,093	38,053	38,053	47
44	6035	Campos Basin	Oil	303	5941	6244	19,691	25,632	25,935	24
45	3702	Greater Sarawak Basin	Gas	3731	78,433	82,164	19,247	97,680	101,411	81
46	3828	South Sumatra Basin	Oil	3125	7080	10,204	18,250	25,330	28,455	36
47	4060	Po Basin	Gas	9527	9360	18,886	17,964	27,324	36,850	51
48	6099	Maracaibo Basin	Oil	979	25,722	26,701	17,576	43,298	44,277	60
49	2054	Trias/Ghadames Basin	Oil	647	24,352	24,999	17,493	41,845	42,492	59
50	2022	Qatar Arch	Gas	1018	464,582	465,600	17,385	481,967	482,985	96
51	1207	Baykit Arch	Oil	0	5052	5052	16,983	22,035	22,035	23
52	6060	Falklands Plateau	Oil	0	0	0	16,581	16,581	16,581	0
53	5304	Saline-Comalcalco Basin	Oil	1023	1223	2246	16,423	17,645	18,669	12
54	8042	Indus	Gas	8385	11,211	19,596	15,964	27,175	35,560	55
55	5305	Villahermosa Uplift	Oil	13,888	27,404	41,292	15,623	43,027	56,915	73
56	3822	North Sumatra Basin	Gas	13,977	11,582	25,559	15,123	26,705	40,682	63
57	2043	Sirte Basin	Oil	5479	32,288	37,767	14,984	47,272	52,752	72
58	4036	Anglo-Dutch Basin	Gas	40,547	30,880	71,427	14,976	45,856	86,402	83
59	6059	Magallanes Basin	Gas	10,053	14,754	24,807	14,040	28,794	38,848	64
60	8043	Bombay	Oil	6228	17,965	24,193	12,944	30,909	37,137	65
61	8025	Sulaiman-Kirthar	Gas	6716	9061	15,777	12,599	21,659	28,376	56
62	6055	Neuquen Basin	Oil	7636	13,068	20,704	12,416	25,483	33,119	63
63	1108	Azov-Kuban Basin	Gas	7492	11,118	18,610	12,252	23,369	30,861	60
64	6107	Lesser Antilles Deformed Belt	Oil	16	7	22	11,957	11,963	11,979	0
65	5243	Alberta Basin	Oil	47,573	31,097	78,670	11,895	42,992	90,565	87
66	6063	Malvinas Basin	Oil	0	75	75	11,327	11,402	11,402	1
67	4035	Northwest German Basin	Gas	76,285	65,440	141,726	10,814	76,254	152,539	93
68	1150	North Ustyurt Basin	Oil	0	2413	2413	10,770	13,183	13,183	18
69	1209	Angara-Lena Terrace	Gas	0	8075	8075	10,705	18,780	18,780	43
70	3142	Sichuan Basin	Gas	6236	6290	12,526	10,311	16,601	22,837	55
71	7183	Gulf of Guinea	Oil	46	1903	1949	10,071	11,974	12,020	16

Rank	Province Code	Province Name	Major Commodity	Cumulative Gas (BCFG)	Remaining Gas (BCFG)	Known Gas (BCFG)	Mean Undiscovered Gas (BCFG)	Future Gas (BCFG)	Gas Endowment (BCFG)	Discovery Maturity, Gas (%)
72	5215	Labrador-Newfoundland Shelf	Oil	0	6630	6630	8893	15,522	15,522	43
73	2058	Grand Erg/Ahnet Basin	Gas	4980	109,053	114,034	8664	117,717	122,697	93
74	3127	Bohaiwan Basin	Oil	2987	12,685	15,672	8514	21,199	24,186	65
75	6029	Sergipe-Alagoas Basin	Oil	348	1059	1407	8334	9393	9741	14
76	2029	Wadi-Surhan Basin	Gas	47	179	226	7873	8052	8099	3
77	3824	Northwest Java Basin	Oil	1026	7028	8055	7453	14,481	15,508	52
78	2009	Masila-Jeza Basin	Oil	0	40	40	6793	6833	6833	1
79	6096	Llanos Basin	Oil	139	15,175	15,314	6217	21,391	15,651	98
80	2004	Ma'Rib-Al Jawf/Masila Basin	Oil	0	16,507	16,507	5944	22,451	22,451	74
81	5302	Veracruz Basin	Oil	335	270	605	5832	6102	6437	9
82	3144	Songliao Basin	Oil	3	1739	1743	5706	7446	7449	23
83	3930	Gippsland Basin	Oil	3501	6274	9775	5658	11,932	15,433	63
84	2010	Ghudun-Khasfeh Flank Province	Oil	0	0	0	5623	5623	5623	0
85	5307	Campeche-Sigsbe Salt Basin	Oil	0	0	0	5565	5565	5565	0
86	2048	Pelagian Basin	Oil	43	17,167	17,210	5556	22,724	22,766	76
87	1211	Cis-Patom Foredeep	Gas	0	100	100	5364	5464	5464	2
88	2014	Ghaba Salt Basin	Oil	138	18,825	18,964	5056	23,881	24,019	79
89	4047	North Carpathian Basin	Oil	8714	7618	16,332	5049	12,667	21,381	76
90	2016	Fahud Salt Basin	Oil	1151	8832	9983	4878	13,710	14,861	67
91	6081	Talara Basin	Oil	1951	941	2892	4795	5736	7687	38
92	2006	Shabwah Basin	Oil	0	573	573	4246	4819	4819	12
93	4048	Pannonian Basin	Oil	6534	7308	13,842	4143	11,452	17,986	77
94	3808	Central Sumatra Basin	Oil	910	2956	3866	4078	7034	7944	49
95	1015	Volga-Ural Region	Oil	16,417	82,750	99,166	3941	86,691	103,108	96
96	6058	San Jorge Basin	Oil	1667	1898	3565	3774	5672	7339	49
97	7303	Orange River Coastal	Gas	0	6015	6015	3603	9618	9618	63
98	6090	Middle Magdalena	Oil	2310	2692	5002	3292	5985	8294	60
99	2015	Central Oman Platform	Oil	0	742	742	3239	3980	3981	19
100	5245	Rocky Mountain Deformed Belt	Gas	8662	6447	15,109	3186	9633	18,295	83
101	4061	Carpathian-Balkanian Basin	Oil	3126	4139	7265	3179	7319	10,445	70
102	2074	Khleisha Uplift	Oil	0	210	210	2580	2790	2790	8
103	4057	Transylvanian Basin	Gas	21,194	9537	30,731	2083	11,620	32,814	94
104	5301	Tampico-Misantla Basin	Oil	2354	4306	6661	2051	6358	8712	76
105	5310	Sierra Madre de Chiapas-Peten Foldbelt	NA	NA	NA	NA	1846	NA	NA	NA
106	6041	Putumayo-Oriente-Maranon Basin	Oil	661	955	1616	1596	2552	3212	50
107	5308	Yucatan Platform	Oil	0	0	0	1496	1496	1496	0
108	8026	Kohat-Potwar	Oil	784	1147	1931	1314	2460	3245	60
109	3115	Junggar Basin	Oil	501	2082	2583	1296	3379	3879	67
110	2025	Hail-Ga'Ara Arch	Gas	0	0	0	1290	1290	1290	0
111	2028	Rutbah Uplift	Gas	0	0	0	1290	1290	1290	0
112	2011	South Oman Salt Basin	Oil	0	188	188	1236	1424	1424	13
113	8034	Assam	Oil	2039	4495	6534	1178	5673	7712	85
114	2031	Zagros Thrust Zone	Gas	0	0	0	966	966	966	0

Rank	Province Code	Province Name	Major Commodity	Cumulative Gas (BCFG)	Remaining Gas (BCFG)	Known Gas (BCFG)	Mean Undiscovered Gas (BCFG)	Future Gas (BCFG)	Gas Endowment (BCFG)	Discovery Maturity, Gas (%)
115	7013	Senegal	Oil	4	45	49	856	901	905	5
116	2075	Euphrates/Mardin	Oil	66	1668	1734	749	2418	2483	70
117	6083	Progreso Basin	Oil	148	350	498	695	1044	1193	42
118	6117	Greater Antilles Deformed Belt	Oil	20	77	96	592	668	688	14
119	5244	Williston Basin, Canada	Oil	10,378	6876	17,254	502	7379	17,756	97
120	2013	Huqf-Haushi Uplift	Oil	0	0	0	404	404	404	0
121	2047	Hamra Basin	Oil	0	0	0	241	241	241	0
122	5306	Macuspana Basin	Oil	8	8	16	210	217	225	7
123	5316	Trans-Mexican Neovolcanic Axis	NA	NA	NA	NA	195	NA	NA	NA
124	5246	Mackenzie Foldbelt	Gas	379	252	631	43	295	674	94
125	2089	Anah Graben	Oil	0	0	0	39	39	39	0
126	5317	Chicontepec Basin	Oil	0	2	2	32	34	34	6
127	2017	Oman Mountains	Oil	0	0	0	22	22	22	0
128	3128	Ordos Basin	Oil	13	5597	5610	20	5617	5630	100
129	1158	Tian Shan Foldbelt	NA	NA	NA	NA	0	NA	NA	NA
130	2033	Sinai Basin	Oil	0	5	5	0	5	5	100
131	2061	Ougarta Uplift	Oil	0	6	6	0	6	6	100
132	3161	Yunnan Guizhou Hubei Fold Belt	NA	NA	NA	NA	0	NA	NA	NA
133	3916	Carnarvon Basin	Oil	32	70	102	0	70	102	100
134	4028	Mid-North Sea High	NA	NA	NA	NA	0	NA	NA	NA
135	4033	German-Polish Basin	Gas	6905	940	7846	0	940	7846	100
136	4038	Munsterland Basin	Gas	121	31	152	0	31	152	100
137	4039	Southwest German Basin	NA	NA	NA	NA	0	NA	NA	NA
138	4046	Bohemia	Gas	114	117	231	0	117	231	100
139	8002	Himalayan	NA	NA	NA	NA	0	NA	NA	NA
140	8023	Central Afghanistan	Gas	27	244	271	0	244	271	100
141	8035	North Burma	Oil	1	10	11	0	10	11	100

APPENDIX 3C. Provinces ranked by mean (expected) undiscovered total petroleum.

[MMBOE = million barrels of oil equivalent. Values given are not necessarily those for the entire province, but only for that part assessed in World Petroleum Assessment 2000 (USGS, 2000). Major commodity is based on the ratio of gas endowment to oil endowment as ft^3 of gas per bbl of oil (CFG/BO). Major commodity is oil if this ratio is less than 20,000 CFG/BO and gas if this ratio is equal to or greater than 20,000 CFG/BO. Total cumulative petroleum is the reported volume of total petroleum that has been produced. Total remaining petroleum is the calculated remaining reserves; the difference between total known and total cumulative petroleum. Total known petroleum is the reported discovered volume (cumulative production plus remaining reserves). Total mean undiscovered petroleum is derived from the mean volumes of oil, gas, and NGL estimated in World Petroleum Assessment 2000 (USGS, 2000). Total future petroleum is the total remaining petroleum plus estimated total mean undiscovered petroleum. Total petroleum endowment is the total known petroleum plus estimated total mean undiscovered petroleum volumes. Discovery maturity is the percentage of total petroleum discovered (total known petroleum) with respect to total petroleum endowment. Calculated values may not be equal to the sums of the component values because numbers have been independently rounded. NA means not applicable; it is shown for all values where there is only one field present, and used where the endowment could not be calculated.] Known volumes from Petroconsultants (1996) and NRG Associates (1995). Note that gas = natural gas.

Rank	Province Code	Province Name	Major Commodity	Total Cumulative Petroleum (MMBOE)	Total Remaining Petroleum (MMBOE)	Total Known Petroleum (MMBOE)	Total Mean Undiscovered Petroleum (MMBOE)	Total Future Petroleum (MMBOE)	Total Petroleum Endowment (MMBOE)	Discovery Maturity, Total Petroleum (%)
1	1174	West Siberian Basin	Oil	83,213	271,517	354,730	182,864	454,380	537,593	66
2	2019	Rub Al Khali Basin	Oil	20,998	101,760	122,757	137,868	239,627	260,625	47
3	2030	Zagros Fold Belt	Oil	51,102	138,437	189,539	90,292	228,729	279,831	68
4	2024	Mesopotamian Foredeep Basin	Oil	58,934	285,059	343,993	78,159	363,217	422,151	81
5	7192	Niger Delta	Oil	17,646	35,353	53,000	68,640	103,994	121,640	44
6	5200	East Greenland Rift Basins	Oil	0	0	0	64,769	64,769	64,769	0
7	2021	Greater Ghawar Uplift	Oil	62,169	129,511	191,680	63,808	193,319	255,488	75
8	1112	South Caspian Basin	Oil	13,334	10,607	23,941	52,725	63,332	76,665	31
9	7203	West-Central Coastal	Oil	5957	9854	15,811	48,665	58,519	64,476	25
10	1016	North Caspian Basin	Oil	1322	44,528	45,850	48,573	93,101	94,423	49
11	2020	Interior Homocline-Central Arch	Oil	84	6516	6601	46,132	52,648	52,733	13
12	2023	Widyan Basin-Interior Platform	Oil	229	18,457	18,686	43,855	62,312	62,541	30
13	4017	Vestford-Helgeland	Oil	76	5897	5973	43,062	48,959	49,035	12
14	6036	Santos Basin	Oil	12	505	517	40,828	41,332	41,344	1
15	6098	East Venezuela Basin	Oil	17,509	34,747	52,257	31,934	66,682	84,191	62
16	1154	Amu-Darya Basin	Gas	13,357	26,975	40,332	31,541	58,516	71,873	56
17	1050	South Barents Basin	Gas	0	11,770	11,770	30,377	42,147	42,147	28
18	6021	Guyana-Suriname Basin	Oil	17	46	63	24,586	24,632	24,649	0
19	4025	North Sea Graben	Oil	28,183	48,322	76,505	21,248	69,570	97,753	78
20	6035	Campos Basin	Oil	1819	9287	11,107	20,128	29,415	31,235	36
21	3154	Tarim Basin	Oil	80	1776	1856	19,250	21,026	21,106	9
22	2071	Red Sea Basin	Oil	6445	5298	11,744	18,533	23,832	30,277	39
23	3817	Kutei Basin	Oil	3691	8040	11,731	18,060	26,100	29,791	39
24	3948	Northwest Shelf	Gas	896	10,589	11,485	16,848	27,437	28,333	41

Rank	Province Code	Province Name	Major Commodity	Total Cumulative Petroleum (MMBOE)	Total Remaining Petroleum (MMBOE)	Total Known Petroleum (MMBOE)	Total Mean Undiscovered Petroleum (MMBOE)	Total Future Petroleum (MMBOE)	Total Petroleum Endowment (MMBOE)	Discovery Maturity, Total Petroleum (%)
25	1008	Timan-Pechora Basin	Oil	5771	14,171	19,942	16,598	30,769	36,539	55
26	1322	North Sakhalin Basin	Oil	785	5292	6077	15,156	20,449	21,234	29
27	1109	Middle Caspian Basin	Oil	6071	8083	14,153	13,171	21,254	27,324	52
28	6099	Maracaibo Basin	Oil	35,745	17,820	53,565	12,086	29,906	65,651	82
29	1060	North Barents Basin	Gas	0	0	0	11,355	11,355	11,355	0
30	4068	Provence Basin	Gas	0	0	0	11,188	11,188	11,188	0
31	5305	Villahermosa Uplift	Oil	17,004	24,993	41,997	10,957	35,950	52,955	79
32	5304	Saline-Comalcalco Basin	Oil	1844	889	2733	10,650	11,539	13,383	20
33	2054	Trias/Ghadames Basin	Oil	7589	12,800	20,389	10,566	23,366	30,955	66
34	6034	Espirito Santo Basin	Oil	100	68	169	10,285	10,354	10,454	2
35	2043	Sirte Basin	Oil	14,929	28,566	43,496	10,000	38,566	53,496	81
36	8047	Ganges-Brahmaputra Delta	Gas	451	1654	2105	9558	11,212	11,663	18
37	6060	Falklands Plateau	Oil	0	0	0	8967	8967	8967	0
38	2022	Qatar Arch	Gas	183	92,332	92,515	8908	101,239	101,423	91
39	6045	Santa Cruz-Tarija Basin	Oil	1655	2045	3699	8629	10,674	12,328	30
40	2056	Illizi Basin	Oil	2485	9593	12,078	8536	18,129	20,614	59
41	3701	Baram Delta/Brunei-Sabah Basin	Oil	5492	7619	13,111	8434	16,054	21,545	61
42	6037	Pelotas Basin	Oil	0	0	0	7846	7846	7846	0
43	1210	Nepa-Botuoba Arch	Oil	1	5974	5976	7643	13,618	13,619	44
44	3703	Malay Basin	Oil	1609	10,346	11,955	6319	16,665	18,274	65
45	3910	Bonaparte Gulf Basin	Gas	123	3046	3169	6281	9327	9450	34
46	3127	Bohaiwan Basin	Oil	10,914	16,339	27,253	6248	22,587	33,501	81
47	1009	Dnieper-Donets Basin	Gas	6137	5523	11,660	5969	11,492	17,629	66
48	1059	Ludlov Saddle	NA	NA	NA	NA	5361	NA	NA	NA
49	3913	Browse Basin	Gas	19	3221	3239	5339	8559	8578	38
50	6022	Foz do Amazonas Basin	NA	NA	NA	NA	5272	NA	NA	NA
51	6096	Llanos Basin	Oil	1202	7203	8405	5004	12,207	13,409	63
52	8048	Irrawaddy	Gas	649	1797	2446	4930	6726	7376	33
53	1207	Baykit Arch	Oil	0	936	936	4897	5833	5833	16
54	8043	Bombay	Oil	4417	8323	12,740	4666	12,989	17,406	73
55	3702	Greater Sarawak Basin	Gas	846	14,044	14,890	4301	18,345	19,191	78
56	6107	Lesser Antilles Deformed Belt	Oil	11	3	14	4110	4113	4124	0
57	5215	Labrador-Newfoundland Shelf	Oil	0	2875	2875	3792	6667	6667	48
58	3828	South Sumatra Basin	Oil	2367	1819	4186	3750	5569	7936	53
59	5307	Campeche-Sigsbe Salt Basin	Oil	0	0	0	3714	3714	3714	0
60	6055	Neuquen Basin	Oil	2883	3276	6159	3562	6838	9721	63
61	6103	Tobago Trough	Gas	0	3867	3867	3551	7418	7418	52
62	4060	Po Basin	Gas	1706	1820	3526	3476	5296	7002	50
63	6041	Putumayo-Oriente-Maranon Basin	Oil	2999	3871	6871	3418	7290	10,289	67
64	1015	Volga-Ural Region	Oil	45,732	35,832	81,564	3403	39,235	84,968	96
65	6059	Magallanes Basin	Gas	2359	3202	5562	3354	6556	8915	62

App-3c_Page 2

Rank	Province Code	Province Name	Major Commodity	Total Cumulative Petroleum (MMBOE)	Total Remaining Petroleum (MMBOE)	Total Known Petroleum (MMBOE)	Total Mean Undiscovered Petroleum (MMBOE)	Total Future Petroleum (MMBOE)	Total Petroleum Endowment (MMBOE)	Discovery Maturity, Total Petroleum (%)
66	8042	Indus	Gas	1512	1970	3482	3328	5299	6811	51
67	6029	Sergipe-Alagoas Basin	Oil	528	402	929	3292	3693	4221	22
68	6063	Malvinas Basin	Oil	0	25	25	3280	3305	3305	1
69	1150	North Ustyurt Basin	Oil	459	2318	2777	3193	5511	5970	47
70	5243	Alberta Basin	Oil	22,342	8760	31,101	3128	11,888	34,229	91
71	7183	Gulf of Guinea	Oil	81	515	596	2965	3480	3561	17
72	3822	North Sumatra Basin	Gas	3409	2451	5860	2914	5365	8774	67
73	6081	Talara Basin	Oil	1715	439	2154	2765	3204	4919	44
74	1108	Azov-Kuban Basin	Gas	2589	1091	3681	2675	3766	6355	58
75	4036	Anglo-Dutch Basin	Gas	7237	5403	12,640	2630	8033	15,270	83
76	2009	Masila-Jeza Basin	Oil	144	496	641	2598	3094	3238	20
77	8025	Sulaiman-Kirthar	Gas	1121	1535	2657	2553	4089	5210	51
78	1209	Angara-Lena Terrace	Gas	0	1409	1409	2526	3935	3935	36
79	2058	Grand Erg/Ahnet Basin	Gas	1783	22,752	24,534	2449	25,201	26,984	91
80	2048	Pelagian Basin	Oil	528	4646	5174	2323	6970	7497	69
81	2004	Ma'Rib-Al Jawf/Masila Basin	Oil	536	3712	4248	2273	5985	6521	65
82	2010	Ghudun-Khasfeh Flank Province	Oil	0	0	0	2258	2258	2258	0
83	5302	Veracruz Basin	Oil	107	125	232	2214	2339	2446	9
84	3142	Sichuan Basin	Gas	1086	1112	2198	2202	3313	4400	50
85	3144	Songliao Basin	Oil	12,399	3466	15,865	2189	5655	18,054	88
86	2016	Fahud Salt Basin	Oil	2754	3535	6288	2008	5543	8297	76
87	3824	Northwest Java Basin	Oil	1771	2914	4685	2007	4922	6692	70
88	2029	Wadi-Surhan Basin	Gas	8	30	38	1942	1972	1979	2
89	4035	Northwest German Basin	Gas	14,589	11,340	25,929	1881	13,221	27,810	93
90	3808	Central Sumatra Basin	Oil	8500	5372	13,872	1663	7034	15,535	89
91	4061	Carpathian-Balkanian Basin	Oil	4624	2571	7195	1659	4230	8854	81
92	2006	Shabwah Basin	Oil	0	210	210	1624	1833	1833	11
93	3930	Gippsland Basin	Oil	3815	2375	6190	1592	3967	7782	80
94	2014	Ghaba Salt Basin	Oil	489	4492	4981	1436	5928	6416	78
95	6090	Middle Magdalena	Oil	2300	1028	3328	1388	2416	4715	71
96	5301	Tampico-Misantla Basin	Oil	6165	1840	8005	1354	3194	9359	86
97	4047	North Carpathian Basin	Oil	2104	1559	3663	1344	2903	5007	73
98	1211	Cis-Patom Foredeep	Gas	0	17	17	1309	1325	1325	1
99	5310	Sierra Madre de Chiapas-Peten Foldbelt	NA	NA	NA	NA	1307	NA	NA	NA
100	6058	San Jorge Basin	Oil	2723	1218	3942	1203	2421	5144	77
101	4048	Pannonian Basin	Oil	2691	1892	4583	1178	3070	5761	80
102	5308	Yucatan Platform	Oil	0	0	0	1092	1092	1092	0
103	2015	Central Oman Platform	Oil	3	198	200	1090	1287	1290	16
104	2075	Euphrates/Mardin	Oil	293	607	900	893	1500	1793	50
105	7303	Orange River Coastal	Gas	0	1003	1003	879	1882	1882	53
106	3115	Junggar Basin	Oil	1035	6206	7241	843	7049	8084	90

Rank	Province Code	Province Name	Major Commodity	Total Cumulative Petroleum (MMBOE)	Total Remaining Petroleum (MMBOE)	Total Known Petroleum (MMBOE)	Total Mean Undiscovered Petroleum (MMBOE)	Total Future Petroleum (MMBOE)	Total Petroleum Endowment (MMBOE)	Discovery Maturity, Total Petroleum (%)
107	2074	Khleisha Uplift	Oil	250	442	692	739	1180	1430	48
108	8034	Assam	Oil	1251	2379	3630	663	3042	4292	85
109	6117	Greater Antilles Deformed Belt	Oil	68	245	313	628	873	940	33
110	5245	Rocky Mountain Deformed Belt	Gas	1916	1199	3115	613	1812	3728	84
111	8026	Kohat-Potwar	Oil	341	343	684	422	765	1106	62
112	6083	Progreso Basin	Oil	147	111	258	385	497	644	40
113	2025	Hail-Ga'Ara Arch	Gas	0	0	0	369	369	369	0
114	2028	Rutbah Uplift	Gas	0	0	0	369	369	369	0
115	2011	South Oman Salt Basin	Oil	203	343	546	357	701	903	60
116	4057	Transylvanian Basin	Gas	3532	1589	5122	354	1943	5475	94
117	7013	Senegal	Oil	1	18	19	343	361	362	5
118	5244	Williston Basin, Canada	Oil	4468	1963	6431	252	2215	6683	96
119	2031	Zagros Thrust Zone	Gas	0	0	0	242	242	242	0
120	5306	Macuspana Basin	Oil	4	5	9	149	154	158	6
121	2047	Hamra Basin	Oil	0	0	0	148	148	148	0
122	3128	Ordos Basin	Oil	83	1535	1618	143	1678	1761	92
123	2013	Huqf-Haushi Uplift	Oil	0	0	0	115	115	115	0
124	5316	Trans-Mexican Neovolcanic Axis	NA	NA	NA	NA	72	NA	NA	NA
125	2089	Anah Graben	Oil	75	80	155	35	115	190	81
126	2017	Oman Mountains	Oil	0	0	0	27	27	27	0
127	5317	Chicontepec Basin	Oil	0	0	0	22	22	22	2
128	5246	Mackenzie Foldbelt	Gas	63	42	105	8	50	113	93
129	1158	Tian Shan Foldbelt	NA	NA	NA	NA	0	NA	NA	NA
130	2033	Sinai Basin	Oil	94	12	106	0	12	106	100
131	2061	Ougarta Uplift	Oil	0	42	42	0	42	42	100
132	3161	Yunnan Guizhou Hubei Fold Belt	NA	NA	NA	NA	0	NA	NA	NA
133	3916	Carnarvon Basin	Oil	21	31	52	0	31	52	100
134	4028	Mid-North Sea High	NA	NA	NA	NA	0	NA	NA	NA
135	4033	German-Polish Basin	Gas	1151	157	1308	0	157	1308	100
136	4038	Munsterland Basin	Gas	20	5	25	0	5	25	100
137	4039	Southwest German Basin	NA	NA	NA	NA	0	NA	NA	NA
138	4046	Bohemia	Gas	21	22	43	0	22	43	100
139	8002	Himalayan	NA	NA	NA	NA	0	NA	NA	NA
140	8023	Central Afghanistan	Gas	5	41	45	0	41	45	100
141	8035	North Burma	Oil	2	2	3	0	2	3	100

App-3c_Page 4

Appendix 3d. Provinces ranked by conventional oil endowment.

[MMBO = million barrels of oil. Values given are not necessarily those for the entire province, but only for that part assessed in World Petroleum Assessment 2000 (USGS, 2000). Major commodity is based on the ratio of gas endowment to oil endowment as ft^3 of gas per bbl of oil (CFG/BO). Major commodity is oil if this ratio is less than 20,000 CFG/BO and gas if this ratio is equal to or greater than 20,000 CFG/BO. Cumulative oil is the reported volume of oil that has been produced. Remaining oil is the calculated remaining reserves; the difference between known and cumulative oil. Known oil is the reported discovered volume (cumulative production plus remaining reserves). Mean undiscovered oil is the mean volume estimated in World Petroleum Assessment 2000 (USGS, 2000). Future oil is the remaining oil plus estimated mean undiscovered oil. Oil endowment is the known oil plus estimated mean undiscovered oil volumes. Discovery maturity is the percentage of oil discovered (known oil) with respect to oil endowment. Calculated values may not be equal to the sums of the component values because numbers have been independently rounded. Negative values result from inconsistent reporting of cumulative and known volumes. Values in parentheses are questionable data. NA means not applicable; it is shown for all values where there is only one field present, and used where the endowment could not be calculated.] Known volumes from Petroconsultants (1996) and NRG Associates (1995). Note that gas = natural gas.

Rank	Province Code	Province Name	Major Commodity	Cumulative Oil (MMBO)	Remaining Oil (MMBO)	Known Oil (MMBO)	Mean Undiscovered Oil (MMBO)	Future Oil (MMBO)	Oil Endowment (MMBO)	Discovery Maturity, Oil (%)
1	2024	Mesopotamian Foredeep Basin	Oil	56,746	235,696	292,442	61,134	296,830	353,576	83
2	1174	West Siberian Basin	Oil	46,530	93,383	139,913	55,235	148,618	195,148	72
3	2030	Zagros Fold Belt	Oil	49,367	72,234	121,601	45,459	117,693	167,060	73
4	2021	Greater Ghawar Uplift	Oil	57,744	83,956	141,700	13,830	97,786	155,530	91
5	2019	Rub Al Khali Basin	Oil	19,320	70,507	89,827	36,647	107,154	126,474	71
6	7192	Niger Delta	Oil	16,598	17,925	34,522	40,487	58,412	75,010	46
7	1015	Volga-Ural Region	Oil	42,690	21,247	63,937	2549	23,796	66,485	96
8	6099	Maracaibo Basin	Oil	35,581	13,491	49,072	8183	21,675	57,256	86
9	4025	North Sea Graben	Oil	22,467	21,427	43,895	13,098	34,526	56,993	77
10	5200	East Greenland Rift Basins	Oil	0	0	0	47,148	47,148	47,148	0
11	2043	Sirte Basin	Oil	13,986	23,087	37,072	6854	29,941	43,927	84
12	7203	West-Central Coastal	Oil	5765	7952	13,717	29,747	37,699	43,464	32
13	5305	Villahermosa Uplift	Oil	14,640	20,382	35,022	7436	27,818	42,457	82
14	6098	East Venezuela Basin	Oil	13,808	16,395	30,203	11,875	28,271	42,078	72
15	2023	Widyan Basin-Interior Platform	Oil	219	17,216	17,435	21,215	38,431	38,650	45
16	1016	North Caspian Basin	Oil	822	9987	10,808	23,320	33,307	34,128	32
17	1112	South Caspian Basin	Oil	11,681	5758	17,438	15,725	21,483	33,163	53
18	3127	Bohaiwan Basin	Oil	10,415	14,139	24,553	4412	18,550	28,965	85
19	6035	Campos Basin	Oil	1769	8287	10,056	16,293	24,580	26,349	38
20	6036	Santos Basin	Oil	6	279	285	23,209	23,488	23,494	1
21	2054	Trias/Ghadames Basin	Oil	7410	7793	15,203	6477	14,270	21,680	70
22	2020	Interior Homocline-Central Arch	Oil	84	4616	4700	16,495	21,111	21,195	22
23	1008	Timan-Pechora Basin	Oil	3346	9774	13,120	5732	15,506	18,852	70
24	3144	Songliao Basin	Oil	12,399	3176	15,575	1026	4203	16,601	94
25	5243	Alberta Basin	Oil	12,821	2458	15,279	818	3276	16,097	95
26	6021	Guyana-Suriname Basin	Oil	17	46	63	15,247	15,293	15,310	0
27	2071	Red Sea Basin	Oil	6280	3580	9861	5187	8767	15,048	66
28	1109	Middle Caspian Basin	Oil	4359	5193	9551	4887	10,080	14,439	66

Rank	Province Code	Province Name	Major Commodity	Cumulative Oil (MMBO)	Remaining Oil (MMBO)	Known Oil (MMBO)	Mean Undiscovered Oil (MMBO)	Future Oil (MMBO)	Oil Endowment (MMBO)	Discovery Maturity, Oil (%)
29	3808	Central Sumatra Basin	Oil	8349	4868	13,217	899	5768	14,117	94
30	8043	Bombay	Oil	3375	5065	8440	2091	7156	10,531	80
31	3701	Baram Delta/Brunei-Sabah Basin	Oil	3953	2945	6898	3390	6335	10,288	67
32	6041	Putumayo-Oriente-Maranon Basin	Oil	2889	3712	6601	3098	6810	9699	68
33	5304	Saline-Comalcalco Basin	Oil	1673	686	2359	7018	7703	9377	25
34	4017	Vestford-Helgeland	Oil	76	2585	2660	6538	9123	9199	29
35	6096	Llanos Basin	Oil	1179	4224	5402	3631	7855	9033	60
36	3817	Kutei Basin	Oil	2195	684	2879	5145	5829	8024	36
37	5301	Tampico-Misantla Basin	Oil	5773	1122	6895	893	2015	7787	89
38	3115	Junggar Basin	Oil	952	5858	6810	563	6422	7373	92
39	4061	Carpathian-Balkanian Basin	Oil	4083	1825	5908	1013	2838	6921	85
40	3154	Tarim Basin	Oil	45	658	704	5948	6606	6652	11
41	2056	Illizi Basin	Oil	2286	1384	3670	2947	4331	6617	55
42	2022	Qatar Arch	Gas	0	1150	1150	5026	6176	6176	19
43	6060	Falklands Plateau	Oil	0	0	0	5833	5833	5833	0
44	2016	Fahud Salt Basin	Oil	2555	1918	4473	1020	2938	5493	81
45	3703	Malay Basin	Oil	1518	2090	3608	1228	3318	4836	75
46	1322	North Sakhalin Basin	Oil	754	1428	2182	2582	4010	4764	46
47	3930	Gippsland Basin	Oil	3077	784	3860	310	1093	4170	93
48	6058	San Jorge Basin	Oil	2445	900	3346	498	1398	3844	87
49	3824	Northwest Java Basin	Oil	1555	1613	3168	658	2271	3825	83
50	5244	Williston Basin, Canada	Oil	2696	809	3506	154	963	3660	96
51	6055	Neuquen Basin	Oil	1532	838	2370	1290	2128	3659	65
52	2048	Pelagian Basin	Oil	521	1719	2239	1268	2986	3507	64
53	3948	Northwest Shelf	Gas	473	647	1119	2381	3028	3500	32
54	6081	Talara Basin	Oil	1346	326	1672	1711	2038	3383	49
55	5215	Labrador-Newfoundland Shelf	Oil	0	1582	1582	1796	3378	3378	47
56	1150	North Ustyurt Basin	Oil	459	1907	2366	954	2861	3320	71
57	1210	Nepa-Botuoba Arch	Oil	1	2005	2006	1208	3213	3214	62
58	6090	Middle Magdalena	Oil	1915	512	2426	709	1220	3135	77
59	6034	Espirito Santo Basin	Oil	93	37	130	2959	2995	3089	4
60	6037	Pelotas Basin	Oil	0	0	0	2938	2938	2938	0
61	8034	Assam	Oil	911	1625	2536	399	2024	2935	86
62	3828	South Sumatra Basin	Oil	1842	587	2429	469	1055	2898	84
63	1009	Dnieper-Donets Basin	Gas	110	1501	1611	1098	2599	2709	59
64	4048	Pannonian Basin	Oil	1567	623	2190	359	982	2550	86
65	5307	Campeche-Sigsbe Salt Basin	Oil	0	0	0	2478	2478	2478	0
66	6045	Santa Cruz-Tarija Basin	Oil	257	63	320	2145	2209	2466	13
67	4035	Northwest German Basin	Gas	1867	413	2280	56	468	2336	98
68	6029	Sergipe-Alagoas Basin	Oil	469	223	692	1487	1711	2179	32
69	2004	Ma'Rib-Al Jawf/Masila Basin	Oil	504	683	1188	959	1643	2147	55
70	6059	Magallanes Basin	Gas	595	620	1216	704	1325	1920	63
71	3910	Bonaparte Gulf Basin	Gas	123	393	517	1286	1679	1803	29

Rank	Province Code	Province Name	Major Commodity	Cumulative Oil (MMBO)	Remaining Oil (MMBO)	Known Oil (MMBO)	Mean Undiscovered Oil (MMBO)	Future Oil (MMBO)	Oil Endowment (MMBO)	Discovery Maturity, Oil (%)
72	2009	Masila-Jeza Basin	Oil	144	488	632	1096	1584	1728	37
73	2014	Ghaba Salt Basin	Oil	466	878	1344	371	1250	1715	78
74	6107	Lesser Antilles Deformed Belt	Oil	8	2	11	1536	1538	1546	1
75	3702	Greater Sarawak Basin	Gas	168	649	817	631	1281	1448	56
76	8048	Irrawaddy	Gas	527	188	715	724	912	1439	50
77	2075	Euphrates/Mardin	Oil	282	312	594	749	1061	1343	44
78	4047	North Carpathian Basin	Oil	649	286	935	391	677	1326	71
79	1207	Baykit Arch	Oil	0	34	34	1213	1246	1246	3
80	7183	Gulf of Guinea	Oil	74	162	236	1004	1166	1240	19
81	3913	Browse Basin	Gas	19	27	46	1055	1083	1101	4
82	2058	Grand Erg/Ahnet Basin	Gas	106	388	494	599	986	1093	45
83	5302	Veracruz Basin	Oil	51	79	130	955	1034	1085	12
84	6063	Malvinas Basin	Oil	0	10	10	1031	1041	1041	1
85	1154	Amu-Darya Basin	Gas	80	686	766	270	957	1037	74
86	1108	Azov-Kuban Basin	Gas	1273	(-750)	524	417	(-332)	941	56
87	2010	Ghudun-Khasfeh Flank Province	Oil	0	0	0	926	926	926	0
88	3822	North Sumatra Basin	Gas	496	178	674	213	391	887	76
89	3128	Ordos Basin	Oil	81	602	683	138	740	821	83
90	2006	Shabwah Basin	Oil	0	113	113	685	798	798	14
91	6117	Greater Antilles Deformed Belt	Oil	65	232	297	494	726	790	38
92	2074	Khleisha Uplift	Oil	250	407	657	105	512	762	86
93	5308	Yucatan Platform	Oil	0	0	0	754	754	754	0
94	4060	Po Basin	Gas	91	254	346	360	615	706	49
95	4036	Anglo-Dutch Basin	Gas	432	159	592	88	247	679	87
96	2011	South Oman Salt Basin	Oil	203	297	500	97	394	596	84
97	2015	Central Oman Platform	Oil	3	68	71	424	493	495	14
98	8026	Kohat-Potwar	Oil	201	120	321	149	268	469	68
99	6083	Progreso Basin	Oil	122	53	175	237	290	412	43
100	4068	Provence Basin	Gas	0	0	0	388	388	388	0
101	8042	Indus	Gas	111	70	180	103	173	283	64
102	1209	Angara-Lena Terrace	Gas	0	0	0	242	242	242	0
103	1050	South Barents Basin	Gas	0	0	0	219	219	219	0
104	5245	Rocky Mountain Deformed Belt	Gas	176	22	198	6	28	204	97
105	2089	Anah Graben	Oil	75	80	155	26	106	181	85
106	7013	Senegal	Oil	0	10	10	157	167	168	6
107	1211	Cis-Patom Foredeep	Gas	0	0	0	161	161	161	0
108	3142	Sichuan Basin	Gas	47	63	110	27	91	138	80
109	1060	North Barents Basin	Gas	0	0	0	135	135	135	0
110	7303	Orange River Coastal	Gas	0	0	0	116	116	116	0
111	5306	Macuspana Basin	Oil	2	4	6	101	105	108	6
112	8047	Ganges-Brahmaputra Delta	Gas	3	3	5	103	105	108	5
113	2033	Sinai Basin	Oil	94	11	105	0	11	105	100
114	2047	Hamra Basin	Oil	0	0	0	89	89	89	0

App-3d_Page 3

Rank	Province Code	Province Name	Major Commodity	Cumulative Oil (MMBO)	Remaining Oil (MMBO)	Known Oil (MMBO)	Mean Undiscovered Oil (MMBO)	Future Oil (MMBO)	Oil Endowment (MMBO)	Discovery Maturity, Oil (%)
115	2025	Hail-Ga'Ara Arch	Gas	0	0	0	53	53	53	0
116	2028	Rutbah Uplift	Gas	0	0	0	53	53	53	0
117	2061	Ougarta Uplift	Oil	0	41	41	0	41	41	100
118	2031	Zagros Thrust Zone	Gas	0	0	0	38	38	38	0
119	3916	Carnarvon Basin	Oil	16	19	35	0	19	35	100
120	2013	Huqf-Haushi Uplift	Oil	0	0	0	30	30	30	0
121	2017	Oman Mountains	Oil	0	0	0	22	22	22	0
122	6103	Tobago Trough	Gas	0	20	20	0	20	20	100
123	5317	Chicontepec Basin	Oil	0	0	0	15	15	15	1
124	8025	Sulaiman-Kirthar	Gas	0	0	0	12	12	12	0
125	4046	Bohemia	Gas	1	0	1	0	0	1	100
126	8035	North Burma	Oil	1	0	1	0	0	1	100
127	2029	Wadi-Surhan Basin	Gas	0	0	0	0	0	0	NA
128	4033	German-Polish Basin	Gas	0	0	0	0	0	0	100
129	4038	Munsterland Basin	Gas	0	0	0	0	0	0	NA
130	4057	Transylvanian Basin	Gas	0	0	0	0	0	0	NA
131	5246	Mackenzie Foldbelt	Gas	0	0	0	0	0	0	NA
132	8023	Central Afghanistan	Gas	0	0	0	0	0	0	NA
NA	1158	Tian Shan Foldbelt	NA	NA	NA	NA	0	NA	NA	NA
NA	1059	Ludlov Saddle	NA	NA	NA	NA	39	NA	NA	NA
NA	3161	Yunnan Guizhou Hubei Fold Belt	NA	NA	NA	NA	0	NA	NA	NA
NA	4028	Mid-North Sea High	NA	NA	NA	NA	0	NA	NA	NA
NA	4039	Southwest German Basin	NA	NA	NA	NA	0	NA	NA	NA
NA	5310	Sierra Madre de Chiapas-Peten Foldbelt	NA	NA	NA	NA	890	NA	NA	NA
NA	5316	Trans-Mexican Neovolcanic Axis	NA	NA	NA	NA	30	NA	NA	NA
NA	6022	Foz do Amazonas Basin	NA	NA	NA	NA	0	NA	NA	NA
NA	8002	Himalayan	NA	NA	NA	NA	0	NA	NA	NA

APPENDIX 3E. Provinces ranked by conventional natural gas endowment.

[BCFG = billion cubic feet of gas. Values given are not necessarily those for the entire province, but only for that part assessed in World Petroleum Assessment 2000 (USGS, 2000). Major commodity is based on the ratio of gas endowment to oil endowment as ft^3 of gas per bbl of oil (CFG/BO). Major commodity is oil if this ratio is less than 20,000 CFG/BO and gas if this ratio is equal to or greater than 20,000 CFG/BO. Cumulative gas is the reported volume of gas that has been produced. Remaining gas is the calculated remaining reserves; the difference between known and cumulative gas. Known gas is the reported discovered volume (cumulative production plus remaining reserves). Mean undiscovered gas is the mean volume estimated in World Petroleum Assessment 2000 (USGS, 2000). Future gas is the remaining gas plus estimated mean undiscovered gas. Gas endowment is the known gas plus estimated mean undiscovered gas volumes. Discovery maturity is the percentage of gas discovered (known gas) with respect to gas endowment. Calculated values may not be equal to the sums of the component values because numbers have been independently rounded. Negative values result from inconsistent reporting of cumulative and known volumes. Values in parentheses are questionable data. NA means not applicable; it is shown for all values where there is only one field present, and used where the endowment could not be calculated.] Known volumes from Petroconsultants (1996) and NRG Associates (1995). Note that gas = natural gas.

Rank	Province Code	Province Name	Major Commodity	Cumulative Gas (BCFG)	Remaining Gas (BCFG)	Known Gas (BCFG)	Mean Undiscovered Gas (BCFG)	Future Gas (BCFG)	Gas Endowment (BCFG)	Discovery Maturity, Gas (%)
1	1174	West Siberian Basin	Oil	218,465	1,051,584	1,270,049	642,934	1,694,518	1,912,983	66
2	2030	Zagros Fold Belt	Oil	10,333	389,111	399,444	212,008	601,119	611,451	65
3	2019	Rub Al Khali Basin	Oil	8303	173,934	182,237	425,712	599,646	607,949	30
4	2022	Qatar Arch	Gas	1018	464,582	465,600	17,385	481,967	482,985	96
5	2021	Greater Ghawar Uplift	Oil	26,493	222,057	248,550	226,969	449,026	475,519	52
6	1154	Amu-Darya Basin	Gas	78,513	151,830	230,343	163,650	315,480	393,993	58
7	2024	Mesopotamian Foredeep Basin	Oil	13,129	285,156	298,285	83,722	368,878	382,007	78
8	1016	North Caspian Basin	Oil	1745	155,161	156,906	119,051	274,212	275,957	57
9	1050	South Barents Basin	Gas	0	70,020	70,020	160,857	230,877	230,877	30
10	7192	Niger Delta	Oil	6232	87,579	93,811	132,716	220,295	226,526	41
11	6098	East Venezuela Basin	Oil	22,053	106,066	128,119	93,561	199,627	221,679	58
12	1112	South Caspian Basin	Oil	8591	27,403	35,994	173,310	200,713	209,304	17
13	4025	North Sea Graben	Oil	31,235	128,603	159,839	37,745	166,348	197,584	81
14	4017	Vestford-Helgeland	Oil	0	15,662	15,662	165,201	180,863	180,863	9
15	4035	Northwest German Basin	Gas	76,285	65,440	141,726	10,814	76,254	152,539	93
16	2020	Interior Homocline-Central Arch	Oil	0	9825	9825	122,074	131,899	131,899	7
17	2058	Grand Erg/Ahnet Basin	Gas	4980	109,053	114,034	8664	117,717	122,697	93
18	3948	Northwest Shelf	Gas	1840	54,405	56,245	64,711	119,116	120,956	47
19	3817	Kutei Basin	Oil	8711	36,762	45,473	64,757	101,519	110,230	41
20	1015	Volga-Ural Region	Oil	16,417	82,750	99,166	3941	86,691	103,108	96
21	2023	Widyan Basin-Interior Platform	Oil	60	7295	7355	94,746	102,041	102,101	7
22	3702	Greater Sarawak Basin	Gas	3731	78,433	82,164	19,247	97,680	101,411	81
23	7203	West-Central Coastal	Oil	1150	11,030	12,180	88,044	99,074	100,223	12
24	5243	Alberta Basin	Oil	47,573	31,097	78,670	11,895	42,992	90,565	87
25	1008	Timan-Pechora Basin	Oil	12,374	24,261	36,635	52,079	76,340	88,714	41
26	4036	Anglo-Dutch Basin	Gas	40,547	30,880	71,427	14,976	45,856	86,402	83
27	1009	Dnieper-Donets Basin	Gas	35,453	23,645	59,098	24,051	47,697	83,149	71
28	6036	Santos Basin	Oil	12	1113	1125	80,547	81,660	81,672	1

Rank	Province Code	Province Name	Major Commodity	Cumulative Gas (BCFG)	Remaining Gas (BCFG)	Known Gas (BCFG)	Mean Undiscovered Gas (BCFG)	Future Gas (BCFG)	Gas Endowment (BCFG)	Discovery Maturity, Gas (%)
29	5200	East Greenland Rift Basins	Oil	0	0	0	80,709	80,709	80 709	0
30	1322	North Sakhalin Basin	Oil	162	22,221	22,383	57,451	79,673	79,835	28
31	3703	Malay Basin	Oil	492	47,657	48,149	27,465	75,122	75,614	64
32	2056	Illizi Basin	Oil	750	44,311	45,061	28,138	72,449	73,199	62
33	2071	Red Sea Basin	Oil	849	8668	9517	60,012	68,680	69,529	14
34	8047	Ganges-Brahmaputra Delta	Gas	2669	9620	12,289	55,091	64,711	67,380	18
35	1109	Middle Caspian Basin	Oil	9984	17,156	27,140	39,685	56,841	66,325	41
36	3154	Tarim Basin	Oil	208	5572	5780	59,939	65,511	65,719	9
37	1060	North Barents Basin	Gas	0	0	0	59,820	59,820	59,320	0
38	3701	Baram Delta/Brunei-Sabah Basin	Oil	9188	27,012	36,200	23,226	50,238	59,426	61
39	5305	Villahermosa Uplift	Oil	13,888	27,404	41,292	15,623	43,027	56,915	73
40	2043	Sirte Basin	Oil	5479	32,288	37,767	14,984	47,272	52,752	72
41	1210	Nepa-Botuoba Arch	Oil	0	22,035	22,035	29,971	52,006	52,006	42
42	4068	Provence Basin	Gas	0	0	0	51,210	51,210	51,210	0
43	6045	Santa Cruz-Tarija Basin	Oil	6928	10,058	16,985	31,107	41,165	48,092	35
44	6099	Maracaibo Basin	Oil	979	25,722	26,701	17,576	43,298	44,277	60
45	6103	Tobago Trough	Gas	0	22,600	22,600	20,103	42,703	42,703	53
46	2054	Trias/Ghadames Basin	Oil	647	24,352	24,999	17,493	41,845	42,492	59
47	6021	Guyana-Suriname Basin	Oil	0	0	0	42,058	42,058	42,058	0
48	3822	North Sumatra Basin	Gas	13,977	11,582	25,559	15,123	26,705	40,682	63
49	6059	Magallanes Basin	Gas	10,053	14,754	24,807	14,040	28,794	38,848	64
50	3913	Browse Basin	Gas	0	17,960	17,960	20,093	38,053	38,053	47
51	8043	Bombay	Oil	6228	17,965	24,193	12,944	30,909	37,137	65
52	4060	Po Basin	Gas	9527	9360	18,886	17,964	27,324	36,850	51
53	3910	Bonaparte Gulf Basin	Gas	0	13,304	13,304	23,492	36,796	36,796	36
54	8042	Indus	Gas	8385	11,211	19,596	15,964	27,175	35,560	55
55	6034	Espirito Santo Basin	Oil	42	180	222	34,287	34,467	34,509	1
56	6055	Neuquen Basin	Oil	7636	13,068	20,704	12,416	25,483	33,119	63
57	4057	Transylvanian Basin	Gas	21,194	9537	30,731	2083	11,620	32,814	94
58	1108	Azov-Kuban Basin	Gas	7492	11,118	18,610	12,252	23,369	30,851	60
59	8048	Irrawaddy	Gas	721	9288	10,009	20,521	29,809	30,530	33
60	3828	South Sumatra Basin	Oil	3125	7080	10,204	18,250	25,330	28,455	36
61	8025	Sulaiman-Kirthar	Gas	6716	9061	15,777	12,599	21,659	28,376	56
62	6035	Campos Basin	Oil	303	5941	6244	19,691	25,632	25,935	24
63	3127	Bohaiwan Basin	Oil	2987	12,685	15,672	8514	21,199	24,186	65
64	2014	Ghaba Salt Basin	Oil	138	18,825	18,964	5056	23,881	24,019	79
65	3142	Sichuan Basin	Gas	6236	6290	12,526	10,311	16,601	22,837	55
66	6037	Pelotas Basin	Oil	0	0	0	22,806	22,806	22,806	0
67	2048	Pelagian Basin	Oil	43	17,167	17,210	5556	22,724	22,766	76
68	2004	Ma'Rib-Al Jawf/Masila Basin	Oil	0	16,507	16,507	5944	22,451	22,451	74
69	1207	Baykit Arch	Oil	0	5052	5052	16,983	22,035	22,035	23
70	4047	North Carpathian Basin	Oil	8714	7618	16,332	5049	12,667	21,381	76
71	1209	Angara-Lena Terrace	Gas	0	8075	8075	10,705	18,780	18,780	43

Rank	Province Code	Province Name	Major Commodity	Cumulative Gas (BCFG)	Remaining Gas (BCFG)	Known Gas (BCFG)	Mean Undiscovered Gas (BCFG)	Future Gas (BCFG)	Gas Endowment (BCFG)	Discovery Maturity, Gas (%)
72	5304	Saline-Comalcalco Basin	Oil	1023	1223	2246	16,423	17,645	18,669	12
73	5245	Rocky Mountain Deformed Belt	Gas	8662	6447	15,109	3186	9633	18,295	83
74	4048	Pannonian Basin	Oil	6534	7308	13,842	4143	11,452	17,986	77
75	5244	Williston Basin, Canada	Oil	10,378	6876	17,254	502	7379	17,756	97
76	6060	Falklands Plateau	Oil	0	0	0	16,581	16,581	16,581	0
77	6096	Llanos Basin	Oil	139	15,175	15,314	6217	21,391	15,651	98
78	5215	Labrador-Newfoundland Shelf	Oil	0	6630	6630	8893	15,522	15,522	43
79	3824	Northwest Java Basin	Oil	1026	7028	8055	7453	14,481	15,508	52
80	3930	Gippsland Basin	Oil	3501	6274	9775	5658	11,932	15,433	63
81	2016	Fahud Salt Basin	Oil	1151	8832	9983	4878	13,710	14,861	67
82	1150	North Ustyurt Basin	Oil	0	2413	2413	10,770	13,183	13,183	18
83	7183	Gulf of Guinea	Oil	46	1903	1949	10,071	11,974	12,020	16
84	6107	Lesser Antilles Deformed Belt	Oil	16	7	22	11,957	11,963	11,979	0
85	6063	Malvinas Basin	Oil	0	75	75	11,327	11,402	11,402	1
86	4061	Carpathian-Balkanian Basin	Oil	3126	4139	7265	3179	7319	10,445	70
87	6029	Sergipe-Alagoas Basin	Oil	348	1059	1407	8334	9393	9741	14
88	7303	Orange River Coastal	Gas	0	6015	6015	3603	9618	9618	63
89	5301	Tampico-Misantla Basin	Oil	2354	4306	6661	2051	6358	8712	76
90	6090	Middle Magdalena	Oil	2310	2692	5002	3292	5985	8294	60
91	2029	Wadi-Surhan Basin	Gas	47	179	226	7873	8052	8099	3
92	3808	Central Sumatra Basin	Oil	910	2956	3866	4078	7034	7944	49
93	4033	German-Polish Basin	Gas	6905	940	7846	0	940	7846	100
94	8034	Assam	Oil	2039	4495	6534	1178	5673	7712	85
95	6081	Talara Basin	Oil	1951	941	2892	4795	5736	7687	38
96	3144	Songliao Basin	Oil	3	1739	1743	5706	7446	7449	23
97	6058	San Jorge Basin	Oil	1667	1898	3565	3774	5672	7339	49
98	2009	Masila-Jeza Basin	Oil	0	40	40	6793	6833	6833	1
99	5302	Veracruz Basin	Oil	335	270	605	5832	6102	6437	9
100	3128	Ordos Basin	Oil	13	5597	5610	20	5617	5630	100
101	2010	Ghudun-Khasfeh Flank Province	Oil	0	0	0	5623	5623	5623	0
102	5307	Campeche-Sigsbe Salt Basin	Oil	0	0	0	5565	5565	5565	0
103	1211	Cis-Patom Foredeep	Gas	0	100	100	5364	5464	5464	2
104	2006	Shabwah Basin	Oil	0	573	573	4246	4819	4819	12
105	2015	Central Oman Platform	Oil	0	742	742	3239	3980	3981	19
106	3115	Junggar Basin	Oil	501	2082	2583	1296	3379	3879	67
107	8026	Kohat-Potwar	Oil	784	1147	1931	1314	2460	3245	60
108	6041	Putumayo-Oriente-Maranon Basin	Oil	661	955	1616	1596	2552	3212	50
109	2074	Khleisha Uplift	Oil	0	210	210	2580	2790	2790	8
110	2075	Euphrates/Mardin	Oil	66	1668	1734	749	2418	2483	70
111	5308	Yucatan Platform	Oil	0	0	0	1496	1496	1496	0
112	2011	South Oman Salt Basin	Oil	0	188	188	1236	1424	1424	13
113	2025	Hail-Ga'Ara Arch	Gas	0	0	0	1290	1290	1290	0
114	2028	Rutbah Uplift	Gas	0	0	0	1290	1290	1290	0

App-3e_Page 3

Rank	Province Code	Province Name	Major Commodity	Cumulative Gas (BCFG)	Remaining Gas (BCFG)	Known Gas (BCFG)	Mean Undiscovered Gas (BCFG)	Future Gas (BCFG)	Gas Endowment (BCFG)	Discovery Maturity, Gas (%)
115	6083	Progreso Basin	Oil	148	350	498	695	1044	1193	42
116	2031	Zagros Thrust Zone	Gas	0	0	0	966	966	966	0
117	7013	Senegal	Oil	4	45	49	856	901	905	5
118	6117	Greater Antilles Deformed Belt	Oil	20	77	96	592	668	688	14
119	5246	Mackenzie Foldbelt	Gas	379	252	631	43	295	674	94
120	2013	Huqf-Haushi Uplift	Oil	0	0	0	404	404	404	0
121	8023	Central Afghanistan	Gas	27	244	271	0	244	271	100
122	2047	Hamra Basin	Oil	0	0	0	241	241	241	0
123	4046	Bohemia	Gas	114	117	231	0	117	231	100
124	5306	Macuspana Basin	Oil	8	8	16	210	217	225	7
125	4038	Munsterland Basin	Gas	121	31	152	0	31	152	100
126	3916	Carnarvon Basin	Oil	32	70	102	0	70	102	100
127	2089	Anah Graben	Oil	0	0	0	39	39	39	0
128	5317	Chicontepec Basin	Oil	0	2	2	32	34	34	6
129	2017	Oman Mountains	Oil	0	0	0	22	22	22	0
130	8035	North Burma	Oil	1	10	11	0	10	11	100
131	2061	Ougarta Uplift	Oil	0	6	6	0	6	6	100
132	2033	Sinai Basin	Oil	0	5	5	0	5	5	100
NA	1158	Tian Shan Foldbelt	NA	NA	NA	NA	0	NA	NA	NA
NA	1059	Ludlov Saddle	NA	NA	NA	NA	28,387	NA	NA	NA
NA	3161	Yunnan Guizhou Hubei Fold Belt	NA	NA	NA	NA	0	NA	NA	NA
NA	4028	Mid-North Sea High	NA	NA	NA	NA	0	NA	NA	NA
NA	4039	Southwest German Basin	NA	NA	NA	NA	0	NA	NA	NA
NA	5310	Sierra Madre de Chiapas-Peten Foldbelt	NA	NA	NA	NA	1846	NA	NA	NA
NA	5316	Trans-Mexican Neovolcanic Axis	NA	NA	NA	NA	195	NA	NA	NA
NA	6022	Foz do Amazonas Basin	NA	NA	NA	NA	29,840	NA	NA	NA
NA	8002	Himalayan	NA	NA	NA	NA	0	NA	NA	NA

APPENDIX 3F. Provinces ranked by conventional total petroleum endowment.

[MMBOE = million barrels of oil equivalent. Values given are not necessarily those for the entire province, but only for that part assessed in World Petroleum Assessment 2000 (USGS, 2000). Major commodity is based on the ratio of gas endowment to oil endowment as ft^3 of gas per bbl of oil (CFG/BO). Major commodity is oil if this ratio is less than 20,000 CFG/BO and gas if this ratio is equal to or greater than 20,000 CFG/BO. Total cumulative petroleum is the reported volume of total petroleum that has been produced. Total remaining petroleum is the calculated remaining reserves; the difference between total known and total cumulative petroleum. Total known petroleum is the reported discovered volume (cumulative production plus remaining reserves). Total mean undiscovered petroleum is derived from the mean volumes of oil, gas, and NGL estimated in World Petroleum Assessment 2000 (USGS, 2000). Total future petroleum is the total remaining petroleum plus estimated total mean undiscovered petroleum. Total petroleum endowment is the total known petroleum plus estimated total mean undiscovered petroleum volumes. Discovery maturity is the percentage of total petroleum discovered (total known petroleum) with respect to total petroleum endowment. Calculated values may not be equal to the sums of the component values because numbers have been independently rounded. NA means not applicable; it is shown for all values where there is only one field present, and used where the endowment could not be calculated.] Known volumes from Petroconsultants (1996) and NRG Associates (1995). Note that gas = natural gas.

Rank	Province Code	Province Name	Major Commodity	Total Cumulative Petroleum (MMBOE)	Total Remaining Petroleum (MMBOE)	Total Known Petroleum (MMBOE)	Total Mean Undiscovered Petroleum (MMBOE)	Total Future Petroleum (MMBOE)	Total Petroleum Endowment (MMBOE)	Discovery Maturity, Total Petroleum (%)
1	1174	West Siberian Basin	Oil	83,213	271,517	354,730	182,864	454,380	537,593	66
2	2024	Mesopotamian Foredeep Basin	Oil	58,934	285,059	343,993	78,159	363,217	422,151	81
3	2030	Zagros Fold Belt	Oil	51,102	138,437	189,539	90,292	228,729	279,831	68
4	2019	Rub Al Khali Basin	Oil	20,998	101,760	122,757	137,868	239,627	260,625	47
5	2021	Greater Ghawar Uplift	Oil	62,169	129,511	191,680	63,808	193,319	255,488	75
6	7192	Niger Delta	Oil	17,646	35,353	53,000	68,640	103,994	121,640	44
7	2022	Qatar Arch	Gas	183	92,332	92,515	8908	101,239	101,423	91
8	4025	North Sea Graben	Oil	28,183	48,322	76,505	21,248	69,570	97,753	78
9	1016	North Caspian Basin	Oil	1322	44,528	45,850	48,573	93,101	94,423	49
10	1015	Volga-Ural Region	Oil	45,732	35,832	81,564	3403	39,235	84,968	96
11	6098	East Venezuela Basin	Oil	17,509	34,747	52,257	31,934	66,682	84,191	62
12	1112	South Caspian Basin	Oil	13,334	10,607	23,941	52,725	63,332	76,665	31
13	1154	Amu-Darya Basin	Gas	13,357	26,975	40,332	31,541	58,516	71,873	56
14	6099	Maracaibo Basin	Oil	35,745	17,820	53,565	12,086	29,906	65,651	82
15	5200	East Greenland Rift Basins	Oil	0	0	0	64,769	64,769	64,769	0
16	7203	West-Central Coastal	Oil	5957	9854	15,811	48,665	58,519	64,476	25
17	2023	Widyan Basin-Interior Platform	Oil	229	18,457	18,686	43,855	62,312	62,541	30
18	2043	Sirte Basin	Oil	14,929	28,566	43,496	10,000	38,566	53,496	81
19	5305	Villahermosa Uplift	Oil	17,004	24,993	41,997	10,957	35,950	52,955	79
20	2020	Interior Homocline-Central Arch	Oil	84	6516	6601	46,132	52,648	52,733	13
21	4017	Vestford-Helgeland	Oil	76	5897	5973	43,062	48,959	49,035	12
22	1050	South Barents Basin	Gas	0	11,770	11,770	30,377	42,147	42,147	28
23	6036	Santos Basin	Oil	12	505	517	40,828	41,332	41,344	1
24	1008	Timan-Pechora Basin	Oil	5771	14,171	19,942	16,598	30,769	36,539	55

Rank	Province Code	Province Name	Major Commodity	Total Cumulative Petroleum (MMBOE)	Total Remaining Petroleum (MMBOE)	Total Known Petroleum (MMBOE)	Total Mean Undiscovered Petroleum (MMBOE)	Total Future Petroleum (MMBOE)	Total Petroleum Endowment (MMBOE)	Discovery Maturity, Total Petroleum (%)
25	5243	Alberta Basin	Oil	22,342	8760	31,101	3128	11,888	34,229	91
26	3127	Bohaiwan Basin	Oil	10,914	16,339	27,253	6248	22,587	33,501	81
27	6035	Campos Basin	Oil	1819	9287	11,107	20,128	29,415	31,235	36
28	2054	Trias/Ghadames Basin	Oil	7589	12,800	20,389	10,566	23,366	30,955	66
29	2071	Red Sea Basin	Oil	6445	5298	11,744	18,533	23,832	30,277	39
30	3817	Kutei Basin	Oil	3691	8040	11,731	18,060	26,100	29,791	39
31	3948	Northwest Shelf	Gas	896	10,589	11,485	16,848	27,437	28,333	41
32	4035	Northwest German Basin	Gas	14,589	11,340	25,929	1881	13,221	27,810	93
33	1109	Middle Caspian Basin	Oil	6071	8083	14,153	13,171	21,254	27,324	52
34	2058	Grand Erg/Ahnet Basin	Gas	1783	22,752	24,534	2449	25,201	26,984	91
35	6021	Guyana-Suriname Basin	Oil	17	46	63	24,586	24,632	24,649	0
36	3701	Baram Delta/Brunei-Sabah Basin	Oil	5492	7619	13,111	8434	16,054	21,545	61
37	1322	North Sakhalin Basin	Oil	785	5292	6077	15,156	20,449	21,234	29
38	3154	Tarim Basin	Oil	80	1776	1856	19,250	21,026	21,106	9
39	2056	Illizi Basin	Oil	2485	9593	12,078	8536	18,129	20,614	59
40	3702	Greater Sarawak Basin	Gas	846	14,044	14,890	4301	18,345	19,191	78
41	3703	Malay Basin	Oil	1609	10,346	11,955	6319	16,665	18,274	65
42	3144	Songliao Basin	Oil	12,399	3466	15,865	2189	5655	18,054	88
43	1009	Dnieper-Donets Basin	Gas	6137	5523	11,660	5969	11,492	17,629	66
44	8043	Bombay	Oil	4417	8323	12,740	4666	12,989	17,406	73
45	3808	Central Sumatra Basin	Oil	8500	5372	13,872	1663	7034	15,535	89
46	4036	Anglo-Dutch Basin	Gas	7237	5403	12,640	2630	8033	15,270	83
47	1210	Nepa-Botuoba Arch	Oil	1	5974	5976	7643	13,618	13,619	44
48	6096	Llanos Basin	Oil	1202	7203	8405	5004	12,207	13,409	63
49	5304	Saline-Comalcalco Basin	Oil	1844	889	2733	10,650	11,539	13,383	20
50	6045	Santa Cruz-Tarija Basin	Oil	1655	2045	3699	8629	10,674	12,328	30
51	8047	Ganges-Brahmaputra Delta	Gas	451	1654	2105	9558	11,212	11,663	18
52	1060	North Barents Basin	Gas	0	0	0	11,355	11,355	11,355	0
53	4068	Provence Basin	Gas	0	0	0	11,188	11,188	11,188	0
54	6034	Espirito Santo Basin	Oil	100	68	169	10,285	10,354	10,454	2
55	6041	Putumayo-Oriente-Maranon Basin	Oil	2999	3871	6871	3418	7290	10,289	67
56	6055	Neuquen Basin	Oil	2883	3276	6159	3562	6838	9721	63
57	3910	Bonaparte Gulf Basin	Gas	123	3046	3169	6281	9327	9450	34
58	5301	Tampico-Misantla Basin	Oil	6165	1840	8005	1354	3194	9359	86
59	6060	Falklands Plateau	Oil	0	0	0	8967	8967	8967	0
60	6059	Magallanes Basin	Gas	2359	3202	5562	3354	6556	8915	62
61	4061	Carpathian-Balkanian Basin	Oil	4624	2571	7195	1659	4230	8854	81
62	3822	North Sumatra Basin	Gas	3409	2451	5860	2914	5365	8774	67
63	3913	Browse Basin	Gas	19	3221	3239	5339	8559	8578	38
64	2016	Fahud Salt Basin	Oil	2754	3535	6288	2008	5543	8297	76
65	3115	Junggar Basin	Oil	1035	6206	7241	843	7049	8084	90

Rank	Province Code	Province Name	Major Commodity	Total Cumulative Petroleum (MMBOE)	Total Remaining Petroleum (MMBOE)	Total Known Petroleum (MMBOE)	Total Mean Undiscovered Petroleum (MMBOE)	Total Future Petroleum (MMBOE)	Total Petroleum Endowment (MMBOE)	Discovery Maturity, Total Petroleum (%)
66	3828	South Sumatra Basin	Oil	2367	1819	4186	3750	5569	7936	53
67	6037	Pelotas Basin	Oil	0	0	0	7846	7846	7846	0
68	3930	Gippsland Basin	Oil	3815	2375	6190	1592	3967	7782	80
69	2048	Pelagian Basin	Oil	528	4646	5174	2323	6970	7497	69
70	6103	Tobago Trough	Gas	0	3867	3867	3551	7418	7418	52
71	8048	Irrawaddy	Gas	649	1797	2446	4930	6726	7376	33
72	4060	Po Basin	Gas	1706	1820	3526	3476	5296	7002	50
73	8042	Indus	Gas	1512	1970	3482	3328	5299	6811	51
74	3824	Northwest Java Basin	Oil	1771	2914	4685	2007	4922	6692	70
75	5244	Williston Basin, Canada	Oil	4468	1963	6431	252	2215	6683	96
76	5215	Labrador-Newfoundland Shelf	Oil	0	2875	2875	3792	6667	6667	48
77	2004	Ma'Rib-Al Jawf/Masila Basin	Oil	536	3712	4248	2273	5985	6521	65
78	2014	Ghaba Salt Basin	Oil	489	4492	4981	1436	5928	6416	78
79	1108	Azov-Kuban Basin	Gas	2589	1091	3681	2675	3766	6355	58
80	1150	North Ustyurt Basin	Oil	459	2318	2777	3193	5511	5970	47
81	1207	Baykit Arch	Oil	0	936	936	4897	5833	5833	16
82	4048	Pannonian Basin	Oil	2691	1892	4583	1178	3070	5761	80
83	4057	Transylvanian Basin	Gas	3532	1589	5122	354	1943	5475	94
84	8025	Sulaiman-Kirthar	Gas	1121	1535	2657	2553	4089	5210	51
85	6058	San Jorge Basin	Oil	2723	1218	3942	1203	2421	5144	77
86	4047	North Carpathian Basin	Oil	2104	1559	3663	1344	2903	5007	73
87	6081	Talara Basin	Oil	1715	439	2154	2765	3204	4919	44
88	6090	Middle Magdalena	Oil	2300	1028	3328	1388	2416	4715	71
89	3142	Sichuan Basin	Gas	1086	1112	2198	2202	3313	4400	50
90	8034	Assam	Oil	1251	2379	3630	663	3042	4292	85
91	6029	Sergipe-Alagoas Basin	Oil	528	402	929	3292	3693	4221	22
92	6107	Lesser Antilles Deformed Belt	Oil	11	3	14	4110	4113	4124	0
93	1209	Angara-Lena Terrace	Gas	0	1409	1409	2526	3935	3935	36
94	5245	Rocky Mountain Deformed Belt	Gas	1916	1199	3115	613	1812	3728	84
95	5307	Campeche-Sigsbe Salt Basin	Oil	0	0	0	3714	3714	3714	0
96	7183	Gulf of Guinea	Oil	81	515	596	2965	3480	3561	17
97	6063	Malvinas Basin	Oil	0	25	25	3280	3305	3305	1
98	2009	Masila-Jeza Basin	Oil	144	496	641	2598	3094	3238	20
99	5302	Veracruz Basin	Oil	107	125	232	2214	2339	2446	9
100	2010	Ghudun-Khasfeh Flank Province	Oil	0	0	0	2258	2258	2258	0
101	2029	Wadi-Surhan Basin	Gas	8	30	38	1942	1972	1979	2
102	7303	Orange River Coastal	Gas	0	1003	1003	879	1882	1882	53
103	2006	Shabwah Basin	Oil	0	210	210	1624	1833	1833	11
104	2075	Euphrates/Mardin	Oil	293	607	900	893	1500	1793	50
105	3128	Ordos Basin	Oil	83	1535	1618	143	1678	1761	92
106	2074	Khleisha Uplift	Oil	250	442	692	739	1180	1430	48

Rank	Province Code	Province Name	Major Commodity	Total Cumulative Petroleum (MMBOE)	Total Remaining Petroleum (MMBOE)	Total Known Petroleum (MMBOE)	Total Mean Undiscovered Petroleum (MMBOE)	Total Future Petroleum (MMBOE)	Total Petroleum Endowment (MMBOE)	Discovery Maturity, Total Petroleum (%)
107	1211	Cis-Patom Foredeep	Gas	0	17	17	1309	1325	1325	1
108	4033	German-Polish Basin	Gas	1151	157	1308	0	157	1308	100
109	2015	Central Oman Platform	Oil	3	198	200	1090	1287	1290	16
110	8026	Kohat-Potwar	Oil	341	343	684	422	765	1106	62
111	5308	Yucatan Platform	Oil	0	0	0	1092	1092	1092	0
112	6117	Greater Antilles Deformed Belt	Oil	68	245	313	628	873	940	33
113	2011	South Oman Salt Basin	Oil	203	343	546	357	701	903	60
114	6083	Progreso Basin	Oil	147	111	258	385	497	644	40
115	2025	Hail-Ga'Ara Arch	Gas	0	0	0	369	369	369	0
116	2028	Rutbah Uplift	Gas	0	0	0	369	369	369	0
117	7013	Senegal	Oil	1	18	19	343	361	362	5
118	2031	Zagros Thrust Zone	Gas	0	0	0	242	242	242	0
119	2089	Anah Graben	Oil	75	80	155	35	115	190	81
120	5306	Macuspana Basin	Oil	4	5	9	149	154	158	6
121	2047	Hamra Basin	Oil	0	0	0	148	148	148	0
122	2013	Huqf-Haushi Uplift	Oil	0	0	0	115	115	115	0
123	5246	Mackenzie Foldbelt	Gas	63	42	105	8	50	113	93
124	2033	Sinai Basin	Oil	94	12	106	0	12	106	100
125	3916	Carnarvon Basin	Oil	21	31	52	0	31	52	100
126	8023	Central Afghanistan	Gas	5	41	45	0	41	45	100
127	4046	Bohemia	Gas	21	22	43	0	22	43	100
128	2061	Ougarta Uplift	Oil	0	42	42	0	42	42	100
129	2017	Oman Mountains	Oil	0	0	0	27	27	27	0
130	4038	Munsterland Basin	Gas	20	5	25	0	5	25	100
131	5317	Chicontepec Basin	Oil	0	0	0	22	22	22	2
132	8035	North Burma	Oil	2	2	3	0	2	3	100
NA	1158	Tian Shan Foldbelt	NA	NA	NA	NA	0	NA	NA	NA
NA	1059	Ludlov Saddle	NA	NA	NA	NA	5361	NA	NA	NA
NA	3161	Yunnan Guizhou Hubei Fold Belt	NA	NA	NA	NA	0	NA	NA	NA
NA	4028	Mid-North Sea High	NA	NA	NA	NA	0	NA	NA	NA
NA	4039	Southwest German Basin	NA	NA	NA	NA	0	NA	NA	NA
NA	5310	Sierra Madre de Chiapas-Peten Foldbelt	NA	NA	NA	NA	1307	NA	NA	NA
NA	5316	Trans-Mexican Neovolcanic Axis	NA	NA	NA	NA	72	NA	NA	NA
NA	6022	Foz do Amazonas Basin	NA	NA	NA	NA	5272	NA	NA	NA
NA	8002	Himalayan	NA	NA	NA	NA	0	NA	NA	NA

Appendix 3g. Provinces ranked by future oil volume.

[MMBO = million barrels of oil. Values given are not necessarily those for the entire province, but only for that part assessed in World Petroleum Assessment 2000 (USGS, 2000). Major commodity is based on the ratio of gas endowment to oil endowment as ft^3 of gas per bbl of oil (CFG/BO). Major commodity is oil if this ratio is less than 20,000 CFG/BO and gas if this ratio is equal to or greater than 20,000 CFG/BO. Cumulative oil is the reported volume of oil that has been produced. Remaining oil is the calculated remaining reserves; the difference between known and cumulative oil. Known oil is the reported discovered volume (cumulative production plus remaining reserves). Mean undiscovered oil is the mean volume estimated in World Petroleum Assessment 2000 (USGS, 2000). Future oil is the remaining oil plus estimated mean undiscovered oil. Oil endowment is the known oil plus estimated mean undiscovered oil volumes. Discovery maturity is the percentage of oil discovered (known oil) with respect to oil endowment. Calculated values may not be equal to the sums of the component values because numbers have been independently rounded. Negative values result from inconsistent reporting of cumulative and known volumes. Values in parentheses are questionable data. NA means not applicable; it is shown for all values where there is only one field present, and used where the endowment could not be calculated.] Known volumes from Petroconsultants (1996) and NRG Associates (1995). Note that gas = natural gas.

Rank	Province Code	Province Name	Major Commodity	Cumulative Oil (MMBO)	Remaining Oil (MMBO)	Known Oil (MMBO)	Mean Undiscovered Oil (MMBO)	Future Oil (MMBO)	Oil Endowment (MMBO)	Discovery Maturity, Oil (%)
1	2024	Mesopotamian Foredeep Basin	Oil	56,746	235,696	292,442	61,134	296,830	353,576	83
2	1174	West Siberian Basin	Oil	46,530	93,383	139,913	55,235	148,618	195,148	72
3	2030	Zagros Fold Belt	Oil	49,367	72,234	121,601	45,459	117,693	167,060	73
4	2019	Rub Al Khali Basin	Oil	19,320	70,507	89,827	36,647	107,154	126,474	71
5	2021	Greater Ghawar Uplift	Oil	57,744	83,956	141,700	13,830	97,786	155,530	91
6	7192	Niger Delta	Oil	16,598	17,925	34,522	40,487	58,412	75,010	46
7	5200	East Greenland Rift Basins	Oil	0	0	0	47,148	47,148	47,148	0
8	2023	Widyan Basin-Interior Platform	Oil	219	17,216	17,435	21,215	38,431	38,650	45
9	7203	West-Central Coastal	Oil	5765	7952	13,717	29,747	37,699	43,464	32
10	4025	North Sea Graben	Oil	22,467	21,427	43,895	13,098	34,526	56,993	77
11	1016	North Caspian Basin	Oil	822	9987	10,808	23,320	33,307	34,128	32
12	2043	Sirte Basin	Oil	13,986	23,087	37,072	6854	29,941	43,927	84
13	6098	East Venezuela Basin	Oil	13,808	16,395	30,203	11,875	28,271	42,078	72
14	5305	Villahermosa Uplift	Oil	14,640	20,382	35,022	7436	27,818	42,457	82
15	6035	Campos Basin	Oil	1769	8287	10,056	16,293	24,580	26,349	38
16	1015	Volga-Ural Region	Oil	42,690	21,247	63,937	2549	23,796	66,485	96
17	6036	Santos Basin	Oil	6	279	285	23,209	23,488	23,494	1
18	6099	Maracaibo Basin	Oil	35,581	13,491	49,072	8183	21,675	57,256	86
19	1112	South Caspian Basin	Oil	11,681	5758	17,438	15,725	21,483	33,163	53
20	2020	Interior Homocline-Central Arch	Oil	84	4616	4700	16,495	21,111	21,195	22
21	3127	Bohaiwan Basin	Oil	10,415	14,139	24,553	4412	18,550	28,965	85
22	1008	Timan-Pechora Basin	Oil	3346	9774	13,120	5732	15,506	18,852	70
23	6021	Guyana-Suriname Basin	Oil	17	46	63	15,247	15,293	15,310	0
24	2054	Trias/Ghadames Basin	Oil	7410	7793	15,203	6477	14,270	21,680	70
25	1109	Middle Caspian Basin	Oil	4359	5193	9551	4887	10,080	14,439	66
26	4017	Vestford-Helgeland	Oil	76	2585	2660	6538	9123	9199	29
27	2071	Red Sea Basin	Oil	6280	3580	9861	5187	8767	15,048	66
28	6096	Llanos Basin	Oil	1179	4224	5402	3631	7855	9033	60

Rank	Province Code	Province Name	Major Commodity	Cumulative Oil (MMBO)	Remaining Oil (MMBO)	Known Oil (MMBO)	Mean Undiscovered Oil (MMBO)	Future Oil (MMBO)	Oil Endowment (MMBO)	Discovery Maturity, Oil (%)
29	5304	Saline-Comalcalco Basin	Oil	1673	686	2359	7018	7703	9377	25
30	8043	Bombay	Oil	3375	5065	8440	2091	7156	10,531	80
31	6041	Putumayo-Oriente-Maranon Basin	Oil	2889	3712	6601	3098	6810	9699	68
32	3154	Tarim Basin	Oil	45	658	704	5948	6606	6652	11
33	3115	Junggar Basin	Oil	952	5858	6810	563	6422	7373	92
34	3701	Baram Delta/Brunei-Sabah Basin	Oil	3953	2945	6898	3390	6335	10,288	67
35	2022	Qatar Arch	Gas	0	1150	1150	5026	6176	6176	19
36	6060	Falklands Plateau	Oil	0	0	0	5833	5833	5833	0
37	3817	Kutei Basin	Oil	2195	684	2879	5145	5829	8024	36
38	3808	Central Sumatra Basin	Oil	8349	4868	13,217	899	5768	14,117	94
39	2056	Illizi Basin	Oil	2286	1384	3670	2947	4331	6617	55
40	3144	Songliao Basin	Oil	12,399	3176	15,575	1026	4203	16,601	94
41	1322	North Sakhalin Basin	Oil	754	1428	2182	2582	4010	4764	46
42	5215	Labrador-Newfoundland Shelf	Oil	0	1582	1582	1796	3378	3378	47
43	3703	Malay Basin	Oil	1518	2090	3608	1228	3318	4836	75
44	5243	Alberta Basin	Oil	12,821	2458	15,279	818	3276	16,097	95
45	1210	Nepa-Botuoba Arch	Oil	1	2005	2006	1208	3213	3214	62
46	3948	Northwest Shelf	Gas	473	647	1119	2381	3028	3500	32
47	6034	Espirito Santo Basin	Oil	93	37	130	2959	2995	3089	4
48	2048	Pelagian Basin	Oil	521	1719	2239	1268	2986	3507	64
49	2016	Fahud Salt Basin	Oil	2555	1918	4473	1020	2938	5493	81
50	6037	Pelotas Basin	Oil	0	0	0	2938	2938	2938	0
51	1150	North Ustyurt Basin	Oil	459	1907	2366	954	2861	3320	71
52	4061	Carpathian-Balkanian Basin	Oil	4083	1825	5908	1013	2838	6921	85
53	1009	Dnieper-Donets Basin	Gas	110	1501	1611	1098	2599	2709	59
54	5307	Campeche-Sigsbe Salt Basin	Oil	0	0	0	2478	2478	2478	0
55	3824	Northwest Java Basin	Oil	1555	1613	3168	658	2271	3825	83
56	6045	Santa Cruz-Tarija Basin	Oil	257	63	320	2145	2209	2466	13
57	6055	Neuquen Basin	Oil	1532	838	2370	1290	2128	3659	65
58	6081	Talara Basin	Oil	1346	326	1672	1711	2038	3383	49
59	8034	Assam	Oil	911	1625	2536	399	2024	2935	86
60	5301	Tampico-Misantla Basin	Oil	5773	1122	6895	893	2015	7787	89
61	6029	Sergipe-Alagoas Basin	Oil	469	223	692	1487	1711	2179	32
62	3910	Bonaparte Gulf Basin	Gas	123	393	517	1286	1679	1803	29
63	2004	Ma'Rib-Al Jawf/Masila Basin	Oil	504	683	1188	959	1643	2147	55
64	2009	Masila-Jeza Basin	Oil	144	488	632	1096	1584	1728	37
65	6107	Lesser Antilles Deformed Belt	Oil	8	2	11	1536	1538	1546	1
66	6058	San Jorge Basin	Oil	2445	900	3346	498	1398	3844	87
67	6059	Magallanes Basin	Gas	595	620	1216	704	1325	1920	63
68	3702	Greater Sarawak Basin	Gas	168	649	817	631	1281	1448	56
69	2014	Ghaba Salt Basin	Oil	466	878	1344	371	1250	1715	78
70	1207	Baykit Arch	Oil	0	34	34	1213	1246	1246	3
71	6090	Middle Magdalena	Oil	1915	512	2426	709	1220	3135	77

Rank	Province Code	Province Name	Major Commodity	Cumulative Oil (MMBO)	Remaining Oil (MMBO)	Known Oil (MMBO)	Mean Undiscovered Oil (MMBO)	Future Oil (MMBO)	Oil Endowment (MMBO)	Discovery Maturity, Oil (%)
72	7183	Gulf of Guinea	Oil	74	162	236	1004	1166	1240	19
73	3930	Gippsland Basin	Oil	3077	784	3860	310	1093	4170	93
74	3913	Browse Basin	Gas	19	27	46	1055	1083	1101	4
75	2075	Euphrates/Mardin	Oil	282	312	594	749	1061	1343	44
76	3828	South Sumatra Basin	Oil	1842	587	2429	469	1055	2898	84
77	6063	Malvinas Basin	Oil	0	10	10	1031	1041	1041	1
78	5302	Veracruz Basin	Oil	51	79	130	955	1034	1085	12
79	2058	Grand Erg/Ahnet Basin	Gas	106	388	494	599	986	1093	45
80	4048	Pannonian Basin	Oil	1567	623	2190	359	982	2550	86
81	5244	Williston Basin, Canada	Oil	2696	809	3506	154	963	3660	96
82	1154	Amu-Darya Basin	Gas	80	686	766	270	957	1037	74
83	2010	Ghudun-Khasfeh Flank Province	Oil	0	0	0	926	926	926	0
84	8048	Irrawaddy	Gas	527	188	715	724	912	1439	50
85	2006	Shabwah Basin	Oil	0	113	113	685	798	798	14
86	5308	Yucatan Platform	Oil	0	0	0	754	754	754	0
87	3128	Ordos Basin	Oil	81	602	683	138	740	821	83
88	6117	Greater Antilles Deformed Belt	Oil	65	232	297	494	726	790	38
89	4047	North Carpathian Basin	Oil	649	286	935	391	677	1326	71
90	4060	Po Basin	Gas	91	254	346	360	615	706	49
91	2074	Khleisha Uplift	Oil	250	407	657	105	512	762	86
92	2015	Central Oman Platform	Oil	3	68	71	424	493	495	14
93	4035	Northwest German Basin	Gas	1867	413	2280	56	468	2336	98
94	2011	South Oman Salt Basin	Oil	203	297	500	97	394	596	84
95	3822	North Sumatra Basin	Gas	496	178	674	213	391	887	76
96	4068	Provence Basin	Gas	0	0	0	388	388	388	0
97	6083	Progreso Basin	Oil	122	53	175	237	290	412	43
98	8026	Kohat-Potwar	Oil	201	120	321	149	268	469	68
99	4036	Anglo-Dutch Basin	Gas	432	159	592	88	247	679	87
100	1209	Angara-Lena Terrace	Gas	0	0	0	242	242	242	0
101	1050	South Barents Basin	Gas	0	0	0	219	219	219	0
102	8042	Indus	Gas	111	70	180	103	173	283	64
103	7013	Senegal	Oil	0	10	10	157	167	168	6
104	1211	Cis-Patom Foredeep	Gas	0	0	0	161	161	161	0
105	1060	North Barents Basin	Gas	0	0	0	135	135	135	0
106	7303	Orange River Coastal	Gas	0	0	0	116	116	116	0
107	2089	Anah Graben	Oil	75	80	155	26	106	181	85
108	5306	Macuspana Basin	Oil	2	4	6	101	105	108	6
109	8047	Ganges-Brahmaputra Delta	Gas	3	3	5	103	105	108	5
110	3142	Sichuan Basin	Gas	47	63	110	27	91	138	80
111	2047	Hamra Basin	Oil	0	0	0	89	89	89	0
112	2025	Hail-Ga'Ara Arch	Gas	0	0	0	53	53	53	0
113	2028	Rutbah Uplift	Gas	0	0	0	53	53	53	0
114	2061	Ougarta Uplift	Oil	0	41	41	0	41	41	100

Rank	Province Code	Province Name	Major Commodity	Cumulative Oil (MMBO)	Remaining Oil (MMBO)	Known Oil (MMBO)	Mean Undiscovered Oil (MMBO)	Future Oil (MMBO)	Oil Endowment (MMBO)	Discovery Maturity, Oil (%)
115	2031	Zagros Thrust Zone	Gas	0	0	0	38	38	38	0
116	2013	Huqf-Haushi Uplift	Oil	0	0	0	30	30	30	0
117	5245	Rocky Mountain Deformed Belt	Gas	176	22	198	6	28	204	97
118	2017	Oman Mountains	Oil	0	0	0	22	22	22	0
119	6103	Tobago Trough	Gas	0	20	20	0	20	20	100
120	3916	Carnarvon Basin	Oil	16	19	35	0	19	35	100
121	5317	Chicontepec Basin	Oil	0	0	0	15	15	15	1
122	8025	Sulaiman-Kirthar	Gas	0	0	0	12	12	12	0
123	2033	Sinai Basin	Oil	94	11	105	0	11	105	100
124	2029	Wadi-Surhan Basin	Gas	0	0	0	0	0	0	NA
125	4033	German-Polish Basin	Gas	0	0	0	0	0	0	100
126	4038	Munsterland Basin	Gas	0	0	0	0	0	0	NA
127	4046	Bohemia	Gas	1	0	1	0	0	1	100
128	4057	Transylvanian Basin	Gas	0	0	0	0	0	0	NA
129	5246	Mackenzie Foldbelt	Gas	0	0	0	0	0	0	NA
130	8023	Central Afghanistan	Gas	0	0	0	0	0	0	NA
131	8035	North Burma	Oil	1	0	1	0	0	1	100
132	1108	Azov-Kuban Basin	Gas	1273	(-750)	524	417	(-332)	941	56
NA	1158	Tian Shan Foldbelt	NA	NA	NA	NA	0	NA	NA	NA
NA	1059	Ludlov Saddle	NA	NA	NA	NA	39	NA	NA	NA
NA	3161	Yunnan Guizhou Hubei Fold Belt	NA	NA	NA	NA	0	NA	NA	NA
NA	4028	Mid-North Sea High	NA	NA	NA	NA	0	NA	NA	NA
NA	4039	Southwest German Basin	NA	NA	NA	NA	0	NA	NA	NA
NA	5310	Sierra Madre de Chiapas-Peten Foldbelt	NA	NA	NA	NA	890	NA	NA	NA
NA	5316	Trans-Mexican Neovolcanic Axis	NA	NA	NA	NA	30	NA	NA	NA
NA	6022	Foz do Amazonas Basin	NA	NA	NA	NA	0	NA	NA	NA
NA	8002	Himalayan	NA	NA	NA	NA	0	NA	NA	NA

APPENDIX 3H. Provinces ranked by future natural gas volume.

[BCFG = billion cubic feet of gas. Values given are not necessarily those for the entire province, but only for that part assessed in World Petroleum Assessment 2000 (USGS, 2000). Major commodity is based on the ratio of gas endowment to oil endowment as ft^3 of gas per bbl of oil (CFG/BO). Major commodity is oil if this ratio is less than 20,000 CFG/BO and gas if this ratio is equal to or greater than 20,000 CFG/BO. Cumulative gas is the reported volume of gas that has been produced. Remaining gas is the calculated remaining reserves; the difference between known and cumulative gas. Known gas is the reported discovered volume (cumulative production plus remaining reserves). Mean undiscovered gas is the mean volume estimated in World Petroleum Assessment 2000 (USGS, 2000). Future gas is the remaining gas plus estimated mean undiscovered gas. Gas endowment is the known gas plus estimated mean undiscovered gas volumes. Discovery maturity is the percentage of gas discovered (known gas) with respect to gas endowment. Calculated values may not be equal to the sums of the component values because numbers have been independently rounded. Negative values result from inconsistent reporting of cumulative and known volumes. Values in parentheses are questionable data. NA means not applicable; it is shown for all values where there is only one field present, and used where the endowment could not be calculated.] Known volumes from Petroconsultants (1996) and NRG Associates (1995). Note that gas = natural gas.

Rank	Province Code	Province Name	Major Commodity	Cumulative Gas (BCFG)	Remaining Gas (BCFG)	Known Gas (BCFG)	Mean Undiscovered Gas (BCFG)	Future Gas (BCFG)	Gas Endowment (BCFG)	Discovery Maturity, Gas (%)
1	1174	West Siberian Basin	Oil	218,465	1,051,584	1,270,049	642,934	1,694,518	1,912,983	66
2	2030	Zagros Fold Belt	Oil	10,333	389,111	399,444	212,008	601,119	611,451	65
3	2019	Rub Al Khali Basin	Oil	8303	173,934	182,237	425,712	599,646	607,949	30
4	2022	Qatar Arch	Gas	1018	464,582	465,600	17,385	481,967	482,985	96
5	2021	Greater Ghawar Uplift	Oil	26,493	222,057	248,550	226,969	449,026	475,519	52
6	2024	Mesopotamian Foredeep Basin	Oil	13,129	285,156	298,285	83,722	368,878	382,007	78
7	1154	Amu-Darya Basin	Gas	78,513	151,830	230,343	163,650	315,480	393,993	58
8	1016	North Caspian Basin	Oil	1745	155,161	156,906	119,051	274,212	275,957	57
9	1050	South Barents Basin	Gas	0	70,020	70,020	160,857	230,877	230,877	30
10	7192	Niger Delta	Oil	6232	87,579	93,811	132,716	220,295	226,526	41
11	1112	South Caspian Basin	Oil	8591	27,403	35,994	173,310	200,713	209,304	17
12	6098	East Venezuela Basin	Oil	22,053	106,066	128,119	93,561	199,627	221,679	58
13	4017	Vestford-Helgeland	Oil	0	15,662	15,662	165,201	180,863	180,863	9
14	4025	North Sea Graben	Oil	31,235	128,603	159,839	37,745	166,348	197,584	81
15	2020	Interior Homocline-Central Arch	Oil	0	9825	9825	122,074	131,899	131,899	7
16	3948	Northwest Shelf	Gas	1840	54,405	56,245	64,711	119,116	120,956	47
17	2058	Grand Erg/Ahnet Basin	Gas	4980	109,053	114,034	8664	117,717	122,697	93
18	2023	Widyan Basin-Interior Platform	Oil	60	7295	7355	94,746	102,041	102,101	7
19	3817	Kutei Basin	Oil	8711	36,762	45,473	64,757	101,519	110,230	41
20	7203	West-Central Coastal	Oil	1150	11,030	12,180	88,044	99,074	100,223	12
21	3702	Greater Sarawak Basin	Gas	3731	78,433	82,164	19,247	97,680	101,411	81
22	1015	Volga-Ural Region	Oil	16,417	82,750	99,166	3941	86,691	103,108	96
23	6036	Santos Basin	Oil	12	1113	1125	80,547	81,660	81,672	1
24	5200	East Greenland Rift Basins	Oil	0	0	0	80,709	80,709	80,709	0
25	1322	North Sakhalin Basin	Oil	162	22,221	22,383	57,451	79,673	79,835	28
26	1008	Timan-Pechora Basin	Oil	12,374	24,261	36,635	52,079	76,340	88,714	41
27	4035	Northwest German Basin	Gas	76,285	65,440	141,726	10,814	76,254	152,539	93
28	3703	Malay Basin	Oil	492	47,657	48,149	27,465	75,122	75,614	64

Rank	Province Code	Province Name	Major Commodity	Cumulative Gas (BCFG)	Remaining Gas (BCFG)	Known Gas (BCFG)	Mean Undiscovered Gas (BCFG)	Future Gas (BCFG)	Gas Endowment (BCFG)	Discovery Maturity, Gas (%)
29	2056	Illizi Basin	Oil	750	44,311	45,061	28,138	72,449	73,199	62
30	2071	Red Sea Basin	Oil	849	8668	9517	60,012	68,680	69,529	14
31	3154	Tarim Basin	Oil	208	5572	5780	59,939	65,511	65,719	9
32	8047	Ganges-Brahmaputra Delta	Gas	2669	9620	12,289	55,091	64,711	67,380	18
33	1060	North Barents Basin	Gas	0	0	0	59,820	59,820	59,820	0
34	1109	Middle Caspian Basin	Oil	9984	17,156	27,140	39,685	56,841	66,825	41
35	1210	Nepa-Botuoba Arch	Oil	0	22,035	22,035	29,971	52,006	52,006	42
36	4068	Provence Basin	Gas	0	0	0	51,210	51,210	51,210	0
37	3701	Baram Delta/Brunei-Sabah Basin	Oil	9188	27,012	36,200	23,226	50,238	59,426	61
38	1009	Dnieper-Donets Basin	Gas	35,453	23,645	59,098	24,051	47,697	83,149	71
39	2043	Sirte Basin	Oil	5479	32,288	37,767	14,984	47,272	52,752	72
40	4036	Anglo-Dutch Basin	Gas	40,547	30,880	71,427	14,976	45,856	86,402	83
41	6099	Maracaibo Basin	Oil	979	25,722	26,701	17,576	43,298	44,277	60
42	5305	Villahermosa Uplift	Oil	13,888	27,404	41,292	15,623	43,027	56,915	73
43	5243	Alberta Basin	Oil	47,573	31,097	78,670	11,895	42,992	90,565	87
44	6103	Tobago Trough	Gas	0	22,600	22,600	20,103	42,703	42,703	53
45	6021	Guyana-Suriname Basin	Oil	0	0	0	42,058	42,058	42,058	0
46	2054	Trias/Ghadames Basin	Oil	647	24,352	24,999	17,493	41,845	42,492	59
47	6045	Santa Cruz-Tarija Basin	Oil	6928	10,058	16,985	31,107	41,165	48,092	35
48	3913	Browse Basin	Gas	0	17,960	17,960	20,093	38,053	38,053	47
49	3910	Bonaparte Gulf Basin	Gas	0	13,304	13,304	23,492	36,796	36,796	36
50	6034	Espirito Santo Basin	Oil	42	180	222	34,287	34,467	34,509	1
51	8043	Bombay	Oil	6228	17,965	24,193	12,944	30,909	37,137	65
52	8048	Irrawaddy	Gas	721	9288	10,009	20,521	29,809	30,530	33
53	6059	Magallanes Basin	Gas	10,053	14,754	24,807	14,040	28,794	38,848	64
54	4060	Po Basin	Gas	9527	9360	18,886	17,964	27,324	36,850	51
55	8042	Indus	Gas	8385	11,211	19,596	15,964	27,175	35,560	55
56	3822	North Sumatra Basin	Gas	13,977	11,582	25,559	15,123	26,705	40,682	63
57	6035	Campos Basin	Oil	303	5941	6244	19,691	25,632	25,935	24
58	6055	Neuquen Basin	Oil	7636	13,068	20,704	12,416	25,483	33,119	63
59	3828	South Sumatra Basin	Oil	3125	7080	10,204	18,250	25,330	28,455	36
60	2014	Ghaba Salt Basin	Oil	138	18,825	18,964	5056	23,881	24,019	79
61	1108	Azov-Kuban Basin	Gas	7492	11,118	18,610	12,252	23,369	30,861	60
62	6037	Pelotas Basin	Oil	0	0	0	22,806	22,806	22,806	0
63	2048	Pelagian Basin	Oil	43	17,167	17,210	5556	22,724	22,766	76
64	2004	Ma'Rib-Al Jawf/Masila Basin	Oil	0	16,507	16,507	5944	22,451	22,451	74
65	1207	Baykit Arch	Oil	0	5052	5052	16,983	22,035	22,035	23
66	8025	Sulaiman-Kirthar	Gas	6716	9061	15,777	12,599	21,659	28,376	56
67	6096	Llanos Basin	Oil	139	15,175	15,314	6217	21,391	15,651	98
68	3127	Bohaiwan Basin	Oil	2987	12,685	15,672	8514	21,199	24,186	65
69	1209	Angara-Lena Terrace	Gas	0	8075	8075	10,705	18,780	18,780	43
70	5304	Saline-Comalcalco Basin	Oil	1023	1223	2246	16,423	17,645	18,669	12
71	3142	Sichuan Basin	Gas	6236	6290	12,526	10,311	16,601	22,837	55

Rank	Province Code	Province Name	Major Commodity	Cumulative Gas (BCFG)	Remaining Gas (BCFG)	Known Gas (BCFG)	Mean Undiscovered Gas (BCFG)	Future Gas (BCFG)	Gas Endowment (BCFG)	Discovery Maturity, Gas (%)
72	6060	Falklands Plateau	Oil	0	0	0	16,581	16,581	16,581	0
73	5215	Labrador-Newfoundland Shelf	Oil	0	6630	6630	8893	15,522	15,522	43
74	3824	Northwest Java Basin	Oil	1026	7028	8055	7453	14,481	15,508	52
75	2016	Fahud Salt Basin	Oil	1151	8832	9983	4878	13,710	14,861	67
76	1150	North Ustyurt Basin	Oil	0	2413	2413	10,770	13,183	13,183	18
77	4047	North Carpathian Basin	Oil	8714	7618	16,332	5049	12,667	21,381	76
78	7183	Gulf of Guinea	Oil	46	1903	1949	10,071	11,974	12,020	16
79	6107	Lesser Antilles Deformed Belt	Oil	16	7	22	11,957	11,963	11,979	0
80	3930	Gippsland Basin	Oil	3501	6274	9775	5658	11,932	15,433	63
81	4057	Transylvanian Basin	Gas	21,194	9537	30,731	2083	11,620	32,814	94
82	4048	Pannonian Basin	Oil	6534	7308	13,842	4143	11,452	17,986	77
83	6063	Malvinas Basin	Oil	0	75	75	11,327	11,402	11,402	1
84	5245	Rocky Mountain Deformed Belt	Gas	8662	6447	15,109	3186	9633	18,295	83
85	7303	Orange River Coastal	Gas	0	6015	6015	3603	9618	9618	63
86	6029	Sergipe-Alagoas Basin	Oil	348	1059	1407	8334	9393	9741	14
87	2029	Wadi-Surhan Basin	Gas	47	179	226	7873	8052	8099	3
88	3144	Songliao Basin	Oil	3	1739	1743	5706	7446	7449	23
89	5244	Williston Basin, Canada	Oil	10,378	6876	17,254	502	7379	17,756	97
90	4061	Carpathian-Balkanian Basin	Oil	3126	4139	7265	3179	7319	10,445	70
91	3808	Central Sumatra Basin	Oil	910	2956	3866	4078	7034	7944	49
92	2009	Masila-Jeza Basin	Oil	0	40	40	6793	6833	6833	1
93	5301	Tampico-Misantla Basin	Oil	2354	4306	6661	2051	6358	8712	76
94	5302	Veracruz Basin	Oil	335	270	605	5832	6102	6437	9
95	6090	Middle Magdalena	Oil	2310	2692	5002	3292	5985	8294	60
96	6081	Talara Basin	Oil	1951	941	2892	4795	5736	7687	38
97	8034	Assam	Oil	2039	4495	6534	1178	5673	7712	85
98	6058	San Jorge Basin	Oil	1667	1898	3565	3774	5672	7339	49
99	2010	Ghudun-Khasfeh Flank Province	Oil	0	0	0	5623	5623	5623	0
100	3128	Ordos Basin	Oil	13	5597	5610	20	5617	5630	100
101	5307	Campeche-Sigsbe Salt Basin	Oil	0	0	0	5565	5565	5565	0
102	1211	Cis-Patom Foredeep	Gas	0	100	100	5364	5464	5464	2
103	2006	Shabwah Basin	Oil	0	573	573	4246	4819	4819	12
104	2015	Central Oman Platform	Oil	0	742	742	3239	3980	3981	19
105	3115	Junggar Basin	Oil	501	2082	2583	1296	3379	3879	67
106	2074	Khleisha Uplift	Oil	0	210	210	2580	2790	2790	8
107	6041	Putumayo-Oriente-Maranon Basin	Oil	661	955	1616	1596	2552	3212	50
108	8026	Kohat-Potwar	Oil	784	1147	1931	1314	2460	3245	60
109	2075	Euphrates/Mardin	Oil	66	1668	1734	749	2418	2483	70
110	5308	Yucatan Platform	Oil	0	0	0	1496	1496	1496	0
111	2011	South Oman Salt Basin	Oil	0	188	188	1236	1424	1424	13
112	2025	Hail-Ga'Ara Arch	Gas	0	0	0	1290	1290	1290	0
113	2028	Rutbah Uplift	Gas	0	0	0	1290	1290	1290	0
114	6083	Progreso Basin	Oil	148	350	498	695	1044	1193	42

Rank	Province Code	Province Name	Major Commodity	Cumulative Gas (BCFG)	Remaining Gas (BCFG)	Known Gas (BCFG)	Mean Undiscovered Gas (BCFG)	Future Gas (BCFG)	Gas Endowment (BCFG)	Discovery Maturity, Gas (%)
115	2031	Zagros Thrust Zone	Gas	0	0	0	966	966	966	0
116	4033	German-Polish Basin	Gas	6905	940	7846	0	940	7846	100
117	7013	Senegal	Oil	4	45	49	856	901	905	5
118	6117	Greater Antilles Deformed Belt	Oil	20	77	96	592	668	688	14
119	2013	Huqf-Haushi Uplift	Oil	0	0	0	404	404	404	0
120	5246	Mackenzie Foldbelt	Gas	379	252	631	43	295	674	94
121	8023	Central Afghanistan	Gas	27	244	271	0	244	271	100
122	2047	Hamra Basin	Oil	0	0	0	241	241	241	0
123	5306	Macuspana Basin	Oil	8	8	16	210	217	225	7
124	4046	Bohemia	Gas	114	117	231	0	117	231	100
125	3916	Carnarvon Basin	Oil	32	70	102	0	70	102	100
126	2089	Anah Graben	Oil	0	0	0	39	39	39	0
127	5317	Chicontepec Basin	Oil	0	2	2	32	34	34	6
128	4038	Munsterland Basin	Gas	121	31	152	0	31	152	100
129	2017	Oman Mountains	Oil	0	0	0	22	22	22	0
130	8035	North Burma	Oil	1	10	11	0	10	11	100
131	2061	Ougarta Uplift	Oil	0	6	6	0	6	6	100
132	2033	Sinai Basin	Oil	0	5	5	0	5	5	100
NA	1158	Tian Shan Foldbelt	NA	NA	NA	NA	0	NA	NA	NA
NA	1059	Ludlov Saddle	NA	NA	NA	NA	28,387	NA	NA	NA
NA	3161	Yunnan Guizhou Hubei Fold Belt	NA	NA	NA	NA	0	NA	NA	NA
NA	4028	Mid-North Sea High	NA	NA	NA	NA	0	NA	NA	NA
NA	4039	Southwest German Basin	NA	NA	NA	NA	0	NA	NA	NA
NA	5310	Sierra Madre de Chiapas-Peten Foldbelt	NA	NA	NA	NA	1846	NA	NA	NA
NA	5316	Trans-Mexican Neovolcanic Axis	NA	NA	NA	NA	195	NA	NA	NA
NA	6022	Foz do Amazonas Basin	NA	NA	NA	NA	29,840	NA	NA	NA
NA	8002	Himalayan	NA	NA	NA	NA	0	NA	NA	NA

APPENDIX 3I. Provinces ranked by total future petroleum volume.

[MMBOE = million barrels of oil equivalent. Values given are not necessarily those for the entire province, but only for that part assessed in World Petroleum Assessment 2000 (USGS, 2000). Major commodity is based on the ratio of gas endowment to oil endowment as ft^3 of gas per bbl of oil (CFG/BO). Major commodity is oil if this ratio is less than 20,000 CFG/BO and gas if this ratio is equal to or greater than 20,000 CFG/BO. Total cumulative petroleum is the reported volume of total petroleum that has been produced. Total remaining petroleum is the calculated remaining reserves; the difference between total known and total cumulative petroleum. Total known petroleum is the reported discovered volume (cumulative production plus remaining reserves). Total mean undiscovered petroleum is derived from the mean volumes of oil, gas, and NGL estimated in World Petroleum Assessment 2000 (USGS, 2000). Total future petroleum is the total remaining petroleum plus estimated total mean undiscovered petroleum. Total petroleum endowment is the total known petroleum plus estimated total mean undiscovered petroleum volumes. Discovery maturity is the percentage of total petroleum discovered (total known petroleum) with respect to total petroleum endowment. Calculated values may not be equal to the sums of the component values because numbers have been independently rounded. NA means not applicable; it is shown for all values where there is only one field present, and used where the endowment could not be calculated.] Known volumes from Petroconsultants (1996) and NRG Associates (1995). Note that gas = natural gas.

Rank	Province Code	Province Name	Major Commodity	Total Cumulative Petroleum (MMBOE)	Total Remaining Petroleum (MMBOE)	Total Known Petroleum (MMBOE)	Total Mean Undiscovered Petroleum (MMBOE)	Total Future Petroleum (MMBOE)	Total Petroleum Endowment (MMBOE)	Discovery Maturity, Total Petroleum (%)
1	1174	West Siberian Basin	Oil	83,213	271,517	354,730	182,864	454,380	537,593	66
2	2024	Mesopotamian Foredeep Basin	Oil	58,934	285,059	343,993	78,159	363,217	422,151	81
3	2019	Rub Al Khali Basin	Oil	20,998	101,760	122,757	137,868	239,627	260,625	47
4	2030	Zagros Fold Belt	Oil	51,102	138,437	189,539	90,292	228,729	279,831	68
5	2021	Greater Ghawar Uplift	Oil	62,169	129,511	191,680	63,808	193,319	255,488	75
6	7192	Niger Delta	Oil	17,646	35,353	53,000	68,640	103,994	121,640	44
7	2022	Qatar Arch	Gas	183	92,332	92,515	8908	101,239	101,423	91
8	1016	North Caspian Basin	Oil	1322	44,528	45,850	48,573	93,101	94,423	49
9	4025	North Sea Graben	Oil	28,183	48,322	76,505	21,248	69,570	97,753	78
10	6098	East Venezuela Basin	Oil	17,509	34,747	52,257	31,934	66,682	84,191	62
11	5200	East Greenland Rift Basins	Oil	0	0	0	64,769	64,769	64,769	0
12	1112	South Caspian Basin	Oil	13,334	10,607	23,941	52,725	63,332	76,665	31
13	2023	Widyan Basin-Interior Platform	Oil	229	18,457	18,686	43,855	62,312	62,541	30
14	7203	West-Central Coastal	Oil	5957	9854	15,811	48,665	58,519	64,476	25
15	1154	Amu-Darya Basin	Gas	13,357	26,975	40,332	31,541	58,516	71,873	56
16	2020	Interior Homocline-Central Arch	Oil	84	6516	6601	46,132	52,648	52,733	13
17	4017	Vestford-Helgeland	Oil	76	5897	5973	43,062	48,959	49,035	12
18	1050	South Barents Basin	Gas	0	11,770	11,770	30,377	42,147	42,147	28
19	6036	Santos Basin	Oil	12	505	517	40,828	41,332	41,344	1
20	1015	Volga-Ural Region	Oil	45,732	35,832	81,564	3403	39,235	84,968	96
21	2043	Sirte Basin	Oil	14,929	28,566	43,496	10,000	38,566	53,496	81
22	5305	Villahermosa Uplift	Oil	17,004	24,993	41,997	10,957	35,950	52,955	79
23	1008	Timan-Pechora Basin	Oil	5771	14,171	19,942	16,598	30,769	36,539	55
24	6099	Maracaibo Basin	Oil	35,745	17,820	53,565	12,086	29,906	65,651	82

Rank	Province Code	Province Name	Major Commodity	Total Cumulative Petroleum (MMBOE)	Total Remaining Petroleum (MMBOE)	Total Known Petroleum (MMBOE)	Total Mean Undiscovered Petroleum (MMBOE)	Total Future Petroleum (MMBOE)	Total Petroleum Endowment (MMBOE)	Discovery Maturity, Total Petroleum (%)
25	6035	Campos Basin	Oil	1819	9287	11,107	20,128	29,415	31,235	36
26	3948	Northwest Shelf	Gas	896	10,589	11,485	16,848	27,437	28,333	41
27	3817	Kutei Basin	Oil	3691	8040	11,731	18,060	26,100	29,791	39
28	2058	Grand Erg/Ahnet Basin	Gas	1783	22,752	24,534	2449	25,201	26,984	91
29	6021	Guyana-Suriname Basin	Oil	17	46	63	24,586	24,632	24,649	0
30	2071	Red Sea Basin	Oil	6445	5298	11,744	18,533	23,832	30,277	39
31	2054	Trias/Ghadames Basin	Oil	7589	12,800	20,389	10,566	23,366	30,955	66
32	3127	Bohaiwan Basin	Oil	10,914	16,339	27,253	6248	22,587	33,501	81
33	1109	Middle Caspian Basin	Oil	6071	8083	14,153	13,171	21,254	27,324	52
34	3154	Tarim Basin	Oil	80	1776	1856	19,250	21,026	21,106	9
35	1322	North Sakhalin Basin	Oil	785	5292	6077	15,156	20,449	21,234	29
36	3702	Greater Sarawak Basin	Gas	846	14,044	14,890	4301	18,345	19,191	78
37	2056	Illizi Basin	Oil	2485	9593	12,078	8536	18,129	20,614	59
38	3703	Malay Basin	Oil	1609	10,346	11,955	6319	16,665	18,274	65
39	3701	Baram Delta/Brunei-Sabah Basin	Oil	5492	7619	13,111	8434	16,054	21,545	61
40	1210	Nepa-Botuoba Arch	Oil	1	5974	5976	7643	13,618	13,619	44
41	4035	Northwest German Basin	Gas	14,589	11,340	25,929	1881	13,221	27,810	93
42	8043	Bombay	Oil	4417	8323	12,740	4666	12,989	17,406	73
43	6096	Llanos Basin	Oil	1202	7203	8405	5004	12,207	13,409	63
44	5243	Alberta Basin	Oil	22,342	8760	31,101	3128	11,888	34,229	91
45	5304	Saline-Comalcalco Basin	Oil	1844	889	2733	10,650	11,539	13,383	20
46	1009	Dnieper-Donets Basin	Gas	6137	5523	11,660	5969	11,492	17,629	66
47	1060	North Barents Basin	Gas	0	0	0	11,355	11,355	11,355	0
48	8047	Ganges-Brahmaputra Delta	Gas	451	1654	2105	9558	11,212	11,663	18
49	4068	Provence Basin	Gas	0	0	0	11,188	11,188	11,188	0
50	6045	Santa Cruz-Tarija Basin	Oil	1655	2045	3699	8629	10,674	12,328	30
51	6034	Espirito Santo Basin	Oil	100	68	169	10,285	10,354	10,454	2
52	3910	Bonaparte Gulf Basin	Gas	123	3046	3169	6281	9327	9450	34
53	6060	Falklands Plateau	Oil	0	0	0	8967	8967	8967	0
54	3913	Browse Basin	Gas	19	3221	3239	5339	8559	8578	38
55	4036	Anglo-Dutch Basin	Gas	7237	5403	12,640	2630	8033	15,270	83
56	6037	Pelotas Basin	Oil	0	0	0	7846	7846	7846	0
57	6103	Tobago Trough	Gas	0	3867	3867	3551	7418	7418	52
58	6041	Putumayo-Oriente-Maranon Basin	Oil	2999	3871	6871	3418	7290	10,289	67
59	3115	Junggar Basin	Oil	1035	6206	7241	843	7049	8084	90
60	3808	Central Sumatra Basin	Oil	8500	5372	13,872	1663	7034	15,535	89
61	2048	Pelagian Basin	Oil	528	4646	5174	2323	6970	7497	69
62	6055	Neuquen Basin	Oil	2883	3276	6159	3562	6838	9721	63
63	8048	Irrawaddy	Gas	649	1797	2446	4930	6726	7376	33
64	5215	Labrador-Newfoundland Shelf	Oil	0	2875	2875	3792	6667	6667	48
65	6059	Magallanes Basin	Gas	2359	3202	5562	3354	6556	8915	62

Rank	Province Code	Province Name	Major Commodity	Total Cumulative Petroleum (MMBOE)	Total Remaining Petroleum (MMBOE)	Total Known Petroleum (MMBOE)	Total Mean Undiscovered Petroleum (MMBOE)	Total Future Petroleum (MMBOE)	Total Petroleum Endowment (MMBOE)	Discovery Maturity, Total Petroleum (%)
66	2004	Ma'Rib-Al Jawf/Masila Basin	Oil	536	3712	4248	2273	5985	6521	65
67	2014	Ghaba Salt Basin	Oil	489	4492	4981	1436	5928	6416	78
68	1207	Baykit Arch	Oil	0	936	936	4897	5833	5833	16
69	3144	Songliao Basin	Oil	12,399	3466	15,865	2189	5655	18,054	88
70	3828	South Sumatra Basin	Oil	2367	1819	4186	3750	5569	7936	53
71	2016	Fahud Salt Basin	Oil	2754	3535	6288	2008	5543	8297	76
72	1150	North Ustyurt Basin	Oil	459	2318	2777	3193	5511	5970	47
73	3822	North Sumatra Basin	Gas	3409	2451	5860	2914	5365	8774	67
74	8042	Indus	Gas	1512	1970	3482	3328	5299	6811	51
75	4060	Po Basin	Gas	1706	1820	3526	3476	5296	7002	50
76	3824	Northwest Java Basin	Oil	1771	2914	4685	2007	4922	6692	70
77	4061	Carpathian-Balkanian Basin	Oil	4624	2571	7195	1659	4230	8854	81
78	6107	Lesser Antilles Deformed Belt	Oil	11	3	14	4110	4113	4124	0
79	8025	Sulaiman-Kirthar	Gas	1121	1535	2657	2553	4089	5210	51
80	3930	Gippsland Basin	Oil	3815	2375	6190	1592	3967	7782	80
81	1209	Angara-Lena Terrace	Gas	0	1409	1409	2526	3935	3935	36
82	1108	Azov-Kuban Basin	Gas	2589	1091	3681	2675	3766	6355	58
83	5307	Campeche-Sigsbe Salt Basin	Oil	0	0	0	3714	3714	3714	0
84	6029	Sergipe-Alagoas Basin	Oil	528	402	929	3292	3693	4221	22
85	7183	Gulf of Guinea	Oil	81	515	596	2965	3480	3561	17
86	3142	Sichuan Basin	Gas	1086	1112	2198	2202	3313	4400	50
87	6063	Malvinas Basin	Oil	0	25	25	3280	3305	3305	1
88	6081	Talara Basin	Oil	1715	439	2154	2765	3204	4919	44
89	5301	Tampico-Misantla Basin	Oil	6165	1840	8005	1354	3194	9359	86
90	2009	Masila-Jeza Basin	Oil	144	496	641	2598	3094	3238	20
91	4048	Pannonian Basin	Oil	2691	1892	4583	1178	3070	5761	80
92	8034	Assam	Oil	1251	2379	3630	663	3042	4292	85
93	4047	North Carpathian Basin	Oil	2104	1559	3663	1344	2903	5007	73
94	6058	San Jorge Basin	Oil	2723	1218	3942	1203	2421	5144	77
95	6090	Middle Magdalena	Oil	2300	1028	3328	1388	2416	4715	71
96	5302	Veracruz Basin	Oil	107	125	232	2214	2339	2446	9
97	2010	Ghudun-Khasfeh Flank Province	Oil	0	0	0	2258	2258	2258	0
98	5244	Williston Basin, Canada	Oil	4468	1963	6431	252	2215	6683	96
99	2029	Wadi-Surhan Basin	Gas	8	30	38	1942	1972	1979	2
100	4057	Transylvanian Basin	Gas	3532	1589	5122	354	1943	5475	94
101	7303	Orange River Coastal	Gas	0	1003	1003	879	1882	1882	53
102	2006	Shabwah Basin	Oil	0	210	210	1624	1833	1833	11
103	5245	Rocky Mountain Deformed Belt	Gas	1916	1199	3115	613	1812	3728	84
104	3128	Ordos Basin	Oil	83	1535	1618	143	1678	1761	92
105	2075	Euphrates/Mardin	Oil	293	607	900	893	1500	1793	50
106	1211	Cis-Patom Foredeep	Gas	0	17	17	1309	1325	1325	1

Rank	Province Code	Province Name	Major Commodity	Total Cumulative Petroleum (MMBOE)	Total Remaining Petroleum (MMBOE)	Total Known Petroleum (MMBOE)	Total Mean Undiscovered Petroleum (MMBOE)	Total Future Petroleum (MMBOE)	Total Petroleum Endowment (MMBOE)	Discovery Maturity, Total Petroleum (%)
107	2015	Central Oman Platform	Oil	3	198	200	1090	1287	1290	16
108	2074	Khleisha Uplift	Oil	250	442	692	739	1180	1430	48
109	5308	Yucatan Platform	Oil	0	0	0	1092	1092	1092	0
110	6117	Greater Antilles Deformed Belt	Oil	68	245	313	628	873	940	33
111	8026	Kohat-Potwar	Oil	341	343	684	422	765	1106	62
112	2011	South Oman Salt Basin	Oil	203	343	546	357	701	903	60
113	6083	Progreso Basin	Oil	147	111	258	385	497	644	40
114	2025	Hail-Ga'Ara Arch	Gas	0	0	0	369	369	369	0
115	2028	Rutbah Uplift	Gas	0	0	0	369	369	369	0
116	7013	Senegal	Oil	1	18	19	343	361	362	5
117	2031	Zagros Thrust Zone	Gas	0	0	0	242	242	242	0
118	4033	German-Polish Basin	Gas	1151	157	1308	0	157	1308	100
119	5306	Macuspana Basin	Oil	4	5	9	149	154	158	6
120	2047	Hamra Basin	Oil	0	0	0	148	148	148	0
121	2013	Huqf-Haushi Uplift	Oil	0	0	0	115	115	115	0
122	2089	Anah Graben	Oil	75	80	155	35	115	190	81
123	5246	Mackenzie Foldbelt	Gas	63	42	105	8	50	113	93
124	2061	Ougarta Uplift	Oil	0	42	42	0	42	42	100
125	8023	Central Afghanistan	Gas	5	41	45	0	41	45	100
126	3916	Carnarvon Basin	Oil	21	31	52	0	31	52	100
127	2017	Oman Mountains	Oil	0	0	0	27	27	27	0
128	4046	Bohemia	Gas	21	22	43	0	22	43	100
129	5317	Chicontepec Basin	Oil	0	0	0	22	22	22	2
130	2033	Sinai Basin	Oil	94	12	106	0	12	106	100
131	4038	Munsterland Basin	Gas	20	5	25	0	5	25	100
132	8035	North Burma	Oil	2	2	3	0	2	3	100
NA	1158	Tian Shan Foldbelt	NA	NA	NA	NA	0	NA	NA	NA
NA	1059	Ludlov Saddle	NA	NA	NA	NA	5361	NA	NA	NA
NA	3161	Yunnan Guizhou Hubei Fold Belt	NA	NA	NA	NA	0	NA	NA	NA
NA	4028	Mid-North Sea High	NA	NA	NA	NA	0	NA	NA	NA
NA	4039	Southwest German Basin	NA	NA	NA	NA	0	NA	NA	NA
NA	5310	Sierra Madre de Chiapas-Peten Foldbelt	NA	NA	NA	NA	1307	NA	NA	NA
NA	5316	Trans-Mexican Neovolcanic Axis	NA	NA	NA	NA	72	NA	NA	NA
NA	6022	Foz do Amazonas Basin	NA	NA	NA	NA	5272	NA	NA	NA
NA	8002	Himalayan	NA	NA	NA	NA	0	NA	NA	NA

APPENDIX 4. Volumes of crude oil, natural gas, and natural gas liquids and geologic attributes of assessment units identified in World Petroleum Assessment 2000 (USGS, 2000).

[BBO = billion barrels of oil; TCFG = trillion cubic feet of gas; BBNGL = billion barrels of natural gas liquids (NGL); BBOE = billion barrels of oil equivalent (BOE). One BOE equals 6000 ft^3 of gas. Total petroleum is the sum of oil, gas in terms of oil equivalents, and NGL. Major commodity is based on the ratio of discovered plus undiscovered gas and discovered and undiscovered oil as ft^3 of gas per bbl of oil (CFG/BO). Major commodity is oil if this ratio is less than 20,000 CFG/BO and gas if this ratio is equal to or greater than 20,000 CFG/BO. Known oil, gas, NGL, and total petroleum are the reported discovered volumes (cumulative production plus remaining reserves). Undiscovered oil, gas, NGL, and total petroleum volumes are the means of probability distributions estimated in World Petroleum Assessment 2000 (USGS, 2000). Numbers have been independently rounded. NA means not applicable; it is shown for all values where there is only one field present, and used where the endowment could not be calculated. Exploration maturity is established if more than 13 fields have been discovered; frontier = 1-13 fields; and hypothetical = 0 fields.] Known volumes from Petroconsultants (1996) and NRG Associates (1995). Note that gas = natural gas.

Province Code	Province Name	Total Petroleum System Code	Total Petroleum System Name	Assessment Unit Code	Assessment Unit Name	Major Commodity	Accumulation Type	Exploration Maturity
1008	Timan-Pechora Basin	100801	Domanik-Paleozoic	10080101	Northwest Izhma-Pechora Depression	Oil	Conventional	Hypothetical
1008	Timan-Pechora Basin	100801	Domanik-Paleozoic	10080102	Main Basin Platform	Oil	Conventional	Established
1008	Timan-Pechora Basin	100801	Domanik-Paleozoic	10080103	Foredeep Basins	Gas	Conventional	Established
1009	Dnieper-Donets Basin	100901	Dnieper-Donets Paleozoic	10090101	Carboniferous-Lower Permian Clastics	Gas	Conventional	Established
1009	Dnieper-Donets Basin	100901	Dnieper-Donets Paleozoic	10090102	Devonian Synrift	Oil	Conventional	Hypothetical
1009	Dnieper-Donets Basin	100901	Dnieper-Donets Paleozoic	10090103	Continuous Basin-Centered Gas Accumulation	Gas	Continuous	Hypothetical
1015	Volga-Ural Region	101501	Volga-Ural Domanik-Paleozoic	10150101	Supra-Domanik Carbonates/Clastics	Oil	Conventional	Established
1015	Volga-Ural Region	101501	Volga-Ural Domanik-Paleozoic	10150102	Lower Volga	Oil	Conventional	Established
1015	Volga-Ural Region	101501	Volga-Ural Domanik-Paleozoic	10150103	Sub-Domanik Devonian Clastics	Oil	Continuous	Established
1015	Volga-Ural Region	101501	Volga-Ural Domanik-Paleozoic	10150104	Self-Sourced Domanik	Oil	Conventional	Hypothetical
1015	Volga-Ural Region	101502	Belsk Basin	10150201	Permian Reefs/Thrust Folds	Gas	Conventional	Frontier
1016	North Caspian Basin	101601	Paleozoic North Caspian	10160101	North and West Margins Subsalt Pinnacle Reefs	Gas	Conventional	Frontier
1016	North Caspian Basin	101601	Paleozoic North Caspian	10160102	North and West Margins Subsalt Barrier Reefs	Oil	Conventional	Frontier
1016	North Caspian Basin	101601	Paleozoic North Caspian	10160103	East and Southeast Margins Subsalt	Oil	Conventional	Frontier
1016	North Caspian Basin	101601	Paleozoic North Caspian	10160104	South Margin Subsalt	Oil	Conventional	Frontier
1016	North Caspian Basin	101601	Paleozoic North Caspian	10160105	Basin Center Subsalt	Gas	Continuous	Hypothetical
1016	North Caspian Basin	101601	Paleozoic North Caspian	10160106	Suprasalt	Oil	Conventional	Established
1050	South Barents Basin	105001	South and North Barents Triassic-Jurassic	10500101	Kolguyev Terrace	Gas	Conventional	Frontier
1050	South Barents Basin	105001	South and North Barents Triassic-Jurassic	10500102	South Barents and Ludlov Saddle	Gas	Conventional	Frontier
1050	South Barents Basin	105001	South and North Barents Triassic-Jurassic	10500103	North Barents	Gas	Conventional	Hypothetical
1108	Azov-Kuban Basin	110801	Azov-Kuban Mesozoic-Cenozoic	11080101	Tertiary Foredeep	Oil	Conventional	Established
1108	Azov-Kuban Basin	110801	Azov-Kuban Mesozoic-Cenozoic	11080102	Foreland Slope	Gas	Conventional	Established
1108	Azov-Kuban Basin	110801	Azov-Kuban Mesozoic-Cenozoic	11080103	Subsalt Jurassic	Gas	Conventional	Frontier
1109	Middle Caspian Basin	110901	Terek-Caspian	11090101	Foldbelt-Foothills	Oil	Conventional	Established
1109	Middle Caspian Basin	110901	Terek-Caspian	11090102	Terek-Sunzha Subsalt Jurassic	Gas	Conventional	Hypothetical
1109	Middle Caspian Basin	110901	Terek-Caspian	11090103	Foreland Slope and Foredeep	Oil	Conventional	Frontier
1109	Middle Caspian Basin	110902	South Mangyshlak	11090201	South Mangyshlak (Entire)	Oil	Conventional	Established
1109	Middle Caspian Basin	110903	Stavropol-Prikumsk	11090301	Offshore Prikumsk Zone	Oil	Conventional	Hypothetical
1109	Middle Caspian Basin	110903	Stavropol-Prikumsk	11090302	Onshore Stavropol-Prikumsk	Oil	Conventional	Established
1109	Middle Caspian Basin	110903	Stavropol-Prikumsk	11090303	Central Caspian Offshore	Oil	Conventional	Hypothetical
1112	South Caspian Basin	111201	Oligocene-Miocene Maykop/Diatom	11120101	Apsheron-Pribalkhan Zone	Oil	Conventional	Established
1112	South Caspian Basin	111201	Oligocene-Miocene Maykop/Diatom	11120102	Lower Kura Depression and Adjacent Shelf	Oil	Conventional	Established
1112	South Caspian Basin	111201	Oligocene-Miocene Maykop/Diatom	11120103	Gograndag-Okarem Zone and Adjacent Shelf	Oil	Conventional	Established
1112	South Caspian Basin	111201	Oligocene-Miocene Maykop/Diatom	11120104	Central Offshore	Oil	Conventional	Hypothetical
1112	South Caspian Basin	111201	Oligocene-Miocene Maykop/Diatom	11120105	Iran Onshore/Nearshore	Oil	Conventional	Hypothetical
1150	North Ustyurt Basin	115001	Buzachi Arch and Surrounding Areas Composite	11500101	Mesozoic Sandstone Reservoirs	Oil	Conventional	Frontier
1150	North Ustyurt Basin	115002	North Ustyurt Jurassic	11500201	Jurassic-Tertiary	Oil	Conventional	Hypothetical
1150	North Ustyurt Basin	115003	North Ustyurt Paleozoic	11500301	Upper Paleozoic Carbonates	Gas	Conventional	Hypothetical
1154	Amu-Darya Basin	115401	Amu-Darya Jurassic-Cretaceous	11540101	Northern and Western Areas	Gas	Conventional	Established
1154	Amu-Darya Basin	115401	Amu-Darya Jurassic-Cretaceous	11540102	Karabil-Badkhyz (Southern Area)	Gas	Conventional	Established

Province Code	Province Name	Total Petroleum System Code	Total Petroleum System Name	Assessment Unit Code	Assessment Unit Name	Major Commodity	Accumulation Type	Exploration Maturity
1154	Amu-Darya Basin	115401	Amu-Darya Jurassic-Cretaceous	11540103	Murgab Depression Suprasalt	Gas	Conventional	Established
1154	Amu-Darya Basin	115401	Amu-Darya Jurassic-Cretaceous	11540104	Murgab Depression Subsalt	Gas	Conventional	Established
1174	West Siberian Basin	117401	Bazhenov-Neocomian	11740101	Upper Jurassic-Cretaceous Sandstones	Oil	Conventional	Established
1174	West Siberian Basin	117401	Bazhenov-Neocomian	11740102	Self-Sourced Bazhenov Fractured Reservoirs	Oil	Continuous	Hypothetical
1174	West Siberian Basin	117402	Togur-Tyumen	11740201	Pre-Upper Jurassic	Oil	Conventional	Established
1174	West Siberian Basin	117403	Northern West Siberian Mesozoic Composite	11740301	Northern West Siberian Onshore Gas	Gas	Conventional	Established
1174	West Siberian Basin	117403	Northern West Siberian Mesozoic Composite	11740302	South Kara Sea Offshore	Gas	Conventional	Frontier
1207	Baykit Arch	120701	Yenisey Foldbelt Riphean-Craton Margin Riphean	12070101	Yenisey Foldbelt Riphean-Craton Margin Riphean	Oil	Conventional	Frontier
1210	Nepa-Botuoba Arch	121001	Baikal-Patom Foldbelt Riphean-Craton Margin Vendian	12100101	Baikal-Patom Foldbelt Riphean-Craton Margin Vendian	Gas	Conventional	Established
1322	North Sakhalin Basin	132201	North Sakhalin Neogene	13220101	Onshore and Offshore Northeastern Shelf	Oil	Conventional	Established
2004	Ma'Rib-Al Jawf/Masila Basin	200401	Madbi Amran/Qishn	20040101	Ma'Rib-Al Jawf/Shabwah/Masila	Oil	Conventional	Established
2014	Ghaba Salt Basin	201401	North Oman Huqf/'Q'-Haushi	20140101	Ghaba-Makarem Combined Structural	Oil	Conventional	Established
2016	Fahud Salt Basin	201601	North Oman Huqf-Shu'aiba	20160101	Fahud-Huqf Combined Structural	Oil	Conventional	Established
2016	Fahud Salt Basin	201602	Middle Cretaceous Natih	20160201	Natih-Fiqa Structural/Stratigraphic	Oil	Conventional	Frontier
2019	Rub Al Khali Basin	201901	Cretaceous Thamama/Wasia	20190101	Cretaceous Reservoirs in Northwest Desert Anticlines	Oil	Conventional	Established
2019	Rub Al Khali Basin	201901	Cretaceous Thamama/Wasia	20190102	Cretaceous Reservoirs in South Gulf Suprasalt Structural	Oil	Conventional	Established
2019	Rub Al Khali Basin	201901	Cretaceous Thamama/Wasia	20190103	Mesozoic/Tertiary Foredeep Fold and Thrust	Gas	Conventional	Established
2019	Rub Al Khali Basin	201902	Jurassic Hanifa/Diyab-Arab	20190201	Jurassic Reservoirs in Northwest Desert Anticlines	Oil	Conventional	Frontier
2019	Rub Al Khali Basin	201902	Jurassic Hanifa/Diyab-Arab	20190202	Jurassic Reservoirs in South Gulf Suprasalt/Qatar Arch Structural	Oil	Conventional	Established
2019	Rub Al Khali Basin	201903	Silurian Qusaiba	20190301	Khuff Carbonates in Salt Structures	Gas	Conventional	Frontier
2019	Rub Al Khali Basin	201903	Silurian Qusaiba	20190302	Paleozoic Reservoirs	Gas	Conventional	Frontier
2021	Greater Ghawar Uplift	202101	Central Arabia Qusaiba Paleozoic	20210101	Central Arch Horst-Block Anticlinal Oil and Gas	Oil	Conventional	Established
2021	Greater Ghawar Uplift	202101	Central Arabia Qusaiba Paleozoic	20210102	North Gulf Salt Basin Structural Gas	Gas	Conventional	Frontier
2021	Greater Ghawar Uplift	202102	Arabian Sub-Basin Tuwaiq/Hanifa-Arab	20210201	Horst-Block Anticlinal Oil	Oil	Conventional	Established
2021	Greater Ghawar Uplift	202102	Arabian Sub-Basin Tuwaiq/Hanifa-Arab	20210202	Salt-Involved Structural Oil	Oil	Conventional	Frontier
2023	Widyan Basin-Interior Platform	202301	Paleozoic Qusaiba /Akkas /Abba /Mudawwara	20230101	Horst/Graben-Related Oil and Gas	Oil	Conventional	Established
2023	Widyan Basin-Interior Platform	202302	Jurassic Gotnia /Barsarin /Sargelu /Najmah	20230201	Platform Horst/Graben-Related Oil	Oil	Conventional	Frontier
2023	Widyan Basin-Interior Platform	202302	Jurassic Gotnia /Barsarin /Sargelu /Najmah	20230202	Basinal Oil and Gas	Oil	Conventional	Hypothetical
2030	Zagros Fold Belt	203001	Zagros-Mesopotamian Cretaceous-Tertiary	20300101	Cretaceous Reservoirs	Oil	Conventional	Established
2030	Zagros Fold Belt	203001	Zagros-Mesopotamian Cretaceous-Tertiary	20300102	Tertiary Reservoirs	Oil	Conventional	Established
2030	Zagros Fold Belt	203002	Paleozoic-Permian/Triassic	20300201	Northern Qatar Arch Extension	Gas	Conventional	Frontier
2043	Sirte Basin	204301	Sirte-Zelten	20430101	Southeast Sirte Clastics	Oil	Conventional	Established
2043	Sirte Basin	204301	Sirte-Zelten	20430102	Central Sirte Carbonates	Oil	Conventional	Established
2043	Sirte Basin	204301	Sirte-Zelten	20430103	Offshore Sirte Hypothetical	Oil	Conventional	Hypothetical
2043	Sirte Basin	204301	Sirte-Zelten	20430104	Southeast Sirte Hypothetical	Oil	Conventional	Hypothetical
2048	Pelagian Basin	204801	Bou Dabbous-Tertiary	20480101	Bou Dabbous-Tertiary Structural/Stratigraphic	Oil	Conventional	Established
2048	Pelagian Basin	204802	Jurassic-Cretaceous Composite	20480201	Jurassic-Cretaceous Structural/Stratigraphic	Oil	Conventional	Established
2054	Trias/Ghadames Basin	205401	Tanezzuft-Oued Mya	20540101	Tanezzuft-Oued Mya Structural/Stratigraphic	Oil	Conventional	Established
2054	Trias/Ghadames Basin	205402	Tanezzuft-Melrhir	20540201	Tanezzuft-Melrhir Structural/Stratigraphic	Oil	Conventional	Frontier
2054	Trias/Ghadames Basin	205403	Tanezzuft-Ghadames	20540301	Tanezzuft-Ghadames Structural/Stratigraphic	Oil	Conventional	Established
2056	Illizi Basin	205601	Tanezzuft-Illizi	20560101	Tanezzuft-Illizi Structural/Stratigraphic	Oil	Conventional	Established
2058	Grand Erg/Ahnet Basin	205801	Tanezzuft-Timimoun	20580101	Tanezzuft-Timimoun Structural/Stratigraphic	Gas	Conventional	Frontier
2058	Grand Erg/Ahnet Basin	205802	Tanezzuft-Ahnet	20580201	Tanezzuft-Ahnet Structural/Stratigraphic	Gas	Conventional	Established
2058	Grand Erg/Ahnet Basin	205803	Tanezzuft-Sbaa	20580301	Tanezzuft-Sbaa Structural/Stratigraphic	Oil	Conventional	Frontier
2058	Grand Erg/Ahnet Basin	205804	Tanezzuft-Mouydir	20580401	Tanezzuft-Mouydir Structural/Stratigraphic	Gas	Conventional	Hypothetical
2058	Grand Erg/Ahnet Basin	205805	Tanezzuft-Benoud	20580501	Tanezzuft-Benoud Structural/Stratigraphic	Gas	Conventional	Frontier
2058	Grand Erg/Ahnet Basin	205806	Tanezzuft-Bechar/Abadla	20580601	Tanezzuft-Bechar/Abadla Structural/Stratigraphic	Gas	Conventional	Hypothetical
2071	Red Sea Basin	207101	Sudr-Nubia	20710101	Gulf of Suez Block-Fault Fairway	Oil	Conventional	Established
2071	Red Sea Basin	207101	Sudr-Nubia	20710102	Gulf of Suez Qaa Plain	Oil	Conventional	Hypothetical
3115	Junggar Basin	207101	Sudr-Nubia	20710103	Southern Gulf of Suez	Oil	Conventional	Frontier
3115	Junggar Basin	207102	Maqna	20710201	Red Sea Coastal Block Faults	Gas	Conventional	Frontier

Province Code	Province Name	Total Petroleum System Code	Total Petroleum System Name	Assessment Unit Code	Assessment Unit Name	Major Commodity	Accumulation Type	Exploration Maturity
3115	Junggar Basin	207102	Maqna	20710202	Red Sea Salt Basin	Oil	Conventional	Hypothetical
3115	Junggar Basin	311501	Lucaogou-Karamay/Ulho/Pindequan	31150101	Upper Paleozoic/Lower Mesozoic Nonmarine Coarse Clastics	Oil	Conventional	Frontier
3115	Junggar Basin	311502	Jurassic Coal-Jurassic /Tertiary	31150201	Jurassic/Tertiary Fluvial and Lacustrine Sandstone	Oil	Conventional	Frontier
3115	Junggar Basin	311503	Lucaogou/Jurassic Coal-Paleozoic/Mesozoic	31150301	Continuous-Type (Basin-Centered) Gas	Gas	Continuous	Hypothetical
3127	Bohaiwan Basin	312701	Shahejie-Shahejie/Guantao/Wumishan	31270101	Tertiary Lacustrine	Oil	Conventional	Established
3127	Bohaiwan Basin	312701	Shahejie-Shahejie/Guantao/Wumishan	31270102	Pre-Tertiary Buried Hills	Oil	Conventional	Established
3127	Bohaiwan Basin	312702	Carboniferous/Permian Coal-Paleozoic	31270201	Coal-Sourced Gas	Gas	Continuous	Hypothetical
3128	Ordos Basin	312801	Yanchang-Yanan	31280101	Jurassic/Triassic Fluvial and Lacustrine Sandstone	Oil	Conventional	Established
3128	Ordos Basin	312802	Taiyuan/Shanxi-Majiagou/Shihezi	31280201	Basin Centered Gas (Continuous) in Ordovician Carbonate and Permian Sandstone	Gas	Continuous	Established
3128	Ordos Basin	312802	Taiyuan/Shanxi-Majiagou/Shihezi	31280202	Carboniferous and Permian Coalbed Methane (Continuous)	Gas	Continuous	Hypothetical
3142	Sichuan Basin	314201	Maokou/Longtang-Jialingjiang /Maokou /Huanglong	31420101	Southeastern Fold Belt	Gas	Conventional	Established
3142	Sichuan Basin	314201	Maokou/Longtang-Jialingjiang /Maokou /Huanglong	31420102	Northwestern Depression/Foldbelt	Gas	Conventional	Frontier
3142	Sichuan Basin	314201	Maokou/Longtang-Jialingjiang /Maokou /Huanglong	31420103	Continuous Gas in Central Uplift/Northwestern Depression	Gas	Continuous	Frontier
3142	Sichuan Basin	314202	Daanzhai-Daanzhai/Lianggaoshan	31420201	Jurassic Lacustrine	Oil	Conventional	Frontier
3142	Sichuan Basin	314203	Xujiahe-Xujiahe/Shaximiao	31420301	Continuous Gas in Northwestern Depression/Central Uplift	Gas	Continuous	Frontier
3142	Sichuan Basin	314204	Cambrian/Silurian Marine Shale-Dengying/Lower Paleozoic	31420401	Leshan-Longnusi Paleohigh	Gas	Conventional	Frontier
3142	Sichuan Basin	314204	Cambrian/Silurian Marine Shale-Dengying/Lower Paleozoic	31420402	Lower Paleozoic of Southeastern Fold Belt	Gas	Conventional	Hypothetical
3144	Songliao Basin	314401	Qingshankou-Putaohua/Shaertu	31440101	Subtle Traps	Oil	Conventional	Established
3144	Songliao Basin	314401	Qingshankou-Putaohua/Shaertu	31440102	Anticlinal	Oil	Conventional	Established
3144	Songliao Basin	314402	Jurassic Coal-Denglouku/Nongan	31440201	Structural Traps	Gas	Conventional	Frontier
3154	Tarim Basin	315401	Ordovician/Jurassic-Phanerozoic	31540101	Tarim Basin Excluding Marginal Foldbelts	Oil	Conventional	Established
3154	Tarim Basin	315401	Ordovician/Jurassic-Phanerozoic	31540102	Kuche (Northern) Foldbelt	Gas	Conventional	Frontier
3154	Tarim Basin	315401	Ordovician/Jurassic-Phanerozoic	31540103	Southwest Foldbelt	Oil	Conventional	Frontier
3701	Baram Delta/Brunei-Sabah Basin	370101	Brunei-Sabah	37010101	Brunei-Sabah Deltaics	Oil	Conventional	Established
3701	Baram Delta/Brunei-Sabah Basin	370101	Brunei-Sabah	37010102	Brunei-Sabah Turbidites	Oil	Conventional	Frontier
3702	Greater Sarawak Basin	370201	Sarawak Basin	37020101	Central Luconia	Gas	Conventional	Established
3702	Greater Sarawak Basin	370201	Sarawak Basin	37020102	Balingian	Oil	Conventional	Established
3702	Greater Sarawak Basin	370202	East Natuna	37020201	East Natuna	Gas	Conventional	Frontier
3703	Malay Basin	370301	Oligocene-Miocene Lacustrine	37030101	South Malay Lacustrine	Oil	Conventional	Established
3703	Malay Basin	370301	Oligocene-Miocene Lacustrine	37030102	North Malay Lacustrine	Oil	Conventional	Frontier
3703	Malay Basin	370302	Miocene Coaly Strata	37030201	South Malay Coaly	Gas	Conventional	Established
3808	Central Sumatra Basin	380801	Brown Shale-Sihapas	38080101	Pematang/Sihapas Siliciclastics	Oil	Conventional	Established
3817	Kutei Basin	381701	Kutei Basin	38170101	Kutei Basin Deltaics	Oil	Conventional	Established
3817	Kutei Basin	381701	Kutei Basin	38170102	Kutei Basin Turbidites	Oil	Conventional	Hypothetical
3817	Kutei Basin	381701	Kutei Basin	38170103	Kutei Basin Fold and Thrust Belt	Oil	Conventional	Frontier
3822	North Sumatra Basin	382201	Bampo-Cenozoic	38220101	North Sumatra	Gas	Conventional	Established
3822	North Sumatra Basin	382201	Bampo-Cenozoic	38220102	Mergui	Gas	Conventional	Hypothetical
3824	Northwest Java Basin	382401	Banuwati-Oligocene/Miocene	38240101	Sunda/Asri	Oil	Conventional	Established
3824	Northwest Java Basin	382402	Jatibarang/Talang Akar-Oligocene/Miocene	38240201	Ardjuna	Oil	Conventional	Established
3824	Northwest Java Basin	382403	Tertiary-Parigi	38240301	Biogenic Gas	Gas	Continuous	Frontier
3824	Northwest Java Basin	382404	Tertiary-Cenozoic	38240401	Billiton Basin Hypothetical			Hypothetical
3828	South Sumatra Basin	382801	Lahat/Talang Akar-Cenozoic	38280101	South Sumatra	Oil	Conventional	Established
3910	Bonaparte Gulf Basin	391001	Milligans-Carboniferous/Permian	39100101	Barnett	Oil	Conventional	Frontier
3910	Bonaparte Gulf Basin	391002	Keyling/Hyland Bay-Permian	39100201	Petrel	Gas	Conventional	Frontier
3910	Bonaparte Gulf Basin	391002	Keyling/Hyland Bay-Permian	39100202	Vulcan Graben	Gas	Conventional	Hypothetical
3910	Bonaparte Gulf Basin	391003	Jurassic/Early Cretaceous-Mesozoic	39100301	Malita	Oil	Conventional	Established
3913	Browse Basin	391301	Late Jurassic/Early Cretaceous-Mesozoic	39130101	Late Jurassic/Early Cretaceous-Mesozoic	Gas	Conventional	Frontier

Province Code	Province Name	Total Petroleum System Code	Total Petroleum System Name	Assessment Unit Code	Assessment Unit Name	Major Commodity	Accumulation Type	Exploration Maturity
3930	Gippsland Basin	393001	Latrobe	39300101	Gippsland	Oil	Conventional	Established
3948	Northwest Shelf	394801	Dingo-Mungaroo/Barrow	39480101	Dingo-Mungaroo/Barrow	Gas	Conventional	Established
3948	Northwest Shelf	394802	Locker-Mungaroo/Barrow	39480201	Locker-Mungaroo/Barrow	Gas	Conventional	Frontier
4017	Vestford-Helgeland	401701	Upper Jurassic Spekk	40170101	Halten Terrace-Trondelag Platform	Oil	Conventional	Established
4017	Vestford-Helgeland	401701	Upper Jurassic Spekk	40170102	Mid-Norway Continental Margin	Gas	Conventional	Hypothetical
4025	North Sea Graben	402501	Kimmeridgian Shales	40250101	Viking Graben	Oil	Conventional	Established
4025	North Sea Graben	402501	Kimmeridgian Shales	40250102	Moray Firth	Oil	Conventional	Established
4025	North Sea Graben	402501	Kimmeridgian Shales	40250103	Central Graben	Oil	Conventional	Established
4036	Anglo-Dutch Basin	403601	Carboniferous-Rotliegend	40360101	Southern Permian Basin-U.K. Onshore	Oil	Conventional	Established
4036	Anglo-Dutch Basin	403601	Carboniferous-Rotliegend	40360102	Southern Permian Basin-Europe Onshore	Gas	Conventional	Established
4036	Anglo-Dutch Basin	403601	Carboniferous-Rotliegend	40360103	Southern Permian Basin-Offshore	Gas	Conventional	Established
4047	North Carpathian Basin	404701	Isotopically Light Gas	40470101	Foreland Basin	Gas		Established
4047	North Carpathian Basin	404702	Mesozoic/Paleogene Composite	40470201	Deformed Belt	Oil	Conventional	Established
4047	North Carpathian Basin	404703	Paleozoic Composite	40470301	Paleozoic Reservoirs	Oil	Conventional	Frontier
4048	Pannonian Basin	404801	Greater Hungarian Plain Neogene	40480101	Greater Hungarian Plain Basins	Oil	Conventional	Established
4048	Pannonian Basin	404802	Zala-Drava-Sava Mesozoic/Neogene	40480201	Zala-Drava-Sava Basins	Oil	Conventional	Established
4048	Pannonian Basin	404803	Danube Neogene	40480301	Danube Basin	Gas	Conventional	Frontier
4048	Pannonian Basin	404804	Transcarpathian Neogene	40480401	Transcarpathian Basin	Gas	Conventional	Frontier
4048	Pannonian Basin	404805	Central Carpathian Paleogene	40480501	Central Carpathian Paleogene Basin	Oil	Conventional	Frontier
4048	Pannonian Basin	404806	Hungarian Paleogene	40480601	Hungarian Paleogene Basin	Oil	Conventional	Frontier
4057	Transylvanian Basin	405701	Transylvanian Composite	40570101	Transylvanian Neogene Suprasalt Gas	Gas		Established
4057	Transylvanian Basin	405701	Transylvanian Composite	40570102	Transylvanian Subsalt			Hypothetical
4060	Po Basin	406001	Porto Garibaldi	40600101	Neogene Flysch Gas	Gas		Established
4060	Po Basin	406002	Meride/Riva di Solto	40600201	Thermal Triassic	Oil	Conventional	Frontier
4061	Carpathian-Balkanian Basin	406101	Moesian Platform Composite	40610101	Moesian Platform	Oil	Conventional	Established
4061	Carpathian-Balkanian Basin	406102	Dysodile Schist-Tertiary	40610201	Romania Flysch Zone	Oil	Conventional	Established
4061	Carpathian-Balkanian Basin	406102	Dysodile Schist-Tertiary	40610202	Romania Ploiesti Zone	Oil	Conventional	Established
4068	Provence Basin	406801	Pre-Messinian	40680101	Subsalt	Gas	Conventional	Hypothetical
5200	East Greenland Rift Basins	520001	Permian/Upper Jurassic Composite	52000101	Northeast Greenland Shelf Rift Systems	Oil	Conventional	Hypothetical
5215	Labrador-Newfoundland Shelf	521501	Egret-Hibernia	52150101	Jeanne d'Arc	Oil	Conventional	Frontier
5215	Labrador-Newfoundland Shelf	521502	Mesozoic Composite	52150201	Graben Systems Exclusive of Jeanne d'Arc	Gas	Conventional	Frontier
5243	Alberta Basin	524301	Keg River-Keg River	52430101	Keg River Gas	Gas	Conventional	Established
5243	Alberta Basin	524301	Keg River-Keg River	52430102	Keg River Oil and Gas	Oil	Conventional	Established
5243	Alberta Basin	524302	Duvernay-Leduc	52430201	Leduc Gas	Gas	Conventional	Established
5243	Alberta Basin	524302	Duvernay-Leduc	52430202	Leduc Oil and Gas	Oil	Conventional	Established
5243	Alberta Basin	524303	Exshaw-Rundle	52430301	Exshaw-Rundle Gas	Gas	Conventional	Established
5243	Alberta Basin	524303	Exshaw-Rundle	52430302	Exshaw-Rundle Oil and Gas	Oil	Conventional	Established
5243	Alberta Basin	524304	Combined Triassic/Jurassic	52430401	Combined Triassic/Jurassic Gas	Gas	Conventional	Established
5243	Alberta Basin	524304	Combined Triassic/Jurassic	52430402	Combined Triassic/Jurassic Oil and Gas	Oil	Conventional	Established
5243	Alberta Basin	524305	Mannville-Upper Mannville	52430501	Mannville Gas	Gas	Conventional	Established
5243	Alberta Basin	524305	Mannville-Upper Mannville	52430502	Mannville Continuous (Basin-Center) Gas	Gas	Continuous	Established
5243	Alberta Basin	524306	Second White Specks-Cardium	52430601	Second White Specks-Cardium Gas	Gas	Conventional	Established
5243	Alberta Basin	524306	Second White Specks-Cardium	52430602	Second White Specks-Cardium Oil and Gas	Oil	Conventional	Established
5243	Alberta Basin	524306	Second White Specks-Cardium	52430603	Upper Cretaceous Eastern Shallow Gas	Gas	Continuous	Established
5244	Williston Basin, Canada	524401	Yeoman	52440101	Yeoman Oil	Oil	Conventional	Frontier
5244	Williston Basin, Canada	524402	Brightholme	52440201	Brightholme Oil	Oil	Conventional	Frontier
5244	Williston Basin, Canada	524403	Bakken	52440301	Bakken Sandstone	Oil	Conventional	Frontier
5244	Williston Basin, Canada	524404	Lodgepole	52440401	Lodgepole Oil	Oil	Conventional	Established
5305	Villahermosa Uplift	530501	Pimienta-Tamabra	53050101	El Abra-Like Reef and Backreef Limestone	Oil	Conventional	Established
5305	Villahermosa Uplift	530501	Pimienta-Tamabra	53050102	Tamabra-Like Debris-Flow-Breccia Limestone Overlying Evaporites	Oil	Conventional	Established

Province Code	Province Name	Total Petroleum System Code	Total Petroleum System Name	Assessment Unit Code	Assessment Unit Name	Major Commodity	Accumulation Type	Exploration Maturity
5305	Villahermosa Uplift	530501	Pimienta-Tamabra	53050103	Tamabra-Like Debris-Flow-Breccia Limestone and Overlying Strata	Oil	Conventional	Established
5305	Villahermosa Uplift	530501	Pimienta-Tamabra	53050104	Tamabra-Like Debris-Flow-Breccia Limestone of the Golden Lane	Oil	Conventional	Established
5305	Villahermosa Uplift	530501	Pimienta-Tamabra	53050105	Tamaulipas-Like Basinal Limestone and Tertiary Strata Without Underlying Evaporites	Oil	Conventional	Established
5305	Villahermosa Uplift	530501	Pimienta-Tamabra	53050106	Tamaulipas-Like Basinal Limestone and Tertiary Strata Overlying Evaporites	Oil	Conventional	Established
5305	Villahermosa Uplift	530501	Pimienta-Tamabra	53050107	Tamabra-Like Debris-Flow-Breccia Limestone North of Campeche	Oil	Conventional	Hypothetical
6021	Guyana-Suriname Basin	602101	Cenomanian-Turonian	60210101	Late Cretaceous-Tertiary Turbidites	Oil	Conventional	Hypothetical
6021	Guyana-Suriname Basin	602101	Cenomanian-Turonian	60210102	Cretaceous Carbonates	Oil	Conventional	Hypothetical
6021	Guyana-Suriname Basin	602101	Cenomanian-Turonian	60210103	Late Cretaceous-Tertiary Nearshore Sandstones	Oil	Conventional	Frontier
6022	Foz do Amazonas Basin	602201	Neogene	60220101	Amazon Delta and Submarine Fan	Gas	Conventional	Frontier
6022	Foz do Amazonas Basin	602201	Neogene	60220102	Offshore Gas Hydrates		Continuous	Hypothetical
6029	Sergipe-Alagoas Basin	602901	Neocomian to Turonian Composite	60290101	Western Pre-Aptian Reservoirs	Oil	Conventional	Established
6029	Sergipe-Alagoas Basin	602901	Neocomian to Turonian Composite	60290102	Late Cretaceous-Tertiary Deep-Water Sandstones	Oil	Conventional	Frontier
6034	Espirito Santo Basin	603401	Cretaceous Composite	60340101	Espirito Santo Shelf	Oil	Conventional	Established
6034	Espirito Santo Basin	603401	Cretaceous Composite	60340102	Late Cretaceous-Tertiary Slide Blocks and Turbidites	Oil	Conventional	Frontier
6034	Espirito Santo Basin	603401	Cretaceous Composite	60340103	Abrolhos Sub-Volcanic Structures	Oil	Conventional	Hypothetical
6035	Campos Basin	603501	Lagoa Feia-Carapebus	60350101	Late Cretaceous-Tertiary Turbidites	Oil	Conventional	Established
6035	Campos Basin	603501	Lagoa Feia-Carapebus	60350102	Cretaceous Carbonates	Oil	Conventional	Frontier
6035	Campos Basin	603501	Lagoa Feia-Carapebus	60350103	Salt Dome Province Tertiary Sandstones	Oil	Conventional	Hypothetical
6036	Santos Basin	603601	Guaratiba-Guaruja (Cretaceous) Composite	60360101	Santos Shelf	Oil	Conventional	Frontier
6036	Santos Basin	603601	Guaratiba-Guaruja (Cretaceous) Composite	60360102	Salt-Structured Deep-Water Sandstones	Oil	Conventional	Hypothetical
6037	Pelotas Basin	603701	Cenomanian-Turonian-Tertiary Composite	60370101	Pelotas Platform and Basin	Oil	Conventional	Hypothetical
6037	Pelotas Basin	603701	Cenomanian-Turonian-Tertiary Composite	60370102	Offshore Gas Hydrates		Continuous	Hypothetical
6041	Putumayo-Oriente-Maranon Basin	604101	Mesozoic-Cenozoic	60410101	Hollin-Napo	Oil	Conventional	Established
6041	Putumayo-Oriente-Maranon Basin	604102	Paleozoic	60410201	Ene	Oil	Conventional	Hypothetical
6045	Santa Cruz-Tarija Basin	604501	Los Monos-Machareti	60450101	Sub-Andean Fold and Thrust Belt	Gas	Conventional	Established
6045	Santa Cruz-Tarija Basin	604501	Los Monos-Machareti	60450102	Foreland Basins	Oil	Conventional	Frontier
6045	Santa Cruz-Tarija Basin	604501	Los Monos-Machareti	60450103	Foreland Central Chaco High	Oil	Conventional	Hypothetical
6055	Neuquen Basin	605501	Neuquen Hybrid	60550101	Neuquen Extensional Structures	Oil	Conventional	Established
6055	Neuquen Basin	605501	Neuquen Hybrid	60550102	Neuquen Foothills Structure	Oil	Conventional	Established
6055	Neuquen Basin	605501	Neuquen Hybrid	60550103	Dorsal de Neuquen Structure	Oil	Conventional	Established
6058	San Jorge Basin	605801	D-129	60580101	San Jorge Extensional Structures	Oil	Conventional	Established
6058	San Jorge Basin	605801	D-129	60580102	San Bernardo Fold Belt Structures	Oil	Conventional	Established
6058	San Jorge Basin	605802	Aguada Bandera	60580201	Hypothetical Basin-Centered Gas	Gas	Continuous	Hypothetical
6059	Magallanes Basin	605901	Lower Inoceramus	60590101	Magallanes Extensional Structures	Gas	Conventional	Established
6059	Magallanes Basin	605901	Lower Inoceramus	60590102	Andean Fold Belt Structures	Oil	Conventional	Frontier
6060	Falklands Plateau	606001	Neocomian Lacustrine	60600101	North Falklands Basin	Oil	Conventional	Hypothetical
6060	Falklands Plateau	606002	Lower Cretaceous	60600201	South Falklands Basin	Oil	Conventional	Hypothetical
6063	Malvinas Basin	606301	Lower Cretaceous Marine	60630101	Malvinas Extensional Structures	Oil	Conventional	Frontier
6081	Talara Basin	608101	Cretaceous-Tertiary	60810101	Cretaceous-Paleogene Basin	Oil	Conventional	Established
6083	Progreso Basin	608301	Neogene	60830101	Neogene Pull-Apart Basin	Oil	Conventional	Frontier
6083	Progreso Basin	608302	Cretaceous-Paleogene	60830201	Cretaceous-Paleogene Santa Elena Block	Oil	Conventional	Frontier
6090	Middle Magdalena	609001	La Luna-La Paz	60900101	Northern	Oil	Conventional	Established
6090	Middle Magdalena	609001	La Luna-La Paz	60900102	Southern	Oil	Conventional	Frontier
6090	Middle Magdalena	609001	La Luna-La Paz	60900103	Eastern	Gas	Conventional	Frontier
6090	Middle Magdalena	609001	La Luna-La Paz	60900104	La Luna and Older	Oil	Conventional	Frontier

Province Code	Province Name	Total Petroleum System Code	Total Petroleum System Name	Assessment Unit Code	Assessment Unit Name	Major Commodity	Accumulation Type	Exploration Maturity
6096	Llanos Basin	609601	Gacheta-Mirador	60960101	Central	Oil	Conventional	Established
6096	Llanos Basin	609601	Gacheta-Mirador	60960102	Peripheral	Oil	Conventional	Established
6098	East Venezuela Basin	609801	Querecual	60980101	East Venezuela Fold and Thrust Belt	Oil	Conventional	Established
6098	East Venezuela Basin	609801	Querecual	60980102	Guarico Sub-Basin	Gas	Conventional	Established
6098	East Venezuela Basin	609801	Querecual	60980103	Maturin Sub-Basin	Oil	Conventional	Established
6098	East Venezuela Basin	609801	Querecual	60980104	Orinoco Heavy Oil and Tar Belt	Oil	Continuous	Frontier
6098	East Venezuela Basin	609802	Upper Cretaceous/Tertiary	60980201	Trinidad Basins	Oil	Conventional	Established
6098	East Venezuela Basin	609802	Upper Cretaceous/Tertiary	60980202	Orinoco Delta and Offshore	Oil	Conventional	Hypothetical
6099	Maracaibo Basin	609901	La Luna/Maracaibo	60990101	Main Maracaibo Basin	Oil	Conventional	Established
6099	Maracaibo Basin	609901	La Luna/Maracaibo	60990102	Southwest Maracaibo Basin Fold Belt	Oil	Conventional	Established
6099	Maracaibo Basin	609901	La Luna/Maracaibo	60990103	Basin-Centered Gas	Gas	Continuous	Hypothetical
6103	Tobago Trough	610301	Lower Cruse	61030101	Carupano Basin Gas	Gas	Conventional	Frontier
6107	Lesser Antilles Deformed Belt	610701	Tobago Trough Paleogene	61070101	Inner Forearc Deformation Belt	Oil	Conventional	Hypothetical
6107	Lesser Antilles Deformed Belt	610701	Tobago Trough Paleogene	61070102	Barbados Accreationary Prism	Oil	Conventional	Established
6117	Greater Antilles Deformed Belt	611701	Upper Jurassic-Neocomian	61170101	North Cuba Fold and Thrust Belt	Oil	Conventional	Established
7013	Senegal	701301	Cretaceous-Tertiary Composite	70130101	Coastal Plain and Offshore	Oil	Conventional	Frontier
7183	Gulf of Guinea	718301	Cretaceous Composite	71830101	Coastal Plain and Offshore	Oil	Conventional	Established
7192	Niger Delta	719201	Tertiary Niger Delta (Agbada/Akata)	71920101	Agbada Reservoirs	Oil	Conventional	Established
7192	Niger Delta	719201	Tertiary Niger Delta (Agbada/Akata)	71920102	Akata Reservoirs	Oil	Conventional	Hypothetical
7203	West-Central Coastal	720301	Melania-Gamba	72030101	Gabon Subsalt	Oil	Conventional	Established
7203	West-Central Coastal	720302	Azile-Senonian	72030201	Gabon Suprasalt	Oil	Conventional	Established
7203	West-Central Coastal	720303	Congo Delta Composite	72030301	Central Congo Delta and Carbonate Platform	Oil	Conventional	Established
7203	West-Central Coastal	720303	Congo Delta Composite	72030302	Central Congo Turbidites	Oil	Conventional	Established
7203	West-Central Coastal	720304	Cuanza Composite	72030401	Cuanza-Namibe	Oil	Conventional	Frontier
7303	Orange River Coastal	730301	Cretaceous Composite	73030101	Offshore	Gas	Conventional	Frontier
8026	Kohat-Potwar	802601	Patala-Namal	80260101	Kohat-Potwar Intrathrust Basin	Oil	Conventional	Established
8034	Assam	803401	Sylhet-Kopili/Barail-Tipam Composite	80340101	Sylhet-Kopili/Barail-Tipam Composite	Oil	Conventional	Established
8042	Indus	804201	Sembar-Goru/Ghazij	80420101	Greater Indus Foreland and Foldbelt	Gas	Conventional	Hypothetical
8042	Indus	804201	Sembar-Goru/Ghazij	80420102	Indus Fan	Gas	Conventional	Hypothetical
8043	Bombay	804301	Eocene-Miocene Composite	80430101	Eocene-Miocene Bombay Shelf	Oil	Conventional	Established
8043	Bombay	804302	Eocene-Miocene Deltaics	80430102	Eocene-Miocene Deltaics	Oil	Conventional	Established
8047	Ganges-Brahmaputra Delta	804701	Permian Coal	80470101	Coalbed Gas	Gas	Continuous	Hypothetical
8047	Ganges-Brahmaputra Delta	804702	Jalangi-Sylhet/Burdwan Composite	80470201	Western Shelf and Slope	Gas	Conventional	Frontier
8047	Ganges-Brahmaputra Delta	804703	Jenam/Bhuban-Bokabil	80470301	Central Basin	Gas	Conventional	Hypothetical
8047	Ganges-Brahmaputra Delta	804703	Jenam/Bhuban-Bokabil	80470302	Eastern Fold Belt	Gas	Conventional	Established
8048	Irrawaddy	804801	Eocene to Miocene Composite	80480101	Central Burma Basin	Oil	Conventional	Established
8048	Irrawaddy	804801	Eocene to Miocene Composite	80480102	Irrawaddy-Andaman	Gas	Conventional	Established

Assessment Unit Code	Assessment Unit Name	Known Oil (BBO)	Known Gas (TCFG)	Known NGL (BBNGL)	Known Total Petroleum (BBOE)	Undiscovered Oil (Mean, BBO)	Undiscovered Gas (Mean, TCFG)	Undiscovered NGL (Mean, BBNGL)	Undiscovered Total Petroleum (Mean, BBOE)
10080101	Northwest Izhma-Pechora Depression	NA	NA	NA	NA	0.018	0.120	0.005	0.043
10080102	Main Basin Platform	12.985	20.515	0.199	16.603	4.938	14.515	0.650	8.006
10080103	Foredeep Basins	0.084	16.117	0.517	3.288	0.214	19.891	0.800	4.329
10090101	Carboniferous-Lower Permian Clastics	1.611	59.098	0.200	11.660	0.251	14.735	0.457	3.163
10090102	Devonian Synrift	0.000	0.000	0.000	0.000	0.848	9.317	0.405	2.806
10090103	Continuous Basin-Centered Gas Accumulation	0.000	0.000	0.000	0.000	NA	NA	NA	NA
10150101	Supra-Domanik Carbonates/Clastics	34.904	77.078	0.857	48.607	1.676	1.141	0.047	1.913
10150102	Lower Volga	2.529	9.932	0.064	4.248	0.148	0.558	0.021	0.262
10150103	Sub-Domanik Devonian Clastics	26.439	9.448	0.175	28.189	0.682	0.695	0.037	0.836
10150104	Self-Sourced Domanik	0.000	0.000	0.000	0.000	NA	NA	NA	NA
10150201	Permian Reefs/Thrust Folds	0.065	2.638	0.005	0.509	0.042	1.547	0.093	0.392
10160101	North and West Margins Subsalt Pinnacle Reefs	NA	NA	NA	NA	1.522	30.544	1.737	8.349
10160102	North and West Margins Subsalt Barrier Reefs	0.134	1.270	0.000	0.346	0.637	4.271	0.107	1.456
10160103	East and Southeast Margins Subsalt	1.217	5.364	0.265	2.376	2.196	6.781	0.385	3.711
10160104	South Margin Subsalt	6.245	98.100	3.400	25.995	15.467	67.521	2.802	29.522
10160105	Basin Center Subsalt	0.000	0.000	0.000	0.000	NA	NA	NA	NA
10160106	Suprasalt	1.830	1.628	0.015	2.116	3.497	9.934	0.382	5.535
10500101	Kolguyev Terrace	0.051	0.003	0.000	0.052	0.563	17.554	0.731	4.219
10500102	South Barents and Ludlov Saddle	0.000	78.140	0.100	13.123	0.257	189.244	3.940	35.738
10500103	North Barents	0.000	0.000	0.000	0.000	0.135	59.820	1.250	11.355
11080101	Tertiary Foredeep	0.498	3.944	0.002	1.157	0.280	3.139	0.069	0.872
11080102	Foreland Slope	0.020	14.595	0.049	2.502	0.048	6.894	0.076	1.272
11080103	Subsalt Jurassic	0.005	0.165	0.005	0.037	0.090	2.219	0.071	0.530
11090101	Foldbelt-Foothills	3.490	7.840	0.029	4.826	2.094	14.512	0.596	5.109
11090102	Terek-Sunzha Subsalt Jurassic	0.000	0.000	0.000	0.000	0.283	11.464	0.688	2.882
11090103	Foreland Slope and Foredeep	0.000	0.000	0.000	0.000	0.452	1.804	0.064	0.816
11090201	South Mangyshlak (Entire)	5.241	5.707	0.046	6.239	0.651	2.262	0.072	1.100
11090301	Offshore Prikumsk Zone	0.000	0.000	0.000	0.000	0.789	3.910	0.141	1.582
11090302	Onshore Stavropol-Prikumsk	0.820	13.499	0.003	3.073	0.287	1.163	0.040	0.521
11090303	Central Caspian Offshore	0.000	0.000	0.000	0.000	0.331	4.572	0.069	1.161
11120101	Apsheron-Pribalkhan Zone	15.093	18.729	0.234	18.448	3.091	7.908	0.306	4.715
11120102	Lower Kura Depression and Adjacent Shelf	2.040	8.224	0.182	3.592	1.275	32.091	1.219	7.843
11120103	Gograndag-Okarem Zone and Adjacent Shelf	0.306	9.042	0.087	1.900	4.764	47.515	2.396	15.079
11120104	Central Offshore	0.000	0.000	0.000	0.000	6.410	82.836	4.051	24.267
11120105	Iran Onshore/Nearshore	0.000	0.000	0.000	0.000	0.185	2.960	0.142	0.821
11500101	Mesozoic Sandstone Reservoirs	2.286	0.943	0.000	2.443	0.814	0.089	0.002	0.831
11500201	Jurassic-Tertiary	0.080	1.450	0.008	0.330	0.139	2.180	0.069	0.572
11500301	Upper Paleozoic Carbonates	NA	NA	NA	NA	0.000	8.500	0.374	1.790
11540101	Northern and Western Areas	0.140	52.412	0.112	8.987	0.070	9.547	0.161	1.822
11540102	Karabil-Badkhyz (Southern Area)	0.111	58.110	0.113	9.909	0.088	7.657	0.127	1.491
11540103	Murgab Depression Suprasalt	0.000	49.516	0.103	8.356	0.000	4.684	0.033	0.813
11540104	Murgab Depression Subsalt	0.516	70.576	0.848	13.126	0.113	141.762	3.675	27.414
11740101	Upper Jurassic-Cretaceous Sandstones	117.972	97.582	0.764	135.000	42.947	96.934	5.749	64.852
11740102	Self-Sourced Bazhenov Fractured Reservoirs	0.000	0.000	0.000	0.000	NA	NA	NA	NA
11740201	Pre-Upper Jurassic	11.756	5.180	0.003	12.623	7.474	5.516	0.316	8.709
11740301	Northern West Siberian Onshore Gas	10.185	1085.588	2.326	193.442	1.304	136.608	3.668	27.741
11740302	South Kara Sea Offshore	0.000	81.700	0.048	13.665	3.510	403.875	10.739	81.561
12070101	Yenisey Foldbelt Riphean-Craton Margin Riphean	0.034	5.052	0.060	0.936	1.213	16.983	0.854	4.897
12100101	Baikal-Patom Foldbelt Riphean-Craton Margin Vendian	2.006	30.210	0.360	7.401	1.611	46.039	2.194	11.478
13220101	Onshore and Offshore Northeastern Shelf	2.182	22.383	0.165	6.077	2.582	57.451	2.999	15.156
20040101	Ma'Rib-Al Jawf/Shabwah/Masila	1.933	17.120	0.312	5.098	2.740	16.982	0.924	6.494
20140101	Ghaba-Makarem Combined Structural	2.028	20.339	0.507	5.925	0.743	10.112	0.443	2.871
20160101	Fahud-Huqf Combined Structural	2.685	7.633	0.058	4.015	1.304	8.206	0.287	2.959
20160201	Natih-Fiqa Structural/Stratigraphic	1.852	3.100	0.100	2.469	0.371	0.363	0.018	0.450
20190101	Cretaceous Reservoirs in Northwest Desert Anticlines	18.981	16.312	0.030	21.729	4.702	2.819	0.127	5.299
20190102	Cretaceous Reservoirs in South Gulf Suprasalt Structural	52.094	67.795	0.643	64.036	3.764	4.073	0.228	4.670
20190103	Mesozoic/Tertiary Foredeep Fold and Thrust	0.156	17.993	1.228	4.382	0.513	15.419	0.728	3.811

Assessment Unit Code	Assessment Unit Name	Known Oil (BBO)	Known Gas (TCFG)	Known NGL (BBNGL)	Known Total Petroleum (BBOE)	Undiscovered Oil (Mean, BBO)	Undiscovered Gas (Mean, TCFG)	Undiscovered NGL (Mean, BBNGL)	Undiscovered Total Petroleum (Mean, BBOE)
20190201	Jurassic Reservoirs in Northwest Desert Anticlines	0.777	6.500	0.200	2.060	10.633	46.058	2.399	20.708
20190202	Jurassic Reservoirs in South Gulf Suprasalt/Qatar Arch Structural	18.236	71.012	0.392	30.463	3.845	12.457	0.684	6.606
20190301	Khuff Carbonates in Salt Structures	0.500	452.000	13.815	89.648	0.000	22.999	1.012	4.845
20190302	Paleozoic Reservoirs	NA	NA	NA	NA	13.235	323.892	25.135	92.352
20210101	Central Arch Horst-Block Anticlinal Oil and Gas	6.226	12.115	0.343	8.588	19.190	230.387	17.861	75.448
20210102	North Gulf Salt Basin Structural Gas	0.150	67.000	0.000	11.317	0.000	152.952	6.727	32.219
20210201	Horst-Block Anticlinal Oil	155.145	229.615	8.400	201.814	26.715	16.055	0.642	30.033
20210202	Salt-Involved Structural Oil	43.850	45.680	0.805	52.268	14.989	17.991	0.719	18.707
20230101	Horst/Graben-Related Oil and Gas	1.213	1.661	0.025	1.515	2.632	64.510	5.080	18.464
20230201	Platform Horst/Graben-Related Oil	0.474	0.100	0.000	0.490	4.682	8.741	0.444	6.583
20230202	Basinal Oil and Gas	0.000	0.000	0.000	0.000	1.886	14.201	0.670	4.923
20300101	Cretaceous Reservoirs	262.917	233.798	1.256	303.139	74.911	104.098	3.157	95.417
20300102	Tertiary Reservoirs	109.310	259.440	0.306	152.856	19.605	93.499	4.804	39.992
20300201	Northern Qatar Arch Extension	0.000	131.220	0.925	22.795	0.000	69.031	3.039	14.544
20430101	Southeast Sirte Clastics	20.895	9.587	0.100	22.593	2.380	0.808	0.048	2.563
20430102	Central Sirte Carbonates	16.177	28.180	0.029	20.902	3.840	9.918	0.413	5.906
20430103	Offshore Sirte Hypothetical	0.000	0.000	0.000	0.000	0.382	4.199	0.183	1.265
20430104	Southeast Sirte Hypothetical	0.000	0.000	0.000	0.000	0.253	0.060	0.004	0.266
20480101	Bou Dabbous-Tertiary Structural/Stratigraphic	2.114	15.509	0.044	4.743	0.667	2.746	0.064	1.189
20480201	Jurassic-Cretaceous Structural/Stratigraphic	0.103	1.301	0.023	0.343	0.403	2.280	0.027	0.810
20540101	Tanezzuft-Oued Mya Structural/Stratigraphic	10.843	8.973	0.000	12.338	0.830	2.341	0.110	1.330
20540201	Tanezzuft-Melrhir Structural/Stratigraphic	0.006	0.125	0.009	0.036	1.875	4.887	0.269	2.959
20540301	Tanezzuft-Ghadames Structural/Stratigraphic	4.538	16.484	1.011	8.296	4.461	12.035	0.908	7.376
20560101	Tanezzuft-Illizi Structural/Stratigraphic	3.670	45.061	0.898	12.078	2.814	27.785	0.873	8.318
20580101	Tanezzuft-Timimoun Structural/Stratigraphic	0.000	4.200	0.000	0.700	0.031	1.128	0.056	0.275
20580201	Tanezzuft-Ahnet Structural/Stratigraphic	0.001	3.117	0.090	0.610	0.034	2.973	0.149	0.678
20580301	Tanezzuft-Sbaa Structural/Stratigraphic	0.284	1.490	0.010	0.542	0.162	0.645	0.011	0.281
20580401	Tanezzuft-Mouydir Structural/Stratigraphic	0.000	0.000	0.000	0.000	0.012	0.292	0.014	0.075
20580501	Tanezzuft-Benoud Structural/Stratigraphic	0.088	105.050	4.935	22.531	0.072	2.541	0.125	0.621
20580601	Tanezzuft-Bechar/Abadla Structural/Stratigraphic	0.000	0.000	0.000	0.000	0.016	0.441	0.022	0.111
20710101	Gulf of Suez Block-Fault Fairway	9.754	5.960	0.041	10.788	2.171	4.172	0.171	3.038
20710102	Gulf of Suez Qaa Plain	0.000	0.000	0.000	0.000	0.093	0.046	0.001	0.102
20710103	Southern Gulf of Suez	0.057	0.035	0.000	0.062	0.344	5.974	0.312	1.651
20710201	Red Sea Coastal Block Faults	0.155	3.527	0.256	0.999	2.136	43.946	2.634	12.094
20710202	Red Sea Salt Basin	0.000	0.000	0.000	0.000	0.443	5.875	0.226	1.649
31150101	Upper Paleozoic/Lower Mesozoic Nonmarine Coarse Clastics	6.624	2.248	0.000	6.998	0.467	0.346	0.021	0.546
31150201	Jurassic/Tertiary Fluvial and Lacustrine Sandstone	0.210	0.342	0.000	0.267	0.096	0.950	0.043	0.297
31150301	Continuous-Type (Basin-Centered) Gas	0.000	0.000	0.000	0.000	NA	NA	NA	NA
31270101	Tertiary Lacustrine	19.937	11.853	0.085	21.998	2.268	4.639	0.223	3.264
31270102	Pre-Tertiary Buried Hills	4.616	3.819	0.002	5.255	2.144	3.875	0.194	2.984
31270201	Coal-Sourced Gas	0.000	0.000	0.000	0.000	NA	NA	NA	NA
31280101	Jurassic/Triassic Fluvial and Lacustrine Sandstone	0.683	0.020	0.000	0.686	0.138	0.020	0.001	0.143
31280201	Basin Centered Gas (Continuous) in Ordovician Carbonate and Permian Sandstone	0.000	5.590	0.000	0.932	NA	NA	NA	NA
31280202	Carboniferous and Permian Coalbed Methane (Continuous)	0.000	0.000	0.000	0.000	NA	NA	NA	NA
31420101	Southeastern Fold Belt	0.000	10.681	0.000	1.781	0.000	4.488	0.197	0.945
31420102	Northwestern Depression/Foldbelt	0.000	0.340	0.000	0.057	0.000	1.650	0.073	0.348
31420103	Continuous Gas in Central Uplift/Northwestern Depression	0.000	0.051	0.000	0.009	NA	NA	NA	NA
31420201	Jurassic Lacustrine	0.109	0.140	0.000	0.132	0.027	0.137	0.008	0.059
31420301	Continuous Gas in Northwestern Depression/Central Uplift	NA	NA	NA	NA	NA	NA	NA	NA
31420401	Leshan-Longnusi Paleohigh	0.001	1.350	0.000	0.226	0.000	2.763	0.122	0.582
31420402	Lower Paleozoic of Southeastern Fold Belt	0.000	0.000	0.000	0.000	0.000	1.273	0.056	0.268
31440101	Subtle Traps	1.150	0.913	0.000	1.302	0.735	1.789	0.083	1.116
31440102	Anticlinal	14.420	0.685	0.000	14.534	0.291	0.705	0.033	0.441
31440201	Structural Traps	0.005	0.145	0.000	0.029	0.000	3.212	0.096	0.632
31540101	Tarim Basin Excluding Marginal Foldbelts	0.655	5.154	0.149	1.664	5.364	41.454	2.478	14.751
31540102	Kuche (Northern) Foldbelt	0.007	0.066	0.002	0.020	0.089	10.712	0.475	2.349

Assessment Unit Code	Assessment Unit Name	Known Oil (BBO)	Known Gas (TCFG)	Known NGL (BBNGL)	Known Total Petroleum (BBOE)	Undiscovered Oil (Mean, BBO)	Undiscovered Gas (Mean, TCFG)	Undiscovered NGL (Mean, BBNGL)	Undiscovered Total Petroleum (Mean, BBOE)
31540103	Southwest Foldbelt	0.041	0.560	0.038	0.172	0.495	7.773	0.359	2.150
37010101	Brunei-Sabah Deltaics	6.652	33.755	0.180	12.457	2.137	15.977	0.799	5.599
37010102	Brunei-Sabah Turbidites	0.246	2.445	0.000	0.654	1.252	7.249	0.375	2.835
37020101	Central Luconia	0.170	35.035	0.366	6.375	0.000	15.067	0.256	2.767
37020102	Balingian	0.628	2.084	0.013	0.988	0.529	2.217	0.116	1.015
37020201	East Natuna	0.020	45.045	0.000	7.528	0.102	1.963	0.090	0.519
37030101	South Malay Lacustrine	3.017	24.073	0.131	7.159	1.003	21.935	0.410	5.069
37030102	North Malay Lacustrine	0.000	0.175	0.006	0.035	0.124	1.762	0.032	0.449
37030201	South Malay Coaly	0.592	23.901	0.186	4.761	0.101	3.768	0.072	0.802
38080101	Pematang/Sihapas Siliciclastics	13.217	3.866	0.011	13.872	0.899	4.078	0.084	1.663
38170101	Kutei Basin Deltaics	2.879	44.473	1.253	11.545	1.258	35.757	1.017	8.234
38170102	Kutei Basin Turbidites	0.000	0.000	0.000	0.000	3.569	25.687	0.991	8.842
38170103	Kutei Basin Fold and Thrust Belt	NA	NA	NA	NA	0.318	3.313	0.114	0.984
38220101	North Sumatra	0.674	25.559	0.926	5.860	0.112	9.117	0.107	1.738
38220102	Mergui	0.000	0.000	0.000	0.000	0.101	6.005	0.073	1.176
38240101	Sunda/Asri	1.259	0.614	0.000	1.362	0.462	0.874	0.019	0.626
38240201	Ardjuna	1.908	7.322	0.175	3.303	0.195	6.579	0.089	1.381
38240301	Biogenic Gas	0.000	0.119	0.000	0.020	NA	NA	NA	NA
38240401	Billiton Basin Hypothetical	0.000	0.000	0.000	0.000	NA	NA	NA	NA
38280101	South Sumatra	2.429	10.204	0.056	4.186	0.469	18.250	0.239	3.750
39100101	Barnett	0.015	0.048	0.000	0.023	0.140	1.151	0.056	0.387
39100201	Petrel	0.000	5.730	0.050	1.005	0.000	8.328	0.367	1.755
39100202	Vulcan Graben	0.000	0.000	0.000	0.000	0.000	2.668	0.117	0.562
39100301	Malita	0.502	7.526	0.385	2.141	1.146	11.346	0.540	3.577
39130101	Late Jurassic/Early Cretaceous-Mesozoic	0.046	17.960	0.200	3.239	1.055	20.093	0.934	5.339
39300101	Gippsland	3.860	9.775	0.701	6.190	0.310	5.658	0.339	1.592
39480101	Dingo-Mungaroo/Barrow	1.149	48.245	0.991	10.181	2.249	56.380	3.314	14.960
39480201	Locker-Mungaroo/Barrow	0.006	8.102	0.000	1.356	0.132	8.331	0.368	1.888
40170101	Halten Terrace-Trondelag Platform	2.660	15.662	0.702	5.973	6.204	57.369	4.202	19.968
40170102	Mid-Norway Continental Margin	0.000	0.000	0.000	0.000	0.334	107.832	4.788	23.095
40250101	Viking Graben	26.681	112.742	3.600	49.072	7.412	22.192	1.036	12.147
40250102	Moray Firth	4.444	5.852	0.322	5.741	0.947	1.752	0.095	1.334
40250103	Central Graben	12.769	40.320	2.049	21.538	4.740	13.801	0.728	7.767
40360101	Southern Permian Basin-U.K. Onshore	0.064	0.099	0.000	0.080	0.007	0.031	0.001	0.014
40360102	Southern Permian Basin-Europe Onshore	2.400	148.227	0.031	27.136	0.074	12.730	0.027	2.223
40360103	Southern Permian Basin-Offshore	0.408	73.834	0.141	12.855	0.062	13.029	0.041	2.274
40470101	Foreland Basin	0.001	10.660	0.003	1.782	0.000	2.321	0.019	0.405
40470201	Deformed Belt	0.923	5.393	0.006	1.828	0.344	2.345	0.075	0.810
40470301	Paleozoic Reservoirs	0.001	0.233	0.000	0.039	0.047	0.383	0.019	0.129
40480101	Greater Hungarian Plain Basins	1.089	9.657	0.024	2.722	0.196	2.395	0.074	0.669
40480201	Zala-Drava-Sava Basins	1.101	3.929	0.063	1.818	0.114	1.151	0.036	0.342
40480301	Danube Basin	0.000	0.247	0.000	0.041	0.000	0.146	0.003	0.027
40480401	Transcarpathian Basin	0.000	0.261	0.000	0.043	0.000	0.112	0.002	0.021
40480501	Central Carpathian Paleogene Basin	NA	NA	NA	NA	NA	NA	NA	NA
40480601	Hungarian Paleogene Basin	0.011	0.016	0.000	0.014	0.049	0.339	0.013	0.118
40570101	Transylvanian Neogene Suprasalt Gas	0.000	30.731	0.000	5.122	0.000	2.083	0.006	0.354
40570102	Transylvanian Subsalt	0.000	0.000	0.000	0.000	NA	NA	NA	NA
40600101	Neogene Flysch Gas	0.014	18.475	0.005	3.098	0.000	16.267	0.034	2.745
40600201	Thermal Triassic	0.332	0.412	0.027	0.428	0.360	1.698	0.087	0.731
40610101	Moesian Platform	1.793	2.217	0.066	2.228	0.171	0.561	0.020	0.285
40610201	Romania Flysch Zone	1.948	3.222	0.011	2.496	0.564	2.177	0.081	1.008
40610202	Romania Ploiesti Zone	2.167	1.826	0.000	2.471	0.278	0.441	0.014	0.366
40680101	Subsalt	0.000	0.000	0.000	0.000	0.388	51.210	2.265	11.188
52000101	Northeast Greenland Shelf Rift Systems	0.000	0.000	0.000	0.000	47.148	80.709	4.170	64.769
52150101	Jeanne d'Arc	1.582	2.406	0.081	2.063	1.796	8.893	0.514	3.792
52150201	Graben Systems Exclusive of Jeanne d'Arc	0.000	4.224	0.107	0.811	NA	NA	NA	NA
52430101	Keg River Gas	0.000	5.812	0.000	0.969	0.000	0.868	0.017	0.162

Assessment Unit Code	Assessment Unit Name	Known Oil (BBO)	Known Gas (TCFG)	Known NGL (BBNGL)	Known Total Petroleum (BBOE)	Undiscovered Oil (Mean, BBO)	Undiscovered Gas (Mean, TCFG)	Undiscovered NGL (Mean, BBNGL)	Undiscovered Total Petroleum (Mean, BBOE)
52430102	Keg River Oil and Gas	1.011	1.261	0.100	1.321	0.084	0.907	0.021	0.256
52430201	Leduc Gas	0.042	9.221	0.334	1.912	0.000	2.442	0.093	0.500
52430202	Leduc Oil and Gas	9.417	5.136	0.692	10.965	0.501	0.782	0.041	0.672
52430301	Exshaw-Rundle Gas	0.191	15.413	0.396	3.156	0.000	2.197	0.044	0.410
52430302	Exshaw-Rundle Oil and Gas	1.537	6.155	0.439	3.002	0.126	0.466	0.016	0.219
52430401	Combined Triassic/Jurassic Gas	0.054	3.965	0.047	0.762	0.000	1.289	0.028	0.243
52430402	Combined Triassic/Jurassic Oil and Gas	1.242	6.615	0.302	2.647	0.096	0.877	0.028	0.270
52430501	Mannville Gas	0.000	24.453	0.265	4.340	0.000	4.597	0.092	0.858
52430502	Mannville Continuous (Basin-Center) Gas	0.000	6.278	0.182	1.228	NA	NA	NA	NA
52430601	Second White Specks-Cardium Gas	0.000	2.071	0.053	0.398	0.000	0.285	0.009	0.056
52430602	Second White Specks-Cardium Oil and Gas	2.423	5.449	0.294	3.625	0.036	0.840	0.026	0.201
52430603	Upper Cretaceous Eastern Shallow Gas	0.265	18.929	0.019	3.439	NA	NA	NA	NA
52440101	Yeoman Oil	0.001	0.000	0.000	0.001	0.005	0.004	0.000	0.006
52440201	Brightholme Oil	0.014	0.011	0.000	0.016	0.008	0.005	0.000	0.009
52440301	Bakken Sandstone	0.007	0.000	0.000	0.007	0.011	0.003	0.000	0.012
52440401	Lodgepole Oil	2.778	0.893	0.037	2.964	0.111	0.063	0.004	0.126
53050101	El Abra-Like Reef and Backreef Limestone	2.063	1.415	0.005	2.303	0.289	0.507	0.028	0.402
53050102	Tamabra-Like Debris-Flow-Breccia Limestone Overlying Evaporites	33.065	35.678	0.072	39.083	9.857	20.450	1.214	14.480
53050103	Tamabra-Like Debris-Flow-Breccia Limestone and Overlying Strata	0.080	0.530	0.001	0.169	0.807	6.147	0.299	2.129
53050104	Tamabra-Like Debris-Flow-Breccia Limestone of the Golden Lane	3.308	3.824	0.000	3.946	0.557	1.113	0.067	0.809
53050105	Tamaulipas-Like Basinal Limestone and Tertiary Strata Without Underlying Evaporites	1.627	1.736	0.000	1.916	0.366	0.804	0.048	0.548
53050106	Tamaulipas-Like Basinal Limestone and Tertiary Strata Overlying Evaporites	4.268	7.640	0.018	5.560	7.823	18.516	0.990	11.900
53050107	Tamabra-Like Debris-Flow-Breccia Limestone North of Campeche	0.000	0.000	0.000	0.000	0.871	1.735	0.104	1.264
60210101	Late Cretaceous-Tertiary Turbidites	0.000	0.000	0.000	0.000	12.745	34.625	1.933	20.449
60210102	Cretaceous Carbonates	0.000	0.000	0.000	0.000	2.200	6.595	0.360	3.660
60210103	Late Cretaceous-Tertiary Nearshore Sandstones	0.063	0.000	0.000	0.063	0.301	0.839	0.036	0.477
60220101	Amazon Delta and Submarine Fan	NA	NA	NA	NA	0.000	29.840	0.298	5.272
60220102	Offshore Gas Hydrates	0.000	0.000	0.000	0.000	NA	NA	NA	NA
60290101	Western Pre-Aptian Reservoirs	0.570	1.171	0.002	0.767	0.031	0.168	0.005	0.063
60290102	Late Cretaceous-Tertiary Deep-Water Sandstones	0.122	0.236	0.001	0.162	1.457	8.166	0.411	3.228
60340101	Espirito Santo Shelf	0.130	0.172	0.000	0.159	0.054	0.065	0.001	0.066
60340102	Late Cretaceous-Tertiary Slide Blocks and Turbidites	NA	NA	NA	NA	2.437	25.677	1.215	7.932
60340103	Abrolhos Sub-Volcanic Structures	0.000	0.000	0.000	0.000	0.468	8.545	0.396	2.288
60350101	Late Cretaceous-Tertiary Turbidites	9.630	5.761	0.010	10.600	10.869	13.917	0.393	13.582
60350102	Cretaceous Carbonates	0.426	0.482	0.000	0.507	0.932	1.743	0.046	1.268
60350103	Salt Dome Province Tertiary Sandstones	0.000	0.000	0.000	0.000	4.493	4.031	0.114	5.278
60360101	Santos Shelf	0.285	1.125	0.044	0.517	13.723	62.874	3.205	27.407
60360102	Salt-Structured Deep-Water Sandstones	0.000	0.000	0.000	0.000	9.486	17.673	0.989	13.420
60370101	Pelotas Platform and Basin	0.000	0.000	0.000	0.000	2.938	22.806	1.107	7.846
60370102	Offshore Gas Hydrates	0.000	0.000	0.000	0.000	NA	NA	NA	NA
60410101	Hollin-Napo	6.601	1.616	0.000	6.871	3.023	0.755	0.015	3.164
60410201	Ene	0.000	0.000	0.000	0.000	0.074	0.841	0.040	0.254
60450101	Sub-Andean Fold and Thrust Belt	0.318	16.818	0.543	3.664	0.181	20.521	0.791	4.392
60450102	Foreland Basins	0.002	0.168	0.005	0.035	1.254	7.375	0.348	2.831
60450103	Foreland Central Chaco High	0.000	0.000	0.000	0.000	0.711	3.211	0.161	1.407
60550101	Neuquen Extensional Structures	1.716	16.699	0.327	4.826	0.862	7.724	0.129	2.278
60550102	Neuquen Foothills Structure	0.297	0.697	0.002	0.416	0.334	3.764	0.061	1.023
60550103	Dorsal de Neuquen Structure	0.357	3.307	0.009	0.917	0.093	0.927	0.013	0.261
60580101	San Jorge Extensional Structures	3.106	2.968	0.000	3.601	0.347	3.038	0.061	0.914
60580102	San Bernardo Fold Belt Structures	0.240	0.597	0.001	0.340	0.151	0.735	0.015	0.288
60580201	Hypothetical Basin-Centered Gas	0.000	0.000	0.000	0.000	NA	NA	NA	NA
60590101	Magallanes Extensional Structures	1.216	24.739	0.211	5.550	0.494	10.386	0.226	2.451
60590102	Andean Fold Belt Structures	0.000	0.069	0.000	0.012	0.210	3.655	0.084	0.903
60600101	North Falklands Basin	0.000	0.000	0.000	0.000	3.603	8.871	0.178	5.259
60600201	South Falklands Basin	0.000	0.000	0.000	0.000	2.229	7.710	0.193	3.707
60630101	Malvinas Extensional Structures	0.010	0.075	0.002	0.025	1.031	11.327	0.362	3.280

Assessment Unit Code	Assessment Unit Name	Known Oil (BBO)	Known Gas (TCFG)	Known NGL (BBNGL)	Known Total Petroleum (BBOE)	Undiscovered Oil (Mean, BBO)	Undiscovered Gas (Mean, TCFG)	Undiscovered NGL (Mean, BBNGL)	Undiscovered Total Petroleum (Mean, BBOE)
60810101	Cretaceous-Paleogene Basin	1.672	2.892	0.000	2.154	1.711	4.795	0.255	2.765
60830101	Neogene Pull-Apart Basin	0.004	0.340	0.000	0.061	0.037	0.471	0.021	0.137
60830201	Cretaceous-Paleogene Santa Elena Block	0.171	0.158	0.000	0.197	0.200	0.224	0.012	0.249
60900101	Northern	1.730	2.791	0.003	2.198	0.191	0.119	0.007	0.218
60900102	Southern	0.681	0.206	0.000	0.715	0.188	0.040	0.002	0.197
60900103	Eastern	0.000	2.000	0.065	0.398	0.099	2.735	0.100	0.654
60900104	La Luna and Older	0.015	0.005	0.000	0.016	0.231	0.399	0.021	0.318
60960101	Central	3.094	15.071	0.450	6.056	1.312	2.180	0.119	1.794
60960102	Peripheral	2.309	0.242	0.001	2.349	2.319	4.036	0.218	3.209
60980101	East Venezuela Fold and Thrust Belt	13.231	52.379	0.522	22.482	4.928	21.298	1.096	9.573
60980102	Guarico Sub-Basin	0.322	47.579	0.005	8.257	0.385	16.995	0.760	3.977
60980103	Maturin Sub-Basin	9.446	12.078	0.002	11.462	1.919	3.258	0.174	2.636
60980104	Orinoco Heavy Oil and Tar Belt	3.757	0.260	0.000	3.801	NA	NA	NA	NA
60980201	Trinidad Basins	3.447	15.822	0.172	6.256	0.416	23.960	1.067	5.477
60980202	Orinoco Delta and Offshore	0.000	0.000	0.000	0.000	4.228	28.049	1.369	10.271
60990101	Main Maracaibo Basin	48.211	24.918	0.026	52.389	7.842	16.085	0.896	11.420
60990102	Southwest Maracaibo Basin Fold Belt	0.861	1.783	0.017	1.175	0.341	1.491	0.077	0.666
60990103	Basin-Centered Gas	0.000	0.000	0.000	0.000	NA	NA	NA	NA
61030101	Carupano Basin Gas	0.020	22.600	0.080	3.867	0.000	20.103	0.201	3.551
61070101	Inner Forearc Deformation Belt	0.000	0.000	0.000	0.000	1.536	11.957	0.582	4.110
61070102	Barbados Accreationary Prism	0.011	0.022	0.000	0.014	NA	NA	NA	NA
61170101	North Cuba Fold and Thrust Belt	0.297	0.096	0.000	0.313	0.494	0.592	0.035	0.628
70130101	Coastal Plain and Offshore	0.010	0.049	0.000	0.019	0.157	0.856	0.043	0.343
71830101	Coastal Plain and Offshore	0.236	1.949	0.036	0.596	1.004	10.071	0.282	2.965
71920101	Agbada Reservoirs	34.522	93.811	2.842	53.000	21.925	85.176	4.078	40.198
71920102	Akata Reservoirs	0.000	0.000	0.000	0.000	18.562	47.540	1.956	28.442
72030101	Gabon Subsalt	1.752	0.843	0.015	1.908	0.727	3.674	0.172	1.511
72030201	Gabon Suprasalt	2.114	1.835	0.010	2.430	5.031	13.230	0.625	7.861
72030301	Central Congo Delta and Carbonate Platform	9.540	9.441	0.039	11.152	4.677	14.314	0.692	7.754
72030302	Central Congo Turbidites	0.205	0.002	0.000	0.205	18.522	55.268	2.682	30.415
72030401	Cuanza-Namibe	0.107	0.059	0.000	0.116	0.790	1.558	0.073	1.123
73030101	Offshore	0.000	6.015	0.000	1.003	0.116	3.603	0.163	0.879
80260101	Kohat-Potwar Intrathrust Basin	0.324	1.981	0.041	0.696	0.145	0.551	0.027	0.264
80340101	Sylhet-Kopili/Barail-Tipam Composite	2.536	6.534	0.005	3.630	0.399	1.178	0.067	0.663
80420101	Greater Indus Foreland and Foldbelt	0.180	35.373	0.063	6.139	0.119	25.405	0.896	5.249
80420102	Indus Fan	0.000	0.000	0.000	0.000	0.000	3.919	0.137	0.790
80430101	Eocene-Miocene Bombay Shelf	6.257	17.528	0.264	9.442	1.612	9.299	0.301	3.462
80430102	Eocene-Miocene Deltaics	2.183	6.665	0.004	3.298	0.479	3.645	0.117	1.204
80470101	Coalbed Gas	0.000	0.000	0.000	0.000	NA	NA	NA	NA
80470201	Western Shelf and Slope	NA	NA	NA	NA	0.064	4.613	0.056	0.888
80470301	Central Basin	0.000	0.000	0.000	0.000	0.015	23.985	0.103	4.115
80470302	Eastern Fold Belt	0.004	12.289	0.051	2.104	0.024	26.494	0.115	4.554
80480101	Central Burma Basin	0.637	0.981	0.000	0.800	0.516	2.008	0.099	0.949
80480102	Irrawaddy-Andaman	0.079	9.040	0.063	1.649	0.209	18.512	0.686	3.981

Assessment Unit Code	Assessment Unit Name	Specific Reservoir Rock Age	General Reservoir Rock Age	Reservoir Rock Lithology	Reservoir Rock Depositional Environment	Seal Rock Lithology	Structural Setting	Trap Type
10080101	Northwest Izhma-Pechora Depression	Lower Paleozoic	Paleozoic	Siliciclastics and Carbonates	Nonmarine to Marine	Shale	Extensional	Structural and Stratigraphic
10080102	Main Basin Platform	Upper Paleozoic	Paleozoic	Siliciclastics and Carbonates	Shallow Marine	Shale	Compressional	Structural
10080103	Foredeep Basins	Permian	Paleozoic	Siliciclastics and Carbonates	Shallow Marine	Evaporites and Shale	Compressional	Structural
10090101	Carboniferous-Lower Permian Clastics	Carboniferous to Permian	Paleozoic	Siliciclastics	Nonmarine to Marine	Evaporites	Extensional	Structural
10090102	Devonian Synrift	Devonian	Paleozoic	Carbonates	Shallow Marine	Evaporites and Shale	Extensional	Structural and Stratigraphic
10090103	Continuous Basin-Centered Gas Accumulation							
10150101	Supra-Domanik Carbonates/Clastics	Upper Paleozoic	Paleozoic	Siliciclastics and Carbonates	Nonmarine to Marine	Evaporites and Shale	Passive	Structural and Stratigraphic
10150102	Lower Volga	Devonian to Carboniferous	Paleozoic	Siliciclastics and Carbonates	Nonmarine to Marine	Shale	Compressional	Structural and Stratigraphic
10150103	Sub-Domanik Devonian Clastics	Devonian	Paleozoic	Siliciclastics	Shallow Marine	Shale	Compressional	Structural
10150104	Self-Sourced Domanik							
10150201	Permian Reefs/Thrust Folds	Carboniferous to Permian	Paleozoic	Carbonates	Shallow Marine	Evaporites	Compressional	Structural and Stratigraphic
10160101	North and West Margins Subsalt Pinnacle Reefs	Upper Paleozoic	Paleozoic	Carbonates	Shallow Marine	Evaporites	Passive	Stratigraphic
10160102	North and West Margins Subsalt Barrier Reefs	Upper Paleozoic	Paleozoic	Carbonates	Shallow Marine	Evaporites	Passive	Stratigraphic
10160103	East and Southeast Margins Subsalt	Carboniferous	Paleozoic	Carbonates	Shallow Marine	Evaporites	Compressional	Structural
10160104	South Margin Subsalt	Carboniferous	Paleozoic	Carbonates	Shallow Marine	Evaporites	Passive	Stratigraphic
10160105	Basin Center Subsalt							
10160106	Suprasalt	Jurassic to Cretaceous	Mesozoic	Siliciclastics	Shallow Marine	Shale	Passive	Structural
10500101	Kolguyev Terrace	Triassic to Jurassic	Mesozoic	Siliciclastics	Nonmarine to Marine	Shale	Compressional	Structural
10500102	South Barents and Ludlov Saddle	Jurassic	Mesozoic	Siliciclastics	Shallow Marine	Shale	Compressional	Structural
10500103	North Barents	Triassic to Cretaceous	Mesozoic	Siliciclastics	Shallow Marine	Shale	Compressional	Structural
11080101	Tertiary Foredeep	Paleogene to Neogene	Cenozoic	Siliciclastics	Deep Marine	Shale	Passive	Structural and Stratigraphic
11080102	Foreland Slope	Cretaceous	Mesozoic	Siliciclastics	Deep Marine	Shale	Compressional	Structural
11080103	Subsalt Jurassic	Jurassic	Mesozoic	Carbonates	Shallow Marine	Evaporites	Compressional	Structural and Stratigraphic
11090101	Foldbelt-Foothills	Cretaceous	Mesozoic	Siliciclastics and Carbonates	Shallow Marine	Shale	Compressional	Structural
11090102	Terek-Sunzha Subsalt Jurassic	Jurassic	Mesozoic	Carbonates	Shallow Marine	Evaporites	Compressional	Structural and Stratigraphic
11090103	Foreland Slope and Foredeep	Cretaceous	Mesozoic	Siliciclastics and Carbonates	Shallow Marine	Shale	Passive	Stratigraphic
11090201	South Mangyshlak (Entire)	Jurassic	Mesozoic	Siliciclastics	Nonmarine to Marine	Shale	Compressional	Structural
11090301	Offshore Prikumsk Zone	Cretaceous	Mesozoic	Siliciclastics	Shallow Marine	Shale	Compressional	Structural
11090302	Onshore Stavropol-Prikumsk	Cretaceous	Mesozoic	Siliciclastics	Shallow Marine	Shale	Compressional	Structural
11090303	Central Caspian Offshore	Jurassic to Cretaceous	Mesozoic	Siliciclastics	Nonmarine to Marine	Shale	Compressional	Structural and Stratigraphic
11120101	Apsheron-Pribalkhan Zone	Neogene	Cenozoic	Siliciclastics	Nonmarine to Marine	Shale	Compressional	Structural
11120102	Lower Kura Depression and Adjacent Shelf	Neogene	Cenozoic	Siliciclastics	Nonmarine to Marine	Shale	Compressional	Structural
11120103	Gograndag-Okarem Zone and Adjacent Shelf	Neogene	Cenozoic	Siliciclastics	Nonmarine to Marine	Shale	Compressional	Structural
11120104	Central Offshore	Neogene	Cenozoic	Siliciclastics	Deep Marine	Shale	Compressional	Structural

Assessment Unit Code	Assessment Unit Name	Specific Reservoir Rock Age	General Reservoir Rock Age	Reservoir Rock Lithology	Reservoir Rock Depositional Environment	Seal Rock Lithology	Structural Setting	Trap Type
11120105	Iran Onshore/Nearshore	Neogene	Cenozoic	Siliciclastics	Nonmarine to Marine	Shale	Compressional	Structural
11500101	Mesozoic Sandstone Reservoirs	Jurassic to Cretaceous	Mesozoic	Siliciclastics	Nonmarine to Marine	Shale	Compressional	Structural
11500201	Jurassic-Tertiary	Jurassic	Mesozoic	Siliciclastics	Nonmarine to Marine	Shale	Compressional	Structural and Stratigraphic
11500301	Upper Paleozoic Carbonates	Upper Paleozoic	Paleozoic	Carbonates	Shallow Marine	Shale	Compressional	Structural and Stratigraphic
11540101	Northern and Western Areas	Jurassic to Cretaceous	Mesozoic	Siliciclastics and Carbonates	Nonmarine to Marine	Evaporites and Shale	Compressional	Structural and Stratigraphic
11540102	Karabil-Badkhyz (Southern Area)	Jurassic to Cretaceous	Mesozoic	Siliciclastics and Carbonates	Nonmarine to Marine	Evaporites and Shale	Passive	Stratigraphic
11540103	Murgab Depression Suprasalt	Cretaceous	Mesozoic	Siliciclastics	Nonmarine to Marine	Shale	Compressional	Structural
11540104	Murgab Depression Subsalt	Jurassic	Mesozoic	Carbonates	Shallow Marine	Evaporites	Compressional	Structural and Stratigraphic
11740101	Upper Jurassic-Cretaceous Sandstones	Cretaceous	Mesozoic	Siliciclastics	Shallow to Deep Marine	Shale	Compressional	Structural and Stratigraphic
11740102	Self-Sourced Bazhenov Fractured Reservoirs							
11740201	Pre-Upper Jurassic	Jurassic	Mesozoic	Siliciclastics	Nonmarine	Shale	Passive	Structural and Stratigraphic
11740301	Northern West Siberian Onshore Gas	Cretaceous	Mesozoic	Siliciclastics	Nonmarine	Shale	Compressional	Structural
11740302	South Kara Sea Offshore	Cretaceous	Mesozoic	Siliciclastics	Nonmarine	Shale	Compressional	Structural
12070101	Yenisey Foldbelt Riphean-Craton Margin Riphean	Precambrian to Cambrian	Precambrian to Paleozoic	Carbonates	Shallow Marine	Evaporites	Passive	Stratigraphic
12100101	Baikal-Patom Foldbelt Riphean-Craton Margin Vendian	Precambrian to Cambrian	Precambrian to Paleozoic	Siliciclastics and Carbonates	Shallow Marine	Evaporites	Compressional	Structural
13220101	Onshore and Offshore Northeastern Shelf	Neogene	Cenozoic	Siliciclastics	Nonmarine to Marine	Shale	Compressional	Structural
20040101	Ma'Rib-Al Jawf/Shabwah/Masila	Jurassic to Cretaceous	Mesozoic	Siliciclastics	Shallow to Deep Marine	Evaporites	Extensional	Structural
20140101	Ghaba-Makarem Combined Structural	Paleozoic	Paleozoic	Siliciclastics	Nonmarine to Marine	Shale	Passive	Structural
20160101	Fahud-Huqf Combined Structural	Cretaceous	Mesozoic	Carbonates	Nonmarine to Marine	Shale	Passive	Structural
20160201	Natih-Fiqa Structural/Stratigraphic	Cretaceous	Mesozoic	Carbonates	Shallow Marine	Shale	Compressional and Extensional	Structural
20190101	Cretaceous Reservoirs in Northwest Desert Anticlines	Cretaceous	Mesozoic	Carbonates	Shallow Marine	Shale	Compressional and Extensional	Structural
20190102	Cretaceous Reservoirs in South Gulf Suprasalt Structural	Cretaceous	Mesozoic	Carbonates	Shallow Marine	Shale	Compressional	Structural
20190103	Mesozoic/Tertiary Foredeep Fold and Thrust	Cretaceous	Mesozoic	Carbonates	Shallow Marine	Shale	Compressional	Structural
20190201	Jurassic Reservoirs in Northwest Desert Anticlines	Jurassic	Mesozoic	Carbonates	Shallow Marine	Evaporites and Shale	Compressional and Extensional	Structural
20190202	Jurassic Reservoirs in South Gulf Suprasalt/Qatar Arch Structural	Jurassic	Mesozoic	Carbonates	Shallow Marine	Evaporites and Shale	Compressional	Structural
20190301	Khuff Carbonates in Salt Structures	Permian	Paleozoic	Carbonates	Shallow Marine	Evaporites	Passive	Structural
20190302	Paleozoic Reservoirs	Permian	Paleozoic	Siliciclastics	Nonmarine	Evaporites	Compressional and Extensional	Structural
20210101	Central Arch Horst-Block Anticlinal Oil and Gas	Paleozoic	Paleozoic	Siliciclastics	Nonmarine to Marine	Evaporites and Shale	Extensional	Structural
20210102	North Gulf Salt Basin Structural Gas	Permian	Paleozoic	Carbonates	Shallow Marine	Evaporites and Shale	Extensional	Structural
20210201	Horst-Block Anticlinal Oil	Jurassic	Mesozoic	Carbonates	Shallow Marine	Evaporites	Extensional	Structural
20210202	Salt-Involved Structural Oil	Jurassic	Mesozoic	Carbonates	Shallow Marine	Evaporites	Extensional	Structural
20230101	Horst/Graben-Related Oil and Gas	Paleozoic	Paleozoic	Siliciclastics	Nonmarine	Evaporites and Shale	Extensional	Structural
20230201	Platform Horst/Graben-Related Oil	Jurassic	Mesozoic	Carbonates	Shallow Marine	Evaporites	Extensional	Structural

Assessment Unit Code	Assessment Unit Name	Specific Reservoir Rock Age	General Reservoir Rock Age	Reservoir Rock Lithology	Reservoir Rock Depositional Environment	Seal Rock Lithology	Structural Setting	Trap Type
20230202	Basinal Oil and Gas	Jurassic	Mesozoic	Carbonates	Shallow Marine	Evaporites	Compressional and Extensional	Structural
20300101	Cretaceous Reservoirs	Cretaceous	Mesozoic	Siliciclastics and Carbonates	Nonmarine to Marine	Shale	Compressional	Structural
20300102	Tertiary Reservoirs	Paleogene to Neogene	Cenozoic	Siliciclastics and Carbonates	Shallow Marine	Evaporites	Compressional	Structural
20300201	Northern Qatar Arch Extension	Permian to Triassic	Paleozoic to Mesozoic	Carbonates	Shallow Marine	Evaporites and Shale	Compressional	Structural
20430101	Southeast Sirte Clastics	Precambrian to Cretaceous	Precambrian to Mesozoic	Siliciclastics	Nonmarine	Evaporites	Extensional	Structural and Stratigraphic
20430102	Central Sirte Carbonates	Cretaceous to Paleogene	Mesozoic to Cenozoic	Carbonates	Shallow Marine	Evaporites	Extensional	Structural and Stratigraphic
20430103	Offshore Sirte Hypothetical	Cretaceous to Paleogene	Mesozoic to Cenozoic	Siliciclastics and Carbonates	Shallow Marine	Evaporites	Compressional and Extensional	Structural
20430104	Southeast Sirte Hypothetical	Cretaceous to Paleogene	Mesozoic to Cenozoic	Siliciclastics	Shallow Marine	Shale	Extensional	Structural and Stratigraphic
20480101	Bou Dabbous-Tertiary Structural/Stratigraphic	Paleogene to Neogene	Cenozoic	Siliciclastics and Carbonates	Shallow Marine	Shale	Compressional and Extensional	Structural and Stratigraphic
20480201	Jurassic-Cretaceous Structural/Stratigraphic	Jurassic to Cretaceous	Mesozoic	Siliciclastics and Carbonates	Shallow to Deep Marine	Shale	Compressional and Extensional	Structural and Stratigraphic
20540101	Tanezzuft-Oued Mya Structural/Stratigraphic	Lower Paleozoic	Paleozoic	Siliciclastics	Nonmarine to Marine	Evaporites and Shale	Compressional and Extensional	Structural and Stratigraphic
20540201	Tanezzuft-Melrhir Structural/Stratigraphic	Ordovician to Triassic	Paleozoic to Mesozoic	Siliciclastics	Nonmarine to Marine	Evaporites and Shale	Compressional and Extensional	Structural and Stratigraphic
20540301	Tanezzuft-Ghadames Structural/Stratigraphic	Cambrian to Triassic	Paleozoic to Mesozoic	Siliciclastics	Nonmarine to Marine	Evaporites and Shale	Compressional and Extensional	Structural and Stratigraphic
20560101	Tanezzuft-Illizi Structural/Stratigraphic	Paleozoic	Paleozoic	Siliciclastics	Nonmarine to Marine	Shale	Compressional and Extensional	Structural and Stratigraphic
20580101	Tanezzuft-Timimoun Structural/Stratigraphic	Paleozoic	Paleozoic	Siliciclastics	Nonmarine to Marine	Shale	Compressional	Structural and Stratigraphic
20580201	Tanezzuft-Ahnet Structural/Stratigraphic	Paleozoic	Paleozoic	Siliciclastics	Nonmarine to Marine	Shale	Compressional	Structural and Stratigraphic
20580301	Tanezzuft-Sbaa Structural/Stratigraphic	Paleozoic	Paleozoic	Siliciclastics	Nonmarine to Marine	Shale	Compressional	Structural and Stratigraphic
20580401	Tanezzuft-Mouydir Structural/Stratigraphic	Paleozoic	Paleozoic	Siliciclastics	Nonmarine to Marine	Shale	Compressional	Structural and Stratigraphic
20580501	Tanezzuft-Benoud Structural/Stratigraphic	Upper Paleozoic to Mesozoic	Paleozoic to Mesozoic	Siliciclastics	Nonmarine to Marine	Evaporites and Shale	Compressional	Structural and Stratigraphic
20580601	Tanezzuft-Bechar/Abadla Structural/Stratigraphic	Paleozoic	Paleozoic	Siliciclastics	Nonmarine to Marine	Shale	Compressional	Structural and Stratigraphic
20710101	Gulf of Suez Block-Fault Fairway	Paleozoic to Mesozoic	Paleozoic to Mesozoic	Siliciclastics	Nonmarine to Marine	Evaporites and Shale	Extensional	Structural
20710102	Gulf of Suez Qaa Plain	Paleozoic to Mesozoic	Paleozoic to Mesozoic	Siliciclastics	Nonmarine to Marine	Evaporites and Shale	Extensional	Structural
20710103	Southern Gulf of Suez	Paleozoic to Mesozoic	Paleozoic to Mesozoic	Siliciclastics	Nonmarine to Marine	Evaporites and Shale	Extensional	Structural
20710201	Red Sea Coastal Block Faults	Neogene	Cenozoic	Siliciclastics and Carbonates	Nonmarine to Marine	Evaporites and Shale	Extensional	Structural
20710202	Red Sea Salt Basin	Neogene	Cenozoic	Siliciclastics and Carbonates	Nonmarine to Marine	Evaporites and Shale	Passive	Structural and Stratigraphic
31150101	Upper Paleozoic/Lower Mesozoic Nonmarine Coarse Clastics	Permian to Triassic	Paleozoic to Mesozoic	Siliciclastics	Nonmarine	Shale	Compressional	Structural
31150201	Jurassic/Tertiary Fluvial and Lacustrine Sandstone	Jurassic to Neogene	Mesozoic to Cenozoic	Siliciclastics	Nonmarine	Shale	Compressional	Structural
31150301	Continuous-Type (Basin-Centered) Gas							
31270101	Tertiary Lacustrine	Paleogene to Neogene	Cenozoic	Siliciclastics	Nonmarine	Shale	Extensional	Structural and Stratigraphic

Assessment Unit Code	Assessment Unit Name	Specific Reservoir Rock Age	General Reservoir Rock Age	Reservoir Rock Lithology	Reservoir Rock Depositional Environment	Seal Rock Lithology	Structural Setting	Trap Type
31270102	Pre-Tertiary Buried Hills	Precambrian to Lower Paleozoic	Precambrian to Paleozoic	Carbonates	Shallow Marine	Shale	Extensional	Structural and Stratigraphic
31270201	Coal-Sourced Gas							
31280101	Jurassic/Triassic Fluvial and Lacustrine Sandstone	Jurassic	Mesozoic	Siliciclastics	Nonmarine	Shale	Passive	Structural and Stratigraphic
31280201	Basin Centered Gas (Continuous) in Ordovician Carbonate and Permian Sandstone							
31280202	Carboniferous and Permian Coalbed Methane (Continuous)							
31420101	Southeastern Fold Belt	Upper Paleozoic to Triassic	Paleozoic to Mesozoic	Carbonates	Shallow Marine	Evaporites and Shale	Compressional	Structural
31420102	Northwestern Depression/Foldbelt	Permian to Triassic	Paleozoic to Mesozoic	Carbonates	Shallow Marine	Evaporites and Shale	Compressional	Structural
31420103	Continuous Gas in Central Uplift/Northwestern Depression							
31420201	Jurassic Lacustrine	Jurassic	Mesozoic	Siliciclastics and Carbonates	Nonmarine	Shale	Compressional	Structural and Stratigraphic
31420301	Continuous Gas in Northwestern Depression/Central Uplift							
31420401	Leshan-Longnusi Paleohigh	Precambrian	Precambrian	Carbonates	Nonmarine to Marine	Shale	Compressional	Structural
31420402	Lower Paleozoic of Southeastern Fold Belt	Lower Paleozoic	Paleozoic	Carbonates	Shallow Marine	Shale	Compressional	Structural
31440101	Subtle Traps	Cretaceous	Mesozoic	Siliciclastics	Nonmarine	Shale	Unknown	Structural and Stratigraphic
31440102	Anticlinal	Cretaceous	Mesozoic	Siliciclastics	Nonmarine	Shale	Compressional and Extensional	Structural
31440201	Structural Traps	Cretaceous	Mesozoic	Siliciclastics	Nonmarine	Shale	Compressional and Extensional	Structural
31540101	Tarim Basin Excluding Marginal Foldbelts	Lower Paleozoic to Neogene	Paleozoic to Cenozoic	Siliciclastics and Carbonates	Nonmarine to Marine	Shale	Compressional	Structural and Stratigraphic
31540102	Kuche (Northern) Foldbelt	Jurassic	Mesozoic	Siliciclastics	Nonmarine	Evaporites and Shale	Compressional	Structural
31540103	Southwest Foldbelt	Neogene	Cenozoic	Siliciclastics	Nonmarine	Shale	Compressional	Structural
37010101	Brunei-Sabah Deltaics	Neogene	Cenozoic	Siliciclastics	Nonmarine to Marine	Shale	Compressional	Structural
37010102	Brunei-Sabah Turbidites	Neogene	Cenozoic	Siliciclastics	Deep Marine	Shale	Compressional	Structural
37020101	Central Luconia	Neogene	Cenozoic	Carbonates	Shallow Marine	Shale	Passive	Stratigraphic
37020102	Balingian	Paleogene to Neogene	Cenozoic	Siliciclastics	Nonmarine to Marine	Shale	Compressional	Structural
37020201	East Natuna	Neogene	Cenozoic	Carbonates	Shallow Marine	Shale	Passive	Stratigraphic
37030101	South Malay Lacustrine	Paleogene to Neogene	Cenozoic	Siliciclastics	Nonmarine to Marine	Shale	Compressional	Structural
37030102	North Malay Lacustrine	Paleogene to Neogene	Cenozoic	Siliciclastics	Nonmarine	Shale	Compressional	Structural
37030201	South Malay Coaly	Neogene	Cenozoic	Siliciclastics	Nonmarine to Marine	Shale	Compressional	Structural
38080101	Pematang/Sihapas Siliciclastics	Paleogene to Neogene	Cenozoic	Siliciclastics	Nonmarine to Marine	Shale	Compressional	Structural
38170101	Kutei Basin Deltaics	Neogene	Cenozoic	Siliciclastics	Nonmarine to Marine	Shale	Compressional	Structural
38170102	Kutei Basin Turbidites	Neogene	Cenozoic	Siliciclastics	Deep Marine	Shale	Compressional	Structural
38170103	Kutei Basin Fold and Thrust Belt	Paleogene	Cenozoic	Carbonates	Shallow Marine	Shale	Compressional	Structural
38220101	North Sumatra	Neogene	Cenozoic	Carbonates	Shallow Marine	Shale	Compressional	Structural and Stratigraphic
38220102	Mergui	Neogene	Cenozoic	Siliciclastics and Carbonates	Shallow to Deep Marine	Shale	Passive	Structural and Stratigraphic

Assessment Unit Code	Assessment Unit Name	Specific Reservoir Rock Age	General Reservoir Rock Age	Reservoir Rock Lithology	Reservoir Rock Depositional Environment	Seal Rock Lithology	Structural Setting	Trap Type
38240101	Sunda/Asri	Paleogene to Neogene	Cenozoic	Siliciclastics	Nonmarine to Marine	Shale	Passive	Structural and Stratigraphic
38240201	Ardjuna	Neogene	Cenozoic	Siliciclastics and Carbonates	Nonmarine to Marine	Shale	Compressional	Structural and Stratigraphic
38240301	Biogenic Gas							
38240401	Billiton Basin Hypothetical							
38280101	South Sumatra	Paleogene to Neogene	Cenozoic	Siliciclastics and Carbonates	Shallow Marine	Shale	Compressional	Structural
39100101	Barnett	Carboniferous	Paleozoic	Siliciclastics	Nonmarine to Marine	Shale	Extensional	Structural
39100201	Petrel	Permian	Paleozoic	Siliciclastics	Nonmarine to Marine	Shale	Compressional	Structural
39100202	Vulcan Graben	Permian	Paleozoic	Siliciclastics	Nonmarine to Marine	Shale	Extensional	Structural
39100301	Malita	Mesozoic	Mesozoic	Siliciclastics	Nonmarine to Marine	Shale	Compressional	Structural
39130101	Late Jurassic/Early Cretaceous-Mesozoic	Mesozoic	Mesozoic	Siliciclastics	Nonmarine to Marine	Shale	Extensional	Structural
39300101	Gippsland	Cretaceous to Paleogene	Mesozoic to Cenozoic	Siliciclastics	Nonmarine to Marine	Shale	Compressional	Structural and Stratigraphic
39480101	Dingo-Mungaroo/Barrow	Jurassic to Cretaceous	Mesozoic	Siliciclastics	Nonmarine to Marine	Shale	Extensional	Structural
39480201	Locker-Mungaroo/Barrow	Mesozoic	Mesozoic	Siliciclastics	Nonmarine to Marine	Shale	Extensional	Structural
40170101	Halten Terrace-Trondelag Platform	Jurassic to Cretaceous	Mesozoic	Siliciclastics	Nonmarine to Marine	Shale	Extensional	Structural
40170102	Mid-Norway Continental Margin	Upper Paleozoic to Paleogene	Paleozoic to Cenozoic	Siliciclastics	Nonmarine to Marine	Shale	Extensional	Structural and Stratigraphic
40250101	Viking Graben	Jurassic	Mesozoic	Siliciclastics	Nonmarine to Marine	Shale	Extensional	Structural and Stratigraphic
40250102	Moray Firth	Jurassic to Cretaceous	Mesozoic	Siliciclastics	Nonmarine to Marine	Shale	Extensional	Structural
40250103	Central Graben	Jurassic to Paleogene	Mesozoic to Cenozoic	Siliciclastics and Carbonates	Nonmarine to Marine	Shale	Extensional	Structural
40360101	Southern Permian Basin-U.K. Onshore	Carboniferous	Paleozoic	Siliciclastics and Carbonates	Nonmarine to Marine	Shale	Unknown	Structural and Stratigraphic
40360102	Southern Permian Basin-Europe Onshore	Permian	Paleozoic	Siliciclastics	Nonmarine	Evaporites	Unknown	Structural and Stratigraphic
40360103	Southern Permian Basin-Offshore	Permian	Paleozoic	Siliciclastics	Nonmarine	Evaporites	Unknown	Structural and Stratigraphic
40470101	Foreland Basin	Neogene	Cenozoic	Siliciclastics	Nonmarine to Marine	Shale	Passive	Stratigraphic
40470201	Deformed Belt	Mesozoic to Cenozoic	Mesozoic to Cenozoic	Siliciclastics and Carbonates	Nonmarine to Marine	Shale	Compressional	Structural
40470301	Paleozoic Reservoirs	Carboniferous	Paleozoic	Carbonates	Shallow Marine	Shale	Passive	Stratigraphic
40480101	Greater Hungarian Plain Basins	Paleozoic to Neogene	Paleozoic to Cenozoic	Siliciclastics and Other	Nonmarine to Marine	Shale	Extensional	Structural
40480201	Zala-Drava-Sava Basins	Paleozoic to Neogene	Paleozoic to Cenozoic	Siliciclastics, Carbonates, and Other	Nonmarine to Marine	Shale	Extensional	Structural
40480301	Danube Basin	Paleozoic to Neogene	Paleozoic to Cenozoic	Siliciclastics and Other	Nonmarine to Marine	Shale	Extensional	Structural
40480401	Transcarpathian Basin	Neogene	Cenozoic	Siliciclastics and Other	Nonmarine to Marine	Shale	Extensional	Structural
40480501	Central Carpathian Paleogene Basin							
40480601	Hungarian Paleogene Basin	Paleogene	Cenozoic	Siliciclastics	Nonmarine to Marine	Shale	Extensional	Structural

Assessment Unit Code	Assessment Unit Name	Specific Reservoir Rock Age	General Reservoir Rock Age	Reservoir Rock Lithology	Reservoir Rock Depositional Environment	Seal Rock Lithology	Structural Setting	Trap Type
40570101	Transylvanian Neogene Suprasalt Gas	Neogene	Cenozoic	Siliciclastics	Nonmarine to Marine	Shale	Passive	Structural
40570102	Transylvanian Subsalt							
40600101	Neogene Flysch Gas	Neogene	Cenozoic	Siliciclastics	Deep Marine	Shale	Compressional	Structural and Stratigraphic
40600201	Thermal Triassic	Mesozoic	Mesozoic	Carbonates	Shallow Marine	Shale	Compressional and Extensional	Structural
40610101	Moesian Platform	Paleogene to Neogene	Cenozoic	Siliciclastics and Carbonates	Nonmarine to Marine	Shale	Unknown	Structural and Stratigraphic
40610201	Romania Flysch Zone	Paleogene to Neogene	Cenozoic	Siliciclastics	Nonmarine to Marine	Shale	Unknown	Structural and Stratigraphic
40610202	Romania Ploiesti Zone	Paleogene to Neogene	Cenozoic	Siliciclastics	Nonmarine to Marine	Shale	Passive	Structural and Stratigraphic
40680101	Subsalt	Cretaceous to Neogene	Mesozoic to Cenozoic	Siliciclastics and Carbonates	Nonmarine to Marine	Evaporites	Extensional	Structural and Stratigraphic
52000101	Northeast Greenland Shelf Rift Systems	Jurassic	Mesozoic	Siliciclastics	Shallow Marine	Shale	Extensional	Structural
52150101	Jeanne d'Arc	Jurassic to Cretaceous	Mesozoic	Siliciclastics	Shallow Marine	Shale	Extensional	Structural
52150201	Graben Systems Exclusive of Jeanne d'Arc							
52430101	Keg River Gas	Devonian	Paleozoic	Carbonates	Shallow Marine	Shale	Passive	Stratigraphic
52430102	Keg River Oil and Gas	Devonian	Paleozoic	Carbonates	Shallow to Deep Marine	Evaporites	Passive	Stratigraphic
52430201	Leduc Gas	Devonian	Paleozoic	Carbonates	Shallow to Deep Marine	Shale	Compressional	Structural and Stratigraphic
52430202	Leduc Oil and Gas	Devonian to Cretaceous	Paleozoic to Mesozoic	Siliciclastics and Carbonates	Nonmarine to Marine	Shale	Passive	Structural and Stratigraphic
52430301	Exshaw-Rundle Gas	Carboniferous	Paleozoic	Siliciclastics and Carbonates	Nonmarine to Marine	Evaporites and Shale	Compressional	Structural and Stratigraphic
52430302	Exshaw-Rundle Oil and Gas	Carboniferous to Cretaceous	Paleozoic to Mesozoic	Siliciclastics and Carbonates	Nonmarine to Marine	Shale	Compressional	Structural and Stratigraphic
52430401	Combined Triassic/Jurassic Gas	Triassic to Jurassic	Mesozoic	Siliciclastics and Carbonates	Nonmarine to Marine	Evaporites and Shale	Compressional	Structural and Stratigraphic
52430402	Combined Triassic/Jurassic Oil and Gas	Triassic to Jurassic	Mesozoic	Siliciclastics and Carbonates	Nonmarine to Marine	Evaporites and Shale	Unknown	Structural and Stratigraphic
52430501	Mannville Gas	Cretaceous	Mesozoic	Siliciclastics	Nonmarine to Marine	Shale	Passive	Stratigraphic
52430502	Mannville Continuous (Basin-Center) Gas							
52430601	Second White Specks-Cardium Gas	Cretaceous	Mesozoic	Siliciclastics	Nonmarine to Marine	Shale	Passive	Stratigraphic
52430602	Second White Specks-Cardium Oil and Gas	Cretaceous	Mesozoic	Siliciclastics	Nonmarine to Marine	Shale	Passive	Stratigraphic
52430603	Upper Cretaceous Eastern Shallow Gas							
52440101	Yeoman Oil	Ordovician	Paleozoic	Carbonates	Shallow Marine	Evaporites	Passive	Structural
52440201	Brightholme Oil	Devonian	Paleozoic	Carbonates	Shallow Marine	Evaporites	Passive	Stratigraphic
52440301	Bakken Sandstone	Devonian to Carboniferous	Paleozoic	Siliciclastics	Shallow to Deep Marine	Shale	Unknown	Structural and Stratigraphic
52440401	Lodgepole Oil	Carboniferous	Paleozoic	Siliciclastics and Carbonates	Shallow Marine	Evaporites and Shale	Unknown	Structural and Stratigraphic
53050101	El Abra-Like Reef and Backreef Limestone	Cretaceous to Neogene	Mesozoic to Cenozoic	Carbonates	Shallow Marine	Shale	Unknown	Structural and Stratigraphic
53050102	Tamabra-Like Debris-Flow-Breccia Limestone Overlying Evaporites	Cretaceous to Neogene	Mesozoic to Cenozoic	Carbonates	Deep Marine	Shale	Unknown	Structural and Stratigraphic
53050103	Tamabra-Like Debris-Flow-Breccia Limestone and Overlying Strata	Cretaceous to Neogene	Mesozoic to Cenozoic	Carbonates	Deep Marine	Shale	Unknown	Structural and Stratigraphic
53050104	Tamabra-Like Debris-Flow-Breccia Limestone of the Golden Lane	Cretaceous to Neogene	Mesozoic to Cenozoic	Carbonates	Deep Marine	Shale	Unknown	Structural and Stratigraphic
53050105	Tamaulipas-Like Basinal Limestone and Tertiary Strata Without Underlying Evaporites	Cretaceous to Neogene	Mesozoic to Cenozoic	Carbonates	Deep Marine	Shale	Unknown	Structural and Stratigraphic

Assessment Unit Code	Assessment Unit Name	Specific Reservoir Rock Age	General Reservoir Rock Age	Reservoir Rock Lithology	Reservoir Rock Depositional Environment	Seal Rock Lithology	Structural Setting	Trap Type
53050106	Tamaulipas-Like Basinal Limestone and Tertiary Strata Overlying Evaporites	Cretaceous to Neogene	Mesozoic to Cenozoic	Carbonates	Deep Marine	Shale	Unknown	Structural and Stratigraphic
53050107	Tamabra-Like Debris-Flow-Breccia Limestone North of Campeche	Cretaceous to Neogene	Mesozoic to Cenozoic	Carbonates	Deep Marine	Shale	Unknown	Structural and Stratigraphic
60210101	Late Cretaceous-Tertiary Turbidites	Cretaceous to Neogene	Mesozoic to Cenozoic	Siliciclastics	Deep Marine	Shale	Unknown	Structural and Stratigraphic
60210102	Cretaceous Carbonates	Cretaceous	Mesozoic	Carbonates	Shallow Marine	Shale	Passive	Structural and Stratigraphic
60210103	Late Cretaceous-Tertiary Nearshore Sandstones	Paleogene to Neogene	Cenozoic	Siliciclastics	Nonmarine to Marine	Shale	Unknown	Structural and Stratigraphic
60220101	Amazon Delta and Submarine Fan	Neogene	Cenozoic	Siliciclastics	Deep Marine	Shale	Extensional	Structural and Stratigraphic
60220102	Offshore Gas Hydrates							
60290101	Western Pre-Aptian Reservoirs	Jurassic to Cretaceous	Mesozoic	Siliciclastics	Nonmarine	Evaporites	Extensional	Structural
60290102	Late Cretaceous-Tertiary Deep-Water Sandstones	Cretaceous to Paleogene	Mesozoic to Cenozoic	Siliciclastics	Deep Marine	Shale	Extensional	Structural and Stratigraphic
60340101	Espirito Santo Shelf	Cretaceous to Neogene	Mesozoic to Cenozoic	Siliciclastics	Nonmarine to Marine	Evaporites and Shale	Extensional	Structural and Stratigraphic
60340102	Late Cretaceous-Tertiary Slide Blocks and Turbidites	Cretaceous to Neogene	Mesozoic to Cenozoic	Siliciclastics	Deep Marine	Evaporites and Shale	Extensional	Structural
60340103	Abrolhos Sub-Volcanic Structures	Cretaceous	Mesozoic	Siliciclastics	Deep Marine	Shale	Compressional	Structural and Stratigraphic
60350101	Late Cretaceous-Tertiary Turbidites	Cretaceous to Neogene	Mesozoic to Cenozoic	Siliciclastics	Deep Marine	Shale	Extensional	Structural and Stratigraphic
60350102	Cretaceous Carbonates	Cretaceous	Mesozoic	Siliciclastics, Carbonates, and Other	Nonmarine to Marine	Shale	Extensional	Structural and Stratigraphic
60350103	Salt Dome Province Tertiary Sandstones	Paleogene to Neogene	Cenozoic	Siliciclastics	Deep Marine	Shale	Passive	Structural
60360101	Santos Shelf	Cretaceous to Neogene	Mesozoic to Cenozoic	Siliciclastics and Carbonates	Shallow to Deep Marine	Shale	Passive	Structural and Stratigraphic
60360102	Salt-Structured Deep-Water Sandstones	Cretaceous to Neogene	Mesozoic to Cenozoic	Siliciclastics	Deep Marine	Evaporites and Shale	Passive	Structural and Stratigraphic
60370101	Pelotas Platform and Basin	Cretaceous to Paleogene	Mesozoic to Cenozoic	Siliciclastics	Shallow to Deep Marine	Shale	Unknown	Structural and Stratigraphic
60370102	Offshore Gas Hydrates							
60410101	Hollin-Napo	Cretaceous	Mesozoic	Siliciclastics	Nonmarine to Marine	Shale	Compressional and Extensional	Structural
60410201	Ene	Permian	Paleozoic	Siliciclastics	Shallow Marine	Evaporites and Shale	Unknown	Structural and Stratigraphic
60450101	Sub-Andean Fold and Thrust Belt	Carboniferous	Paleozoic	Siliciclastics	Nonmarine to Marine	Shale	Compressional	Structural
60450102	Foreland Basins	Carboniferous	Paleozoic	Siliciclastics	Nonmarine to Marine	Shale	Extensional	Structural and Stratigraphic
60450103	Foreland Central Chaco High	Paleogene to Neogene	Cenozoic	Siliciclastics	Nonmarine to Marine	Shale	Extensional	Structural and Stratigraphic
60550101	Neuquen Extensional Structures	Jurassic to Cretaceous	Mesozoic	Siliciclastics and Carbonates	Nonmarine to Marine	Shale	Extensional	Structural and Stratigraphic
60550102	Neuquen Foothills Structure	Cretaceous to Paleogene	Mesozoic to Cenozoic	Siliciclastics	Nonmarine to Marine	Shale	Compressional	Structural
60550103	Dorsal de Neuquen Structure	Jurassic	Mesozoic	Siliciclastics	Nonmarine to Marine	Shale	Compressional and Extensiona	Structural
60580101	San Jorge Extensional Structures	Cretaceous to Paleogene	Mesozoic to Cenozoic	Siliciclastics	Nonmarine to Marine	Shale	Extensional	Structural
60580102	San Bernardo Fold Belt Structures	Cretaceous to Paleogene	Mesozoic to Cenozoic	Siliciclastics	Nonmarine to Marine	Shale	Compressional	Structural
60580201	Hypothetical Basin-Centered Gas							

Assessment Unit Code	Assessment Unit Name	Specific Reservoir Rock Age	General Reservoir Rock Age	Reservoir Rock Lithology	Reservoir Rock Depositional Environment	Seal Rock Lithology	Structural Setting	Trap Type
60590101	Magallanes Extensional Structures	Jurassic to Cretaceous	Mesozoic	Siliciclastics	Nonmarine to Marine	Shale	Extensional	Structural
60590102	Andean Fold Belt Structures	Paleogene	Cenozoic	Siliciclastics	Nonmarine to Marine	Shale	Compressional	Structural
60600101	North Falklands Basin	Jurassic to Cretaceous	Mesozoic	Siliciclastics	Nonmarine to Marine	Shale	Extensional	Structural
60600201	South Falklands Basin	Jurassic to Cretaceous	Mesozoic	Siliciclastics	Nonmarine to Marine	Shale	Compressional and Extensional	Structural
60630101	Malvinas Extensional Structures	Jurassic to Cretaceous	Mesozoic	Siliciclastics	Nonmarine to Marine	Shale	Compressional and Extensional	Structural
60810101	Cretaceous-Paleogene Basin	Cretaceous to Paleogene	Mesozoic to Cenozoic	Siliciclastics	Nonmarine to Marine	Shale	Extensional	Structural
60830101	Neogene Pull-Apart Basin	Paleogene to Neogene	Cenozoic	Siliciclastics	Nonmarine to Marine	Shale	Extensional	Structural
60830201	Cretaceous-Paleogene Santa Elena Block	Cretaceous to Paleogene	Mesozoic to Cenozoic	Siliciclastics	Nonmarine to Marine	Shale	Extensional	Structural
60900101	Northern	Cretaceous to Neogene	Mesozoic to Cenozoic	Siliciclastics	Nonmarine to Marine	Shale	Compressional	Structural
60900102	Southern	Paleogene to Neogene	Cenozoic	Siliciclastics	Nonmarine	Shale	Compressional	Structural
60900103	Eastern	Cretaceous to Neogene	Mesozoic to Cenozoic	Siliciclastics	Nonmarine to Marine	Shale	Compressional	Structural
60900104	La Luna and Older	Cretaceous	Mesozoic	Siliciclastics and Carbonates	Shallow Marine	Shale	Compressional and Extensional	Structural
60960101	Central	Cretaceous to Neogene	Mesozoic to Cenozoic	Siliciclastics	Nonmarine to Marine	Shale	Compressional and Extensional	Structural
60960102	Peripheral	Cretaceous to Neogene	Mesozoic to Cenozoic	Siliciclastics	Nonmarine to Marine	Shale	Compressional and Extensional	Structural
60980101	East Venezuela Fold and Thrust Belt	Neogene	Cenozoic	Siliciclastics	Nonmarine to Marine	Shale	Compressional	Structural
60980102	Guarico Sub-Basin	Paleogene to Neogene	Cenozoic	Siliciclastics	Nonmarine to Marine	Shale	Compressional	Structural
60980103	Maturin Sub-Basin	Cretaceous to Neogene	Mesozoic to Cenozoic	Siliciclastics	Nonmarine to Marine	Shale	Compressional	Structural and Stratigraphic
60980104	Orinoco Heavy Oil and Tar Belt							
60980201	Trinidad Basins	Neogene	Cenozoic	Siliciclastics	Nonmarine to Marine	Shale	Compressional and Extensional	Structural
60980202	Orinoco Delta and Offshore	Paleogene to Neogene	Cenozoic	Siliciclastics	Deep Marine	Shale	Extensional	Structural and Stratigraphic
60990101	Main Maracaibo Basin	Paleogene to Neogene	Cenozoic	Siliciclastics	Nonmarine to Marine	Shale	Compressional	Structural
60990102	Southwest Maracaibo Basin Fold Belt	Paleogene to Neogene	Cenozoic	Siliciclastics	Nonmarine to Marine	Shale	Compressional	Structural
60990103	Basin-Centered Gas							
61030101	Carupano Basin Gas	Neogene	Cenozoic	Siliciclastics	Nonmarine to Marine	Shale	Compressional and Extensional	Structural
61070101	Inner Forearc Deformation Belt	Paleogene	Cenozoic	Siliciclastics	Deep Marine	Shale	Compressional	Structural
61070102	Barbados Accreationary Prism							
61170101	North Cuba Fold and Thrust Belt	Jurassic to Cretaceous	Mesozoic	Siliciclastics and Carbonates	Nonmarine to Marine	Shale	Compressional	Structural
70130101	Coastal Plain and Offshore	Cretaceous to Paleogene	Mesozoic to Cenozoic	Siliciclastics and Carbonates	Shallow Marine	Shale	Passive	Structural and Stratigraphic
71830101	Coastal Plain and Offshore	Cretaceous	Mesozoic	Siliciclastics	Shallow to Deep Marine	Shale	Extensional	Structural and Stratigraphic
71920101	Agbada Reservoirs	Paleogene to Neogene	Cenozoic	Siliciclastics	Nonmarine to Marine	Shale	Extensional	Structural
71920102	Akata Reservoirs	Paleogene to Neogene	Cenozoic	Siliciclastics	Deep Marine	Shale	Passive	Stratigraphic

Assessment Unit Code	Assessment Unit Name	Specific Reservoir Rock Age	General Reservoir Rock Age	Reservoir Rock Lithology	Reservoir Rock Depositional Environment	Seal Rock Lithology	Structural Setting	Trap Type
72030101	Gabon Subsalt	Cretaceous	Mesozoic	Siliciclastics	Nonmarine to Marine	Evaporites and Shale	Extensional	Structural
72030201	Gabon Suprasalt	Cretaceous to Neogene	Mesozoic to Cenozoic	Siliciclastics	Deep Marine	Shale	Passive	Structural and Stratigraphic
72030301	Central Congo Delta and Carbonate Platform	Cretaceous to Neogene	Mesozoic to Cenozoic	Siliciclastics and Carbonates	Shallow Marine	Shale	Extensional	Structural and Stratigraphic
72030302	Central Congo Turbidites	Paleogene to Neogene	Cenozoic	Siliciclastics	Deep Marine	Shale	Passive	Stratigraphic
72030401	Cuanza-Namibe	Cretaceous to Neogene	Mesozoic to Cenozoic	Siliciclastics and Carbonates	Shallow to Deep Marine	Shale	Extensiona	Structural and Stratigraphic
73030101	Offshore	Cretaceous	Mesozoic	Siliciclastics	Deep Marine	Shale	Extensional	Structural and Stratigraphic
80260101	Kohat-Potwar Intrathrust Basin	Cretaceous to Paleogene	Mesozoic to Cenozoic	Siliciclastics	Nonmarine to Marine	Shale	Unknown	Structural and Stratigraphic
80340101	Sylhet-Kopili/Barail-Tipam Composite	Paleogene to Neogene	Cenozoic	Siliciclastics and Carbonates	Nonmarine to Marine	Shale	Compressional	Structural
80420101	Greater Indus Foreland and Foldbelt	Permian to Paleogene	Paleozoic to Cenozoic	Siliciclastics and Carbonates	Nonmarine to Marine	Shale	Compressional	Structural and Stratigraphic
80420102	Indus Fan	Neogene	Cenozoic	Siliciclastics	Deep Marine	Shale	Passive	Stratigraphic
80430101	Eocene-Miocene Bombay Shelf	Paleogene to Neogene	Cenozoic	Siliciclastics, Carbonates, and Other	Shallow Marine	Shale	Extensional	Structural and Stratigraphic
80430102	Eocene-Miocene Deltaics	Paleogene to Neogene	Cenozoic	Siliciclastics	Nonmarine to Marine	Shale	Extensional	Structural and Stratigraphic
80470101	Coalbed Gas							
80470201	Western Shelf and Slope	Paleogene	Cenozoic	Siliciclastics and Carbonates	Nonmarine to Marine	Shale	Passive	Stratigraphic
80470301	Central Basin	Paleogene to Neogene	Cenozoic	Siliciclastics	Nonmarine to Marine	Shale	Passive	Stratigraphic
80470302	Eastern Fold Belt	Paleogene to Neogene	Cenozoic	Siliciclastics	Nonmarine to Marine	Shale	Compressional	Structural
80480101	Central Burma Basin	Paleogene to Neogene	Cenozoic	Siliciclastics	Nonmarine to Marine	Shale	Compressional and Extensional	Structural
80480102	Irrawaddy-Andaman	Paleogene to Neogene	Cenozoic	Siliciclastics	Nonmarine to Marine	Shale	Compressional and Extensional	Structural

APPENDIX 5. Volumes of crude oil, natural gas, and natural gas liquids and geologic attributes of total petroleum systems containing assessment units identified in World Petroleum Assessment 2000 (USGS, 2000).

[BBO = billion barrels of oil; TCFG = trillion cubic feet of gas; BBNGL = billion barrels of natural gas liquids (NGL); BBOE = billion barrels of oil equivalent (BOE). One BOE equals 6000 ft^3 of gas. Total petroleum is the sum of oil, gas in terms of oil equivalents, and NGL. Major commodity is based on the ratio of discovered plus undiscovered gas and discovered and undiscovered oil as ft^3 of gas per bbl of oil (CFG/BO). Major commodity is oil if this ratio is less than 20,000 CFG/BO and gas if this ratio is equal to or greater than 20,000 CFG/BO. Known oil, gas, NGL, and total petroleum are the reported discovered volumes (cumulative production plus remaining reserves). Undiscovered oil, gas, NGL, and total petroleum volumes are the means of probability distributions estimated in World Petroleum Assessment 2000 (USGS, 2000). Numbers have been independently rounded. NA means not applicable; it is shown for all values where there is only one field present, and used where the endowment could not be calculated.] Known volumes from Petroconsultants (1996) and NRG Associates (1995). Note that gas = natural gas.

Province Code	Province Name	Total Petroleum System Code	Total Petroleum System Name	Number of Assess-ment Units	Number of Fields	Known Oil (BBO)	Known Gas (TCFG)	Known NGL (BBNGL)	Known Total Petroleum (BBOE)
1008	Timan-Pechora Basin	100801	Domanik-Paleozoic	3	254	13.069	36.632	0.716	19.890
1009	Dnieper-Donets Basin	100901	Dnieper-Donets Paleozoic	3	222	1.611	59.098	0.200	11.660
1015	Volga-Ural Region	101501	Volga-Ural Domanik-Paleozoic	4	1365	63.872	96.458	1.096	81.044
1015	Volga-Ural Region	101502	Belsk Basin	1	42	0.065	2.638	0.005	0.509
1016	North Caspian Basin	101601	Paleozoic North Caspian	6	173	10.808	156.976	8.890	45.861
1050	South Barents Basin	105001	South and North Barents Triassic-Jurassic	3	6	0.051	78.143	0.100	13.175
1108	Azov-Kuban Basin	110801	Azov-Kuban Mesozoic-Cenozoic	3	178	0.524	18.704	0.056	3.696
1109	Middle Caspian Basin	110901	Terek-Caspian	3	66	3.490	7.840	0.029	4.826
1109	Middle Caspian Basin	110902	South Mangyshlak	1	42	5.241	5.707	0.046	6.239
1109	Middle Caspian Basin	110903	Stavropol-Prikumsk	3	147	0.820	13.499	0.003	3.073
1112	South Caspian Basin	111201	Oligocene-Miocene Maykop/Diatom	5	116	17.438	35.994	0.503	23.941
1150	North Ustyurt Basin	115001	Buzachi Arch and Surrounding Areas Composite	1	9	2.286	0.943	0.000	2.443
1150	North Ustyurt Basin	115002	North Ustyurt Jurassic	1	13	0.080	1.450	0.008	0.330
1150	North Ustyurt Basin	115003	North Ustyurt Paleozoic	1	1	NA	NA	NA	NA
1154	Amu-Darya Basin	115401	Amu-Darya Jurassic-Cretaceous	4	250	0.766	230.614	1.175	40.377
1174	West Siberian Basin	117401	Bazhenov-Neocomian	2	475	117.972	97.582	0.764	135.000
1174	West Siberian Basin	117402	Togur-Tyumen	1	84	11.756	5.180	0.003	12.623
1174	West Siberian Basin	117403	Northern West Siberian Mesozoic Composite	2	107	10.185	1167.288	2.375	207.107
1207	Baykit Arch	120701	Yenisey Foldbelt Riphean-Craton Margin Riphean	1	6	0.034	5.052	0.060	0.936
1210	Nepa-Botuoba Arch	121001	Baikal-Patom Foldbelt Riphean-Craton Margin Vendian	1	33	2.006	30.210	0.360	7.401
1322	North Sakhalin Basin	132201	North Sakhalin Neogene	1	73	2.182	22.383	0.165	6.077
2004	Ma'Rib-Al Jawf/Masila Basin	200401	Madbi Amran/Qishn	1	54	1.933	17.120	0.312	5.098
2014	Ghaba Salt Basin	201401	North Oman Huqf/'Q'-Haushi	1	57	2.028	20.339	0.507	5.925
2016	Fahud Salt Basin	201601	North Oman Huqf-Shu'aiba	1	33	2.685	7.633	0.058	4.015
2016	Fahud Salt Basin	201602	Middle Cretaceous Natih	1	8	1.852	3.100	0.100	2.469
2019	Rub Al Khali Basin	201901	Cretaceous Thamama/Wasia	3	78	71.231	102.100	1.901	90.148
2019	Rub Al Khali Basin	201902	Jurassic Hanifa/Diyab-Arab	2	51	19.013	77.512	0.592	32.523
2019	Rub Al Khali Basin	201903	Silurian Qusaiba	2	4	0.555	452.030	13.815	89.708
2021	Greater Ghawar Uplift	202101	Central Arabia Qusaiba Paleozoic	2	20	6.376	79.115	0.343	19.905
2021	Greater Ghawar Uplift	202102	Arabian Sub-Basin Tuwaiq/Hanifa-Arab	2	37	198.995	275.295	9.205	254.083
2023	Widyan Basin-Interior Platform	202301	Paleozoic Qusaiba /Akkas /Abba /Mudawwara	1	30	1.213	1.661	0.025	1.515
2023	Widyan Basin-Interior Platform	202302	Jurassic Gotnia /Barsarin /Sargelu /Najmah	2	14	0.474	0.100	0.000	0.490
2030	Zagros Fold Belt	203001	Zagros-Mesopotamian Cretaceous-Tertiary	2	319	372.226	493.238	1.562	455.995
2030	Zagros Fold Belt	203002	Paleozoic-Permian/Triassic	1	10	0.000	131.220	0.925	22.795
2043	Sirte Basin	204301	Sirte-Zelten	4	252	37.072	37.767	0.129	43.496
2048	Pelagian Basin	204801	Bou Dabbous-Tertiary	1	46	2.114	15.509	0.044	4.743
2048	Pelagian Basin	204802	Jurassic-Cretaceous Composite	1	26	0.103	1.301	0.023	0.343
2054	Trias/Ghadames Basin	205401	Tanezzuft-Oued Mya	1	36	10.843	8.973	0.000	12.338
2054	Trias/Ghadames Basin	205402	Tanezzuft-Melrhir	1	4	0.006	0.125	0.009	0.036
2054	Trias/Ghadames Basin	205403	Tanezzuft-Ghadames	1	93	4.538	16.484	1.011	8.296

Province Code	Province Name	Total Petroleum System Code	Total Petroleum System Name	Number of Assess-ment Units	Number of Fields	Known Oil (BBO)	Known Gas (TCFG)	Known NGL (BBNGL)	Known Total Petroleum (BBOE)
2056	Illizi Basin	205601	Tanezzuft-Illizi	1	102	3.670	45.061	0.898	12.078
2058	Grand Erg/Ahnet Basin	205801	Tanezzuft-Timimoun	1	11	0.000	4.200	0.000	0.700
2058	Grand Erg/Ahnet Basin	205802	Tanezzuft-Ahnet	1	24	0.001	3.117	0.090	0.610
2058	Grand Erg/Ahnet Basin	205803	Tanezzuft-Sbaa	1	15	0.284	1.490	0.010	0.542
2058	Grand Erg/Ahnet Basin	205804	Tanezzuft-Mouydir	1	0	0.000	0.000	0.000	0.000
2058	Grand Erg/Ahnet Basin	205805	Tanezzuft-Benoud	1	4	0.088	105.050	4.935	22.531
2058	Grand Erg/Ahnet Basin	205806	Tanezzuft-Bechar/Abadla	1	0	0.000	0.000	0.000	0.000
2071	Red Sea Basin	207101	Sudr-Nubia	3	110	9.810	5.995	0.041	10.850
2071	Red Sea Basin	207102	Maqna	2	17	0.155	3.527	0.256	0.999
3115	Junggar Basin	311501	Lucaogou-Karamay/Ulho/Pindequan	1	12	6.624	2.248	0.000	6.998
3115	Junggar Basin	311502	Jurassic Coal-Jurassic/Tertiary	1	9	0.210	0.342	0.000	0.267
3115	Junggar Basin	311503	Lucaogou/Jurassic Coal-Paleozoic/Mesozoic	1	0	0.000	0.000	0.000	0.000
3127	Bohaiwan Basin	312701	Shahejie-Shahejie/Guantao/Wumishan	2	158	24.553	15.672	0.088	27.253
3127	Bohaiwan Basin	312702	Carboniferous/Permian Coal-Paleozoic	1	0	0.000	0.000	0.000	0.000
3128	Ordos Basin	312801	Yanchang-Yanan	1	26	0.683	0.020	0.000	0.686
3128	Ordos Basin	312802	Taiyuan/Shanxi-Majiagou/Shihezi	2	21	0.000	5.590	0.000	0.932
3142	Sichuan Basin	314201	Maokou/Longtang-Jialingjiang/Maokou/Huanglong	3	67	0.000	11.072	0.000	1.846
3142	Sichuan Basin	314202	Daanzhai-Daanzhai/Lianggaoshan	1	17	0.109	0.140	0.000	0.132
3142	Sichuan Basin	314203	Xujiahe-Xujiahe/Shaximiao	1	1	NA	NA	NA	NA
3142	Sichuan Basin	314204	Cambrian/Silurian Marine Shale-Dengying/Lower Paleozoic	2	2	0.001	1.350	0.000	0.226
3144	Songliao Basin	314401	Qingshankou-Putaohua/Shaertu	2	73	15.570	1.598	0.000	15.836
3144	Songliao Basin	314402	Jurassic Coal-Denglouku/Nongan	1	8	0.005	0.145	0.000	0.029
3154	Tarim Basin	315401	Ordovician/Jurassic-Phanerozoic	3	26	0.704	5.780	0.189	1.856
3701	Baram Delta/Brunei-Sabah Basin	370101	Brunei-Sabah	2	82	6.898	36.200	0.180	13.111
3702	Greater Sarawak Basin	370201	Sarawak Basin	2	62	0.797	37.119	0.379	7.363
3702	Greater Sarawak Basin	370202	East Natuna	1	3	0.020	45.045	0.000	7.528
3703	Malay Basin	370301	Oligocene-Miocene Lacustrine	2	58	3.017	24.248	0.136	7.194
3703	Malay Basin	370302	Miocene Coaly Strata	1	33	0.592	23.901	0.186	4.761
3808	Central Sumatra Basin	380801	Brown Shale-Sihapas	1	188	13.217	3.866	0.011	13.872
3817	Kutei Basin	381701	Kutei Basin	3	73	2.879	45.473	1.273	11.731
3822	North Sumatra Basin	382201	Bampo-Cenozoic	2	87	0.674	25.559	0.926	5.860
3824	Northwest Java Basin	382401	Banuwati-Oligocene/Miocene	1	59	1.259	0.614	0.000	1.362
3824	Northwest Java Basin	382402	Jatibarang/Talang Akar-Oligocene/Miocene	1	142	1.908	7.322	0.175	3.303
3824	Northwest Java Basin	382403	Tertiary-Parigi	1	8	0.000	0.119	0.000	0.020
3824	Northwest Java Basin	382404	Tertiary-Cenozoic	1	0	0.000	0.000	0.000	0.000
3828	South Sumatra Basin	382801	Lahat/Talang Akar-Cenozoic	1	193	2.429	10.204	0.056	4.186
3910	Bonaparte Gulf Basin	391001	Milligans-Carboniferous/Permian	1	7	0.015	0.048	0.000	0.023
3910	Bonaparte Gulf Basin	391002	Keyling/Hyland Bay-Permian	2	3	0.000	5.730	0.050	1.005
3910	Bonaparte Gulf Basin	391003	Jurassic/Early Cretaceous-Mesozoic	1	17	0.502	7.526	0.385	2.141
3913	Browse Basin	391301	Late Jurassic/Early Cretaceous-Mesozoic	1	8	0.046	17.960	0.200	3.239
3930	Gippsland Basin	393001	Latrobe	1	48	3.860	9.775	0.701	6.190
3948	Northwest Shelf	394801	Dingo-Mungaroo/Barrow	1	64	1.149	48.245	0.991	10.181
3948	Northwest Shelf	394802	Locker-Mungaroo/Barrow	1	7	0.006	8.102	0.000	1.356
4017	Vestford-Helgeland	401701	Upper Jurassic Spekk	2	16	2.660	15.662	0.702	5.973
4025	North Sea Graben	402501	Kimmeridgian Shales	3	470	43.895	158.914	5.971	76.351
4036	Anglo-Dutch Basin	403601	Carboniferous-Rotliegend	3	793	2.872	222.159	0.172	40.071
4047	North Carpathian Basin	404701	Isotopically Light Gas	1	142	0.001	10.660	0.003	1.782
4047	North Carpathian Basin	404702	Mesozoic/Paleogene Composite	1	188	0.923	5.393	0.006	1.828
4047	North Carpathian Basin	404703	Paleozoic Composite	1	3	0.001	0.233	0.000	0.039
4048	Pannonian Basin	404801	Greater Hungarian Plain Neogene	1	321	1.089	9.657	0.024	2.722
4048	Pannonian Basin	404802	Zala-Drava-Sava Mesozoic/Neogene	1	165	1.101	3.929	0.063	1.818
4048	Pannonian Basin	404803	Danube Neogene	1	16	0.000	0.247	0.000	0.041
4048	Pannonian Basin	404804	Transcarpathian Neogene	1	15	0.000	0.261	0.000	0.043
4048	Pannonian Basin	404805	Central Carpathian Paleogene	1	1	NA	NA	NA	NA
4048	Pannonian Basin	404806	Hungarian Paleogene	1	8	0.011	0.016	0.000	0.014
4057	Transylvanian Basin	405701	Transylvanian Composite	2	119	0.000	30.731	0.000	5.122
4060	Po Basin	406001	Porto Garibaldi	1	244	0.014	18.475	0.005	3.098

App-5_Page 2

Province Code	Province Name	Total Petroleum System Code	Total Petroleum System Name	Number of Assess-ment Units	Number of Fields	Known Oil (BBO)	Known Gas (TCFG)	Known NGL (BBNGL)	Known Total Petroleum (BBOE)
4060	Po Basin	406002	Meride/Riva di Solto	1	12	0.332	0.412	0.027	0.428
4061	Carpathian-Balkanian Basin	406101	Moesian Platform Composite	1	202	1.793	2.217	0.066	2.228
4061	Carpathian-Balkanian Basin	406102	Dysodile Schist-Tertiary	2	185	4.115	5.048	0.011	4.968
4068	Provence Basin	406801	Pre-Messinian	1	0	0.000	0.000	0.000	0.000
5200	East Greenland Rift Basins	520001	Permian/Upper Jurassic Composite	1	0	0.000	0.000	0.000	0.000
5215	Labrador-Newfoundland Shelf	521501	Egret-Hibernia	1	15	1.582	2.406	0.081	2.063
5215	Labrador-Newfoundland Shelf	521502	Mesozoic Composite	1	6	0.000	4.224	0.107	0.811
5243	Alberta Basin	524301	Keg River-Keg River	2	840	1.011	7.073	0.100	2.290
5243	Alberta Basin	524302	Duvernay-Leduc	2	1666	9.459	14.357	1.025	12.877
5243	Alberta Basin	524303	Exshaw-Rundle	2	1633	1.728	21.568	0.836	6.158
5243	Alberta Basin	524304	Combined Triassic/Jurassic	2	1308	1.296	10.580	0.350	3.409
5243	Alberta Basin	524305	Mannville-Upper Mannville	2	2901	0.000	30.731	0.447	5.569
5243	Alberta Basin	524306	Second White Specks-Cardium	3	1432	2.688	26.449	0.365	7.462
5244	Williston Basin, Canada	524401	Yeoman	1	2	0.001	0.000	0.000	0.001
5244	Williston Basin, Canada	524402	Brightholme	1	7	0.014	0.011	0.000	0.016
5244	Williston Basin, Canada	524403	Bakken	1	7	0.007	0.000	0.000	0.007
5244	Williston Basin, Canada	524404	Lodgepole	1	282	2.778	0.893	0.037	2.964
5305	Villahermosa Uplift	530501	Pimienta-Tamabra	7	418	44.412	50.822	0.095	52.977
6021	Guyana-Suriname Basin	602101	Cenomanian-Turonian	3	3	0.063	0.000	0.000	0.063
6022	Foz do Amazonas Basin	602201	Neogene	2	1	NA	NA	NA	NA
6029	Sergipe-Alagoas Basin	602901	Neocomian to Turonian Composite	2	80	0.692	1.407	0.003	0.929
6034	Espirito Santo Basin	603401	Cretaceous Composite	3	46	0.130	0.222	0.001	0.169
6035	Campos Basin	603501	Lagoa Feia-Carapebus	3	73	10.056	6.244	0.010	11.107
6036	Santos Basin	603601	Guaratiba-Guaruja (Cretaceous) Composite	2	11	0.285	1.125	0.044	0.517
6037	Pelotas Basin	603701	Cenomanian-Turonian-Tertiary Composite	2	0	0.000	0.000	0.000	0.000
6041	Putumayo-Oriente-Maranon Basin	604101	Mesozoic-Cenozoic	1	171	6.601	1.616	0.000	6.871
6041	Putumayo-Oriente-Maranon Basin	604102	Paleozoic	1	0	0.000	0.000	0.000	0.000
6045	Santa Cruz-Tarija Basin	604501	Los Monos-Machareti	3	100	0.320	16.985	0.548	3.699
6055	Neuquen Basin	605501	Neuquen Hybrid	3	297	2.370	20.704	0.338	6.159
6058	San Jorge Basin	605801	D-129	2	182	3.346	3.565	0.002	3.942
6058	San Jorge Basin	605802	Aguada Bandera	1	0	0.000	0.000	0.000	0.000
6059	Magallanes Basin	605901	Lower Inoceramus	2	252	1.216	24.807	0.211	5.562
6060	Falklands Plateau	606001	Neocomian Lacustrine	1	0	0.000	0.000	0.000	0.000
6060	Falklands Plateau	606002	Lower Cretaceous	1	0	0.000	0.000	0.000	0.000
6063	Malvinas Basin	606301	Lower Cretaceous Marine	1	3	0.010	0.075	0.002	0.025
6081	Talara Basin	608101	Cretaceous-Tertiary	1	42	1.672	2.892	0.000	2.154
6083	Progreso Basin	608301	Neogene	1	6	0.004	0.340	0.000	0.061
6083	Progreso Basin	608302	Cretaceous-Paleogene	1	13	0.171	0.158	0.000	0.197
6090	Middle Magdalena	609001	La Luna-La Paz	4	79	2.426	5.002	0.068	3.328
6096	Llanos Basin	609601	Gacheta-Mirador	2	93	5.402	15.314	0.451	8.405
6098	East Venezuela Basin	609801	Querecual	4	311	26.756	112.296	0.528	46.001
6098	East Venezuela Basin	609802	Upper Cretaceous/Tertiary	2	97	3.447	15.822	0.172	6.256
6099	Maracaibo Basin	609901	La Luna/Maracaibo	3	80	49.072	26.701	0.043	53.565
6103	Tobago Trough	610301	Lower Cruse	1	10	0.020	22.600	0.080	3.867
6107	Lesser Antilles Deformed Belt	610701	Tobago Trough Paleogene	2	27	0.011	0.022	0.000	0.014
6117	Greater Antilles Deformed Belt	611701	Upper Jurassic-Neocomian	1	21	0.297	0.096	0.000	0.313
7013	Senegal	701301	Cretaceous-Tertiary Composite	1	7	0.010	0.049	0.000	0.019
7183	Gulf of Guinea	718301	Cretaceous Composite	1	34	0.236	1.949	0.036	0.596
7192	Niger Delta	719201	Tertiary Niger Delta (Agbada/Akata)	2	619	34.522	93.811	2.842	53.000
7203	West-Central Coastal	720301	Melania-Gamba	1	26	1.752	0.843	0.015	1.908
7203	West-Central Coastal	720302	Azile-Senonian	1	104	2.114	1.835	0.010	2.430
7203	West-Central Coastal	720303	Congo Delta Composite	2	224	9.745	9.443	0.039	11.357
7203	West-Central Coastal	720304	Cuanza Composite	1	25	0.107	0.059	0.000	0.116
7303	Orange River Coastal	730301	Cretaceous Composite	1	3	0.000	6.015	0.000	1.003
8026	Kohat-Potwar	802601	Patala-Namal	1	22	0.324	1.981	0.041	0.696

Province Code	Province Name	Total Petroleum System Code	Total Petroleum System Name	Number of Assess-ment Units	Number of Fields	Known Oil (BBO)	Known Gas (TCFG)	Known NGL (BBNGL)	Known Total Petroleum (BBOE)
8034	Assam	803401	Sylhet-Kopili/Barail-Tipam Composite	1	46	2.536	6.534	0.005	3.630
8042	Indus	804201	Sembar-Goru/Ghazij	2	92	0.180	35.373	0.063	6.139
8043	Bombay	804301	Eocene-Miocene Composite	2	162	8.440	24.193	0.267	12.739
8047	Ganges-Brahmaputra Delta	804701	Permian Coal	1	0	0.000	0.000	0.000	0.000
8047	Ganges-Brahmaputra Delta	804702	Jalangi-Sylhet/Burdwan Composite	1	1	NA	NA	NA	NA
8047	Ganges-Brahmaputra Delta	804703	Jenam/Bhuban-Bokabil	2	35	0.004	12.289	0.051	2.104
8048	Irrawaddy	804801	Eocene to Miocene Composite	2	60	0.716	10.020	0.053	2.449

Province Code	Province Name	Total Petroleum System Code	Total Petroleum System Name	Number of Assess-ment Units	Number of Fields	Undiscovered Oil (Mean, BBO)	Undiscovered Gas (Mean, TCFG)	Undiscovered NGL (Mean, BBNGL)	Undiscovered Total Petroleum (Mean, BBOE)
1008	Timan-Pechora Basin	100801	Domanik-Paleozoic	3	254	5.170	34.525	1.454	12.378
1009	Dnieper-Donets Basin	100901	Dnieper-Donets Paleozoic	3	222	1.098	24.051	0.862	5.969
1015	Volga-Ural Region	101501	Volga-Ural Domanik-Paleozoic	4	1365	2.507	2.394	0.105	3.011
1015	Volga-Ural Region	101502	Belsk Basin	1	42	0.042	1.547	0.093	0.392
1016	North Caspian Basin	101601	Paleozoic North Caspian	6	173	23.320	119.051	5.412	48.573
1050	South Barents Basin	105001	South and North Barents Triassic-Jurassic	3	6	0.955	266.618	5.921	51.312
1108	Azov-Kuban Basin	110801	Azov-Kuban Mesozoic-Cenozoic	3	178	0.417	12.252	0.215	2.675
1109	Middle Caspian Basin	110901	Terek-Caspian	3	66	2.829	27.779	1.348	8.807
1109	Middle Caspian Basin	110902	South Mangyshlak	1	42	0.651	2.262	0.072	1.100
1109	Middle Caspian Basin	110903	Stavropol-Prikumsk	3	147	1.407	9.644	0.250	3.264
1112	South Caspian Basin	111201	Oligocene-Miocene Maykop/Diatom	5	116	15.725	173.310	8.115	52.725
1150	North Ustyurt Basin	115001	Buzachi Arch and Surrounding Areas Composite	1	9	0.814	0.089	0.002	0.831
1150	North Ustyurt Basin	115002	North Ustyurt Jurassic	1	13	0.139	2.180	0.069	0.572
1150	North Ustyurt Basin	115003	North Ustyurt Paleozoic	1	1	0.000	8.500	0.374	1.790
1154	Amu-Darya Basin	115401	Amu-Darya Jurassic-Cretaceous	4	250	0.270	163.650	3.995	31.541
1174	West Siberian Basin	117401	Bazhenov-Neocomian	2	475	42.947	96.934	5.749	64.852
1174	West Siberian Basin	117402	Togur-Tyumen	1	84	7.474	5.516	0.316	8.709
1174	West Siberian Basin	117403	Northern West Siberian Mesozoic Composite	2	107	4.814	540.483	14.407	109.302
1207	Baykit Arch	120701	Yenisey Foldbelt Riphean-Craton Margin Riphean	1	6	1.213	16.983	0.854	4.897
1210	Nepa-Botuoba Arch	121001	Baikal-Patom Foldbelt Riphean-Craton Margin Vendian	1	33	1.611	46.039	2.194	11.478
1322	North Sakhalin Basin	132201	North Sakhalin Neogene	1	73	2.582	57.451	2.999	15.156
2004	Ma'Rib-Al Jawf/Masila Basin	200401	Madbi Amran/Qishn	1	54	2.740	16.982	0.924	6.494
2014	Ghaba Salt Basin	201401	North Oman Huqf/'Q'-Haushi	1	57	0.743	10.112	0.443	2.871
2016	Fahud Salt Basin	201601	North Oman Huqf-Shu'aiba	1	33	1.304	8.206	0.287	2.959
2016	Fahud Salt Basin	201602	Middle Cretaceous Natih	1	8	0.371	0.363	0.018	0.450
2019	Rub Al Khali Basin	201901	Cretaceous Thamama/Wasia	3	78	8.979	22.311	1.083	13.780
2019	Rub Al Khali Basin	201902	Jurassic Hanifa/Diyab-Arab	2	51	14.478	58.515	3.083	27.314
2019	Rub Al Khali Basin	201903	Silurian Qusaiba	2	4	13.235	346.892	26.147	97.197
2021	Greater Ghawar Uplift	202101	Central Arabia Qusaiba Paleozoic	2	20	19.190	383.339	24.587	107.667
2021	Greater Ghawar Uplift	202102	Arabian Sub-Basin Tuwaiq/Hanifa-Arab	2	37	41.703	34.046	1.362	48.739
2023	Widyan Basin-Interior Platform	202301	Paleozoic Qusaiba /Akkas /Abba /Mudawwara	1	30	2.632	64.510	5.080	18.464
2023	Widyan Basin-Interior Platform	202302	Jurassic Gotnia /Barsarin /Sargelu /Najmah	2	14	6.569	22.942	1.114	11.506
2030	Zagros Fold Belt	203001	Zagros-Mesopotamian Cretaceous-Tertiary	2	319	94.516	197.597	7.960	135.409
2030	Zagros Fold Belt	203002	Paleozoic-Permian/Triassic	1	10	0.000	69.031	3.039	14.544
2043	Sirte Basin	204301	Sirte-Zelten	4	252	6.854	14.984	0.648	10.000
2048	Pelagian Basin	204801	Bou Dabbous-Tertiary	1	46	0.667	2.746	0.064	1.189
2048	Pelagian Basin	204802	Jurassic-Cretaceous Composite	1	26	0.403	2.280	0.027	0.810
2054	Trias/Ghadames Basin	205401	Tanezzuft-Oued Mya	1	36	0.830	2.341	0.110	1.330
2054	Trias/Ghadames Basin	205402	Tanezzuft-Melrhir	1	4	1.875	4.887	0.269	2.959
2054	Trias/Ghadames Basin	205403	Tanezzuft-Ghadames	1	93	4.461	12.035	0.908	7.376
2056	Illizi Basin	205601	Tanezzuft-Illizi	1	102	2.814	27.785	0.873	8.318
2058	Grand Erg/Ahnet Basin	205801	Tanezzuft-Timimoun	1	11	0.031	1.128	0.056	0.275
2058	Grand Erg/Ahnet Basin	205802	Tanezzuft-Ahnet	1	24	0.034	2.973	0.149	0.678
2058	Grand Erg/Ahnet Basin	205803	Tanezzuft-Sbaa	1	15	0.162	0.645	0.011	0.281
2058	Grand Erg/Ahnet Basin	205804	Tanezzuft-Mouydir	1	0	0.012	0.292	0.014	0.075
2058	Grand Erg/Ahnet Basin	205805	Tanezzuft-Benoud	1	4	0.072	2.541	0.125	0.621
2058	Grand Erg/Ahnet Basin	205806	Tanezzuft-Bechar/Abadla	1	0	0.016	0.441	0.022	0.111
2071	Red Sea Basin	207101	Sudr-Nubia	3	110	2.608	10.192	0.484	4.791
2071	Red Sea Basin	207102	Maqna	2	17	2.579	49.820	2.860	13.743
3115	Junggar Basin	311501	Lucaogou-Karamay/Ulho/Pindequan	1	12	0.467	0.346	0.021	0.546
3115	Junggar Basin	311502	Jurassic Coal-Jurassic/Tertiary	1	9	0.096	0.950	0.043	0.297
3115	Junggar Basin	311503	Lucaogou/Jurassic Coal-Paleozoic/Mesozoic	1	0	NA	NA	NA	NA
3127	Bohaiwan Basin	312701	Shahejie-Shahejie/Guantao/Wumishan	2	158	4.412	8.514	0.417	6.248
3127	Bohaiwan Basin	312702	Carboniferous/Permian Coal-Paleozoic	1	0	NA	NA	NA	NA
3128	Ordos Basin	312801	Yanchang-Yanan	1	26	0.138	0.020	0.001	0.143
3128	Ordos Basin	312802	Taiyuan/Shanxi-Majiagou/Shihezi	2	21	NA	NA	NA	NA

Province Code	Province Name	Total Petroleum System Code	Total Petroleum System Name	Number of Assess-ment Units	Number of Fields	Undiscovered Oil (Mean, BBO)	Undiscovered Gas (Mean, TCFG)	Undiscovered NGL (Mean, BBNGL)	Undiscovered Total Petroleum (Mean, BBOE)
3142	Sichuan Basin	314201	Maokou/Longtang-Jialingjiang/Maokou/Huanglong	3	67	0.000	6.137	0.270	1.293
3142	Sichuan Basin	314202	Daanzhai-Daanzhai/Lianggaoshan	1	17	0.027	0.137	0.008	0.059
3142	Sichuan Basin	314203	Xujiahe-Xujiahe/Shaximiao	1	1	NA	NA	NA	NA
3142	Sichuan Basin	314204	Cambrian/Silurian Marine Shale-Dengying/Lower Paleozoic	2	2	0.000	4.036	0.178	0.850
3144	Songliao Basin	314401	Qingshankou-Putaohua/Shaertu	2	73	1.026	2.494	0.115	1.557
3144	Songliao Basin	314402	Jurassic Coal-Denglouku/Nongan	1	8	0.000	3.212	0.096	0.632
3154	Tarim Basin	315401	Ordovician/Jurassic-Phanerozoic	3	26	5.948	59.939	3.312	19.250
3701	Baram Delta/Brunei-Sabah Basin	370101	Brunei-Sabah	2	82	3.390	23.226	1.174	8.434
3702	Greater Sarawak Basin	370201	Sarawak Basin	2	62	0.529	17.284	0.372	3.782
3702	Greater Sarawak Basin	370202	East Natuna	1	3	0.102	1.963	0.090	0.519
3703	Malay Basin	370301	Oligocene-Miocene Lacustrine	2	58	1.127	23.696	0.442	5.518
3703	Malay Basin	370302	Miocene Coaly Strata	1	33	0.101	3.768	0.072	0.802
3808	Central Sumatra Basin	380801	Brown Shale-Sihapas	1	188	0.899	4.078	0.084	1.663
3817	Kutei Basin	381701	Kutei Basin	3	73	5.145	64.757	2.122	18.060
3822	North Sumatra Basin	382201	Bampo-Cenozoic	2	87	0.213	15.123	0.181	2.914
3824	Northwest Java Basin	382401	Banuwati-Oligocene/Miocene	1	59	0.462	0.874	0.019	0.626
3824	Northwest Java Basin	382402	Jatibarang/Talang Akar-Oligocene/Miocene	1	142	0.195	6.579	0.089	1.381
3824	Northwest Java Basin	382403	Tertiary-Parigi	1	8	NA	NA	NA	NA
3824	Northwest Java Basin	382404	Tertiary-Cenozoic	1	0	NA	NA	NA	NA
3828	South Sumatra Basin	382801	Lahat/Talang Akar-Cenozoic	1	193	0.469	18.250	0.239	3.750
3910	Bonaparte Gulf Basin	391001	Milligans-Carboniferous/Permian	1	7	0.140	1.151	0.056	0.387
3910	Bonaparte Gulf Basin	391002	Keyling/Hyland Bay-Permian	2	3	0.000	10.995	0.484	2.316
3910	Bonaparte Gulf Basin	391003	Jurassic/Early Cretaceous-Mesozoic	1	17	1.146	11.346	0.540	3.577
3913	Browse Basin	391301	Late Jurassic/Early Cretaceous-Mesozoic	1	8	1.055	20.093	0.934	5.339
3930	Gippsland Basin	393001	Latrobe	1	48	0.310	5.658	0.339	1.592
3948	Northwest Shelf	394801	Dingo-Mungaroo/Barrow	1	64	2.249	56.380	3.314	14.960
3948	Northwest Shelf	394802	Locker-Mungaroo/Barrow	1	7	0.132	8.331	0.368	1.888
4017	Vestford-Helgeland	401701	Upper Jurassic Spekk	2	16	6.538	165.201	8.990	43.062
4025	North Sea Graben	402501	Kimmeridgian Shales	3	470	13.098	37.745	1.859	21.248
4036	Anglo-Dutch Basin	403601	Carboniferous-Rotliegend	3	793	0.143	25.790	0.069	4.510
4047	North Carpathian Basin	404701	Isotopically Light Gas	1	142	0.000	2.321	0.019	0.405
4047	North Carpathian Basin	404702	Mesozoic/Paleogene Composite	1	188	0.344	2.345	0.075	0.810
4047	North Carpathian Basin	404703	Paleozoic Composite	1	3	0.047	0.383	0.019	0.129
4048	Pannonian Basin	404801	Greater Hungarian Plain Neogene	1	321	0.196	2.395	0.074	0.669
4048	Pannonian Basin	404802	Zala-Drava-Sava Mesozoic/Neogene	1	165	0.114	1.151	0.036	0.342
4048	Pannonian Basin	404803	Danube Neogene	1	16	0.000	0.146	0.003	0.027
4048	Pannonian Basin	404804	Transcarpathian Neogene	1	15	0.000	0.112	0.002	0.021
4048	Pannonian Basin	404805	Central Carpathian Paleogene	1	1	NA	NA	NA	NA
4048	Pannonian Basin	404806	Hungarian Paleogene	1	8	0.049	0.339	0.013	0.118
4057	Transylvanian Basin	405701	Transylvanian Composite	2	119	0.000	2.083	0.006	0.354
4060	Po Basin	406001	Porto Garibaldi	1	244	0.000	15.267	0.034	2.745
4060	Po Basin	406002	Meride/Riva di Solto	1	12	0.360	1.698	0.087	0.731
4061	Carpathian-Balkanian Basin	406101	Moesian Platform Composite	1	202	0.171	0.561	0.020	0.285
4061	Carpathian-Balkanian Basin	406102	Dysodile Schist-Tertiary	2	185	0.842	2.618	0.096	1.374
4068	Provence Basin	406801	Pre-Messinian	1	0	0.388	51.210	2.265	11.188
5200	East Greenland Rift Basins	520001	Permian/Upper Jurassic Composite	1	0	47.148	80.709	4.170	64.769
5215	Labrador-Newfoundland Shelf	521501	Egret-Hibernia	1	15	1.796	8.893	0.514	3.792
5215	Labrador-Newfoundland Shelf	521502	Mesozoic Composite	1	6	NA	NA	NA	NA
5243	Alberta Basin	524301	Keg River-Keg River	2	840	0.084	1.776	0.039	0.419
5243	Alberta Basin	524302	Duvernay-Leduc	2	1666	0.501	3.224	0.133	1.172
5243	Alberta Basin	524303	Exshaw-Rundle	2	1633	0.126	2.663	0.060	0.629
5243	Alberta Basin	524304	Combined Triassic/Jurassic	2	1308	0.096	2.166	0.056	0.514
5243	Alberta Basin	524305	Mannville-Upper Mannville	2	2901	0.000	4.597	0.092	0.858
5243	Alberta Basin	524306	Second White Specks-Cardium	3	1432	0.036	1.126	0.034	0.257
5244	Williston Basin, Canada	524401	Yeoman	1	2	0.005	0.004	0.000	0.006
5244	Williston Basin, Canada	524402	Brightholme	1	7	0.008	0.005	0.000	0.009

Province Code	Province Name	Total Petroleum System Code	Total Petroleum System Name	Number of Assess-ment Units	Number of Fields	Undiscovered Oil (Mean, BBO)	Undiscovered Gas (Mean, TCFG)	Undiscovered NGL (Mean, BBNGL)	Undiscovered Total Petroleum (Mean, BBOE)
5244	Williston Basin, Canada	524403	Bakken	1	7	0.011	0.003	0.000	0.012
5244	Williston Basin, Canada	524404	Lodgepole	1	282	0.111	0.063	0.004	0.126
5305	Villahermosa Uplift	530501	Pimienta-Tamabra	7	418	20.569	49.272	2.751	31.531
6021	Guyana-Suriname Basin	602101	Cenomanian-Turonian	3	3	15.247	42.058	2.330	24.586
6022	Foz do Amazonas Basin	602201	Neogene	2	1	0.000	29.840	0.298	5.272
6029	Sergipe-Alagoas Basin	602901	Neocomian to Turonian Composite	2	80	1.487	8.334	0.415	3.292
6034	Espirito Santo Basin	603401	Cretaceous Composite	3	46	2.959	34.287	1.612	10.285
6035	Campos Basin	603501	Lagoa Feia-Carapebus	3	73	16.293	19.691	0.553	20.128
6036	Santos Basin	603601	Guaratiba-Guaruja (Cretaceous) Composite	2	11	23.209	80.547	4.194	40.828
6037	Pelotas Basin	603701	Cenomanian-Turonian-Tertiary Composite	2	0	2.938	22.806	1.107	7.846
6041	Putumayo-Oriente-Maranon Basin	604101	Mesozoic-Cenozoic	1	171	3.023	0.755	0.015	3.164
6041	Putumayo-Oriente-Maranon Basin	604102	Paleozoic	1	0	0.074	0.841	0.040	0.254
6045	Santa Cruz-Tarija Basin	604501	Los Monos-Machareti	3	100	2.145	31.107	1.300	8.629
6055	Neuquen Basin	605501	Neuquen Hybrid	3	297	1.290	12.416	0.203	3.562
6058	San Jorge Basin	605801	D-129	2	182	0.498	3.774	0.075	1.203
6058	San Jorge Basin	605802	Aguada Bandera	1	0	NA	NA	NA	NA
6059	Magallanes Basin	605901	Lower Inoceramus	2	252	0.704	14.040	0.310	3.354
6060	Falklands Plateau	606001	Neocomian Lacustrine	1	0	3.603	8.871	0.178	5.259
6060	Falklands Plateau	606002	Lower Cretaceous	1	0	2.229	7.710	0.193	3.707
6063	Malvinas Basin	606301	Lower Cretaceous Marine	1	3	1.031	11.327	0.362	3.280
6081	Talara Basin	608101	Cretaceous-Tertiary	1	42	1.711	4.795	0.255	2.765
6083	Progreso Basin	608301	Neogene	1	6	0.037	0.471	0.021	0.137
6083	Progreso Basin	608302	Cretaceous-Paleogene	1	13	0.200	0.224	0.012	0.249
6090	Middle Magdalena	609001	La Luna-La Paz	4	79	0.709	3.292	0.130	1.388
6096	Llanos Basin	609601	Gacheta-Mirador	2	93	3.631	6.217	0.337	5.004
6098	East Venezuela Basin	609801	Querecual	4	311	7.231	41.552	2.030	16.186
6098	East Venezuela Basin	609802	Upper Cretaceous/Tertiary	2	97	4.644	52.009	2.436	15.748
6099	Maracaibo Basin	609901	La Luna/Maracaibo	3	80	8.183	17.576	0.973	12.086
6103	Tobago Trough	610301	Lower Cruse	1	10	0.000	20.103	0.201	3.551
6107	Lesser Antilles Deformed Belt	610701	Tobago Trough Paleogene	2	27	1.536	11.957	0.582	4.110
6117	Greater Antilles Deformed Belt	611701	Upper Jurassic-Neocomian	1	21	0.494	0.592	0.035	0.628
7013	Senegal	701301	Cretaceous-Tertiary Composite	1	7	0.157	0.856	0.043	0.343
7183	Gulf of Guinea	718301	Cretaceous Composite	1	34	1.004	10.071	0.282	2.965
7192	Niger Delta	719201	Tertiary Niger Delta (Agbada/Akata)	2	619	40.487	132.716	6.034	68.640
7203	West-Central Coastal	720301	Melania-Gamba	1	26	0.727	3.674	0.172	1.511
7203	West-Central Coastal	720302	Azile-Senonian	1	104	5.031	13.230	0.625	7.861
7203	West-Central Coastal	720303	Congo Delta Composite	2	224	23.198	69.582	3.374	38.169
7203	West-Central Coastal	720304	Cuanza Composite	1	25	0.790	1.558	0.073	1.123
7303	Orange River Coastal	730301	Cretaceous Composite	1	3	0.116	3.603	0.163	0.879
8026	Kohat-Potwar	802601	Patala-Namal	1	22	0.145	0.551	0.027	0.264
8034	Assam	803401	Sylhet-Kopili/Barail-Tipam Composite	1	46	0.399	1.178	0.067	0.663
8042	Indus	804201	Sembar-Goru/Ghazij	2	92	0.119	29.325	1.033	6.039
8043	Bombay	804301	Eocene-Miocene Composite	2	162	2.091	12.944	0.418	4.666
8047	Ganges-Brahmaputra Delta	804701	Permian Coal	1	0	NA	NA	NA	NA
8047	Ganges-Brahmaputra Delta	804702	Jalangi-Sylhet/Burdwan Composite	1	1	0.064	4.613	0.056	0.888
8047	Ganges-Brahmaputra Delta	804703	Jenam/Bhuban-Bokabil	2	35	0.039	50.479	0.218	8.670
8048	Irrawaddy	804801	Eocene to Miocene Composite	2	60	0.724	20.521	0.785	4.930

Province Code	Province Name	Total Petroleum System Code	Number of Assessment Units	Number of Fields	Major Commodity	Specific Source Rock Age	General Source Rock Age	Source Rock Depositional Environment	Age of Peak Maturation
1008	Timan-Pechora Basin	100801	3	254	Oil	Devonian	Paleozoic	Deep Marine	Paleozoic to Mesozoic
1009	Dnieper-Donets Basin	100901	3	222	Gas	Devonian to Carboniferous	Paleozoic	Deep Marine	Paleozoic to Mesozoic
1015	Volga-Ural Region	101501	4	1365	Oil	Devonian	Paleozoic	Deep Marine	Paleozoic to Mesozoic
1015	Volga-Ural Region	101502	1	42	Gas	Permian	Paleozoic	Deep Marine	Paleozoic to Mesozoic
1016	North Caspian Basin	101601	6	173	Oil	Upper Paleozoic	Paleozoic	Deep Marine	Paleozoic to Mesozoic
1050	South Barents Basin	105001	3	6	Gas	Triassic	Mesozoic	Coaly and Shallow Marine	Mesozoic
1108	Azov-Kuban Basin	110801	3	178	Gas	Jurassic to Neogene	Mesozoic to Cenozoic	Coaly and Shallow Marine	Cenozoic
1109	Middle Caspian Basin	110901	3	66	Oil	Jurassic to Neogene	Mesozoic to Cenozoic	Shallow and Deep Marine	Cenozoic
1109	Middle Caspian Basin	110902	1	42	Oil	Triassic	Mesozoic	Deep Marine	Mesozoic to Cenozoic
1109	Middle Caspian Basin	110903	3	147	Oil	Mesozoic to Cenozoic	Mesozoic to Cenozoic	Shallow and Deep Marine	Cenozoic
1112	South Caspian Basin	111201	5	116	Oil	Paleogene to Neogene	Cenozoic	Deep Marine	Cenozoic
1150	North Ustyurt Basin	115001	1	9	Oil	Paleozoic to Mesozoic	Paleozoic to Mesozoic	Deep Marine	Mesozoic to Cenozoic
1150	North Ustyurt Basin	115002	1	13	Oil	Jurassic	Mesozoic	Coaly and Lacustrine	Mesozoic to Cenozoic
1150	North Ustyurt Basin	115003	1	1	Gas	Carboniferous	Paleozoic	Deep Marine	Mesozoic
1154	Amu-Darya Basin	115401	4	250	Gas	Jurassic	Mesozoic	Coaly and Deep Marine	Cenozoic
1174	West Siberian Basin	117401	2	475	Oil	Jurassic	Mesozoic	Deep Marine	Cenozoic
1174	West Siberian Basin	117402	1	84	Oil	Jurassic	Mesozoic	Lacustrine	Mesozoic to Cenozoic
1174	West Siberian Basin	117403	2	107	Gas	Jurassic to Cretaceous	Mesozoic	Coaly and Shallow Marine	Mesozoic to Cenozoic
1207	Baykit Arch	120701	1	6	Oil	Precambrian	Precambrian	Deep Marine	Paleozoic
1210	Nepa-Botuoba Arch	121001	1	33	Gas	Precambrian	Precambrian	Deep Marine	Paleozoic
1322	North Sakhalin Basin	132201	1	73	Oil	Paleogene to Neogene	Cenozoic	Deep Marine	Cenozoic
2004	Ma'Rib-Al Jawf/Masila Basin	200401	1	54	Oil	Jurassic	Mesozoic	Deep Marine	Cenozoic
2014	Ghaba Salt Basin	201401	1	57	Oil	Precambrian	Precambrian	Shallow Marine	Paleozoic to Mesozoic
2016	Fahud Salt Basin	201601	1	33	Oil	Precambrian	Precambrian	Shallow Marine	Paleozoic to Mesozoic
2016	Fahud Salt Basin	201602	1	8	Oil	Cretaceous	Mesozoic	Shallow Marine	Cenozoic
2019	Rub Al Khali Basin	201901	3	78	Oil	Cretaceous	Mesozoic	Deep Marine	Cenozoic
2019	Rub Al Khali Basin	201902	2	51	Oil	Jurassic	Mesozoic	Shallow Marine	Mesozoic to Cenozoic
2019	Rub Al Khali Basin	201903	2	4	Gas	Silurian	Paleozoic	Shallow Marine	Mesozoic to Cenozoic
2021	Greater Ghawar Uplift	202101	2	20	Oil	Silurian	Paleozoic	Shallow Marine	Mesozoic
2021	Greater Ghawar Uplift	202102	2	37	Oil	Jurassic to Cretaceous	Mesozoic	Shallow Marine	Cenozoic
2023	Widyan Basin-Interior Platform	202301	1	30	Oil	Silurian	Paleozoic	Shallow Marine	Mesozoic
2023	Widyan Basin-Interior Platform	202302	2	14	Oil	Jurassic	Mesozoic	Shallow Marine	Mesozoic to Cenozoic
2030	Zagros Fold Belt	203001	2	319	Oil	Jurassic to Cretaceous	Mesozoic	Shallow Marine	Cenozoic
2030	Zagros Fold Belt	203002	1	10	Gas	Silurian	Paleozoic	Shallow Marine	Mesozoic to Cenozoic
2043	Sirte Basin	204301	4	252	Oil	Cretaceous	Mesozoic	Deep Marine	Cenozoic
2048	Pelagian Basin	204801	1	46	Oil	Paleogene	Cenozoic	Shallow and Deep Marine	Cenozoic
2048	Pelagian Basin	204802	1	26	Oil	Jurassic to Cretaceous	Mesozoic	Shallow and Deep Marine	Cenozoic
2054	Trias/Ghadames Basin	205401	1	36	Oil	Silurian to Devonian	Paleozoic	Shallow and Deep Marine	Mesozoic to Cenozoic
2054	Trias/Ghadames Basin	205402	1	4	Oil	Silurian	Paleozoic	Shallow and Deep Marine	Mesozoic to Cenozoic
2054	Trias/Ghadames Basin	205403	1	93	Oil	Silurian to Devonian	Paleozoic	Shallow and Deep Marine	Mesozoic
2056	Illizi Basin	205601	1	102	Oil	Silurian to Devonian	Paleozoic	Shallow and Deep Marine	Mesozoic
2058	Grand Erg/Ahnet Basin	205801	1	11	Gas	Silurian to Devonian	Paleozoic	Shallow and Deep Marine	Paleozoic to Mesozoic
2058	Grand Erg/Ahnet Basin	205802	1	24	Gas	Silurian to Devonian	Paleozoic	Shallow and Deep Marine	Paleozoic to Mesozoic
2058	Grand Erg/Ahnet Basin	205803	1	15	Oil	Silurian to Devonian	Paleozoic	Shallow and Deep Marine	Paleozoic
2058	Grand Erg/Ahnet Basin	205804	1	0	Gas	Silurian to Devonian	Paleozoic	Shallow and Deep Marine	Paleozoic
2058	Grand Erg/Ahnet Basin	205805	1	4	Gas	Silurian	Paleozoic	Shallow and Deep Marine	Mesozoic to Cenozoic
2058	Grand Erg/Ahnet Basin	205806	1	0	Gas	Silurian to Devonian	Paleozoic	Shallow and Deep Marine	Paleozoic to Mesozoic
2071	Red Sea Basin	207101	3	110	Oil	Cretaceous	Mesozoic	Shallow Marine	Cenozoic
2071	Red Sea Basin	207102	2	17	Oil	Neogene	Cenozoic	Shallow Marine	Cenozoic
3115	Junggar Basin	311501	1	12	Oil	Permian	Paleozoic	Lacustrine	Mesozoic
3115	Junggar Basin	311502	1	9	Oil	Jurassic	Mesozoic	Coaly	Cenozoic
3115	Junggar Basin	311503	1	0	Gas				
3127	Bohaiwan Basin	312701	2	158	Oil	Paleogene	Cenozoic	Lacustrine	Cenozoic
3127	Bohaiwan Basin	312702	1	0	Gas				
3128	Ordos Basin	312801	1	26	Oil	Triassic	Mesozoic	Lacustrine	Mesozoic
3128	Ordos Basin	312802	2	21	Gas				
3142	Sichuan Basin	314201	3	67	Gas	Permian	Paleozoic	Coaly and Shallow Marine	Mesozoic

Province Code	Province Name	Total Petroleum System Code	Number of Assess-ment Units	Number of Fields	Major Com-modity	Specific Source Rock Age	General Source Rock Age	Source Rock Depositional Environment	Age of Peak Maturation
3142	Sichuan Basin	314202	1	17	Oil	Jurassic	Mesozoic	Lacustrine	Mesozoic
3142	Sichuan Basin	314203	1	1	Gas				
3142	Sichuan Basin	314204	2	2	Gas	Lower Paleozoic	Paleozoic	Shallow Marine	Mesozoic
3144	Songliao Basin	314401	2	73	Oil	Cretaceous	Mesozoic	Lacustrine	Mesozoic
3144	Songliao Basin	314402	1	8	Gas	Jurassic	Mesozoic	Coaly	Mesozoic
3154	Tarim Basin	315401	3	26	Oil	Paleozoic to Mesozoic	Paleozoic to Mesozoic	Coaly, Lacustrine, and Shallow Marine	Paleozoic to Cenozoic
3701	Baram Delta/Brunei-Sabah Basin	370101	2	82	Oil	Neogene	Cenozoic	Coaly, Shallow, and Deep Marine	Cenozoic
3702	Greater Sarawak Basin	370201	2	62	Gas	Paleogene to Neogene	Cenozoic	Coaly and Shallow Marine	Cenozoic
3702	Greater Sarawak Basin	370202	1	3	Gas	Neogene	Cenozoic	Coaly and Shallow Marine	Cenozoic
3703	Malay Basin	370301	2	58	Oil	Paleogene to Neogene	Cenozoic	Lacustrine	Cenozoic
3703	Malay Basin	370302	1	33	Gas	Neogene	Cenozoic	Coaly	Cenozoic
3808	Central Sumatra Basin	380801	1	188	Oil	Paleogene	Cenozoic	Lacustrine	Cenozoic
3817	Kutei Basin	381701	3	73	Oil	Paleogene to Neogene	Cenozoic	Coaly, Shallow, and Deep Marine	Cenozoic
3822	North Sumatra Basin	382201	2	87	Gas	Paleogene to Neogene	Cenozoic	Lacustrine and Shallow Marine	Cenozoic
3824	Northwest Java Basin	382401	1	59	Oil	Paleogene	Cenozoic	Lacustrine	Cenozoic
3824	Northwest Java Basin	382402	1	142	Oil	Paleogene to Neogene	Cenozoic	Coaly	Cenozoic
3824	Northwest Java Basin	382403	1	8	Gas				
3824	Northwest Java Basin	382404	1	0					
3828	South Sumatra Basin	382801	1	193	Oil	Paleogene	Cenozoic	Lacustrine and Shallow Marine	Cenozoic
3910	Bonaparte Gulf Basin	391001	1	7	Oil	Carboniferous	Paleozoic	Shallow Marine	Paleozoic
3910	Bonaparte Gulf Basin	391002	2	3	Gas	Permian	Paleozoic	Coaly and Shallow Marine	Mesozoic
3910	Bonaparte Gulf Basin	391003	1	17	Oil	Jurassic to Cretaceous	Mesozoic	Coaly and Shallow Marine	Mesozoic to Cenozoic
3913	Browse Basin	391301	1	8	Gas	Jurassic to Cretaceous	Mesozoic	Shallow Marine	Cenozoic
3930	Gippsland Basin	393001	1	48	Oil	Cretaceous to Paleogene	Mesozoic to Cenozoic	Coaly	Mesozoic to Cenozoic
3948	Northwest Shelf	394801	1	64	Gas	Jurassic to Cretaceous	Mesozoic	Shallow Marine	Mesozoic
3948	Northwest Shelf	394802	1	7	Gas	Triassic	Mesozoic	Shallow Marine	Mesozoic
4017	Vestford-Helgeland	401701	2	16	Oil	Jurassic	Mesozoic	Deep Marine	Cenozoic
4025	North Sea Graben	402501	3	470	Oil	Jurassic	Mesozoic	Deep Marine	Cenozoic
4036	Anglo-Dutch Basin	403601	3	793	Gas	Carboniferous	Paleozoic	Coaly and Shallow Marine	Paleozoic to Mesozoic
4047	North Carpathian Basin	404701	1	142	Gas	Neogene	Cenozoic	Coaly and Shallow Marine	Not mature
4047	North Carpathian Basin	404702	1	188	Oil	Paleogene	Cenozoic	Shallow Marine	Cenozoic
4047	North Carpathian Basin	404703	1	3	Oil	Paleozoic	Paleozoic	Shallow and Deep Marine	Cenozoic
4048	Pannonian Basin	404801	1	321	Oil	Neogene	Cenozoic	Lacustrine and Deep Marine	Cenozoic
4048	Pannonian Basin	404802	1	165	Oil	Triassic to Neogene	Mesozoic to Cenozoic	Lacustrine and Deep Marine	Cenozoic
4048	Pannonian Basin	404803	1	16	Gas	Neogene	Cenozoic	Lacustrine and Deep Marine	Cenozoic
4048	Pannonian Basin	404804	1	15	Gas	Neogene	Cenozoic	Lacustrine and Deep Marine	Cenozoic
4048	Pannonian Basin	404805	1	1	Gas				
4048	Pannonian Basin	404806	1	8	Oil	Paleogene	Cenozoic	Deep Marine	Cenozoic
4057	Transylvanian Basin	405701	2	119	Gas	Neogene	Cenozoic	Coaly and Lacustrine and Shallow Marine	Not mature
4060	Po Basin	406001	1	244	Gas	Neogene	Cenozoic	Deep Marine	Cenozoic
4060	Po Basin	406002	1	12	Oil	Triassic	Mesozoic	Shallow and Deep Marine	Cenozoic
4061	Carpathian-Balkanian Basin	406101	1	202	Oil	Devonian to Neogene	Paleozoic to Cenozoic	Lacustrine and Deep Marine	Mesozoic to Cenozoic
4061	Carpathian-Balkanian Basin	406102	2	185	Oil	Paleogene to Neogene	Cenozoic	Shallow Marine	Cenozoic
4068	Provence Basin	406801	1	0	Gas	Cretaceous to Neogene	Mesozoic to Cenozoic	Shallow and Deep Marine	Mesozoic to Cenozoic
5200	East Greenland Rift Basins	520001	1	0	Oil	Jurassic	Mesozoic	Shallow and Deep Marine	Mesozoic to Cenozoic
5215	Labrador-Newfoundland Shelf	521501	1	15	Oil	Jurassic	Mesozoic	Deep Marine	Mesozoic to Cenozoic
5215	Labrador-Newfoundland Shelf	521502	1	6	Gas				
5243	Alberta Basin	524301	2	840	Oil	Devonian	Paleozoic	Deep Marine	Mesozoic to Cenozoic
5243	Alberta Basin	524302	2	1666	Oil	Devonian	Paleozoic	Deep Marine	Cenozoic
5243	Alberta Basin	524303	2	1633	Oil	Devonian to Carboniferous	Paleozoic	Deep Marine	Cenozoic
5243	Alberta Basin	524304	2	1308	Oil	Triassic to Jurassic	Mesozoic	Shallow and Deep Marine	Cenozoic
5243	Alberta Basin	524305	2	2901	Gas	Jurassic to Cretaceous	Mesozoic	Coaly and Shallow Marine	Cenozoic
5243	Alberta Basin	524306	3	1432	Oil	Cretaceous	Mesozoic	Shallow and Deep Marine	Cenozoic

Province Code	Province Name	Total Petroleum System Code	Number of Assessment Units	Number of Fields	Major Commodity	Specific Source Rock Age	General Source Rock Age	Source Rock Depositional Environment	Age of Peak Maturation
5244	Williston Basin, Canada	524401	1	2	Oil	Ordovician	Paleozoic	Shallow Marine	Paleozoic
5244	Williston Basin, Canada	524402	1	7	Oil	Devonian	Paleozoic	Deep Marine	Mesozoic
5244	Williston Basin, Canada	524403	1	7	Oil	Devonian to Carboniferous	Paleozoic	Deep Marine	Mesozoic to Cenozoic
5244	Williston Basin, Canada	524404	1	282	Oil	Carboniferous	Paleozoic	Deep Marine	Mesozoic to Cenozoic
5305	Villahermosa Uplif	530501	7	418	Oil	Jurassic	Mesozoic	Deep Marine	Cenozoic
6021	Guyana-Suriname Basin	602101	3	3	Oil	Cretaceous	Mesozoic	Deep Marine	Cenozoic
6022	Foz do Amazonas Basin	602201	2	1	Gas	Neogene	Cenozoic	Shallow and Deep Marine	Cenozoic
6029	Sergipe-Alagoas Basin	602901	2	80	Oil	Cretaceous	Mesozoic	Lacustrine and Shallow Marine	Cenozoic
6034	Espirito Santo Basin	603401	3	46	Oil	Cretaceous	Mesozoic	Shallow and Deep Marine	Mesozoic to Cenozoic
6035	Campos Basin	603501	3	73	Oil	Cretaceous	Mesozoic	Lacustrine	Cenozoic
6036	Santos Basin	603601	2	11	Oil	Cretaceous	Mesozoic	Deep Marine	Mesozoic to Cenozoic
6037	Pelotas Basin	603701	2	0	Oil	Cretaceous to Paleogene	Mesozoic to Cenozoic	Deep Marine	Cenozoic
6041	Putumayo-Oriente-Maranon Basin	604101	1	171	Oil	Cretaceous	Mesozoic	Shallow Marine	Cenozoic
6041	Putumayo-Oriente-Maranon Basin	604102	1	0	Oil	Permian	Paleozoic	Shallow Marine	Cenozoic
6045	Santa Cruz-Tarija Basin	604501	3	100	Oil	Silurian to Devonian	Paleozoic	Shallow Marine	Paleozoic to Cenozoic
6055	Neuquen Basin	605501	3	297	Oil	Jurassic to Cretaceous	Mesozoic	Shallow and Deep Marine	Mesozoic to Cenozoic
6058	San Jorge Basin	605801	2	182	Oil	Cretaceous	Mesozoic	Lacustrine	Mesozoic
6058	San Jorge Basin	605802	1	0	Gas				
6059	Magallanes Basin	605901	2	252	Gas	Cretaceous	Mesozoic	Shallow Marine	Cenozoic
6060	Falklands Plateau	606001	1	0	Oil	Cretaceous	Mesozoic	Lacustrine	Cenozoic
6060	Falklands Plateau	606002	1	0	Oil	Cretaceous	Mesozoic	Lacustrine and Shallow Marine	Cenozoic
6063	Malvinas Basin	606301	1	3	Oil	Cretaceous	Mesozoic	Shallow Marine	Cenozoic
6081	Talara Basin	608101	1	42	Oil	Cretaceous to Paleogene	Mesozoic to Cenozoic	Lacustrine and Shallow Marine	Cenozoic
6083	Progreso Basin	608301	1	6	Oil	Paleogene to Neogene	Cenozoic	Shallow Marine	Cenozoic
6083	Progreso Basin	608302	1	13	Oil	Cretaceous	Mesozoic	Shallow Marine	Cenozoic
6090	Middle Magdalena	609001	4	79	Oil	Cretaceous	Mesozoic	Deep Marine	Cenozoic
6096	Llanos Basin	609601	2	93	Oil	Cretaceous	Mesozoic	Shallow Marine	Cenozoic
6098	East Venezuela Basin	609801	4	311	Oil	Cretaceous	Mesozoic	Shallow Marine	Cenozoic
6098	East Venezuela Basin	609802	2	97	Oil	Cretaceous to Neogene	Mesozoic to Cenozoic	Shallow and Deep Marine	Cenozoic
6099	Maracaibo Basin	609901	3	80	Oil	Cretaceous	Mesozoic	Shallow Marine	Cenozoic
6103	Tobago Trough	610301	1	10	Gas	Neogene	Cenozoic	Shallow Marine	Cenozoic
6107	Lesser Antilles Deformed Belt	610701	2	27	Oil	Cretaceous to Paleogene	Mesozoic to Cenozoic	Shallow Marine	Cenozoic
6117	Greater Antilles Deformed Belt	611701	1	21	Oil	Jurassic to Cretaceous	Mesozoic	Shallow Marine	Cenozoic
7013	Senegal	701301	1	7	Oil	Cretaceous	Mesozoic	Deep Marine	Cenozoic
7183	Gulf of Guinea	718301	1	34	Oil	Cretaceous	Mesozoic	Shallow and Deep Marine	Mesozoic to Cenozoic
7192	Niger Delta	719201	2	619	Oil	Paleogene to Neogene	Cenozoic	Shallow and Deep Marine	Cenozoic
7203	West-Central Coastal	720301	1	26	Oil	Cretaceous	Mesozoic	Lacustrine	Cenozoic
7203	West-Central Coastal	720302	1	104	Oil	Cretaceous	Mesozoic	Deep Marine	Cenozoic
7203	West-Central Coastal	720303	2	224	Oil	Cretaceous to Neogene	Mesozoic to Cenozoic	Lacustrine and Deep Marine	Mesozoic to Cenozoic
7203	West-Central Coastal	720304	1	25	Oil	Cretaceous	Mesozoic	Lacustrine and Deep Marine	Mesozoic to Cenozoic
7303	Orange River Coastal	730301	1	3	Gas	Cretaceous	Mesozoic	Lacustrine and Deep Marine	Cenozoic
8026	Kohat-Potwar	802601	1	22	Oil	Paleogene	Cenozoic	Lacustrine and Shallow Marine	Cenozoic
8034	Assam	803401	1	46	Oil	Paleogene to Neogene	Cenozoic	Coaly and Shallow Marine	Cenozoic
8042	Indus	804201	2	92	Gas	Permian to Neogene	Paleozoic to Cenozoic	Shallow and Deep Marine	Cenozoic
8043	Bombay	804301	2	162	Oil	Paleogene to Neogene	Cenozoic	Shallow Marine	Cenozoic
8047	Ganges-Brahmaputra Delta	804701	1	0	Gas				
8047	Ganges-Brahmaputra Delta	804702	1	1	Gas	Paleogene	Cenozoic	Coaly	Cenozoic
8047	Ganges-Brahmaputra Delta	804703	2	35	Gas	Paleogene to Neogene	Cenozoic	Coaly	Cenozoic
8048	Irrawaddy	804801	2	60	Gas	Paleogene to Neogene	Cenozoic	Coaly and Shallow Marine	Cenozoic

Appendix 6. Assessment units ranked by estimated size of their largest undiscovered oil field.

[MMBO = million barrels of oil. Size of largest field is the mean of the distribution for size of the largest undiscovered oil field. Hypothetical = no discovered fields in the assessment unit of at least minimum field size; frontier = 1 to 13 fields; established = more than 13 fields.]

Rank	Assessment Unit Code	Assessment Unit Name	Largest Field (MMBO)	Assessment Unit Maturity
1	20300101	Cretaceous Reservoirs	11,984	Established
2	52000101	Northeast Greenland Shelf Rift Systems	6107	Hypothetical
3	10160104	South Margin Subsalt	5070	Frontier
4	20190302	Paleozoic Reservoirs	4234	Established
5	20300102	Tertiary Reservoirs	3426	Established
6	20210201	Horst-Block Anticlinal Oil	2728	Established
7	72030302	Central Congo Turbidites	2698	Established
8	60210101	Late Cretaceous-Tertiary Turbidites	2654	Hypothetical
9	20190201	Jurassic Reservoirs in Northwest Desert Anticlines	2581	Frontier
10	20210202	Salt-Involved Structural Oil	2204	Frontier
11	20210101	Central Arch Horst-Block Anticlinal Oil and Gas	2141	Established
12	60360101	Santos Shelf	1989	Frontier
13	60350103	Salt Dome Province Tertiary Sandstones	1872	Hypothetical
14	60360102	Salt-Structured Deep-Water Sandstones	1624	Hypothetical
15	71920102	Akata Reservoirs	1586	Hypothetical
16	11740101	Upper Jurassic-Cretaceous Sandstones	1577	Established
17	60350101	Late Cretaceous-Tertiary Turbidites	1571	Established
18	60980101	East Venezuela Fold and Thrust Belt	1412	Established
19	11120104	Central Offshore	1397	Hypothetical
20	53050106	Tamaulipas-Like Basinal Limestone and Tertiary Strata Overlying Evaporites	1264	Established
21	72030201	Gabon Suprasalt	1112	Established
22	11120103	Gogr态ndag-Okarem Zone and Adjacent Shelf	1106	Established
23	53050102	Tamabra-Like Debris-Flow-Breccia Limestone Overlying Evaporites	1105	Established
24	20190101	Cretaceous Reservoirs in Northwest Desert Anticlines	1067	Established
25	60980202	Orinoco Delta and Offshore	1048	Hypothetical
26	61070101	Inner Forearc Deformation Belt	1048	Hypothetical
27	71920101	Agbada Reservoirs	1013	Established
28	60990101	Main Maracaibo Basin	994	Established
29	60600201	South Falklands Basin	961	Hypothetical
30	60600101	North Falklands Basin	955	Hypothetical
31	40170101	Halten Terrace-Trondelag Platform	897	Established
32	11740302	South Kara Sea Offshore	891	Frontier
33	60370101	Pelotas Platform and Basin	824	Hypothetical
34	20190102	Cretaceous Reservoirs in South Gulf Suprasalt Structural	820	Established
35	20540301	Tanezzuft-Ghadames Structural/Stratigraphic	817	Established
36	38170102	Kutei Basin Turbidites	815	Hypothetical
37	11120101	Apsheron-Pribalkhan Zone	765	Established

Rank	Assessment Unit Code	Assessment Unit Name	Largest Field (MMBO)	Assessment Unit Maturity
38	31540101	Tarim Basin Excluding Marginal Foldbelts	717	Hypothetical
39	40250101	Viking Graben	709	Established
40	60340102	Late Cretaceous-Tertiary Slide Blocks and Turbidites	708	Established
41	60210102	Cretaceous Carbonates	701	Hypothetical
42	10080102	Main Basin Platform	589	Established
43	60960102	Peripheral	568	Established
44	20230101	Horst/Graben-Related Oil and Gas	563	Established
45	72030301	Central Congo Delta and Carbonate Platform	557	Established
46	20540201	Tanezzuft-Melrhir Structural/Stratigraphic	550	Frontier
47	20190202	Jurassic Reservoirs in South Gulf Suprasalt/Qatar Arch Structural	534	Established
48	13220101	Onshore and Offshore Northeastern Shelf	520	Established
49	53050107	Tamabra-Like Debris-Flow-Breccia Limestone North of Campeche	506	Hypothetical
50	40250103	Central Graben	466	Established
51	11740201	Pre-Upper Jurassic	446	Established
52	12100101	Baikal-Patom Foldbelt Riphean-Craton Margin Vendian	444	Established
53	10160103	East and Southeast Margins Subsalt	430	Frontier
54	20430102	Central Sirte Carbonates	426	Established
55	60960101	Central	426	Established
56	20230201	Platform Horst/Graben-Related Oil	418	Frontier
57	39130101	Late Jurassic/Early Cretaceous-Mesozoic	408	Frontier
58	10160101	North andWest Margins Subsalt Pinnacle Reefs	405	Established
59	60290102	Late Cretaceous-Tertiary Deep-Water Sandstones	390	Frontier
60	60450103	Foreland Central Chaco High	387	Hypothetical
61	11090101	Foldbelt-Foothills	375	Established
62	12070101	Yenisey Foldbelt Riphean-Craton Margin Riphean	374	Frontier
63	20430101	Southeast Sirte Clastics	363	Established
64	20710101	Gulf of Suez Block-Fault Fairway	358	Established
65	20160101	Fahud-Huqf Combined Structural	349	Established
66	31270102	Pre-Tertiary Buried Hills	349	Established
67	11120102	Lower Kura Depression and Adjacent Shelf	348	Established
68	60410101	Hollin-Napo	332	Established
69	60340103	Abrolhos Sub-Volcanic Structures	331	Hypothetical
70	20230202	Basinal Oil and Gas	315	Hypothetical
71	72030401	Cuanza-Namibe	315	Frontier
72	11500101	Mesozoic Sandstone Reservoirs	298	Frontier
73	40680101	Subsalt	297	Hypothetical
74	60450102	Foreland Basins	295	Frontier
75	31270101	Tertiary Lacustrine	285	Established
76	60810101	Cretaceous-Paleogene Basin	283	Established
77	37010101	Brunei-Sabah Deltaics	276	Established
78	37010102	Brunei-Sabah Turbidites	272	Frontier
79	20040101	Ma'Rib-Al Jawf/Shabwah/Masila	267	Established
80	52150101	Jeanne d'Arc	263	Established

Rank	Assessment Unit Code	Assessment Unit Name	Largest Field (MMBO)	Assessment Unit Maturity
81	39480101	Dingo-Mungaroo/Barrow	257	Established
82	39100301	Malita	255	Established
83	10160106	Suprasalt	253	Established
84	11740301	NorthernWest Siberian Onshore Gas	244	Established
85	38170101	Kutei Basin Deltaics	243	Established
86	20430103	Offshore Sirte Hypothetical	223	Hypothetical
87	20560101	Tanezzuft-Illizi Structural/Stratigraphic	222	Established
88	10090102	Devonian Synrift	210	Hypothetical
89	60630101	Malvinas Extensional Structures	208	Frontier
90	80430101	Eocene-Miocene Bombay Shelf	203	Established
91	71830101	Coastal Plain and Offshore	201	Established
92	53050103	Tamabra-Like Debris-Flow-Breccia Limestone and Overlying Strata	200	Established
93	20710201	Red Sea Coastal Block Faults	192	Frontier
94	53050104	Tamabra-Like Debris-Flow-Breccia Limestone of the Golden Lane	188	Established
95	11090301	Offshore Prikumsk Zone	182	Hypothetical
96	40170102	Mid-Norway Continental Margin	181	Hypothetical
97	20190103	Mesozoic/Tertiary Foredeep Fold and Thrust	177	Established
98	72030101	Gabon Subsalt	173	Established
99	31540103	Southwest Foldbelt	171	Established
100	10500101	Kolguyev Terrace	143	Frontier
101	37030101	South Malay Lacustrine	143	Established
102	60350102	Cretaceous Carbonates	141	Frontier
103	40250102	Moray Firth	140	Established
104	60980103	Maturin Sub-Basin	134	Established
105	40600201	Thermal Triassic	133	Frontier
106	60550101	Neuquen Extensional Structures	128	Established
107	20430104	Southeast Sirte Hypothetical	127	Hypothetical
108	80480101	Central Burma Basin	123	Established
109	20160201	Natih-Fiqa Structural/Stratigraphic	119	Frontier
110	20710202	Red Sea Salt Basin	118	Hypothetical
111	10500102	South Barents and Ludlov Saddle	115	Frontier
112	20540101	Tanezzuft-Oued Mya Structural/Stratigraphic	114	Established
113	11090201	South Mangyshlak (Entire)	114	Established
114	20710103	Southern Gulf of Suez	112	Frontier
115	20140101	Ghaba-Makarem Combined Structural	111	Established
116	20480101	Bou Dabbous-Tertiary Structural/Stratigraphic	110	Established
117	11090102	Terek-Sunzha Subsalt Jurassic	109	Hypothetical
118	60980201	Trinidad Basins	109	Established
119	53050105	Tamaulipas-Like Basinal Limestone and Tertiary Strata Without Underlying Evaporites	107	Established
120	80430102	Eocene-Miocene Cambay Deltaic	106	Established
121	38080101	Pematang/Sihapas Siliciclastics	105	Established
122	11120105	Iran Onshore/Nearshore	97	Hypothetical
123	31150101	Upper Paleozoic/Lower Mesozoic Nonmarine Coarse Clastics	93	Frontier

Rank	Assessment Unit Code	Assessment Unit Name	Largest Field (MMBO)	Assessment Unit Maturity
124	61170101	North Cuba Fold and Thrust Belt	93	Established
125	38170103	Kutei Basin Fold and Thrust Belt	93	Established
126	38240101	Sunda/Asri	91	Established
127	40610202	Romania Ploiesti Zone	85	Established
128	10150103	Sub-Domanik Devonian Clastics	82	Established
129	38280101	South Sumatra	82	Established
130	60550102	Neuquen Foothills Structure	77	Established
131	10080103	Foredeep Basins	74	Established
132	10500103	North Barents	74	Hypothetical
133	31440101	Subtle Traps	73	Established
134	60210103	Late Cretaceous-Tertiary Nearshore Sandstones	73	Established
135	60990102	Southwest Maracaibo Basin Fold Belt	73	Established
136	37020102	Balingian	72	Established
137	20710102	Gulf of Suez Qaa Plain	71	Hypothetical
138	73030101	Offshore	69	Established
139	10160102	North andWest Margins Subsalt Barrier Reefs	66	Frontier
140	20480201	Jurassic-Cretaceous Structural/Stratigraphic	66	Established
141	70130101	Coastal Plain and Offshore	66	Frontier
142	60980102	Guarico Sub-Basin	65	Established
143	10150101	Supra-Domanik Carbonates/Clastics	64	Established
144	20580301	Tanezzuft-Sbaa Structural/Stratigraphic	63	Frontier
145	10090101	Carboniferous-Lower Permian Clastics	63	Established
146	60580101	San Jorge Extensional Structures	62	Established
147	80480102	Irrawaddy-Andaman	61	Established
148	11090303	Central Caspian Offshore	61	Hypothetical
149	60830201	Cretaceous-Paleogene Santa Elena Block	60	Frontier
150	53050101	El Abra-Like Reef and Backreef Limestone	60	Established
151	60590101	Magallanes Extensional Structures	59	Established
152	60900102	Southern	57	Frontier
153	60590102	Andean Fold Belt Structures	55	Frontier
154	37030102	North Malay Lacustrine	55	Frontier
155	11080101	Tertiary Foredeep	55	Established
156	11540104	Murgab Depression Subsalt	54	Established
157	31440102	Anticlinal	53	Established
158	80340101	Sylhet-Kopili/Barail-Tipam Composite	52	Established
159	60900104	La Luna and Older	52	Frontier
160	37020201	East Natuna	51	Frontier
161	31540102	Kuche (Northern) Foldbelt	49	Frontier
162	39300101	Gippsland	49	Established
163	40610201	Romania Flysch Zone	48	Established
164	39480201	Locker-Mungaroo/Barrow	47	Frontier
165	11090103	Foreland Slope and Foredeep	43	Frontier
166	38220102	Mergui	42	Hypothetical

Rank	Assessment Unit Code	Assessment Unit Name	Largest Field (MMBO)	Assessment Unit Maturity
167	38240201	Ardjuna	42	Established
168	11500201	Jurassic-Tertiary Reservoirs	42	Frontier
169	60900101	Northern	41	Established
170	40470201	Deformed Belt	39	Established
171	60410201	Ene	37	Hypothetical
172	31150201	Jurassic/Tertiary Fluvial and Lacustrine Sandstone	34	Frontier
173	20580501	Tanezzuft-Benoud Structural/Stratigraphic	34	Frontier
174	11080103	Subsalt Jurassic	34	Frontier
175	80260101	Kohat-Potwar Intrathrust Basin	33	Established
176	60900103	Eastern	33	Established
177	11090302	Onshore Stavropol-Prikumsk	33	Established
178	39100101	Barnett	31	Frontier
179	60450101	Sub-Andean Fold and Thrust Belt	28	Established
180	38220101	North Sumatra	26	Established
181	10150102	Lower Volga	26	Established
182	60580102	San Bernardo Fold Belt Structures	26	Established
183	11080102	Foreland Slope	23	Established
184	20580201	Tanezzuft-Ahnet Structural/Stratigraphic	23	Established
185	11540102	Karabil-Badkhyz (Southern Area)	23	Established
186	10080101	Northwest Izhma-Pechora Depression	23	Hypothetical
187	40610101	Moesian Platform	22	Established
188	37030201	South Malay Coaly	21	Established
189	20580101	Tanezzuft-Timimoun Structural/Stratigraphic	21	Frontier
190	52430202	Leduc Oil and Gas	21	Established
191	60550103	Dorsal de Neuquen Structure	21	Established
192	20580401	Tanezzuft-Mouydir Structural/Stratigraphic	21	Hypothetical
193	40480101	Greater Hungarian Plain Basins	21	Established
194	80470201	Western Shelf and Slope	19	Established
195	60830101	Neogene Pull-Apart Basin	19	Frontier
196	31280101	Jurassic/Triassic Fluvial and Lacustrine Sandstone	19	Established
197	80420101	Greater Indus Foreland and Foldbelt	18	Established
198	40360102	Southern Permian Basin-Europe Onshore	18	Established
199	11540101	Northern andWestern Areas	18	Established
200	40480201	Zala-Drava-Sava Basins	18	Established
201	20580601	Tanezzuft-Bechar/Abadla Structural/Stratigraphic	18	Hypothetical
202	10150201	Permian Reefs/Thrust Folds	16	Frontier
203	40480601	Hungarian Paleogene Basin	15	Frontier
204	52430102	Keg River Oil and Gas	15	Established
205	40470301	Paleozoic Reservoirs	12	Established
206	52430302	Exshaw-Rundle Oil and Gas	11	Established
207	40360103	Southern Permian Basin-Offshore	9	Established
208	80470302	Eastern Fold Belt	9	Established
209	52430402	Combined Triassic/Jurassic Oil and Gas	8	Established

Rank	Assessment Unit Code	Assessment Unit Name	Largest Field (MMBO)	Assessment Unit Maturity
210	60290101	Western Pre-Aptian Reservoirs	8	Established
211	60340101	Espirito Santo Shelf	7	Established
212	52440401	Lodgepole Oil	7	Established
213	80470301	Central Basin	7	Hypothetical
214	31420201	Jurassic Lacustrine	6	Frontier
215	52430602	Second White Specks-Cardium Oil and Gas	5	Established
216	40360101	Southern Permian Basin-U.K. Onshore	3	Established
217	52440101	Yeoman Oil	2	Established
218	52440301	Bakken Sandstone	2	Frontier
219	52440201	Brighthoime Oil	2	Frontier

Appendix 7. Assessment units ranked by estimated size of their largest undiscovered natural gas field.

[BCFG = billion cubic feet of gas. Size of largest field is the mean of the distribution for size of the largest undiscovered gas field. Hypothetical = no discovered fields in the assessment unit of at least minimum field size; frontier = 1 to 13 fields; established = more than 13 fields.]

Rank	Assessment Unit Code	Assessment Unit Name	Largest Field (BCFG)	Assessment Unit Maturity
1	11740302	South Kara Sea Offshore	88,058	Frontier
2	20190302	Paleozoic Reservoirs	47,103	Established
3	11540104	Murgab Depression Subsalt	35,323	Established
4	40170102	Mid-Norway Continental Margin	31,943	Hypothetical
5	10500102	South Barents and Ludlov Saddle	29,826	Frontier
6	20210102	North Gulf Salt Basin Structural Gas	22,624	Frontier
7	20210101	Central Arch Horst-Block Anticlinal Oil and Gas	22,212	Established
8	11740301	NorthernWest Siberian Onshore Gas	18,754	Established
9	10160104	South Margin Subsalt	13,408	Frontier
10	40680101	Subsalt	12,709	Hypothetical
11	20300201	Northern Qatar Arch Extension	12,597	Frontier
12	10500103	North Barents	11,967	Hypothetical
13	10160101	North andWest Margins Subsalt Pinnacle Reefs	10,292	Established
14	11120104	Central Offshore	9982	Hypothetical
15	13220101	Onshore and Offshore Northeastern Shelf	8217	Established
16	20300102	Tertiary Reservoirs	7312	Established
17	52000101	Northeast Greenland Shelf Rift Systems	7112	Hypothetical
18	11120102	Lower Kura Depression and Adjacent Shelf	6692	Established
19	20190201	Jurassic Reservoirs in Northwest Desert Anticlines	6635	Frontier
20	11120103	Gograndag-Okarem Zone and Adjacent Shelf	6578	Established
21	60360101	Santos Shelf	6507	Frontier
22	61070101	Inner Forearc Deformation Belt	5567	Hypothetical
23	38170101	Kutei Basin Deltaics	5557	Established
24	40170101	Halten Terrace-Trondelag Platform	5473	Established
25	20300101	Cretaceous Reservoirs	5401	Established
26	60340102	Late Cretaceous-Tertiary Slide Blocks and Turbidites	5058	Established
27	20230101	Horst/Graben-Related Oil and Gas	4993	Established
28	12100101	Baikal-Patom Foldbelt Riphean-Craton Margin Vendian	4954	Established
29	39130101	Late Jurassic/Early Cretaceous-Mesozoic	4935	Frontier
30	39480101	Dingo-Mungaroo/Barrow	4727	Established
31	20710201	Red Sea Coastal Block Faults	4611	Frontier
32	20190301	Khuff Carbonates in Salt Structures	4446	Frontier
33	60370101	Pelotas Platform and Basin	4244	Hypothetical
34	38170102	Kutei Basin Turbidites	4196	Hypothetical
35	31540101	Tarim Basin Excluding Marginal Foldbelts	3965	Hypothetical
36	60980202	Orinoco Delta and Offshore	3965	Hypothetical
37	80420101	Greater Indus Foreland and Foldbelt	3810	Established

Rank	Assessment Unit Code	Assessment Unit Name	Largest Field (BCFG)	Assessment Unit Maturity
38	71920101	Agbada Reservoirs	3778	Established
39	72030302	Central Congo Turbidites	3483	Established
40	60340103	Abrolhos Sub-Volcanic Structures	3430	Hypothetical
41	10080103	Foredeep Basins	3373	Established
42	80470302	Eastern Fold Belt	3312	Established
43	61030101	Carupano Basin Gas	3292	Frontier
44	80480102	Irrawaddy-Andaman	3276	Established
45	60220101	Amazon Delta and Submarine Fan	3241	Established
46	60210101	Late Cretaceous-Tertiary Turbidites	3141	Hypothetical
47	10500101	Kolguyev Terrace	2910	Frontier
48	60980101	East Venezuela Fold and Thrust Belt	2900	Established
49	12070101	Yenisey Foldbelt Riphean-Craton Margin Riphean	2709	Frontier
50	80470301	Central Basin	2645	Hypothetical
51	20190103	Mesozoic/Tertiary Foredeep Fold and Thrust	2604	Established
52	60980201	Trinidad Basins	2558	Established
53	60450101	Sub-Andean Fold and Thrust Belt	2533	Established
54	39480201	Locker-Mungaroo/Barrow	2475	Frontier
55	11500301	Upper Paleozoic Carbonates	2429	Hypothetical
56	38280101	South Sumatra	2407	Established
57	53050106	Tamaulipas-Like Basinal Limestone and Tertiary Strata Overlying Evaporites	2323	Established
58	39100301	Malita	2281	Established
59	37030101	South Malay Lacustrine	2266	Established
60	40250101	Viking Graben	2185	Established
61	10080102	Main Basin Platform	2125	Established
62	31540102	Kuche (Northern) Foldbelt	2108	Frontier
63	60980102	Guarico Sub-Basin	2056	Established
64	11090102	Terek-Sunzha Subsalt Jurassic	2054	Hypothetical
65	72030201	Gabon Suprasalt	2022	Established
66	20140101	Ghaba-Makarem Combined Structural	2006	Established
67	80420102	Indus Fan	1994	Hypothetical
68	37020101	Central Luconia	1923	Established
69	20230202	Basinal Oil and Gas	1891	Hypothetical
70	60360102	Salt-Structured Deep-Water Sandstones	1886	Hypothetical
71	20560101	Tanezzuft-Illizi Structural/Stratigraphic	1842	Established
72	60290102	Late Cretaceous-Tertiary Deep-Water Sandstones	1813	Frontier
73	37010101	Brunei-Sabah Deltaics	1729	Established
74	38220102	Mergui	1715	Hypothetical
75	11090101	Foldbelt-Foothills	1710	Established
76	39100201	Petrel	1560	Frontier
77	10090102	Devonian Synrift	1525	Hypothetical
78	60630101	Malvinas Extensional Structures	1524	Frontier
79	20430102	Central Sirte Carbonates	1522	Established
80	71920102	Akata Reservoirs	1513	Hypothetical

Rank	Assessment Unit Code	Assessment Unit Name	Largest Field (BCFG)	Assessment Unit Maturity
81	11120101	Apsheron-Pribalkhan Zone	1489	Established
82	31540103	Southwest Foldbelt	1457	Established
83	40250103	Central Graben	1452	Established
84	60600101	North Falklands Basin	1448	Hypothetical
85	38220101	North Sumatra	1438	Established
86	60600201	South Falklands Basin	1435	Hypothetical
87	40360102	Southern Permian Basin-Europe Onshore	1374	Established
88	20430103	Offshore Sirte Hypothetical	1338	Hypothetical
89	73030101	Offshore	1247	Established
90	10090101	Carboniferous-Lower Permian Clastics	1207	Established
91	37010102	Brunei-Sabah Turbidites	1183	Frontier
92	60590101	Magallanes Extensional Structures	1176	Established
93	71830101	Coastal Plain and Offshore	1138	Established
94	40600101	Neogene Flysch Gas	1097	Established
95	80470201	Western Shelf and Slope	1090	Established
96	11080102	Foreland Slope	1061	Established
97	20160101	Fahud-Huqf Combined Structural	1033	Established
98	20540301	Tanezzuft-Ghadames Structural/Stratigraphic	1014	Established
99	39100202	Vulcan Graben	1009	Hypothetical
100	60210102	Cretaceous Carbonates	1000	Hypothetical
101	60990101	Main Maracaibo Basin	995	Established
102	20040101	Ma'Rib-Al Jawf/Shabwah/Masila	991	Established
103	72030301	Central Congo Delta and Carbonate Platform	983	Established
104	20710103	Southern Gulf of Suez	947	Frontier
105	20190202	Jurassic Reservoirs in South Gulf Suprasalt/Qatar Arch Structural	946	Established
106	38240201	Ardjuna	926	Established
107	60350101	Late Cretaceous-Tertiary Turbidites	909	Established
108	60450102	Foreland Basins	903	Frontier
109	39300101	Gippsland	888	Established
110	53050103	Tamabra-Like Debris-Flow-Breccia Limestone and Overlying Strata	885	Established
111	11540102	Karabil-Badkhyz (Southern Area)	835	Established
112	11120105	Iran Onshore/Nearshore	823	Hypothetical
113	20580501	Tanezzuft-Benoud Structural/Stratigraphic	823	Frontier
114	60450103	Foreland Central Chaco High	817	Hypothetical
115	20710202	Red Sea Salt Basin	804	Hypothetical
116	20710101	Gulf of Suez Block-Fault Fairway	800	Established
117	10160103	East and Southeast Margins Subsalt	788	Frontier
118	80430101	Eocene-Miocene Bombay Shelf	774	Established
119	60550101	Neuquen Extensional Structures	765	Established
120	38080101	Pematang/Sihapas Siliciclastics	763	Established
121	31270101	Tertiary Lacustrine	756	Established
122	31420401	Leshan-Longnusi Paleohigh	750	Frontier
123	11540101	Northern andWestern Areas	719	Established

Rank	Assessment Unit Code	Assessment Unit Name	Largest Field (BCFG)	Assessment Unit Maturity
124	60900103	Eastern	718	Established
125	20230201	Platform Horst/Graben-Related Oil	700	Frontier
126	20540201	Tanezzuft-Melrhir Structural/Stratigraphic	686	Frontier
127	11540103	Murgab Depression Suprasalt	683	Established
128	60960102	Peripheral	651	Established
129	60550102	Neuquen Foothills Structure	634	Established
130	31270102	Pre-Tertiary Buried Hills	633	Established
131	11740101	Upper Jurassic-Cretaceous Sandstones	632	Established
132	38170103	Kutei Basin Fold and Thrust Belt	629	Established
133	40360103	Southern Permian Basin-Offshore	626	Established
134	60350103	Salt Dome Province Tertiary Sandstones	620	Hypothetical
135	37030201	South Malay Coaly	611	Established
136	11080103	Subsalt Jurassic	607	Frontier
137	53050102	Tamabra-Like Debris-Flow-Breccia Limestone Overlying Evaporites	583	Established
138	60590102	Andean Fold Belt Structures	580	Frontier
139	31420101	Southeastern Fold Belt	580	Established
140	37020201	East Natuna	578	Frontier
141	20580201	Tanezzuft-Ahnet Structural/Stratigraphic	557	Established
142	72030101	Gabon Subsalt	550	Established
143	80430102	Eocene-Miocene Cambay Deltaic	524	Established
144	11090303	Central Caspian Offshore	517	Hypothetical
145	60580101	San Jorge Extensional Structures	474	Established
146	60410201	Ene	469	Hypothetical
147	11090301	Offshore Prikumsk Zone	465	Hypothetical
148	72030401	Cuanza-Namibe	450	Frontier
149	31420102	Northwestern Depression/Foldbelt	450	Frontier
150	60960101	Central	449	Established
151	60810101	Cretaceous-Paleogene Basin	446	Established
152	52150101	Jeanne d'Arc	433	Established
153	40600201	Thermal Triassic	423	Frontier
154	40570101	Transylvanian Neogene Suprasalt Gas	417	Established
155	11080101	Tertiary Foredeep	413	Established
156	20190102	Cretaceous Reservoirs in South Gulf Suprasalt Structural	408	Established
157	20540101	Tanezzuft-Oued Mya Structural/Stratigraphic	405	Established
158	31440201	Structural Traps	393	Frontier
159	31420402	Lower Paleozoic of Southeastern Fold Belt	389	Hypothetical
160	10160106	Suprasalt	389	Established
161	10160102	North andWest Margins Subsalt Barrier Reefs	382	Frontier
162	11500201	Jurassic-Tertiary Reservoirs	362	Frontier
163	60980103	Maturin Sub-Basin	350	Established
164	37030102	North Malay Lacustrine	321	Frontier
165	20580101	Tanezzuft-Timimoun Structural/Stratigraphic	309	Frontier
166	52430201	Leduc Gas	304	Established

Rank	Assessment Unit Code	Assessment Unit Name	Largest Field (BCFG)	Assessment Unit Maturity
167	20580601	Tanezzuft-Bechar/Abadla Structural/Stratigraphic	303	Hypothetical
168	10150201	Permian Reefs/Thrust Folds	297	Frontier
169	11090201	South Mangyshlak (Entire)	293	Established
170	52430301	Exshaw-Rundle Gas	288	Established
171	20480201	Jurassic-Cretaceous Structural/Stratigraphic	285	Established
172	20480101	Bou Dabbous-Tertiary Structural/Stratigraphic	283	Established
173	60350102	Cretaceous Carbonates	260	Frontier
174	20580301	Tanezzuft-Sbaa Structural/Stratigraphic	259	Frontier
175	31150201	Jurassic/Tertiary Fluvial and Lacustrine Sandstone	254	Frontier
176	60990102	Southwest Maracaibo Basin Fold Belt	237	Established
177	11740201	Pre-Upper Jurassic	233	Established
178	20580401	Tanezzuft-Mouydir Structural/Stratigraphic	230	Hypothetical
179	10150103	Sub-Domanik Devonian Clastics	220	Established
180	80480101	Central Burma Basin	217	Established
181	40480101	Greater Hungarian Plain Basins	217	Established
182	70130101	Coastal Plain and Offshore	208	Frontier
183	39100101	Barnett	190	Frontier
184	60210103	Late Cretaceous-Tertiary Nearshore Sandstones	188	Established
185	40470101	Foreland Basin	187	Established
186	38240101	Sunda/Asri	184	Established
187	40250102	Moray Firth	181	Established
188	31440101	Subtle Traps	174	Established
189	40610201	Romania Flysch Zone	171	Established
190	60830101	Neogene Pull-Apart Basin	170	Frontier
191	10150101	Supra-Domanik Carbonates/Clastics	169	Established
192	40480201	Zala-Drava-Sava Basins	168	Established
193	37020102	Balingian	166	Established
194	60550103	Dorsal de Neuquen Structure	165	Established
195	11090103	Foreland Slope and Foredeep	162	Frontier
196	40470201	Deformed Belt	152	Established
197	52430401	Combined Triassic/Jurassic Gas	143	Established
198	52430501	Mannville Gas	136	Established
199	10080101	Northwest Izhma-Pechora Depression	135	Hypothetical
200	52430102	Keg River Oil and Gas	132	Established
201	10150102	Lower Volga	116	Established
202	80260101	Kohat-Potwar Intrathrust Basin	108	Established
203	60900104	La Luna and Older	104	Frontier
204	52430101	Keg River Gas	99	Established
205	20160201	Natih-Fiqa Structural/Stratigraphic	98	Frontier
206	11090302	Onshore Stavropol-Prikumsk	98	Established
207	31440102	Anticlinal	96	Established
208	80340101	Sylhet-Kopili/Barail-Tipam Composite	91	Established
209	40610101	Moesian Platform	89	Established

Rank	Assessment Unit Code	Assessment Unit Name	Largest Field (BCFG)	Assessment Unit Maturity
210	53050101	El Abra-Like Reef and Backreef Limestone	87	Established
211	60580102	San Bernardo Fold Belt Structures	85	Established
212	40480601	Hungarian Paleogene Basin	75	Frontier
213	40470301	Paleozoic Reservoirs	74	Established
214	60830201	Cretaceous-Paleogene Santa Elena Block	60	Frontier
215	40480301	Danube Basin	59	Frontier
216	40610202	Romania Ploiesti Zone	53	Established
217	52430602	Second White Specks-Cardium Oil and Gas	48	Established
218	52430202	Leduc Oil and Gas	47	Established
219	40480401	Transcarpathian Basin	46	Frontier
220	52430402	Combined Triassic/Jurassic Oil and Gas	44	Established
221	60290101	Western Pre-Aptian Reservoirs	40	Established
222	52430601	Second White Specks-Cardium Gas	28	Established
223	52430302	Exshaw-Rundle Oil and Gas	24	Established
224	40360101	Southern Permian Basin-U.K. Onshore	12	Established
225	52440401	Lodgepole Oil	8	Established

APPENDIX 8. Assessment units ranked by estimated number of undiscovered oil fields larger than 1024 MMBO.

[MMBO = million barrels of oil. Both percents and numbers of fields larger than 1024 MMBO are for undiscovered fields only.] Note that gas = natural gas.

Rank	Assessment Unit Code	Assessment Unit Name	Percent of Fields > 1024 MMBO	Number of Fields > 1024 MMBO
1	20300101	Cretaceous Reservoirs	9.79	15.31
2	52000101	Northeast Greenland Shelf Rift Systems	5.15	12.89
3	20210201	Horst-Block Anticlinal Oil	3.40	4.08
4	20300102	Tertiary Reservoirs	4.70	3.74
5	10160104	South Margin Subsalt	11.39	3.34
6	72030302	Central Congo Turbidites	2.89	3.01
7	11740101	Upper Jurassic-Cretaceous Sandstones	0.27	2.91
8	20210101	Central Arch Horst-Block Anticlinal Oil and Gas	2.83	2.61
9	20190302	Paleozoic Reservoirs	5.30	2.60
10	20210202	Salt-Involved Structural Oil	4.85	2.50
11	60210101	Late Cretaceous-Tertiary Turbidites	3.57	2.29
12	20190201	Jurassic Reservoirs in Northwest Desert Anticlines	4.70	2.03
13	60360101	Santos Shelf	1.45	1.89
14	71920102	Akata Reservoirs	0.69	1.74
15	60350101	Late Cretaceous-Tertiary Turbidites	1.76	1.32
16	60360102	Salt-Structured Deep-Water Sandstones	2.14	1.23
17	11120104	Central Offshore	2.07	0.86
18	53050106	Tamaulipas-Like Basinal Limestone and Tertiary Strata Overlying Evaporites	0.87	0.83
19	60350103	Salt Dome Province Tertiary Sandstones	3.82	0.78
20	60980101	East Venezuela Fold and Thrust Belt	1.41	0.76
21	53050102	Tamabra-Like Debris-Flow-Breccia Limestone Overlying Evaporites	0.56	0.71
22	71920101	Agbada Reservoirs	0.12	0.70
23	72030201	Gabon Suprasalt	0.56	0.60
24	11120103	Gograndag-Okarem Zone and Adjacent Shelf	1.10	0.55
25	60990101	Main Maracaibo Basin	0.51	0.52
26	20190101	Cretaceous Reservoirs in Northwest Desert Anticlines	1.21	0.52
27	60980202	Orinoco Delta and Offshore	0.72	0.44
28	40170101	Halten Terrace-Trondelag Platform	1.07	0.41
29	60600101	North Falklands Basin	1.00	0.40
30	11740302	South Kara Sea Offshore	1.70	0.38
31	20540301	Tanezzuft-Ghadames Structural/Stratigraphic	0.43	0.35
32	20190102	Cretaceous Reservoirs in South Gulf Suprasalt Structural	0.67	0.32
33	38170102	Kutei Basin Turbidites	0.86	0.31
34	60370101	Pelotas Platform and Basin	1.20	0.30
35	11120101	Apsheron-Pribalkhan Zone	0.86	0.27
36	60340102	Late Cretaceous-Tertiary Slide Blocks and Turbidites	0.90	0.25
37	60600201	South Falklands Basin	1.00	0.25
38	31540101	Tarim Basin Excluding Marginal Foldbelts	0.43	0.24

Rank	Assessment Unit Code	Assessment Unit Name	Percent of Fields > 1024 MMBO	Number of Fields > 1024 MMBO
39	61070101	Inner Forearc Deformation Belt	2.81	0.23
40	40250101	Viking Graben	0.13	0.21
41	60210102	Cretaceous Carbonates	0.87	0.18
42	60960102	Peripheral	0.19	0.17
43	20540201	Tanezzuft-Melrhir Structural/Stratigraphic	0.43	0.15
44	20230101	Horst/Graben-Related Oil and Gas	0.32	0.13
45	10080102	Main Basin Platform	0.10	0.11
46	13220101	Onshore and Offshore Northeastern Shelf	0.25	0.11
47	60960101	Central	0.19	0.09
48	20190202	Jurassic Reservoirs in South Gulf Suprasalt/Qatar Arch Structural	0.14	0.09
49	39130101	Late Jurassic/Early Cretaceous-Mesozoic	0.60	0.08
50	12100101	Baikal-Patom Foldbelt Riphean-Craton Margin Vendian	0.32	0.08
51	72030301	Central Congo Delta and Carbonate Platform	0.03	0.07
52	12070101	Yenisey Foldbelt Riphean-Craton Margin Riphean	0.28	0.06
53	60290102	Late Cretaceous-Tertiary Deep-Water Sandstones	0.23	0.06
54	72030401	Cuanza-Namibe	0.18	0.06
55	53050107	Tamabra-Like Debris-Flow-Breccia Limestone North of Campeche	0.55	0.06
56	10160103	East and Southeast Margins Subsalt	0.13	0.05
57	60340103	Abrolhos Sub-Volcanic Structures	0.84	0.05
58	60450103	Foreland Central Chaco High	0.30	0.04
59	11500101	Mesozoic Sandstone Reservoirs	0.22	0.04
60	20160101	Fahud-Huqf Combined Structural	0.13	0.04
61	11120102	Lower Kura Depression and Adjacent Shelf	0.16	0.03
62	40680101	Subsalt	0.60	0.02
63	60450102	Foreland Basins	0.04	0.02
64	10160101	North andWest Margins Subsalt Pinnacle Reefs	0.11	0.01
65	11090101	Foldbelt-Foothills	0.03	0.01

Appendix 9. Assessment units ranked by estimated number of undiscovered natural gas fields larger than 1024 million BOE.

[MMBOE = million barrels of oil equivalent. Both percents and numbers of fields larger than 1024 MMBOE (6144 bcf of gas) are for undiscovered fields only. One barrel of oil equivalent (BOE) equals 6000 ft^3 of natural gas.] Note that gas = natural gas.

Rank	Assessment Unit Code	Assessment Unit Name	Percent of Fields > 1024 MMBOE	Number of Fields > 1024 MMBOE
1	11740302	South Kara Sea Offshore	17.18	12.71
2	20190302	Paleozoic Reservoirs	5.30	9.31
3	20210101	Central Arch Horst-Block Anticlinal Oil and Gas	5.13	6.15
4	10500102	South Barents and Ludlov Saddle	4.58	5.93
5	20210102	North Gulf Salt Basin Structural Gas	6.52	5.07
6	11540104	Murgab Depression Subsalt	5.06	4.63
7	40170102	Mid-Norway Continental Margin	12.28	4.10
8	11740301	Northern West Siberian Onshore Gas	2.94	3.55
9	20300201	Northern Qatar Arch Extension	4.01	1.89
10	10500103	North Barents	2.42	1.48
11	40680101	Subsalt	3.40	1.29
12	11120104	Central Offshore	2.07	1.28
13	10160101	North and West Margins Subsalt Pinnacle Reefs	13.53	1.11
14	10160104	South Margin Subsalt	3.66	1.07
15	13220101	Onshore and Offshore Northeastern Shelf	1.04	0.90
16	20300102	Tertiary Reservoirs	1.52	0.81
17	52000101	Northeast Greenland Shelf Rift Systems	1.45	0.73
18	60360101	Santos Shelf	0.87	0.59
19	11120102	Lower Kura Depression and Adjacent Shelf	1.40	0.57
20	11120103	Gograndag-Okarem Zone and Adjacent Shelf	1.10	0.55
21	20190201	Jurassic Reservoirs in Northwest Desert Anticlines	2.55	0.52
22	38170101	Kutei Basin Deltaics	0.39	0.44
23	40170101	Halten Terrace-Trondelag Platform	1.11	0.43
24	20300101	Cretaceous Reservoirs	1.03	0.38
25	20230101	Horst/Graben-Related Oil and Gas	0.20	0.35
26	60340102	Late Cretaceous-Tertiary Slide Blocks and Turbidites	0.90	0.35
27	12100101	Baikal-Patom Foldbelt Riphean-Craton Margin Vendian	0.28	0.33
28	39130101	Late Jurassic/Early Cretaceous-Mesozoic	1.02	0.31
29	20710201	Red Sea Coastal Block Faults	0.15	0.29
30	39480101	Dingo-Mungaroo/Barrow	0.24	0.28
31	20190301	Khuff Carbonates in Salt Structures	0.50	0.26
32	38170102	Kutei Basin Turbidites	0.86	0.23
33	60370101	Pelotas Platform and Basin	0.87	0.21
34	61070101	Inner Forearc Deformation Belt	2.33	0.19
35	31540101	Tarim Basin Excluding Marginal Foldbelts	0.43	0.19
36	80420101	Greater Indus Foreland and Foldbelt	0.16	0.18
37	60980202	Orinoco Delta and Offshore	0.28	0.17
38	72030302	Central Congo Turbidites	0.68	0.15

Rank	Assessment Unit Code	Assessment Unit Name	Percent of Fields > 1024 MMBOE	Number of Fields > 1024 MMBOE
39	80480102	Irrawaddy-Andaman	0.14	0.14
40	61030101	Carupano Basin Gas	0.16	0.13
41	60210101	Late Cretaceous-Tertiary Turbidites	0.80	0.13
42	60340103	Abrolhos Sub-Volcanic Structures	0.84	0.12
43	10080103	Foredeep Basins	0.20	0.11
44	60980101	East Venezuela Fold and Thrust Belt	0.28	0.10
45	12070101	Yenisey Foldbelt Riphean-Craton Margin Riphean	0.28	0.10
46	80470302	Eastern Fold Belt	0.07	0.09
47	39480201	Locker-Mungaroo/Barrow	0.39	0.09
48	10500101	Kolguyev Terrace	0.18	0.09
49	71920101	Agbada Reservoirs	0.03	0.08
50	60220101	Amazon Delta and Submarine Fan	0.09	0.08
51	53050106	Tamaulipas-Like Basinal Limestone and Tertiary Strata Overlying Evaporites	0.30	0.07
52	20190103	Mesozoic/Tertiary Foredeep Fold and Thrust	0.20	0.07
53	72030201	Gabon Suprasalt	0.14	0.05
54	11500301	Upper Paleozoic Carbonates	0.18	0.05
55	39100301	Malita	0.21	0.05
56	60290102	Late Cretaceous-Tertiary Deep-Water Sandstones	0.23	0.04
57	60360102	Salt-Structured Deep-Water Sandstones	0.36	0.03
58	80420102	Indus Fan	0.18	0.02
59	20140101	Ghaba-Makarem Combined Structural	0.10	0.02
60	38220102	Mergui	0.08	0.02
61	10080102	Main Basin Platform	0.04	0.01
62	11120101	Apsheron-Pribalkhan Zone	0.10	0.01
63	31540102	Kuche (Northern) Foldbelt	0.05	0.01
64	73030101	Offshore	0.09	0.01
65	60600101	North Falklands Basin	0.03	0.01
66	60600201	South Falklands Basin	0.03	0.00
67	53050102	Tamabra-Like Debris-Flow-Breccia Limestone Overlying Evaporites	0.10	0.00

APPENDIX 10. Discovered and undiscovered petroleum volumes in assessed portions of countries.

[MMBO = million barrels of oil. BCFG = billion cubic feet of gas. MMBNGL = million barrels of natural gas liquids (NGL). Values are not necessarily those for the entire country, but only for that part assessed in World Petroleum Assessment 2000 (USGS, 2000). Cumulative = cumulative production; Known = cumulative production plus remaining reserves; Mean undiscovered = volumes estimated in World Petroleum Assessment 2000 (USGS, 2000).] Known volumes from Petroconsultants (1996) and NRG Associates (1995). Note that gas = natural gas.

Country Name	Cumulative Oil (MMBO)	Known Oil (MMBO)	Mean Undiscovered Oil (MMBO)	Cumulative Gas (BCFG)	Known Gas (BCFG)	Mean Undiscovered Gas (BCFG)	Cumulative NGL (MMBNGL)	Known NGL (MMBNGL)	Mean Undiscovered NGL (MMBNGL)
Members of OPEC, Organization of the Petroleum Exporting Countries									
Algeria	9145	18,277	7732	5798	178,727	48,982	991	6751	2135
Indonesia	14,452	22,420	7435	27,749	138,847	107,710	676	2446	2777
Iran	33,728	104,984	53,114	9772	631,042	314,567	10	4575	13,955
Iraq	22,414	100,066	45,099	5411	74,806	120,006	0	195	6225
Kuwait	25,840	80,100	3840	2796	61,478	5915	0	812	191
Kuwait/Saudi Arabia Neutral	5149	13,511	0	901	11,715	0	0	100	0
Libya	14,160	38,897	8271	5479	56,668	21,109	30	313	919
Nigeria	15,803	33,195	37,616	6232	89,756	123,209	5	2796	5572
Qatar	5007	9203	3617	2949	371,890	41,100	14	10,700	1821
Saudi Arabia	72,776	283,508	87,093	24,960	312,848	680,996	0	9998	48,872
United Arab Emirates	15,741	72,931	7695	7994	147,676	44,524	294	1991	2365
Venezuela	46,439	76,044	19,664	19,369	154,448	101,240	1	649	4579
Members of OECD, Organisation for Economic Co-operation and Development									
Australia	3708	5556	4975	5373	91,370	109,418	272	1968	5833
Austria	0	0	0	107	222	70	1	3	1
Canada	15,693	20,564	2774	66,991	118,293	24,519	1930	3347	932
Czech Republic	1	1	9	0	11	534	0	0	7
Denmark	495	1285	95	1987	9603	777	82	241	40
France	0	0	311	0	0	20,827	0	0	927
Germany	1614	1895	156	22,571	33,891	12,993	5	18	71
Hungary	547	680	146	5444	8954	2508	0	12	78
Italy	91	371	371	9527	18,247	27,272	27	32	564
Netherlands	658	997	319	73,393	141,848	8554	10	47	155
Norway	8792	22,319	12,881	14,675	110,934	183,036	217	2864	9853
Spain	0	0	78	0	0	17,795	0	0	785
Turkey	559	752	749	5	453	749	0	1	19
United Kingdom	13,283	22,931	6329	42,474	100,459	23,377	251	3675	799

Country Name	Cumulative Oil (MMBO)	Known Oil (MMBO)	Mean Undiscovered Oil (MMBO)	Cumulative Gas (BCFG)	Known Gas (BCFG)	Mean Undiscovered Gas (BCFG)	Cumulative NGL (MMBNGL)	Known NGL (MMBNGL)	Mean Undiscovered NGL (MMBNGL)
Part OPEC, part OECD									
Australia/Indonesia Zone of	0	21	57	0	6016	4536	0	360	202
Other countries									
Afghanistan	0	111	88	1497	5989	14,653	3	3	347
Angola	2600	6780	14,516	873	7140	42,739	0	16	2070
Argentina	4289	6628	3218	15,606	46,051	36,695	219	677	867
Azerbaijan	9210	13,622	6306	5360	17,348	67,449	186	278	2943
Bahrain	897	1050	899	4725	12,500	16,516	9	75	711
Bangladesh	1	1	37	2616	11,784	33,581	3	51	153
Barbados	8	11	1152	16	22	8968	0	0	436
Benin	19	30	70	0	60	705	0	1	20
Bolivia	232	279	1203	4399	10,460	25,013	130	345	1016
Bosnia-	0	1	0	0	0	0	0	0	0
Brazil	2337	11,163	46,746	705	9878	194,408	5	58	8127
Brunei	2448	4091	1802	8468	23,086	12,412	8	141	627
Bulgaria	38	46	9	153	195	28	1	47	1
Cambodia	0	0	124	0	175	1762	0	6	32
Cameroon	794	1241	1533	0	4193	5581	0	10	263
Chile	308	355	337	6278	9615	6405	60	80	143
China	23,939	48,458	12,115	9966	43,967	85,786	1	277	4461
Colombia	3596	8134	5120	3277	21,600	10,101	0	521	491
Congo	883	2498	5800	6	2255	17,395	0	23	843
Congo (Kinshasa)	185	257	326	5	99	982	0	0	48
Côte d'Ivoire	51	134	603	46	1702	6043	0	35	169
Croatia	587	813	86	680	3733	1718	30	61	30
Cuba	65	297	494	20	96	592	0	0	35
Ecuador	2034	5448	970	487	1621	552	0	0	22
Egypt	6374	9866	3119	850	6162	20,435	24	44	1088
Equatorial Guinea	0	125	2352	0	897	7538	5	45	350
Eritrea	0	0	559	0	5	10,928	0	0	633
Falkland Islands (Islas Malvinas)	0	0	5833	0	0	16,581	0	0	370
French Guiana	0	0	12	0	0	34	0	0	1
Gabon	2097	4165	8184	265	1660	24,307	0	16	1160
Ghana	3	50	201	0	177	2014	0	0	56
Greenland	0	0	47,148	0	0	80,709	0	0	4170
Grenada	0	0	0	0	0	804	0	0	8

Country Name	Cumulative Oil (MMBO)	Known Oil (MMBO)	Mean Undiscovered Oil (MMBO)	Cumulative Gas (BCFG)	Known Gas (BCFG)	Mean Undiscovered Gas (BCFG)	Cumulative NGL (MMBNGL)	Known NGL (MMBNGL)	Mean Undiscovered NGL (MMBNGL)
Guinea-Bissau	0	0	38	0	0	205	0	0	10
Guyana	0	0	2206	0	0	6022	0	0	333
India	4287	10,980	2556	8320	31,355	30,279	4	273	622
Joint Thailand/Malaysia	0	0	91	0	5200	2091	0	26	39
Jordan	0	0	0	47	226	2422	0	0	194
Kazakhstan	3683	18,327	21,094	1271	70,388	72,255	231	5696	3586
Malaysia	3176	7179	3042	4757	87,963	50,174	62	651	1329
Malta	0	0	36	0	0	205	0	0	2
Mauritania	0	0	49	0	0	265	0	0	13
Mexico	22,139	44,412	20,569	17,609	50,822	49,272	51	95	2751
Moldova	0	2	0	0	0	0	0	0	0
Morocco	0	0	4	0	0	106	0	0	5
Myanmar	528	716	730	722	10,020	27,144	2	63	814
Namibia	0	0	89	0	6000	1534	0	0	70
Oman	3552	7279	3451	1342	33,229	33,746	7	716	1761
Pakistan	314	505	258	15,908	37,231	28,606	15	103	1016
Paraguay	0	0	902	0	11	4506	0	0	221
Peru	2047	2655	3316	1951	3025	6337	44	0	316
Poland	59	125	206	1946	5232	2822	0	0	61
Romania	4319	6264	1073	24,334	38,227	5416	22	36	128
Russia	96,448	225,955	77,382	265,367	1,675,579	1,168,735	1038	8996	37,960
Senegal	0	10	50	4	49	274	0	0	14
Serbia and Montenegro	163	297	49	215	1150	271	0	5	9
Slovakia	0	0	10	0	363	256	0	0	6
Slovenia	5	6	0	35	55	10	1	1	0
South Africa	0	0	35	0	15	2085	0	0	93
Sudan	0	0	771	0	855	15,491	0	123	914
Suriname	17	63	13,029	0	0	36,003	0	0	1995
Syria	1652	3365	1311	322	2469	5097	2	26	332
Thailand	0	0	95	185	4400	4674	3	53	65
The Gambia	0	0	6	0	0	34	0	0	2
Togo	0	3	30	0	0	302	0	0	8
Trinidad and Tobago	2722	3292	758	3156	22,047	31,790	26	172	1178
Tunisia	1004	1519	2180	622	3681	7141	0	84	324
Turkmenistan	2572	3997	6836	59,231	146,476	207,692	186	590	6774
Ukraine	694	2429	1340	42,215	70,138	27,536	121	180	929
Uruguay	0	0	147	0	0	1140	0	0	55
Uzbekistan	79	614	140	20,559	82,894	15,038	37	791	437
Vietnam	0	0	44	0	0	777	0	0	14

Country Name	Cumulative Oil (MMBO)	Known Oil (MMBO)	Mean Undiscovered Oil (MMBO)	Cumulative Gas (BCFG)	Known Gas (BCFG)	Mean Undiscovered Gas (BCFG)	Cumulative NGL (MMBNGL)	Known NGL (MMBNGL)	Mean Undiscovered NGL (MMBNGL)
Western Sahara	0	0	14	0	0	77	0	0	4
Yemen	649	1933	3307	0	17,120	21,901	31	312	1265

XI: Glossary

Selected terms used in this report are defined here. The definitions are intended to be explanatory rather than strictly technical. Additional definitions are given in the glossary of USGS (2000) by Klett et al. (2000b), and Klett (2004).

Assessment unit (AU): A mappable part of a total petroleum system in which discovered and undiscovered oil and natural gas accumulations constitute a single, relatively homogeneous population such that the methodology of resource assessment is appropriate.

Continuous accumulations or continuous petroleum accumulations: Those accumulations that are areally extensive reservoirs of petroleum not necessarily related to conventional structural or stratigraphic traps. These accumulations lack well-defined downdip petroleum/ water contacts, and thus are not localized by the buoyancy of oil and natural gas in water. This definition does not incorporate factors such as deep water, remoteness, harsh climate, regulatory status, or engineering techniques.

Conventional accumulations or conventional petroleum accumulations: Those accumulations associated with structural and stratigraphic traps, commonly bounded by a downdip water contact that is significantly controlled by the buoyancy of petroleum in water.

Crude oil (oil): A mixture of hydrocarbons and small quantities of various nonhydrocarbons that exists in a liquid phase in natural underground reservoirs and remains liquid at atmospheric pressure after passing through surface separation facilities (API, 1995).

Cumulative petroleum production: Reported total volume of petroleum that has been produced to date. Cumulative oil, cumulative natural gas, and cumulative production are sometimes used.

Discovery maturity: Percentage of the known (discovered) volumes with respect to conventional endowment.

Endowment of conventional petroleum: The sum of the known petroleum volume (cumulative production plus remaining reserves), the mean of the undiscovered volume. Endowment, oil endowment, and natural gas endowment are sometimes used as abbreviated forms of this term.

Field: A production unit consisting of a collection of oil and natural gas pools that, when projected to the surface, form an approximately contiguous area that can be circumscribed.

Future petroleum volume: The sum of the remaining reserves and the mean of the undiscovered volume. Cumulative production does not contribute to the future petroleum volume. Future oil volume and future natural gas volume are sometimes used as variations of this term.

Future grown petroleum volume: The sum of the remaining reserves, the mean of the undiscovered volume, and additions to reserves by reserve growth. Cumulative production does not contribute to the future grown petroleum volume.

Grown conventional petroleum endowment: The sum of the known petroleum volume (cumulative production plus remaining reserves), the mean of the undiscovered volume, and additions to reserves by reserve growth.

Grown petroleum volume: Known petroleum volume adjusted upward to account for future reserve growth. For this assessment, 30 yr of reserve growth is considered.

Known petroleum volume: The sum of cumulative production and remaining reserves as reported in the databases used in this assessment. Also called estimated

 DOI:10.1306/1061772M863175

total recoverable volume, ultimate recoverable reserves, or estimated ultimate recovery.

Natural gas: A mixture of hydrocarbon compounds and small quantities of various nonhydrocarbons existing in the gaseous phase or in solution with oil in underground reservoirs at reservoir temperature and pressure conditions as a gas at standard temperature and pressure conditions (60°F or 15°C and 1 atmosphere –14.7 pounds per square inch or 101.325 kilopascals; API, 1995).

Natural gas field: A field with a natural gas-to-oil-ratio of 20,000 ft^3 per bbl or greater.

Natural gas liquids (NGL): Petroleum that occurs naturally as a vapor phase in the reservoir, but as a liquid at surface conditions. Natural gas liquids are commonly reported separately from oil and were considered to be 55° API or greater oil gravity for assessment purposes.

Oil field: A field with a natural gas-to-oil-ratio less than 20,000 ft^3 per bbl.

Remaining petroleum reserves: Volume of petroleum in discovered fields that has not yet been produced. For this assessment, remaining reserves were calculated by subtracting cumulative production from known volumes. Remaining reserves are typically reported as proven plus probable reserves for most of the world except Canada and the U.S., where remaining reserves are reported as only proven. An abbreviated form of this term is remaining reserves.

Reserve growth (also known as field growth): That part of the identified resources over and above measured reserves, estimated to be added to existing accumulations through delineation of additional in-place oil and natural gas, improved recovery efficiency, and revisions resulting from recalculation of viable reserves in dynamically changing economic and operating conditions. Reserve growth is expressed as increases in successive estimates of recoverable crude oil, natural gas, natural gas liquids, and condensates in existing (discovered) accumulations.

Stranded gas: A discovered natural gas accumulation that has not been produced or developed because, for whatever reason, it is currently uneconomic to deliver to market.

Total petroleum system (TPS): All genetically related petroleum that occurs, or is estimated to occur, in shows and accumulations, both discovered and undiscovered, generated by a body or closely related bodies of mature source rock.

XII: References

Adams, T. D., and M. A. Kirkby, 1975, Estimate of world gas reserves, *in* Proceedings of the 9th World Petroleum Congress, Tokyo, May 1975: John Wiley and Sons, Chichester, v. 3, no. 9, p. 3–9.

Ahlbrandt, T. S., 2001, The Sirte Basin Province of Libya—Sirte Zelten Total Petroleum System: U.S. Geological Survey Bulletin 2202F, 29 p.

Ahlbrandt, T. S., 2002, Future petroleum energy resources of the world: International Geology Review, v. 44, no. 12, p. 1092–1104.

Ahlbrandt, T. S., R. R. Charpentier, T. R. Klett, J. W. Schmoker, and C. J. Schenk, 2000, Analysis of Assessment Results, Chapter AR, *in* USGS World Petroleum Assessment 2000—Description and Results, U. S. Geological Survey Digital Data Series DDS-60, 323 p.

Ahlbrandt, T. S., T. R. Klett, and R. M. Pollastro, 1998, Estimates of undiscovered oil and gas in petroleum systems of the world: North Africa and Middle East examples: Proceedings of the 15th World Petroleum Congress, Beijing, China, John Wiley and Sons, London, v. 2, p. 180–183.

API (American Petroleum Institute), 1995, Standard definitions for petroleum statistics: American Petroleum Institute, 5th edition, 62 p.

Arrington, J. R., 1960, Predicting the size of crude reserves is key to evaluating exploration programs: Oil and Gas Journal, v. 58, no. 9 (February 19), p. 130–134.

Attanasi, E. D., and D. H. Root, 1994, The enigma of oil and gas field growth: AAPG Bulletin, v. 78, p. 321–332.

Attanasi, E. D., and J. W. Schmoker, 1997, Long-term implications of new United States gas estimates: Nonrenewable Resources, March, 1997, v. 6, no. 1, p. 53–62.

Attanasi, E. D., R. F. Mast, and D. H. Root, 1999, Oil and gas field growth projections: wishful thinking or reality?: Oil and Gas Journal, v. 97, no. 4, p. 79–81.

Ausubel, J. H., 1998, Resources and environment in the 21st century: Seeing past the phantoms: World Energy Council Journal, July 1998, p. 8–16.

Ausubel, J. H., 2000, Where is energy going?: Industrial Physicist, v. 6, p. 16–19.

Ausubel, J. H., 2001, Some ways to lessen worries about climate change: The Electricity Journal, v. 14, p. 24–33.

Ausubel, J. H., and C. Marchetti, 2001, The evolution of transport: The Industrial Physicist, v. 7, p. 20–24.

Barthel, F., P. Kehrer, J. Koch, F. K. Miscus, and D. Weigel, 1976, Die kunftige Entwicklung der Energienachfrage und deren Deckung: in Perspektiven bis zum Jahr 2000. Section 3, Das Angebot von Energie-Rohstoffen; Hannover, Federal Republic of Germany: Bundesanstalt für Geowissenschaften und Rohshtoffe, p. 89–189.

Bauquis, P. R., R. Brasseur, and J. Masseron, 1972, Les réserves de pétrole et les perspectives de production a moyen et long terme. Petroleum reserves and production possibilities in the medium and long range: Revue de l'Institut Français du Pétrole et Annales des Combustibles Liquides, v. 27, no. 4, p. 631–658.

BGR (Bundesministerium fur Wirtschaft und Arbeit or Federal Institute for Geosciences and Natural Resources, Hannover), 2002, Reserves, resources and availability of energy resources 2002: Number 15, E. Schwiezbart'sche Verlagsbuchhandlung (Nagele u. Obermiller), Johannes strasse 70174 Stuttgart, Germany 41 p. (Note: this is a BWMA-Documentation ISSN 0342-9288 published by the Federal Ministry of Economics and Labor).

Bjørkum, P. A., and P. H. Nadeau, 1998, Temperature controlled porosity/permeability reduction, fluid migration, and petroleum exploration in sedimentary basins: Australian Petroleum Production and Exploration Association Journal, v. 38, part 1, p. 453–464.

Bois, C., P. Bouche, and R. Pelet, 1979, Histoire Géologique et Repartition des reserves dans le Monde. Paris, Institut Français du Pétrole, Report no. 27.542. (Also reported in Revue d l'Institut Français du Petrole et Annales des Combustibles Liquides, v. 35, no. 2, p. 275–298.

Bookout, J. F., 1989, Two centuries of fossil fuel energy: Episodes, v. 12, no. 4, p. 257–262.

BP (British Petroleum), 1998, BP statistical review of world energy, London. (Also available at http://www.bp.com/centres/energy).

BP (British Petroleum), 2001, BP Statistical Review of World Energy June 2002. BP, London. (Also available at http://www.bp.com/centres/energy).

Campbell, C. J., 1989, Oil price leap in the early nineties: Noroil, v. 17, no. 12, p. 35–38.

Campbell C. J., 1991, The golden century of oil 1950–2050: The depletion of a resource: Kluwer Academic Publishers, Dordrecht, The Netherlands, 345 p.

Campbell, C. J., 1992, The depletion of oil: Marine and Petroleum Geology, v. 9, December 1992, p. 666–671.

Campbell, C. J., 1993, The depletion of the world's oil: Petrole et technique. No. 383, Paris, France, p. 5–12.

Campbell, C. J., 1994, The imminent end of cheap oil-based energy: SunWorld v. 18, no. 4, p. 7–19.

Campbell, C. J., 1995, The next oil price shock: The world's remaining oil and its depletion: Energy Exploration and Exploitation, v. 13, no. 1, p. 19–46.

Campbell, C. J., 1997, The coming oil crisis: Multi-Science Publications Co., Essex, England, 210 p.

Cavallo, A. J., 2002, Predicting the peak in world oil production: Natural Resources Research, v. 11, no. 3, p. 187–195.

CERIGAZ, 2001, Natural gas in the world: 2001 survey: Paris: Institut Français du Pétrole, (Cited in IEA, 2002).

Charpentier, R. R., and T. R. Klett, 2000, Monte Carlo simulation method, Chapter MC, *in* U.S. Geological Survey Digital Data Series 60: U.S. Geological Survey World Petroleum Assessment 2000, 15 p.

Charpentier, R. R., T. R. Klett, and J. W. Schmoker, 2000, Procedures for allocation and aggregation of resources: Chapter AA, *in* U.S. Geological Survey Digital Data Series 60: U.S. Geological Survey World Petroleum Assessment 2000, 82 p.

DeBruyne, M. D., 1980, Financing problems in the oil industry, *in* Proceedings of the 10th World Petroleum Congress, v. 1, p. 51–54.

Deffeyes, K. S., 2001, Hubbert's peak: The impending world oil shortage: Princeton University Press, Princeton, New Jersey, 208 p.

Desprairies, P., 1978, Worldwide petroleum supply limits, *in* World Energy Conference, Guildford, United Kingdom, IPC Science and Technology Press, New York, p. 1–30.

Doust, H., 2003, Placing petroleum systems and plays in their basin history context: A means of assist in the identification of new opportunities: First Break, v. 21, p. 73–83.

Downey, M. W., 1994, Hydrocarbon seal rocks, *in* L. B. Magoon and W. G. Dow, eds., The petroleum system from source to trap: AAPG Memoir 60, p. 159–164.

Duce, J. T., 1946, Post-war oil supply areas: The Petroleum Times (London), April 13, v. 50, p. 382–389.

Dyman, T. S., R. A. Crovelli, C. E. Bartberger, and K. I. Takahashi, 2002, Worldwide estimates of deep natural gas resources based on the U. S. Geological Survey World Petroleum Assessment 2000: Natural Resources Research (International Association for Mathematical Geology), September 2002, v. 11, no. 3, p. 207–218.

Edwards, J. D., 1997, Crude oil and alternative energy production forecasts for the twenty-first century: The end of the hydrocarbon era: AAPG Bulletin, v. 81, p. 1291–1305.

Edwards, J. S., 2001, Twenty-first century energy: Decline of fossil fuel, increase of renewable nonpolluting energy sources, *in* M. W. Downey, J. C. Threet, and W. A. Morgan, eds., Petroleum provinces of the twenty-first century: AAPG Memoir 74, p. 21–34.

Edwards, J. S., 2002a, Twenty-first century energy: Transition from fossil fuels to renewable, nonpolluting energy sources, *in* L. C. Gerhard, P. P. Leahy, and V. J. Yannacone, eds., Sustainability of Energy and Water through the 21st Century: Published jointly by the Kansas Geological Survey and the AAPG and the AAPG Division of Environmental Geosciences, Proceedings of the Arbor Day Farm Conference, p. 37–48.

Edwards, J. S., 2002b, Twenty-first century energy transition from fossil fuels to renewable, nonpolluting energy sources: 25th Annual International Conference of the International Association of Energy Economists, Aberdeen, Scotland, PowerPoint® file and abstract, CD-ROM, 24 slides.

EIA (Energy Information Administration), 2000, Annual energy review 1999: Department of Energy, Energy Information Administration Publication DOE/EIA-0384(99), 374 p.

EIA (Energy Information Administration), 2001a, Annual energy review 2000: Department of Energy, Energy Information Administration Publication DOE/EIA-0384 (2000), 379 p.

EIA (Energy Information Administration), 2001b, International energy outlook: Department of Energy, Energy Information Administration Publication DOE/EIA-0484(2002), 273 p.

Folinsbee, R., 1977, World view from Alph to Zipf: Geological Society of America, v. 88, p. 897–907.

Gautier, D. L., G. L. Dolton, K. I. Takahashi, and K. L. Varnes, eds., 1996, 1995, National Assessment of United States Oil and Gas Resources—Results, Methodology, and Supporting Data: U.S. Geological Survey Digital Data Series DDS-30 (version 2). (Also available at http://energy.cr.usgs.gov/oilgas/noga/nogaindex 95.htm [accessed March 23, 2005].)

Greene, D. L., J. L. Hopson, and J. Li, 2003, Running out of and into oil: Analyzing global oil depletion and transition through 2050: Oak Ridge National Laboratory, TM-2003/259, October 2003, 124 p.

Grossling, B. F., 1976, Window on oil: A survey of world petroleum resources: London, The Financial Times, 140 p.

Halbouty, M. T., and J. D. Moody, 1979, World ultimate reserves of crude oil, *in* Proceedings of the 10th World Petroleum Congress, Bucharest, Hungary, v. 2, p. 291–301.

Halbouty, M. T., 1981, Prospectos futuros de exploración y las reservas petrolíferas mundiales (Future exploration prospects and the world petroleum reserves): Boletín de la Asociación Mexicana de Geólogos Petroleros, v. 33, no. 2, p. 3–22.

Hendricks, T. A., 1965, Resources of oil, gas, and natural-gas liquids in the United States and the world: U. S. Geological Survey Circular 522, 20 p.

Henry, M. E., and R. R. Charpentier, 2001, Comparison of recent natural gas estimates for the western Canada sedimentary basin (abs): AAPG Annual Meeting, Denver, Colorado, June 3–6, 2001, p. A85.

Hubbert, M. K., 1956, Nuclear energy and the fossil fuels: American Petroleum Institute Drilling and Production Practice, Proceedings of the Spring Meeting, San Antonio, Texas. p. 7–25.

Hubbert, M. K., 1969, Energy resources, *in* Resources and man: Committee on resources and man, National Academy of Science—National Research Council, San Francisco, W. H. Freeman, p. 157–242.

Hubbert, M. K., 1973, Survey of world energy resources: Canadian Mining and Metallurgy Bulletin, Montreal, v. 66, no. 735, p. 37–53.

Hubbert, M. K., 1974, U.S. energy resources: A review as of 1972: U.S. Senate Committee on Interior and Insular Affairs, U.S. energy resources, a review as of 1972, a background paper: U.S. Congress, 93rd, 2nd session, Committee Print, serial no. 93-40 (92-74), pt. 1, p. 1–201.

IEA (International Energy Agency), 1998, World Energy Outlook: International Energy Agency/OECD, 2000 Edition, Paris, 475 p.

IEA (International Energy Agency), 2000, World Energy Outlook: International Energy Agency/OECD, 2000 Edition, Paris, 457 p.

IEA (International Energy Agency), 2001, World Energy Outlook—2001 Insights: International Energy Agency/ OECD, 2001 Edition, Paris, 421 p.

IEA (International Energy Agency), 2002, World Energy Outlook: International Energy Agency/OECD, 2002 Edition, Paris, 530 p.

IFP (Institut Français du Pétrole), 1978, p. 24. (Cited in Odell and Rosing, 1983).

IHS Energy Group, 2001, International petroleum exploration and production database, Petroconsultants, Inc. data: Englewood, Colorado, IHS Energy Group. [Database available from IHS Energy Group, 15 Inverness Way East, Englewood, Colorado, 80112, U.S.A.].

International Gas Union, 1977, Discussed in "Survey of energy resources. Hydrocarbons: Part A2 by E. Schubert": 1980 11th World Energy Conference, Munich, Germany, p. 75–169.

Kirkby, M. A., and T. D. Adams, 1974, The search for oil around the world up to 1999: Petroleum Times, London, November 1, 1974, p. 25–29.

Klemme, H. D., 1977, World oil and gas reserves from analysis of giant fields and petroleum basins (provinces), *in* R. F. Meyer, ed., The future supply of nature-made petroleum and gas — UNITAR Conference on Energy and the Future, 1st, Laxenburg, Austria, 1976, Technical Reports: Pergamon Press, New York, p. 217–260.

Klemme, H. D., 1978, Worldwide petroleum exploration and prospects, *in* M. L. Landwehr, ed., New ideas, new methods, new developments: Exploration and Economics of the Petroleum Industry, v. 16, p. 39–101.

Klemme, H. D., 1986, Field size distribution related to basin characteristics, *in* D. D. Rice, ed., Oil and gas assessment—methods and applications: AAPG Studies in Geology, 21, p. 85–89.

Klemme, H. D., and G. F. Ulmishek, 1991, Effective petroleum source rocks of the world; stratigraphic distribution and controlling factors: AAPG Bulletin, v. 75, no. 12, p. 1809–1851.

Klett, T. R., 2003, Graphical comparison of reserve-growth models for conventional oil and gas accumulations: U.S. Geological Survey Bulletin 2172, 98 p., 1 table, Appendices (3 discovery tables), 11 figures, CD-ROM. (Also available at http://pubs.usgs.gov/bul/b2172-f/ [accessed March 23, 2005].)

Klett, T. R., 2004, Oil and natural gas resource assessment: Classification and terminology: Elsevier Encyclopedia of Energy, v. 4, Elsevier, San Diego, 11 p.

Klett, T. R., and D. L. Gautier, 2003, Characteristics of reserve growth in oil fields of the North Sea graben area (abs.): 65th EAGE Conference and Exhibiton, European Association of Geoscientists and Engineers, Stavanger, Norway, June 2-5, 2003, 4 p., 10 figures, 1 table.

Klett, T. R., and J. W. Schmoker, 2003, Reserve growth in the world's giant oil fields *in* M. T. Halbouty, Giant oil and gas fields of the decade 1990-1999, AAPG Memoir p. 78, p. 107–122.

Klett, T. R., J. W. Schmoker, and T. S., Ahlbrandt, 2000a, Assessment hierarchy and initial province ranking, Chapter RH, *in* U.S. Geological Survey Digital Data Series 60: U.S. Geological Survey World Petroleum Assessment 2000, 27 p.

Klett, T. R., T. S. Ahlbrandt, J. W. Schmoker, and G. L. Dolton, 1997, Ranking of the world's oil and gas provinces by known petroleum volumes: U.S. Geological Survey Open File Report 97-463, CD-ROM.

Klett, T. R., J. W. Schmoker, R. R. Charpentier, T. S. Ahlbrandt, and G. F. Ulmishek, 2000b, Glossary, Chapter GL, *in* U.S. Geological Survey Digital Data Series 60: U.S. Geological Survey World Petroleum Assessment 2000, 8 p.

Klett, T. R., D. L. Gautier, and T. S. Ahlbrandt, *in press*, Additions of crude oil and natural gas to reserves: AAPG Bulletin, v. 89, n. 8.

Laherrère, J. H., 1993, Le pétrole, une ressource sure, des réserves incertaines: Pétrole et Techniques, v. 383, p. 13–19.

Law, B. E., and C. W. Spencer, 1993, Gas in tight reservoirs—an emerging major source of energy, *in* D. Howell, ed., The Future of Energy Gases: U.S. Geological Survey Professional Paper 1570, p. 233–252.

Law, B. E., G. F. Ulmishek, and V. I. Slavin, eds., 1998, Abnormal pressures in hydrocarbon environments: AAPG Memoir 70, 264 p.

Levorsen A. I., 1950, Estimates of undiscovered petroleum reserves, *in* Proceedings of the United Nations Scientific Conference on the Conservation and Utilization of Resources, v. 1, Plenary meetings, United Nations, p. 107–110.

MacKay, I. H., and F. K. North, 1975, Undiscovered oil reserves, *in* J. D. Haun, ed., Proceedings of AAPG Con-

ference on Methods of Estimating the Volume of Undiscovered Oil and Gas Resources, Stanford University, Palo Alto, California, August 21-23, 1974: AAPG Studies in Geology 1, p. 76–86.
MacKenzie, J. J., 1996, Oil is a finite resource, when is global production likely to peak?: Washington D.C., World Resources Institute, 22 p.
Magoon, L. B., and W. G. Dow, eds., 1994, The petroleum system—from source to trap: AAPG Memoir 60, 655 p.
Magoon, L. B., and J. W. Schmoker, 2000, The total petroleum system—the natural fluid network that constrains the assessment unit, Chapter PS, *in* U.S. Geological Survey World Petroleum Assessment 2000-Description and Results: U.S. Geological Survey Digital Data Series DDS-60, 4 CD-ROMs.
Martin, A. J., 1985, Prediction of strategic reserves in prospect for the world oil industry, *in* T. Niblock and R. Lawless, eds.: University of Durham, p. 16–39.
Masters, C. D., 1987, Global oil assessments and the search for non-OPEC oil: OPEC Review, Summer 1987, p. 153–169.
Masters, C. D., E. D. Attanasi, and D. H. Root, 1994, World petroleum assessment and analysis: Proceedings of the 14th World Petroleum Congress: John Wiley and Sons, London, p. 529–541.
Masters, C. D., D. H. Root, and E. D. Attanasi, 1991, World resources of crude oil and natural gas: Proceedings of the 13th World Petroleum Congress, Chichester, England, John Wiley and Sons, p. 51–64
Masters, C. D., D. H. Root, and W. D. Dietzman, 1983, Distribution and quantitative assessment of world crude oil reserves and resources; Open-File Report—U. S. Geological Survey, Report OF 83-0728, 23 p.
Masters, C. D., D. H. Root, and W. D. Dietzman, 1984, Distribution and quantitative assessment of world crude oil reserves and resources: Proceedings of the 11th World Petroleum Congress, Chichester, England, John Wiley and Sons, v. 2, p. 229–237.
Masters, C. D., D. H. Root, and R. M. Turner, 1997, World resource statistics geared for electronic access: Oil & Gas Journal, v. 95, no. 41, p. 98–104.
McCabe, P. J., 1998, Energy resources—cornucopia or empty barrel: AAPG Bulletin, v. 82, p. 2110–2134.
McCormick, W. T. Jr., L. W. Fish, R. B. Kalisch, and T. J. Wander, 1978, The future for world natural gas supply: World Energy resources 1985–2020, Oil and gas resources, the full reports to the conservation commission of the World's Energy Conference: Guildford, United Kingdom, IPC Science and Technology Press for the World Energy Conference, p. 49–56.
Meyerhoff, A., 1979, Proved and ultimate reserves of natural gas and natural gas liquids in the world: World reserves of oil and gas, *in* Proceedings of the 10th World Petroleum Congress, Bucharest, v. 2, p. 303–343.
MMS (Minerals Management Service), 1996, An assessment of the undiscovered hydrocarbon potential of the Nation's Outer Continental Shelf: Minerals Management Service OCS Report MMS 96-0034, 40 p.
Moody, J. D., 1970, Petroleum demand of future decades: AAPG Bulletin, v. 54, p. 2239–2245.
Moody, J. D., 1978, The world hydrocarbon resource base and related problems, *in* G. M. Philip and K. L. Williams, eds., Australia's mineral resources assessment and potential: University of Sydney Earth Resources Foundation Occasional Publication 1, p. 63–69.
Moody, J. D., and R. W. Esser, 1975, An estimate of the world's recoverable crude oil resource, *in* Proceedings of the 9th World Petroleum Congress, Tokyo, Japan, May 1975, Applied Science Publishers Ltd., (London), v. 3, p. 11–20.
Moody, J. D., and R. E. Geiger, 1975, Petroleum Resources: How much oil and where?: Technology Review, Boston, Massachusetts Institute of Technology, v. 77, April 1975, p. 38–45.
Nakicenovic, N., 1993, The methane age and beyond, *in* D. Howell, ed., The Future of Energy Gases: U.S. Geological Survey Professional Paper 1570, p. 661–676.
National Academy of Science, 1975, Reported in "Survey of energy resources. Hydrocarbons: Part A2 by E. Schubert": 1980 11th World Energy Conference, Munich, Germany, p. 75–169.
National Petroleum Council, 2003, Balancing natural gas policy: Fueling the demands of a growing economy: National Petroleum Council, v. 1, Summary of findings and recommendations, September, 2003, Washington, D.C., 68 p., 2 appendices.
Nehring, R., 1978, Giant oil fields and world oil resources: CIA report R-2284-CIA, Santa Monica, California, Rand Corporation, 162 p.
Nehring, R., 1979, The outlook for conventional petroleum resources: Paper P-6413 Rand Corporation, 21 p.
NRG Associates, Inc., 1995, The significant oil and gas fields of Canada database: Colorado Springs, Colorado, NRG Associates, Inc. [Database available from NRG Associates, Inc., P.O. Box 1655, Colorado Springs, Colorado, 80901, U.S.A.].
Odell, P. R., 1973, The future of oil: a rejoinder: Geographical Journal, v. 139, p. 436–454.
Odell, P. R., 1998, Oil and gas reserves: Retrospect and prospect: Energy Exploration and Exploitation, v. 16, p. 117–124.
Odell, P. R., and K. E. Rosing, 1975, Estimating world oil discoveries up to 1999: The question and method: Petroleum Times, London, February 1975, p. 26–29.
Odell, P. R., and K. E. Rosing, 1983, The future of oil, world oil resources and use: Nichols Publishing, New York, 224 p.
Parent, J. D, 1977, A survey of United States and total world production, proved reserves and remaining recoverable resources of fossil fuels and uranium as of December 31, 1975: Chicago, Institute of Gas Technology, 50 p.
Parent, J. D, 1980, A survey of United States and total world production, proved reserves and remaining recoverable resources of fossil fuels and uranium as of December 31, 1978: Chicago, Institute of Gas Technology, 140 p.
Parent, J. D, and H. R. Linden, 1974, Analysis of world

energy supplies, *in* Proceedings of the 9th World Energy Conference, Detroit, September 1974, ch. 1.2, p. 25–29.

Perrodon, A., 1992, Petroleum systems: Models and applications: Journal of Petroleum Geology, v. 15, no. 3, p. 319–326.

Petroconsultants, 1996, Petroleum exploration and production database: Petroconsultants, Inc. [Database available from IHS Energy Group, 15 Inverness Way East, D205, Englewood, Colorado, 80112, U.S.A.]

Pitman, J. K., D. Steinshouer, and M. D. Lewan, 2004, Petroleum generation and migration in the Mesopotamian Basin and Zagros fold belt of Iraq: Results from a basin-modeling study: GeoArabia, v. 9, no. 4, p. 41–72.

Pogue, J. E., 1946, Oil in the world: Yale Review, New Haven Connecticut, Yale University, v. 35, p. 623–632.

Pratt, W. E., 1952, Toward a philosophy of oil finding: AAPG Bulletin, v. 36, no. 12, p. 2231–2236.

Riva, J. P., 1995, World production after year 2000: Business as usual or crises?: Congressional Research Service, Library of Congress, Report for Congress, August 18, 20 p.

Root, D. H., and R. F. Mast, 1993, Future growth in known oil and gas fields: AAPG Bulletin, v. 77, no. 3, p. 479–484.

Root, D. H., E. D. Attanasi, R. F. Mast, and D. L. Gautier, 1995, Estimates of inferred reserves for the 1995 National Oil and Gas Resources Assessment: U.S. Geological Survey Open-File Report 95-75-L, 29 p.

Ryman, W. P., 1967, Discussed by Hubbert, M. K., "Energy resources": Resources and man, Committee on resources and man, National Academy of Sciences—National Research Council, San Francisco, W. H. Freeman and Co., 1968, p. 157–242.

Schlee, J. S., 1984, Interregional unconformities and hydrocarbon accumulation: AAPG Memoir 36, 184 p.

Schmoker, J. W., and E. D. Attanasi, 1996, The importance of reserve growth to the nation's supply of natural gas: U.S. Geological Survey Fact Sheet FS-202-96, 2 p.

Schmoker, J. W., and E. D. Attanasi, 1997, Reserve growth important to U.S. gas supply: Oil & Gas Journal, v. 95, no. 4 (January 27), p. 95–96.

Schmoker, J. W., and T. R. Klett, 2000, Estimating potential reserve growth of known (discovered) fields: A component of the USGS World Petroleum Assessment 2000, Chapter RG, *in* U.S. Geological Survey World Petroleum Assessment 2000 — Description and Results: U.S. Geological Survey Digital Data Series DDS-60, 183 p.

Schollnberger, W. E., 1998, Projections of the world's hydrocarbon resources and reserve depletion in the 21st century: Houston Geological Society Bulletin, November, p. 31–37.

Schubert, E., 1980, Survey of energy resources: Hydrocarbons: Part A2, 11th World Energy Conference, Munich, Germany, p. 75–169.

Schweinfurth, S. P., 1973, Crude oil and natural gas liquids resources of the United States and the world: Special reports, U. S. Federal Power Commission, Washington, D. C., United States, p. 55–72.

Seidl, R., 1977, Oil: The picture is changing: Options, International Institute for Applied Systems Analysis, Laxenburg, Austria, Winter, p. 4.

Shell, 1967, Reported in "Survey of energy resources. Hydrocarbons: Part A2 by E. Schubert": 1980, 11th World Energy Conference, Munich, Germany, p. 75–169.

Shell, 2001, Energy needs, choices and Possibilities scenarios to 2050, http://www.shell.com/files/media-en/scenarios.pdf. (Also available at http://hubbertpeak.com/laherrere/shell2001/ [accessed March 23, 2005].)

Stark, P. R., 2003, Report finds most new oil around the world comes from old fields: First Break (news feature), v. 21, p. 11–13 (also available from IHS Energy 2003 World Petroleum Trends).

Styrikovich, M.A., 1977, The long range energy perspective: Natural Resources Forum, v. 1, no. 3, p. 252–253.

Tissot, B. P., R. Pelet, and P. Ungever, 1987, Thermal history of sedimentary basins, maturation indices, and kinetics of oil and gas generation: AAPG Bulletin, v. 71, p. 1445–1466.

USGS (U.S. Geological Survey), 1995, National assessment of United States oil and gas resources: U.S. Geological Survey Circular 1118, 20 p.

USGS (U.S. Geological Survey), 2000, U.S. Geological Survey World Petroleum Assessment 2000—Description and Results by U.S. Geological Survey World Energy Assessment Team, USGS Digital Data Series DDS-60, a 4 CD-ROM set.

Verma, M. K., 2000, The significance of field growth and the role of enhanced recovery: U.S. Geological Survey Fact Sheet FS-115-00, 4 p.

Verma, M. K., and G. F. Ulmishek, 2003, Reserve growth in oil fields of West Siberian Basin, Russia: Natural Resources Research, v. 12, no. 3, p. 105–119.

Verma, M. K., T. S. Ahlbrandt, and M. Al-Gailani, 2004, Petroleum reserves and undiscovered resources in the total petroleum systems of Iraq: Reserve growth and production implications: GeoArabia, v. 9, no. 3, p. 51–74.

Verma, M. K., G. F. Ulmishek, and A. P. Gilberstein, 2000, Oil and gas reserve growth—a model for the Volga-Ural province, Russia: Society of Petroleum Engineers (SPE) Publication 62616 presented at the 2000 SPE/AAPG Western Regional Meeting, Long Beach, California, June 19-23, 2000, 7 p.

Warman, H. R., 1972, The future of oil: Geographical Journal, London, v. 138/3, p. 287–297.

Weeks L. G., 1948, Highlights on 1947 developments in foreign petroleum fields: AAPG Bulletin, v. 32, p. 1093– 1160.

Weeks, L. G., 1950, Discussion of "Estimates of undiscovered petroleum reserves by A.I. Levorsen", *in* Proceedings of the United Nations Scientific Conference on the Conservation and Utilization of Resources, v. 1, p. 107–110.

Weeks, L. G., 1958, Fuel reserves of the future: AAPG Bulletin, v. 42, no. 2, p. 431–441.

Weeks, L. G., 1959, Where will energy come from in 2059?: The Petroleum Engineer, Dallas, August 1959, p. A-24–A-31.

Weeks, L. G., 1965, World offshore petroleum resources, AAPG Bulletin, v. 49, no. 10, p. 1680–1693.

Weeks, L. G., 1968, The gas, oil, and sulfur potential of the sea: Ocean Industry, Houston, June 1968, p. 43–51.

Weeks, L. G., 1971, Marine geology and petroleum resources, *in* Proceedings of the 8th World Petroleum Congress, Moscow, v. 2, p. 99–106.

Williams, R., 2003, Debate over peak-oil issue boiling over, with major implications for industry, society: Oil & Gas Journal, v. 101 (July 14, 2002), no. 27, p. 18–37.